PHYSICAL PROPERTIES OF FOODS

ift Basic Symposium Series

Edited by
INSTITUTE OF FOOD TECHNOLOGISTS
221 N. LaSalle St.
Chicago, Illinois

Other Books in This Series

FOOD PROTEINS
Whitaker and Tannenbaum
POSTHARVEST BIOLOGY AND BIOTECHNOLOGY
Hultin and Milner
IMPACT OF TOXICOLOGY ON FOOD PROCESSING
Ayres and Kirschman
FOOD CARBOHYDRATES
Lineback and Inglett

Other IFT-AVI Books

CARBOHYDRATES AND HEALTH
Hood, Wardrip and Bollenback
EVALUATION OF PROTEINS FOR HUMANS
Bodwell

PHYSICAL PROPERTIES OF FOODS

Edited by

Micha Peleg
Department of Food Engineering
University of Massachusetts
Amherst, Massachusetts

Edward B. Bagley
Northern Regional Research Center
U.S. Department of Agriculture
Agricultural Research Service
Peoria, Illinois

AVI PUBLISHING COMPANY, INC.
Westport, Connecticut

Library of Congress Cataloging in Publication Data

Main entry under title:

Physical properties of foods.

(IFT basic symposium series)
Papers of the IFT–IUFoST Basic Symposium held in 1982 at Las Vegas, Nevada.
Bibliography: p.
Includes index.
1. Food—Analysis. I. Peleg, Micha. II. Bagley, Edward B. III. IFT–IUFoST Basic Symposium (1982: Las Vegas, Nev.) IV. Institute of Food Technologists. V. International Union of Food Science and Technology. VI. Series.
TX541.P48 1983 664 83–8744
ISBN 0-87055-418-2

Printed in the United States of America

Contents

Contributors

BAGLEY, E.B., Ph.D., Northern Regional Research Center, Agricultural Research Service, U.S. Department of Agriculture, Peoria, IL 61604 and Adjunct Professor of Food Science, Department of Food Science, College of Agriculture, University of Illinois, Urbana-Champaign, Urbana, IL 61801

BAIRD, DONALD G., Ph.D., Department of Chemical Engineering, Virginia Polytechnic Institute and State University, Blacksburg, VA 24061

BOURNE, MALCOLM C., Ph.D., New York State Agricultural Experiment Station and Institute of Food Science, Cornell University, Geneva, NY 14456

deMAN, J.M., Ph.D., Department of Food Science, University of Guelph, Guelph, Ontario, Canada NlG 2Wl

EL-NOKALY, MAGDA, Ph.D., Department of Chemistry, University of Missouri, Rolla, MO 65401

FLINK, JAMES M., Ph.D., Department for the Technology of Plant Food Products, The Royal Veterinary and Agricultural University, Copenhagen, Denmark

FRANCIS, F.J., Ph.D., Department of Food Science and Nutrition, University of Massachusetts, Amherst, MA 01003

FRIBERG, STIG E., Ph.D., Department of Chemistry, University of Missouri, Rolla, MO 65401

HAMANN, D.D., Ph.D., Food Science Department, North Carolina State University, Raleigh, NC 27650

KALAB, MILOSLAV, Ph.D., Food Research Institute, Research Branch, Agriculture Canada, Ottawa, Ontario, Canada KIA 0C6

KING, C. JUDSON, Sc.D., Department of Chemical Engineering, University of California, Berkeley, CA 94720

KRIEGER, IRVIN M., Ph.D., Departments of Chemistry and Macromolecular Science, Case Western Reserve University, Cleveland, OH 44106

LUND, DARYL B., Ph.D., Department of Food Science, University of Wisconsin-Madison, Madison, WI 53706

PELEG, MICHA, D.Sc., Department of Food Engineering, University of Massachusetts, Amherst, MA 01003

SCHWARTZBERG, HENRY G., Ph.D., Department of Food Engineering, University of Massachusetts, Amherst, MA 01003

STANLEY, D.W., Ph.D., Department of Food Science, University of Guelph, Ontario, Canada. NIG 2WI

SZCZESNIAK, ALINA SURMACKA, Sc.D., Central Research, General Foods Corp., White Plains, New York 10625

TARANTO, M.V., Ph.D., ITT Continental Baking Company, Research & Development Laboratories, Rye, NY 10580.

Preface

In 1980 the IFT Basic Symposium Committee selected the topic "Physical Properties of Foods" as the theme of the 1982 IFT–IUFoST Basic Symposium, to be held immediately preceding the 42nd Annual IFT Meeting in Las Vegas, Nevada. This was a result of the growing importance of this field to the food industry as well as to government and academia. The term "physical properties" itself means many things to many people. It covers a large variety of physical phenomena that are dealt with by a multitude of scientific disciplines. Because of the vastness of the domain, the choice of topics to be discussed in the symposium was extremely difficult because of the inevitable result that, whatever was included for the limited time available, other important and interesting topics would be left out. The selection of topics was also limited by two other considerations. Instrumental aspects will be covered in a separate symposium in 1983. Engineering properties (e.g., thermal, electrical) were discussed at the 41st Annual IFT Meeting in Atlanta, Georgia (1981) during a symposium sponsored by the IFT Food Engineering Division, and the proceedings were published as an OverView in the February, 1982, issue of *Food Technology*. It was not possible to eliminate all reference to either instrumental aspects of engineering properties from discussions of physical properties in general, but the format and topics were chosen to minimize overlap with the 1981 and 1983 symposia. The final 1982 program, covered in four half-day sessions, was devoted to the following subject areas.

Session I: Principles and methods of measuring physical properties including electron microscopy, colorimetry, and differential scanning calorimetry.

Session II: Structure and other characteristics of plant, animal, synthetic, baked, and particulate foods.

Session III: Rheology of foods, doughs, and emulsions.

Session IV: Volatility, expressibility, stickiness, and phase transitions in carbohydrates.

The symposium took place on June 20 and 21, 1982, at Caesars Palace Hotel, in Las Vegas, Nevada. The success of the program is attested to by the excellent response, with 160 participants attending from the United States and abroad. The timeliness of the topic is apparent from this result and thanks are due to the Basic Symposia Committee for their foresight in choosing this topic area. The committee consisted of the following people: Ernest J. Briskey (Chairman), Oregon State University; John R. Whitaker (Past Chairman), University of California-Davis; Wassef Nawar (1982), University of Massachusetts; John D. Sink (1982), West Virginia University; Darrell E. Goll (1983), University of Arizona; Richard V. Lechowich (1983), University of Arizona; Larry R. Beuchat (1984), University of Georgia; and Thomas Richardson (1984), University of Wisconsin-Madison.

There are many details to be arranged in such a meeting, from publicity to meeting room arrangements. Thanks are therefore extended here to the IFT staff, under the direction of Calvert L. Willey, Executive Director of IFT, for the very smooth and trouble-free operation. Special thanks are accorded to John B. Klis, Director of Publications, and Anna May Schenck, JFS Assistant Scientific Editor, for their invaluable help, not only before and during the meeting, but for their patience, persistence, and expertise in bringing this volume to completion.

We also thank the management and secretaries of the USDA Northern Regional Research Center and the Department of Food Engineering at the University of Massachusetts for their assistance. Without their cooperation this symposium and book would not have been possible.

MICHA PELEG
EDWARD B. BAGLEY

1

Physical Properties of Foods: What They Are and Their Relation to Other Food Properties[1]

Alina Surmacka Szczesniak[2]

INTRODUCTION

In a broad sense, the physical properties of foods may be defined as those properties that lend themselves to description and quantification by physical rather than chemical means. Their importance stretches from product handling, to processing, to consumer acceptance. Most of the early advances in this area have been made in the context of agricultural products, with much credit due to agricultural engineers, quality control personnel, and agriculture-oriented universities. More recently, food process engineers and food scientists interested in this area have made significant contributions. Although much still needs to be researched and understood, these efforts have resulted in a sizeable body of organized knowledge.

Much of this knowledge has been developed against the objective of determining and quantifying quality factors that govern consumer acceptance, suitability of the product for specific uses and, thus, its economic value. Kramer defined quality of foods "as the composite of those characteristics that differentiate individual units of a product, and have significance in determining the degree of acceptability of that unit by the buyer" (Kramer and Twigg 1970). He classified quality attributes as quantitative (e.g., proportion of ingredients, drained weight); hidden

[1]This chapter is dedicated to Prof. Amihud Kramer whose death on Dec. 8, 1981, prevented him from discussing this subject within this Symposium.

[2]Central Research, General Foods Corp., White Plains, New York 10625.

(e.g., nutritive value, adulterants, toxic substances); and sensory (e.g., appearance, kinesthetics, flavor). Physical properties classified as sensory attributes are of specific interest to this discussion and will be treated in some detail.

Another area of technical needs that gave impetus to research on physical properties of foods were problems in handling and processing of foods. Important in handling are the engineering parameters of shape, size, volume, density, and surface area. Furthermore, storage of grains and seeds in silos, mechanical harvesting, and transport of fruits and vegetables over long distances require that the products withstand static and dynamic loading, the latter often of impact type. Thus, the engineering parameters having a bearing on the behavior of foods on handling must encompass the stress–strain–time relationships. Friction, as in silos, both against the surface of the grain and against the surface of the construction material is another physical force of importance.

Processing may involve thermal, mechanical, rheological, electrical, and other physical properties of foods. Thermal properties, such as specific heat and conductivity, are important in food preservation processing involving adding or removing heat as in canning, pasteurization, and freezing. They are also important in dehydration, although mass transfer and other physical properties of the food, such as its porosity, may be more important.

Within the group of electrical properties are electrical conductivity and capacitance, dielectric properties, and reaction to electromagnetic radiation. The ability of a particle to hold a surface charge, which is determined by its electrical conductivity, has been used to separate similar seed varieties and to make hydrocolloids more dispersible in water. Dielectric properties govern the behavior of the food in dielectric and microwave heating. The most important application of microwave energy is rapid thawing–heating–cooking both in the home and in institutional feeding establishments. Its applications to food processing have been limited by cost comparisons to other heat sources and by cost–quality relationships. Key industrial applications have been in finish drying, e.g., of potato chips.

Mechanical properties and rheological properties govern the behavior of solid materials during size reduction processes such as grinding or pureeing of fluids during flow through pipes and orifices, and very importantly, they affect the consumer acceptance through the sensory property of texture. Included in this concept are also the terms "consistency" and "mouthfeel."

Other physical properties of foods important in handling and processing are aero- and hydrodynamic characteristics affecting transfer through air or water, and surface properties: surface tension, contact angle and surface rheology.

Let us now examine some selected physical properties of foods in detail and address their bearing on food quality.

GEOMETRICAL PROPERTIES

Historically, geometrical properties encompass the parameters of size, shape, volume, density, and surface area as related to homogenous food units. Included in this group of physical properties should also be the geometrical characteristics of texture (Szczesniak 1963A) which, for the most part, refer to structural geometry and structurally heterogenous foodstuffs. The geometrical texture characteristics can be divided into two classes: those referring to particle size and shape (gritty, grainy), and those referring to particle shape and orientation (fibrous, cellular). These are best treated within a discussion of textural parameters and will not be discussed here.

Size and Shape

We can differentiate here two general cases: (1) food products, such as agricultural commodities, in which the shape and size can be differentiated with the naked eye, and (2) food powders, such as ground coffee, salt, and milk powder, in which the differentiation of shape and size can be best done with the aid of magnifying lenses.

The size and shape of an agricultural commodity, or of a processed product, not only have hedonic connotations (i.e., affect the degree of consumer acceptance) but in many cases also influence packaging, distribution of stresses when forces are applied, and processability. Round tomatoes are difficult to harvest mechanically and to pack, and the concentration of stresses in small areas results in easy bruising. Thus, considerable efforts were expanded by plant breeders to produce a "square" tomato (Stevens *et al.* 1976).

In the area of food powders, serious processing problems are encountered in soluble coffee manufacturing owing to the presence of fines in ground coffee. These will clog extraction columns causing machinery downtime. Similarly, fines will clog filters in home brewing equipment. Particle size and, especially, particle size distribution in porous media have been shown to govern permeability to water and, thus, will affect extractability and solubility characteristics.

As pointed out by Medalia (1980), "to define the shape of a body fully, one must specify the location of all points on the external surface." This is a very time-consuming process that also poses mathematical difficulties, especially when applied to powders whose shape tends to be much more irregular than that of agricultural commodities. This is especially true of agglomerated powders.

For these reasons, qualitative shape descriptions are the most popular with food graders. The shapes of fruits and vegetables have been classified into 13 categories such as round, oblate, oblong, conic, elliptical, truncated, ribbed, etc. (Mohsenin 1965). Standard charts have been prepared such as the one shown in Fig. 1.1 for apples. They enable the grader to characterize the shape either by a number on the chart or by word description.

The most prevailing method for quantitative shape description involves calculations of similarity to a sphere:

$$\text{sphericity} = d_e/d_c$$

where d_e is the diameter of a sphere of the same volume as the test object and d_c is the diameter of the smallest circumscribing sphere (usually the longest diameter of the test object). Other equations for calculating sphericity have been given by Mohsenin (1980). Published values for the sphericity of fruits are of the order 89–97. These values are ex-

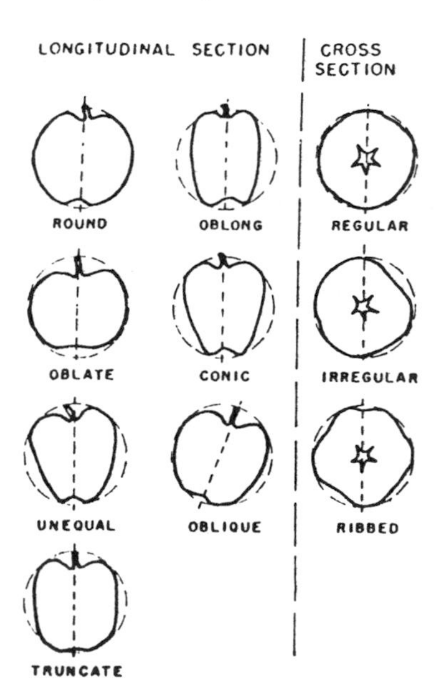

Fig. 1.1. Charted standard for describing the shape of apple fruit.
From Mohsenin (1980).

pressed as the percentage; the higher the number, the greater the similarity to a sphere. Oblong-shaped products, such as rice, would exhibit a low value of sphericity.

Other ways of describing the shape of an object involve estimations of roundness, i.e., sharpness of the corners, measurement of axial dimensions, resemblance to geometric bodies (Mohsenin 1980), and angle of curvature (as in pickles).

Size is usually characterized in practice by determining the opening, as in a sieve or screen, through which the product will or will not pass (Table 1.1). This method is used with both whole and ground materials. Although simple and widely used for grading agricultural commodities (e.g., peas, cherries, potatoes), the method has considerable disadvantages, the main one being the fact that sieving separates the product according to the narrowest dimension. It is most appropriate for products that are approximately spherical in shape. For oblong materials, such as rice, slit screens are used.

TABLE 1.1. GRADED SIZE/DIAMETER RELATIONSHIP IN CANNED PEAS

	Diameter of circular openings (in.)	
Size designation	Will not pass through	Will pass through
1	—	9/32
2	9/32	10/32
3	10/32	11/32
4	11/32	12/32
5	12/32	13/32
6	13/32	14/32
7	14/32	—

Source: Anon. (1981).

Other practical methods of determining the size of fruits and vegetables are diameter or length measurements, and count per weight or volume. With fruits, the larger size is most desirable for the fresh market. With vegetables, which are most preferred when immature, the smaller size is more desirable.

It is only recently that serious attention began to be paid to devising more adequate methods for characterizing the three-dimensional shape and size of food powders. This came with the realization that these physical properties affect aero- and hydrodynamic characteristics, mixing, segregation, and other types of behavior that often lead to processing problems, as well as detract from quality in the hands of the consumer. The physical properties of food powders are discussed in greater detail in Chapter 10 by Peleg.

With powders, both the average particle size and the particle size distribution are very important. For example, using a system of glass beads and water, Pall and Mohsenin (1980) developed an experimental model that indicates that intrinsic gas permeability is a straight line function of the particle size index squared and porosity. Intrinsic permeability describes the rate at which a fluid passes through a porous medium. The particle size index is defined as the product of the volume surface mean diameter and the uniformity coefficient. The uniformity coefficient is calculated as the ratio of the maximum diameter of the smallest 60% of the particles to the maximum diameter of the smallest 10% of the particles.

The most popular method of determining the shape and size of dry particles involves microscopic examination under a suitable magnification of particles spread on a glass slide. A commercial particle size analyzer facilitates calculation of particle sizes and their distribution from the photographs. The method has inherent biases, especially when used to determine shape parameters, because of the manner in which the particles fall on a solid support, and thus more adequate methods need to be developed.

Sphericity (designated as ψ and defined as the surface area of a sphere having a volume equal to that of the particle divided by the surface area of the particle) is an important physical property of powders which, together with porosity, affects the values of the Reynolds number and the friction factor, or drag coefficient. These influence drying processes, especially fluidized bed drying, as well as handling and conveying of the powders. For an equal porosity, the Reynolds number will increase with decreasing sphericity. For an equal Reynolds number, the friction factor will increase with decreasing sphericity (Charm 1978).

Another important effect of particle size and shape is that when the particles are present in a liquid or semi-liquid food product their size and shape exert an important influence on viscosity and other rheological parameters of the system. When present in a liquid suspending medium (as in fruit and vegetable juices), the particles increase viscosity as their size increases. When present as globules in an emulsion, the particles increase viscosity as their size decreases.

Working with commercial tomato juices, Surak *et al.* (1979) have found that the bulk viscosity increased approximately linearly with increasing volume mean diameter of the suspended particles. They also stated that the particle shape, in addition to particle diameter, influences the viscosity. Kattan *et al.* (1956) have shown that tomato juice containing elongated particles exhibits a higher viscosity than that containing spherical particles. The shape of the particles is influenced by the operating speed of the puddle finisher, the higher rpm's resulting in a greater occurrence of elongated particles.

Working with ice cream mixes, Sherman (1961) has demonstrated a linear relationship between viscosity and the reciprocal of the globule mean volume diameter. The higher the dispersed phase volume concentration (i.e., the larger the number of globules), the more pronounced was the viscosity increase with decreasing globule diameter.

The above remarks refer to the dispersed systems of relatively low concentrations (<20%) and containing deformable particles, such as is the situation with food products. At high concentrations (above 50% by volume) and with rigid spherical particles, the viscosity has been reported to be independent of the particle size (Chong *et al.* 1971). It is interesting to note that rheological investigations with rigid spheres at solids concentrations above 17.5% by volume showed a reduction in viscosity when a certain proportion (up to 40%) of the dispersed phase was made up of very small particles (Eveson 1959; Chong *et al.* 1971; Goto and Kuno 1982). The effect, attributable to the "ball bearing" action of the small particles around the large particles, indicates the importance of particle size distribution when fractions composing the dispersed phase are widely different in size. Translated into liquid food suspensions, this would suggest that the presence of small particles would be detrimental to the viscosity of products such as fruit and vegetable pastes, purees, and some juices.

The size of particles suspended in a liquid food product can be determined microscopically, by wet sieving techniques, or by means of a Coulter Counter Particle Size Analyzer. The latter method requires that the particles be greater than 0.5 Mm in diameter and that the liquid medium be a conductor. With emulsions, microscopic counts represent the best technique. The theory of the Coulter Counter is based on the principle that particles passing through an aperture change the resistance between the electrodes. This produces a current pulse of short duration, the magnitude of which is proportional to the volume of the particles. The pulses are scaled and counted electronically.

Volume, Density and Surface Area

Volume and density measurements of liquid foods present no special problems, other than the proper control of temperature at which the measurements are made. Standard volumetric methods (e.g., graduated cylinder) for volume quantification and pycnometer or commercial density meters for density measurements are simple and straightforward.

The situation is more complicated with solid materials, especially those of a porous nature. Volume of agricultural products, especially those exhibiting an irregular shape, is usually determined by water displacement. The product is weighed in air and in water (using an analytical balance, or a special gravity balance) and the volume is calculated according to

$$\text{volume} = \frac{\text{weight in air} - \text{weight in water}}{\text{weight density of water}}$$

Density of solids can be calculated as the ratio of weight to volume or can be determined by flotation in liquids (usually salt solutions) of different densities. The density of the liquid in which the product will neither sink nor float is equivalent to the density of the product. This method is a popular technique for separating peas into maturity grades. Low density peas represent a higher quality product since they are more tender and contain more sugar. Higher density peas represent a lower quality product since they are more mature, less tender, and contain more starch. Thus, the density of certain agricultural products (peas, lima beans, potatoes) is an indirect measure of their texture. Separation by density in flotation is also used with many agricultural commodities to remove defective materials and extraneous matter.

Again, food powders present problems in volume and density measurements because of their packing characteristics. Generally, two types of measurements are useful: free flow density and tapped density, the difference being in the manner of filling the volumetric container. Tapped density gives a higher number than free flow density because of the displacement of air from between the particles. Free flow and tapped density relate to container fill and settling during shipment and handling.

Surface area values are of greater concern to plant scientists than to food scientists. Nevertheless, they do have a bearing on heat transfer in heating and cooling processes and, thus, are of interest to food and agricultural engineers. A number of methods were developed for calculating the surface area of products such as fruits and eggs based on shape factor measurements (e.g., areas of transverse cross sections, transverse diameters, areas of axial or longitudinal cross sections). Equations were developed relating weight to surface area. These resulted in excellent values of the correlation coefficient and can be used to estimate the surface area by knowing the weight of the fruit or egg. Methods for surface area determination have been discussed in depth by Mohsenin (1980). Typical values for the surface area of fruits are apples,

17.2–25.2 in.^{2}, plums, 5.4–7.0 in.^{2}, pears 22.2–23.0 in.^{2} (Mohsenin 1980).

Defects

In the sense of quality control, most defects (defined in the broad sense as factors detracting from high quality) are detected by physical means and, thus, in the context of this discussion can be classified as physical properties of foods.

Notable here are structural defects resulting in misshapened crop; off-color caused by genetic disturbance, microbial action, or enzymatic changes; holes and scars produced by chewing insects; mechanical damage to tissue resulting in bruising; and presence of extraneous or foreign matter. These are usually detected through visual inspection using aids such as white backgrounds, special lights, dilution, and reference standards to facilitate the description of the nature and severity of the defect.

OPTICAL PROPERTIES

The most important optical properties from the quality standpoint are the visual color and surface appearance (or gloss) of the product. These involve reflected light, although some spectrophotometers measure light in both reflectance and transmission modes. Transmitted light may be used for detecting internal defects such as blood spots in eggs and water cores in apples (Fig. 1.2).

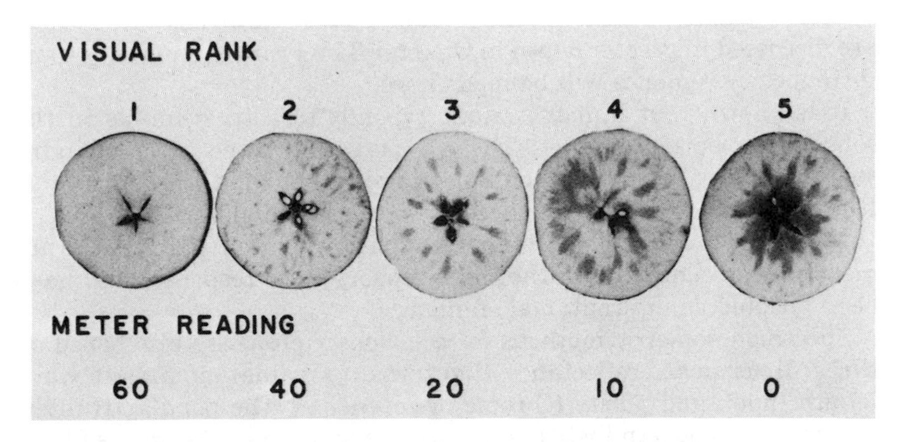

Fig. 1.2. Use of transmitted light measurement (difference meter) to detect water-core defect in apples.
From Birth and Olsen (1964).

Color

Although only a few definitive studies have been done on the effect of color on consumer acceptance, there is much experimental evidence that it is one of the most important quality attributes. It can signal a high quality product (such as the golden yellow of a table orange) or can alert the consumer to a potential physiological danger (such as green processed meat). Furthermore, it is well recognized that color has important psychological connotations that can influence the mood and emotional state of the human being. This fact has not been fully utilized by the manufacturers of compounded food products. Examples of research papers dealing with the effect of color on consumer acceptance of foodstuffs are the work by Czerkaskyj (1971) on color of canned cling peaches, research by Maga (1973) on color of potato chips, the publication by von Elbe and Johnson (1971) on color of canned peas, and the paper by DuBose *et al.* (1980) on color of fruit-flavored beverages and cakes.

There is also convincing evidence that color influences flavor recognition in products such as beverages and dessert gels and that it affects consumer perception of flavor intensity. The darker the brown color of chocolate-flavored products, for example, the more chocolaty they will be rated by consumers. The published evidence on the effect of color on the perceived flavor intensity has been reviewed by Kostyla and Clydesdale (1978) and DuBose *et al.* (1980). The latter authors also have shown that atypical colors in fruit-flavored beverages evoked incorrect flavor responses that were characteristically associated with the atypical color.

Techniques for measuring the color of food are quite advanced. They are discussed in greater depth in Chapter 3 by Francis; thus only some introductory remarks will be made here.

It is known that human vision responds to a tri-stimulus in the sensory perception of color. The eye possesses three types of light-sensing devices each corresponding to a different band of wavelengths. It is generally accepted that these are red, green and blue. The Y, X, Z values of the International Committee on Illumination (CIE) are the numerical representatives of the three sensory color responses and have been adopted as international standards.

Spectrophotometric methods for color description are also based on three dimensions: reflectance (lightness, or value), dominant wave length (hue), and purity (chrome or intensity). The popular Hunter, Gardner, or Macbeth instruments use the opponent color scales (L a b type). The L dimension defines the lightness; the a dimension refers to

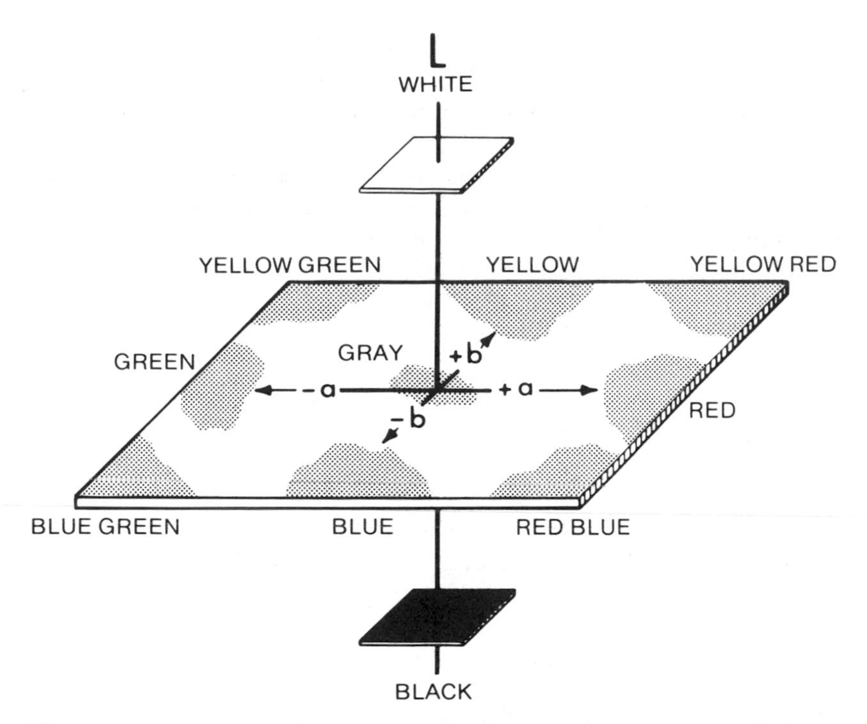

Fig. 1.3. *L a b* system for color description and quantification.

the red–green hues; and the *b* dimension refers to the blue–yellow hues. As shown in Fig. 1.3, a sample with a positive *a* value and a positive *b* value will be in the yellow–red quadrant. A sample with a positive *a* value and a negative *b* value will be in the red–blue quadrant. Values attached to these dimensions will indicate the chrome or intensity of color. Data obtained from the spectrophotometer can be converted to the CIE X, Y, Z values by means of established equations, a task that has been greatly facilitated with computerized spectrophotometers. An excellent source of detailed information on food color measurements is the book by Francis and Clydesdale (1975).

Gloss

The appearance of a surface, whether it is glossy or dull, is an important physical aspect of food quality detected by human vision. In general, shiny appearances are preferred, although this varies with the product and appears to be a learned response.

Typical of the products where a shiny surface is valued are apples, cucumbers, cherries, rice, and fruit pie fillings. On the other hand, oranges, green beans, mushrooms, and similar products are expected to have a dull surface.

In contrast to color, the appearance of a surface, which is associated with the special distribution of light by the object, cannot be defined in any organized coordinate measurement. We recognize reflected light (i.e., light leaving the object from the illuminated side) and transmitted light (i.e., light passing through the object and leaving from the other side). These may be further subdivided into diffused and nondiffused light.

The four types of light distribution from objects have the following general relationship to surface appearance (Hunter 1973, 1975): (1) diffuse reflection—shiny; (2) specular reflection—glossy, mirror-like; (3) diffuse transmission—cloudy, opaque; (4) specular transmission—translucent. Figure 1.4 illustrates these four types of light distribution.

Since gloss is the psychological attribute of surfaces associated with the specular reflection and since this can vary from surface to surface, no well-developed methods exist for quantifying this property. Hunter (1973, 1975) discusses this area in detail.

Fig. 1.4. Surface appearance of different objects. *(Left to right)* Yellow vase illustrates diffuse reflection—the yellow light is reflecting in all directions and the surface looks the same from any direction. Brass bell illustrates specular reflection—the reflection is highly directed and the position of the highlights moves with the light source of our eye. Plastic cup illustrates diffuse transmission—the transmitted light leaves the surface in all directions making the object appear cloudy. Bottle of oil illustrates specular transmission—the light passes through making the object appear transluscent.
From Hunter (1973).

THERMAL PROPERTIES

Thermal properties of foods are of considerable importance to food engineers and processors since they govern the change of temperature in food processing involving heat transfer such as heating, cooling, drying, and freezing. Together with the other physical characteristics such as density, viscosity, shape, and size, they govern the design of the equipment and the profile of the process. Included here are such properties as specific heat, thermal conductivity, thermal diffusivity, surface conductance, and emissivity.

Specific Heat

By definition, specific heat of a substance is the ratio of its heat capacity to that of water, which is given a value of 1. It denotes the amount of heat in calories required to change the temperature of 1 gram of the substance 1°C. In the British system, the corresponding units will be Btu, pound, and °F. In freezing, we are also concerned about the latent heat of fusion, i.e., the quantity of heat that must be removed at (or near) 0°C to freeze the product.

Values for specific heat and latent heat of fusion are known for most foods and typical values are given in Table 1.2. It will be noticed that the

TABLE 1.2. THERMAL PROPERTIES OF SELECTED FRESH FOODS

	Approx. % H_2O	Specific heat (cal/g/°C)		Latent heat of fusion (cal/g)
		Above freezing	Below freezing	
Asparagus	92	0.94	0.48	74.5
Bananas	76	0.80	0.42	60.0
Beef, fresh	48–72	0.70–0.84	0.38–0.43	49.5–61.2
Blueberries	83	0.86	0.45	65.6
Bread	36	0.70	0.34	25.6–29.5
Cauliflower	91	0.93	0.47	73.4
Cheese	40	0.50	0.31	30.0
Cherries	84	0.87	0.45	66.7
Dates	22	0.36	0.26	53.4
Eggs, dried, whole	4	0.25	0.21	5.0
Figs, dried	23	0.39	0.27	18.9
fresh	78	0.82	0.43	62.3
Honey	17	0.35	0.26	14.5
Ice Cream	63	0.80	0.45	16.1
Lettuce	95	0.96	0.48	75.6
Mushrooms	90	0.93	0.47	72.3
Peas, green	78	0.79	0.42	58.9
Spinach	91	0.94	0.48	73.4
Tomatoes	94	0.95	0.48	74.5
Water	100	1.00	0.48	80.0

Source: Potter (1978).

specific heat of a given food is higher above its freezing point and lower below its freezing point. Also, the higher the moisture content of the food, the higher and closer is its specific heat to that of water. Thus, high moisture products require more energy to heat to a given temperature. The latent heat of fusion bears a similar relationship to the moisture content of the foodstuff and to the value of pure water.

Thermal Conductivity

Thermal conductivity of a food depends on its porosity, structure, and chemical constituents, primarily on the properties of air, fat, and water. Since the thermal conductivity of fat is lower than that of water, and the thermal conductivity of air is even lower, high levels of fat or entrapped air will decrease the thermal conductivity of a foodstuff. This has an important impact on the rates of heating and cooling and, thus, on process efficiency and potential flavor deterioration.

Since the thermal conductivity of ice is greater than that of water, frozen foods will exhibit greater thermal conductivity than their unfrozen counterparts. On either side of the freezing point, thermal conductivity is highly temperature dependent. It is often assumed that the thermal conductivity of foodstuffs increases with increasing temperature above freezing in the same manner as the conductivity of water increases, and increases with decreasing temperature below freezing in the same manner as the conductivity of ice increases. However, this assumption does not appear to be valid in all cases.

The thermal conductivity of liquids was also found to increase with pressure, and decrease with increasing concentration of solutes (Cuevas and Cheryan 1979).

Reviews on thermal conductivity of foods and food products and methods used for measurement have been published by Woodams and Nowrey (1968), Qashou *et al.* (1972), and Cuevas and Cheryan (1979). The latter publication deals exclusively with liquid products.

Experimental methods of determining thermal conductivity are most applicable to fluids. They involve measuring the temperature differential across the fluid placed between two surfaces, one of which is heated and the other cooled. Several geometries may be used: concentric cylinders, concentric spheres, or parallel plates (Woodams and Nowrey 1968). Typical values cited in the literature are shown in Table 1.3.

Thermal conductivity is difficult to determine experimentally, and food engineers have studied the mathematical approach to its estimation. Various predictive equations have been proposed based on approximations of the foodstuff to a model. These have been discussed in detail

TABLE 1.3. THERMAL CONDUCTIVITY OF SELECTED FOODS

Food	Temp. (°F)	Thermal Conductivity (Btu/hr·ft·°F)
Apple juice	68	0.323
	176	0.365
Milk		
fresh, whole, 3.6% fat	68	0.318
	176	0.355
skim, fat <0.1%	68	0.328
Milk, condensed, 50% water	78.8	0.1875
Olive oil	84	0.0970
Honey	70	0.031
Pork		
fat, 93% fat	50	0.12
lean	39	0.278
Orange	32–59	0.24
Strawberries	57–76	0.39
Oats, 9.9% water	0–104	0.037
	39	0.290
Turkey breast	5	0.702
Water	50	0.338
	176	0.338
Ice	5	1.35
	−13	1.40

Source: Adapted from Woodams and Nowrey (1968).

by Heldman and Singh (1981) and Cuevas and Cheryan (1979). The simplest model considers the food to be a two-phase system composed of water and solids. Others take into account the structure of the food and its anisotropic nature.

Thermal Diffusivity

When heat transfer occurs by conduction rather than convection, the property of thermal diffusivity becomes important. It is related to thermal conductivity, density, and specific heat of the product and determines the rate of heat propagation through the food. It may be calculated from the equation

$$\alpha = k/(\rho \cdot c_p)$$

where α = thermal diffusivity; k = thermal conductivity; ρ = density; and c_p = specific heat. It may also be determined experimentally under transient heat transfer conditions (Dickerson 1965). This technique assumes that the sample is homogeneous, an assumption that is not valid for many foodstuffs, especially agricultural products.

Thermal diffusivity values for foods are in the range of $1-2 \times 10^{-7} m^2 s^{-1}$ and increase with increasing temperature. Typical values are shown in Table 1.4. A comprehensive review of the importance of thermal diffusivity in food processing and methods of determining this thermal property has been published recently by Singh (1982).

TABLE 1.4. THERMAL DIFFUSIVITY OF SELECTED FOODS

Product	% H_2O	°C	Thermal diffusivity 10^{-7} m^2/sec
Apples, Red Delicious	85	0–30	1.37
Applesauce	37	5	1.05
		65	1.12
Banana	76	5	1.18
		65	1.42
Beans, baked	—	4–122	1.68
Squash, whole	—	48	1.34
Strawberry	92	5	1.27
Codfish	81	5	1.22
		65	1.42
Beef, round	71	40–65	1.33
Ham, smoked	64	40–65	1.38
Water	100	30	1.48
		65	1.60

Source: Different authors as compiled by Singh (1982).

ELECTRICAL PROPERTIES

Electrical properties of foods have not been studied extensively and only limited data are available in the literature. One reason is that electrical properties of foods depend on the strength of the electromagnetic field, on one hand, and the exact composition, packing density, and temperature on the other. Thus, wide variations in values can be expected for any given product depending on the samples used and the test conditions employed. A good source of information are publications by Jason and Jowitt (1968), Bengtsson *et al.* (1963), Ede and Haddow (1951), and Mudgett (1982). Values for a large number of nonfood materials have been compiled by von Hippel (1954).

Foods can be classified as conductors (materials that conduct electromagnetic energy) and dielectrics (materials that neutralize and dissipate electromagnetic energy).

Electrical Conductivity

With foods that are conductors (e.g., sugar, whole egg, beef, salt, dried milk), the electrical conductivity is significantly dependent on the frequency of the electromagnetic field. Most foods, however, are poor conductors and their conductivity is essentially independent of the electromagnetic field. Ede and Haddow (1951) reported that, for most foods, the conductivity decreases with temperature and shows a sharp drop at the freezing point.

Typical values of electrical conductivity and the effect of temperature are shown in Table 1.5 for selected foodstuffs.

TABLE 1.5. ELECTRICAL CONDUCTIVITY OF SELECTED FOOD MATERIALS[a]

	20°C	70°C	100°C
Lean beef	6	12	16
Lean pork	5	13	17
Lean bacon	22	60	60
Herring	4	10	14
Potato, raw	6	11	14
Apple, raw	1.3	2.8	4
Milk	4	10	12
Dough	0.7	1.3	1.6
Salt	2.6	2.6	2.5
Sugar	2.3	2.5	2.6

Source: Ede and Haddow (1951).
[a]In millimho/cm^3 at 0.2–20 MHz frequencies.

Dielectric Properties

The principle of applying dielectric (1–150 MHz) and microwave (915–2450 MHz) energy to foods is based on their nonconducting, or dielectric, character and the fact that the alternating current of the electromagnetic field will cause the charged asymmetric molecules (such as water) to oscillate around their axes at, or near, the frequency of the field. This produces intermolecular friction that is dissipated as heat. The dielectric region of the radiation spectrum is of longer wavelength than the microwave region, the frequency being inversely related to wavelength.

Three main parameters are important here: the dielectric constant (ε_r'), the dielectric loss factor (ε_r''), and the dielectric loss tangent ($\tan \delta = \varepsilon_r''/\varepsilon_r'$).

The dielectric constant is a measure of a substance's ability to neutralize the attraction between electric charges. Water has a very high dielectric constant since, due to its high dipole moment, the molecules tend to orient themselves so as to neutralize an electric field (Paul and Palmer 1972). The value of the dielectric constant for water is about 80 (at 1000 MHz and room temperature); it decreases drastically with increasing frequency (above 3000 MHz) and also decreases with increasing temperature.

The dielectric loss tangent (tan δ) is the ratio of the dielectric loss factor to the dielectric constant and is a measure of the dissipation factor. The higher the value of the dielectric loss factor, the faster the material will heat when subjected to microwave, and the smaller will be the depth of microwave penetration. In general, the loss tangent increases with increasing frequency and decreases with increasing temperature. The specific effect of frequency on the dielectric properties depends on the frequency range and the energy absorption process.

The effect of temperature on dielectric properties of foods may be positive or negative depending on the temperature, frequency range, and state of dielectric dispersion in the food material. A marked increase in the dielectric constant and loss factor is observed between frozen and thawed foods (Nelson 1978).

Dielectric properties of selected food materials are shown in Tables 1.6 and 1.7. The effects of microwave frequency on the dielectric loss tangent of water are summarized in Table 1.8. Additional data for specific food products may by found in the papers by Pace *et al.* (1968), Nelson (1973A, B; 1978, 1979), Mudgett (1982), and others. As pointed out by Nelson (1978), a considerable body of published information exists on dielectric properties of many foodstuffs. However, because of the different factors in the dielectric field and in the food that affect these properties, precise values for a particular product under a specific set of conditions can be obtained only by actual measurements.

Determinations of the dielectric properties are usually based on measurements of electrical impedance, attenuation, or reflection coefficient as a function of sample thickness (de Loor and Meijboom 1966). Commercial instruments are available, e.g., Marconi Q-meter, Boonton RX-meter (Bengtsson *et al.* 1963), or Rhode and Schwarz (Nelson 1973B).

Principles and industrial applications of microwave energy have been discussed by Goldblith (1967), Copson (1975), Potter (1978), and Mudgett (1982). The review by Mudgett covers, in addition to product heat-

TABLE 1.6. DIELECTRIC CONSTANT AND ELECTRICAL CONDUCTIVITY OF SELECTED FOOD MATERIALS

Material	ε'_r	Conductivity (mMho/cm)
Agar gel		
4%	73.5	0.85
8%	66	1.70
Potato, raw	63	0.33
Potato liquor	77.5	11
Milk	69.5	4.1

Source: de Loor and Meijboom (1966).

ing characteristics in microwave food processing, factors affecting the electrical properties of foodstuffs and suggestions for future research.

Since the volumetric displacement of fluids and the binding of water and ions by colloidal solids depress the values for the dielectric constant and the dielectric loss of foods (Mudgett 1982), attempts have been made to use these measurements as nondestructive means of predicting fruit texture and maturity. However, so far experimental evidence has failed to establish a valid correlation with soluble solids and fruit maturity (Nelson 1980, 1982). A correlation does exist between dielectric properties and tissue density, and between dielectric properties and moisture content. The latter was shown to depend on the microwave frequency (Nelson 1978). The strong correlation between the dielectric constant and free moisture in grains has been used as the basis for quick moisture determinations employing specific frequencies (Nelson 1973B).

TABLE 1.7. DIELECTRIC CONSTANTS FOR SELECTED MUSCLE FOODS AND FRESH PRODUCE

Beef and Codfish[a] (+2°C and 200 MHz)	ε′	tan δ	
Beef			
rib	64.9	0.61	
brisket	64.9	0.62	
top round	68.5	0.59	
bottom round	65.7	0.60	
thick flank	66.0	0.62	
Cod fish			
back	67.7	0.70	
belly flap	68.0	0.72	
Fresh Produce[b] (+23°C and 2500 MHz)	**ε′**	**tan δ**	**%H_2O**
Apples	43	0.21	86
Cantaloupes	49–55	0.25–0.37	90–92
Carrots	57	0.32	86
Cucumbers	63.5	0.20	96
Peaches	51–62	0.31	89–91
Potatoes	52–54	0.32–0.37	80
Sweet potatoes	41–58	0.28–0.32	74–84
Watermelon	59	0.26	91

[a]Source: Bengtsson *et al.* (1963).
[b]Source: Nelson (1980, 1982).

TABLE 1.8. EFFECT OF TEMPERATURE AND MICROWAVE FREQUENCY ON THE DIELECTRIC LOSS FACTOR (tan δ) OF WATER

°C	900 MHz	2450 MHz
15	0.07	0.17
55	0.03	0.07
95	0.02	0.04

Source: Jeppson (1964).

MECHANICAL PROPERTIES

Of all the physical properties of foods, color and mechanical properties have received the greatest attention, with research on mechanical properties having gained significantly in importance in the last one to two decades. The reason for this activity is the fact that, in addition to affecting the mechanical behavior of the product during transport, handling, and processing, the mechanical properties form the basis for the sensory property of texture (how the food behaves on mastication in the mouth).

In both cases, that is, during consumption and handling by food or agricultural engineers, the foodstuff is subjected to various forces, and its mechanical properties will govern how it reacts to these imposed stresses. During mastication and during industrial size reduction processes (e.g., slicing, grinding, pureeing) it is desirable to have a "weak" product—one that will disintegrate in the proper manner when forces are applied. On the other hand, during transport and industrial handling, it is desirable to have a "strong" product—one which will not suffer any substantial damage when impact (as in mechanical harvesting) or static compressive forces (as in silos storage) are applied.

Relationship to Rheology and Texture

Mohsenin (1980) defines the mechanical properties "as those having to do with the behavior of the material under applied forces." Rheology has been defined as "a science devoted to the study of deformation and flow," and more recently as the "the study of those properties of materials that govern the relationship between stress and strain." (Dealy 1982). "Stress" is defined as the intensity of force components acting on a body and is expressed in units of force per unit area. "Strain" is the change in size or shape of a body in response to the applied force. It is a nondimensional parameter (reported as a ratio or percentage) and is expressed as the change in relation to the original size or shape. Researchers use two modes of strain expression: engineering strain, which expresses the change in shape at any given time with respect to the original shape at $t = 0$, and true strain, which expresses the continuous change in shape as the stress is being applied. The former is less cumbersome to calculate and, thus, more prevalent. However, as pointed out by Calzada and Peleg (1978) differently shaped stress–strain plots may be obtained and, thus, different conclusions may be drawn depending on how strain is expressed.

Rheologically, the behavior of a material is expressed in terms of stress, strain, and time effects.

Some researchers use interchangeably the terms "mechanical properties" and "rheological properties." This is only approximately correct. One can say that all rheological properties are mechanical properties, but not all mechanical properties are rheological properties. There are certain mechanical properties that do not involve deformation and, thus, cannot be classified as rheological properties. These usually deal with the motion of the material as a result of the applied force and involve properties such as drag coefficient, rebound coefficient in impact, and flow of material in bulk (Mohsenin 1980).

The relationship between the mechanical properties of foods and the textural parameters has been pointed out by Szczesniak (1963A) and is shown in Table 1.9. The developed classification of the textural parameters was intended to form a bridge between the popular sensory terms used to describe texture and the rheological principles of stress, strain, and time effects. It has been well accepted both by the sensory evaluation people and by food technologists interested in instrumental texture description in a manner correlatable with sensory perception (see, e.g., Amerine *et al.* 1965; Pomeranz and Meloan, 1971; and Brennan 1980).

Since rheological tests express the behavior of the material in terms of stress–strain relationships, they must be conducted on samples of known, and usually simple, geometry. This is done reasonably easily

TABLE 1.9. MECHANICAL CHARACTERISTICS OF FOOD TEXTURE

Primary Parameters	Secondary Parameters	Popular Terms
Hardness[a]		Soft → Firm → Hard
Cohesiveness[b]	Fracturability[f]	Crumby → Crunchy → Brittle
	Chewiness[g]	Tender → Chewy → Tough
	Gumminess[h]	Short → Mealy → Pasty → Gummy
Viscosity[c]		Thin → Viscous
Springiness[d]		Plastic → Elastic
Adhesiveness[e]		Sticky → Tacky → Gooey

Source: Szczesniak (1963A).
[a]Hardness—the force necessary to attain a given deformation.
[b]Cohesiveness—the strength of the internal bonds making up the body of the product.
[c]Viscosity—the rate of flow per unit force.
[d]Springiness—the rate at which a deformed material goes back to its undeformed condition after the deforming force is removed.
[e]Adhesiveness—the work necessary to overcome the attractive forces between the surface of the food and the surface of other materials with which the food comes in contact (e.g., tongue, teeth, palate, etc.)
[f]Fracturability—the force with which the material fractures. It is related to the primary parameters of hardness and cohesiveness. In fracturable materials, cohesiveness is low and hardness can vary from low to high.
[g]Chewiness—the energy required to masticate a solid food product to a state ready for swallowing. It is related to the primary parameters of hardness, cohesiveness, and springiness.
[h]Gumminess—the energy required to disintegrate a semisolid food product to a state ready for swallowing. It is related to the primary parameters of hardness and cohesiveness. With semisolid food products, hardness is low.

when dealing with synthetic polymers, which can be molded or cut to the desired size and shape. The large majority of foods, however, do not lend themselves to this treatment. In the case of fresh produce, for example, cutting will injure the tissue and cause changes in mechanical properties due to enzymatic action. Furthermore, engineers concerned with the handling behavior of foods are interested in the mechanical properties of the naturally shaped product and the distribution of stresses over the spherical bodies of foodstuffs such as oranges and apples. To this end, agricultural engineers have adapted to foods (Arnold and Mohsenin 1971; ASAE 1976) the Hertz and Boussinesq equations for the solution of contact stresses so that the observed forces and deformations can be used in constitutive equations. This has been discussed at length by Mohsenin (1980).

In addition, classical rheological tests of solids usually involve small strains, while the mechanical parameters of food important to sensory texture and to behavior on processing manifest themselves under large stresses (Bourne 1975). It is still subject to debate whether the behavior under small strains can be extrapolated to that under large strains.

From the rheological standpoint, materials can be divided into those that deform and those that flow. The former are solids and the latter are liquids. Each category can be further subdivided into ideal, i.e., strain-rate independent and nonideal, i.e., strain-rate dependent materials. Ideal solids deform in an elastic, Hookean manner, while ideal liquids flow in a viscous, Newtonian manner; in each case the behavior is independent of the strain rate. Most foodstuffs are nonideal and thus, strain-rate dependent. Furthermore, they usually contain some solid and some liquid attributes and, rheologically, are termed viscoelastic bodies. In addition, many possess well-developed structural elements which "yield" or rupture when forces are applied, thus changing the stress–strain behavior not only with the applied rate of strain, but also with the applied amount of strain. Thus, the rheological behavior of foods is extremely complicated. Because foods are often anisotropic in nature, their rheological properties will vary with the direction of stress application. When, in addition, one considers nature-introduced variations and time-related changes due to enzymatic activities and moisture transfer, the task of describing the behavior of foods in the sense of classical rheology becomes a formidable and frustrating endeavor. Nevertheless, the recent revival of interest in the application of rheological principles and rheological models to the characterization of mechanical–textural properties of foods has found support and is yielding useful and interesting information. Good treatment of the mechanical, rheological, and textural properties of foodstuffs can be found in books by Mohsenin (1980), Muller (1973), Kramer and Szczesniak (1973), and Sherman (1970), as well as in chapters by Brennan (1980) and deMan (1976).

Simplified Rheological Principles

As has already been stated, rheology defines the behavior of a material in terms of stress, strain, and time effects. The latter can refer to the rate of strain (or stress), the effect of time on behavior under constant stress (or strain), or the effect of time on the return to the original state when stress is removed.

There are three commonly applied types of stress: *compressive* (directed toward the material), *tensile* (directed away from the material), and *shearing* (directed tangentially to the material). These are depicted in Fig. 1.5. Shear stress is most prevalent with fluids. Action of pure shear occurs only rarely with solid foods; so-called "shear measurements" usually involve a combination of forces (compression, shear, and

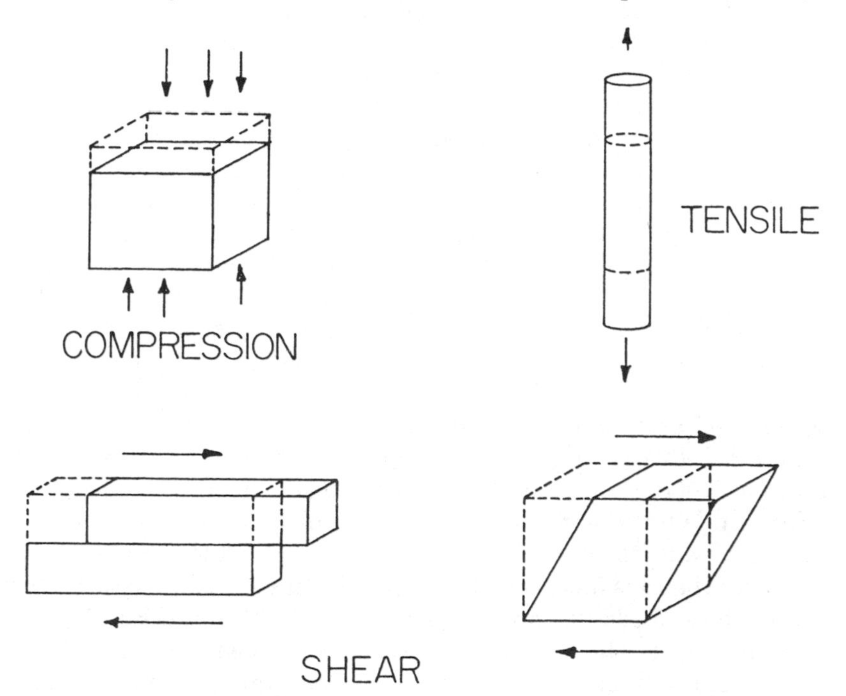

Fig. 1.5. Fundamental types of stress acting on a body.

tensile). Since strain is the response of a material to stress, there are also three types of strain: *compressive, tensile, and shear*. The ratio of stress to strain is called the "modulus"—compression modulus, tensile modulus, or shear modulus. When an elastic material is compressed, the stress–strain plot is a straight line starting at the origin and its slope is called "Young's modulus of elasticity." Many foods exhibit this behavior under small compressions before the so-called "limit of elasticity" is reached. As a general rule, the shear modulus is usually three times as great as the modulus of elasticity. The maximum stress that the material is capable of sustaining before rupture is called "strength"—compressive strength, tensile strength, or shear strength.

Rheologically, a material may deform in three ways: elastic, plastic, or viscous. In an ideal *elastic body*, deformation (or strain) occurs instantly the moment stress is applied, is directly proportional to stress, and disappears instantly and completely when stress is removed. In an ideal *plastic body*, deformation does not begin until a certain value of stress (called yield stress) is reached. Deformation is permanent and no recovery occurs when the stress is removed. In an ideal viscous body, deformation occurs instantly the moment stress is applied, but—in contrast with an elastic body—it is proportional to the rate of strain and is not recovered when stress is removed.

These three types of behavior are denoted by a spring, a friction element, and a dashpot, respectively, in rheological models. They can be arranged in series or in parallel and in any number to depict the fact that a real material usually exhibits a combination of rheological behavior types. Figure 1.6 shows some typical rheological models. Recently, Peleg (1976, 1977A) has introduced contact and fracture elements to account more closely for the nonlinear behavior of food products and its dependency on the deformation history.

As has already been mentioned, food products usually behave as a combination of elastic and viscous elements; such materials are called "viscoelastic" and their stress–strain relationship depends on the rate of strain that introduces the time dependency. When the stress–strain ratio is a function of time alone, the material is said to be linearly viscoelastic. When the stress–strain ratio is also a function of stress, the material is said to be nonlinearly viscoelastic. Unfortunately, most rheological theories have been developed for linear viscoelastic materials, while most foodstuffs are nonlinear viscoelastic materials. Because of the absence of rheological theories for nonlinear viscoelastic materials, certain liberties and approximations must be taken in the rheological characterization and interpretation of food behavior.

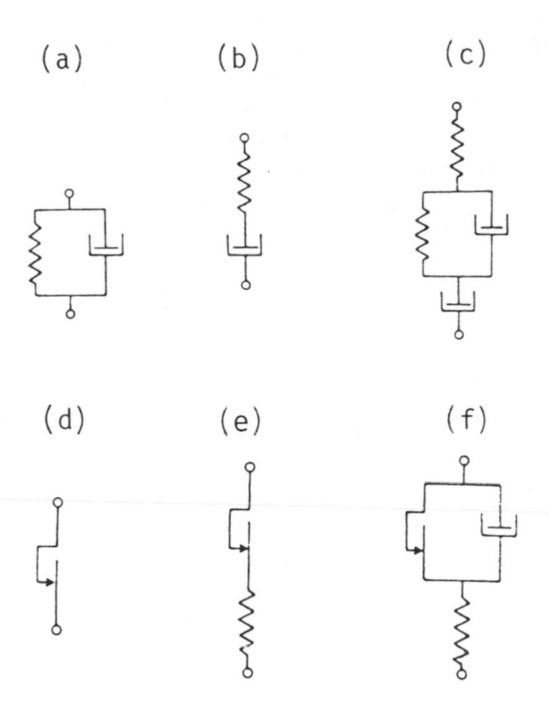

Fig. 1.6. Rheological models. (a) Voight-Kelvin, (b) Maxwell, (c) Burgers, (d) St. Venant, (e) plasto-elastic body, (f) plasto-viscoelastic or Bingham body. Springs represent elastic elements, dashpots represent viscous elements, and slides represent friction elements.

The manner in which stress is to act on a body can involve static (constant stress or strain), dynamic (varying stress or strain), or impact (stress is exerted and removed in a very short period of time) situations. Impact during mechanical handling is the most common cause of mechanical damage to foodstuffs. Behavior under static stress governs the extent of potential mechanical injury in situations involving storage in bulk, containerized shipping, etc. Working with oranges, Chuma *et al.* (1978) found that bruising of the fruit by impact was three times as great as that by static loading.

When constant stress is applied to a body, the increase in strain as a function of time is called "creep." The ratio of strain at any time t to the constant stress is called "creep compliance."

When constant strain is applied to a body, the decrease in stress as a function of time is called "stress relaxation." It is caused by the inner flow of the material, which is governed by its viscosity. The ratio of viscosity to the shear modulus is called the "relaxation time."

When stress is applied sinusoidally with time, the strain response will also vary sinusoidally, but in general will be out of phase with the stress. The strain response can be vectorially divided into two components: one which is in phase with the applied stress (and corresponds to the elastic components of the tested system) and one which lags 90° behind the applied stress (and corresponds to the viscous components of the tested system). This forms the basis for the *dynamic viscoelastic measurements*. For a more extensive treatment of rheological principles the reader is referred to the text by Sherman (1970).

Behavior of Fluids

The most important mechanical–rheological behavior of fluid foods is their flow behavior and the accompanying parameters. Figure 1.7 shows the three types of flow most common among liquid foodstuffs: *Newtonian*, where the shear stress increases linearly with the shear rate, *pseudoplastic*, where the shear stress increases curvilinearly down with the shear rate, and *Bingham*, where the flow is Newtonian following a yield point (τ_0), i.e., the level of stress needed to initiate flow. Since viscosity is the ratio of shear stress to shear rate

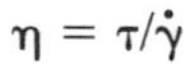

$$\eta = \tau/\dot{\gamma}$$

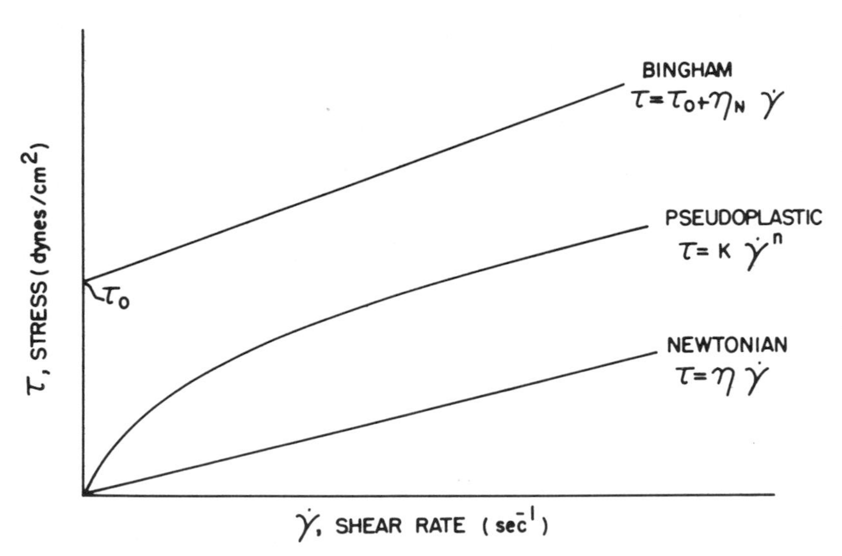

Fig. 1.7. Typical flow curves for liquid food products.

Newtonian and Bingham behaviors indicate that the viscosity of the material is shear-rate independent. Pseudoplastic (or shear-thinning) behavior indicates that the viscosity decreases with increasing shear rate. Newtonian flow is exhibited by water, sugar solutions, and vegetable oils. Pseudoplastic flow is exhibited by most hydrocolloid solutions, ketchup, fruit purees, condensed milk, mustard, and other materials possessing a shear-sensitive structure. Bingham flow is shown by certain dessert sauces and chocolate toppings. Not shown in Fig. 1.7 are two other flow types: dilatant (shear thickening) and plastic (pseudoplastic flow preceded by a yield stress); these are not very common among food materials.

A number of equations have been developed for the characterization of non-Newtonian flow. Most pseudoplastic liquid foods obey the empirical equation known as the "power law." Since pseudoplastic flow is characterized by

$$\tau = k\dot{\gamma}^n$$

its logarithmic form will be

$$\log \tau = \log k + n \log \dot{\gamma}$$

For most liquid foods the plot of τ vs $\dot{\gamma}$ on log–log coordinates gives a straight line whose slope is n and intercept is $\log k$. The slope denotes the deviation from Newtonian behavior ($n = 1$ for Newtonian flow, $n < 1$ for pseudoplastic flow, and $n > 1$ for dilatant flow), and k is usually called the consistency index. Both parameters are affected by temperature and the solids content, the effect being greater on the consistency index. Values of these parameters for various fluid foods were compiled by Holdsworth (1971), Higgs and Norrington (1971), Rao (1977A, B) and others. Some selected values are illustrated in Table 1.10.

TABLE 1.10. POWER LAW CONSTANTS FOR SELECTED FLUID FOODSTUFFS AT 25°C

	n	K (dynes s^n/cm^2)
Tomato puree	0.55	10.80
Apricot puree	0.31	200
Apple sauce	0.64	5.0
French mustard	0.40	334
Tomato ketchup	0.28	187
Guar gum, 1.5%	0.16	464
Xanthan gum, 1.2%	0.26	39.8
Sweetened condensed milk	0.83	36

Source: Adapted from Higgs and Norrington (1971), Rao (1977A), and Holdsworth (1971).

In cases of materials exhibiting a yield point, the above equations are modified to the form

$$\tau - \tau_0 = k\,\dot{\gamma}^{\,n} \qquad \text{and} \qquad \log(\tau - \tau_0) = \log k + n \log \dot{\gamma}$$

Another flow model that has been widely used for foods, especially for molten chocolate (Chevalley 1975), is Casson's model

$$\tau^{0.5} = K_0 + K_1\dot{\gamma}^{0.5}$$

where K_0 and K_1 are constants, and $K_0{}^2$ can be considered as the yield stress.

Flow properties influence the mouthfeel of liquid foods and their processibility. Working with aqueous solutions of food gums, Szczesniak and Farkas (1962) have shown that sensory sliminess decreases with increasing deviation from Newtonian behavior, and that the greater the pseudoplasticity of the solution, the less slimy it felt in the mouth and the easier it was to manipulate and swallow. This finding was subsequently confirmed by Stone and Oliver (1966).

In processing, the flow characteristics influence pumping requirements, flow of the fluid through pipes, extrudability, behavior on pasteurization, etc. These are dealt with in detail in the texts by Heldman and Singh (1981) and Charm (1978), and in a recent review article by Rao and Anantheswaran (1982).

Flow properties can be determined using any of the variety of available viscometer and rheometers. Their types can be classified as rotational, tube (capillary, orifice, and pipe), and other systems (e.g., falling ball, rising bubble, sliding plate). An excellent source of information on the different types of viscometers, their principles and operation is the text by Van Wazer *et al.* (1963). The most serious issues in flow properties determination are (a) avoiding errors caused by equipment geometry, and (b) in the case of non-Newtonian flow, ascertaining that the characterization is either complete or is done under conditions similar to those to be used in practice. This is necessary to insure that meaningful processing data are obtained.

Behavior of Solids

Compression. The behavior of foods in compression is one of the easiest and most important mechanical tests to perform. With the growing popularity of the universal testing machines such as the Instron, which provide means of applying, detecting, and recording the forces of sample resistance, the test requires only two flat parallel plates, one fixed (on which the sample is placed) and the other moving at a predetermined rate to impose strain on the test sample. A typical compression force–deformation curve for a foodstuff is shown in Fig. 1.8. As indicated, a number of mechanical parameters can be quantified from such a curve. The slope of the initial straight line portion is taken as the elastic modulus and is often considered to be a measure of firmness. Figure 1.8 also shows two yield points. The first point of inflection indicates a bioyield—failure of some structural elements in the tested biological material. The final yield point is the rupture yield—massive failure of the specimen. It is equivalent to the compressive strength.

Because of the presence of a viscous element in the rheological structure of most foods, the shape and position of the compression force–deformation curves are often highly influenced by the rate of the imposed strain (Culioli and Sherman 1976). This is due to the fact that the material relaxes (or flows) while being compressed, the extent of this flow being dependent on the nature of the viscous element. The lower the viscosity of the material (i.e., the lower the internal friction), the faster and more complete will be the relaxation, and the greater the strain-rate sensitivity. Because of this phenomenon, usually the faster the rate of deformation, the higher the forces needed to compress the material. Compression testing of foods is usually done at deformation rates of 2–50 cm/min.

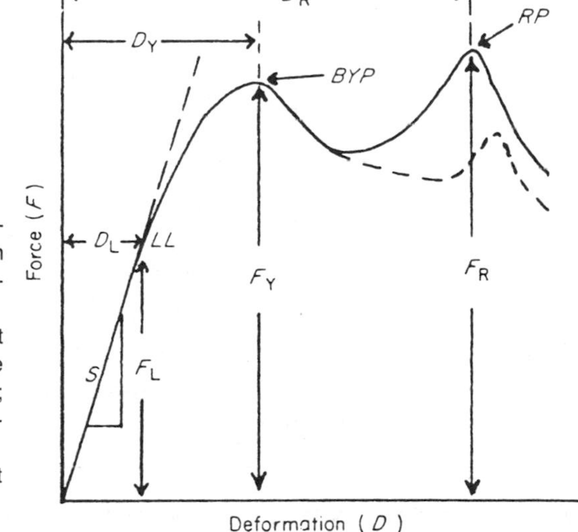

Fig. 1.8. Generalized compressive force-deformation curve. S = modulus of elasticity; BYP = bioyield point; RP = rupture point; LL = limit of elasticity; F_L and D_Y = force and deformation at elastic limit; F_Y and D_Y = force and deformation at bioyield; F_R and D_R = force and deformation at rupture.
From Brennan (1980).

Impact testing involves very high deformation rates of the order of 9000 cm/min and upward. Holt and Schoorl (1977) and Schoorl and Holt (1977) reported on impact testing of apples and described the employed testing techniques. They found a linear relationship between bruising of the fruit and the level of impact energy absorbed and pointed out implications to package design and materials handling equipment. An in-depth treatment of impact testing can be found in the text by Mohsenin (1980).

Tensile. In general, tensile measurements are difficult to perform on foods because of their geometry and because of gripping problems with materials whose geometry lends itself to this type of a test (e.g., noodles, meat). It is important that failure occurs between and not at the grips. For this reason it is preferable to use dumbbell-shaped specimens.

Figure 1.9 shows a tensile stress–strain curve for stick chewing gum and indicates the different parameters that can be quantified. As pointed out by Voisey and deMan (1976), tensile parameters are important in mastication since the wedging action of teeth imposes tensile stresses.

Shear. Initially developed for meat, shear measurements are still the most prevalent test in meat research. As has already been pointed out, pure shear forces are only rarely encountered in food products. Figure 1.10 shows typical "shear" force–deformation curves for cooked beef muscle obtained with a recording Warner–Bratzler shear appara-

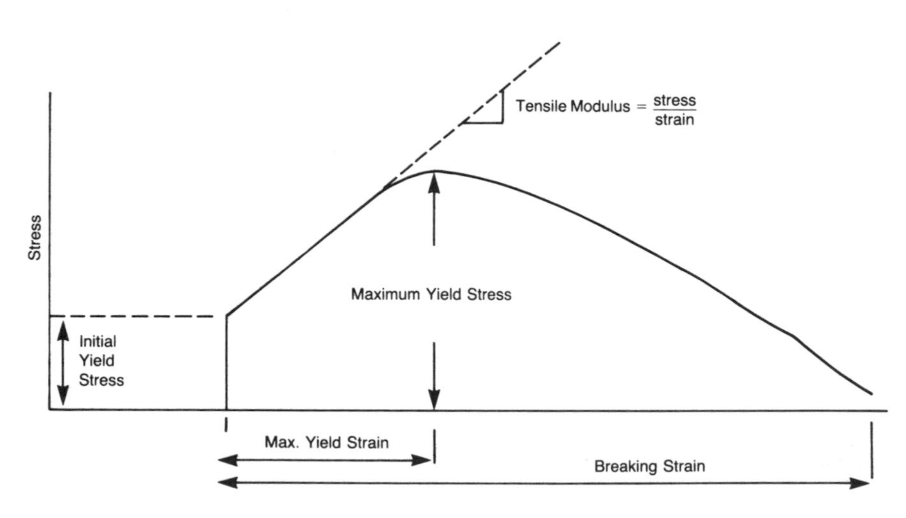

Fig. 1.9. Generalized tensile stress–strain curve for stick chewing gum.

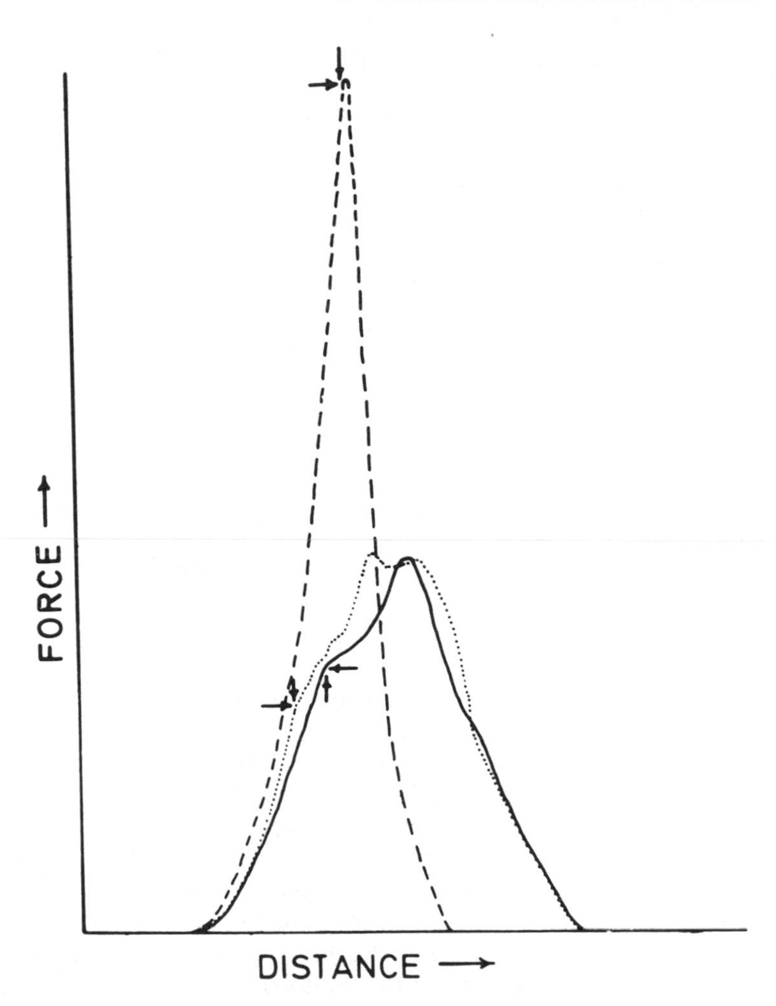

Fig. 1.10. Shear force-deformation curves for cooked beef deep pectoral muscles. ——control; ··· stretched; - - - cold-shortened; vectorial arrows indicate first yield points.
From Bouton et al. (1975).

tus. It illustrates that the curves can assume different shapes depending on the myofibrillar contraction states and, thus, structure of the muscle fibers. Typical shear deformation curves obtained with the Kramer shear press for a variety of foodstuffs have been published by Szczesniak *et al.* (1970).

Stress Relaxation. A typical stress relaxation curve for an agricultural product is shown in Fig. 1.11. It illustrates that the rate of relaxa-

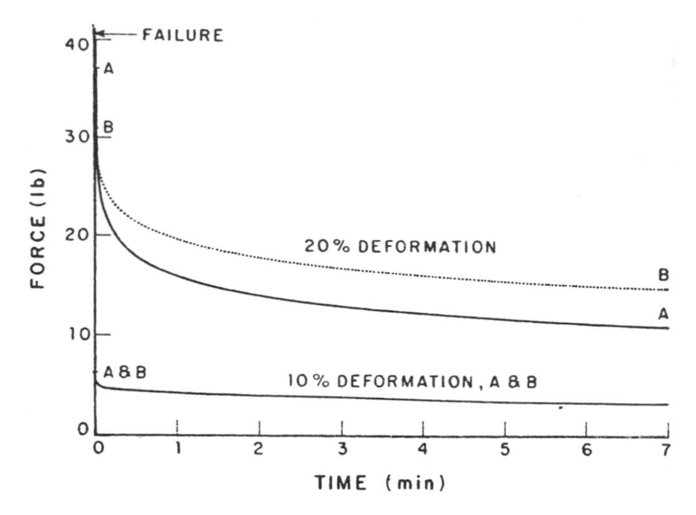

Fig. 1.11. Stress relaxation curve for raw potato tissue. Curve A, 10 cm/min; curve B, 1 cm/min.
From Peleg and Calzada (1976).

tion and residual stress at any given time depend on the initial force and deformation applied and the rate at which the test material was initially deformed. The greater the extent of initial deformation and of applied force, the higher the residual stress, and the faster the relaxation rate. The slower the rate of initial deformation (or force application), the slower the material appears to relax. This is due to the fact that the material already had started to relax during the initial force application stage, as has been already mentioned in the discussion of the compression test. For these reasons, the initial load should be applied as quickly as possible, being careful to avoid errors caused by pen overshoot.

A convenient way of expressing stress relaxation is to measure the time required for the force to relax to a given percentage (e.g., 60%) of its initial value. A preferred, but a more time-consuming, approach is to follow the equation for a viscoelastic body

$$\tau(t) = \tau_{\text{initial}} e^{-t/\alpha}$$

where $\tau(t)$ is stress at time t and α is the "relaxation time" that characterizes the material response. This quantity is determined by plotting $\log \tau(t)/\tau_{\text{initial}}$ vs time. The slope of the linear portion of the curve is $-1/\alpha$.

As pointed out by Peleg (1977B) and Peleg and Calzada (1976), the stress relaxation test is a very useful supplement to the compression test. The shape of the curve can be related to the elements in the rheological model. The test can also yield information about the structural changes that might have occurred on previous deformation and about the material's rheological memory and past history.

In applying the stress relaxation test to foods one should keep in mind that viscous flow may not be the only mechanism responsible for relaxation, and that the decrease in force may also be caused by structural failure (e.g., development of internal cracks). This invalidates the test from the theoretical viewpoint but may still provide valuable empirical information.

Creep. In contrast to the stress relaxation test, which can be easily performed with an ordinary Instron or a similar apparatus, the creep test requires a special instrument or a special capability with the Instron to keep the load (or force) constant and record the change in deformation (creep) as a function of time. However, rheologically, this test yields significant data and was very popular in the early days of food rheology. It is still very popular among the European researchers.

Figure 1.12 shows a typical creep curve. As discussed in detail by Sherman (1970), the creep curve has three main regions: (a) an instan-

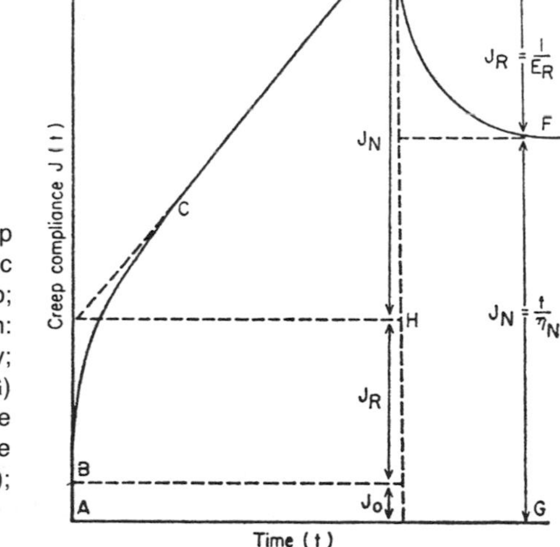

Fig. 1.12. Model creep curve. Creep region: (A-B) instantaneous elastic creep; (B-C) retarded elastic creep; (C-D) Newtonian flow. Recovery region: (D-E) instantaneous elastic recovery; (E-F) retarded elastic recovery; (F-G) unrecovered deformation. The three creep regions correspond to the three recovery regions: (A-B) = (D-E); (B-C) = (E-F); (C-D) =(F-G).
From Sherman (1970).

taneous elastic region in which the bonds are stretched elastically and complete recovery occurs on removal of stress; (b) a retarded elastic region in which the bonds break and reform; (c) a linear region of Newtonian flow in which the units flow past one another since the time for the bonds to reform is longer than the test period. Similarly, three regions are found when stress is removed and recovery occurs: (a) instantaneous elastic recovery; (b) retarded elastic recovery; (c) unrecovered deformation.

Rheological Model Building. Stress relaxation and, especially, creep tests can be used to define the rheological elements operating in the material and to construct models portraying its rheological behavior. This approach is illustrated by the classic work of Sherman on ice cream (Shama and Sherman 1966). The observed creep behavior was characterized in terms of a six-element viscoelastic model composed of a Maxwell and two Voight–Kelvin models connected in series. As illustrated in Fig. 1.13, the specific elements in the model were related to the structural units in the frozen ice cream. Similar models were constructed for the ice cream mix and the melted product.

Instrumentation. A number of publications are available that discuss instrumentation used for quantifying the mechanical properties of foods important to their rheological and textural characteristics (Szczesniak 1963B; Voisey and deMan 1976). The text by Kramer and Szczesniak (1973) contains listings of commercially available instruments and of named instruments described in the literature.

All the texture testing instruments have a number of elements in

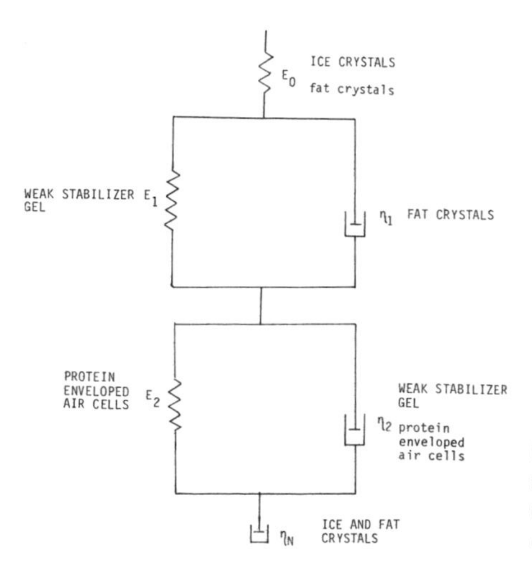

Fig. 1.13. Rheological model for frozen ice cream showing the relationship between rheological and structural elements.
From Shama and Sherman (1966).

common, as shown in Table 1.11. Of these, the probe contacting the test sample is the key feature, since it governs the type of forces that act on the specimen and their distribution. The current trend is to use a universal testing machine (e.g., the Instron) for the driving mechanism, the sensory element, and the readout system and to equip it with different probes depending on the type of test desired.

Texture testing instruments can be classified in a number of ways. Classification according to the measured variable is shown in Table 1.11. The Instron is a multiple-measuring instrument as is the GF Texturometer.

Texture Profiling. The common use of recorders has facilitated another current trend and that is to quantify a number of different parameters from one test. This is illustrated in Figs. 1.8, 1.9, 1.10 and 1.12. This approach to food texture evaluation is termed "texture profiling." In the broad sense, it involves multiparameter interpretation of stress–strain curves. In a narrower sense, and when originally developed, it

TABLE 1.11. CHARACTERISTIC FEATURES OF TEXTURE TESTING INSTRUMENTS

Basic Elements[a]

Element	Examples
Probe	Flat plunger, plate, piercing rod, penetrating cone, cutting blade, shearing jaws, cutting wires
Driving mechanism	Weight/pulley arrangement, hydraulic system, variable drive electric motor
Sensing element	Simple spring, strain gauge, load cell
Readout system	Max. force dial, oscilloscope, $X-Y$ recorder

Classification According to Measured Variable[b,c]

Method	Measured variable	Dimensional units	Examples
1. Force-measuring	Force (F)	ml/t^{-2}	Tenderometer
2. Distance-measuring	distance	l	Penetrometers
	area	l^2	Grawemeyer Consistometer
	volume	l^3	Seed Displacement (bread volume)
3. Time-measuring	Time (T)	t	Ostwald viscometer
4. Energy-measuring	Work (F × D)	ml^2/t^2	Farinograph
5. Ratio-measuring	F, D, or T, or F × D, measured twice	Dimensionless	Cohesiveness (G.F. Texture Profile)
6. Multiple-measuring	F, D, and T, and F × D	$m\ l/t^2$, l, t, ml^2/t^2	G.F. Texturometer
7. Multiple-variable	F, D, or T (all vary)	Unclear	Durometer
8. Chemical analysis	Concentration	Dimensionless (% or ppm)	Alcohol insoluble solids
9. Miscellaneous	Anything	Anything	Optical density of fish homogenate

[a]After Szczesniak (1966).
[b]After Bourne (1966).
[c]Abbreviations are as follows: m, mass; l, length; t, time.

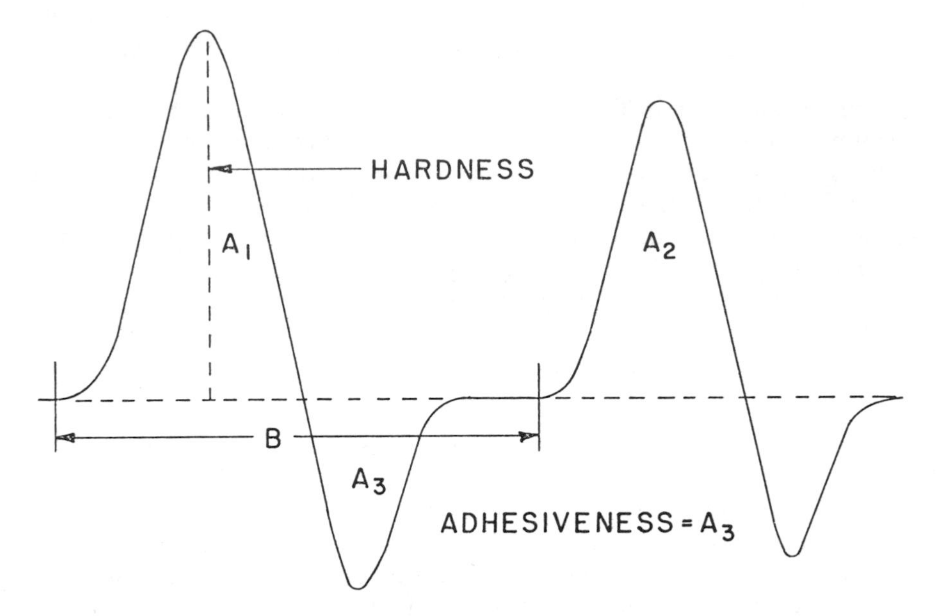

Fig. 1.14. Generalized texture profile analysis curve obtained with the General Foods texturometer. Cohesiveness = A_2/A_1; springiness = $C - B$; C = time constant for clay.
From Friedman et al. (1963).

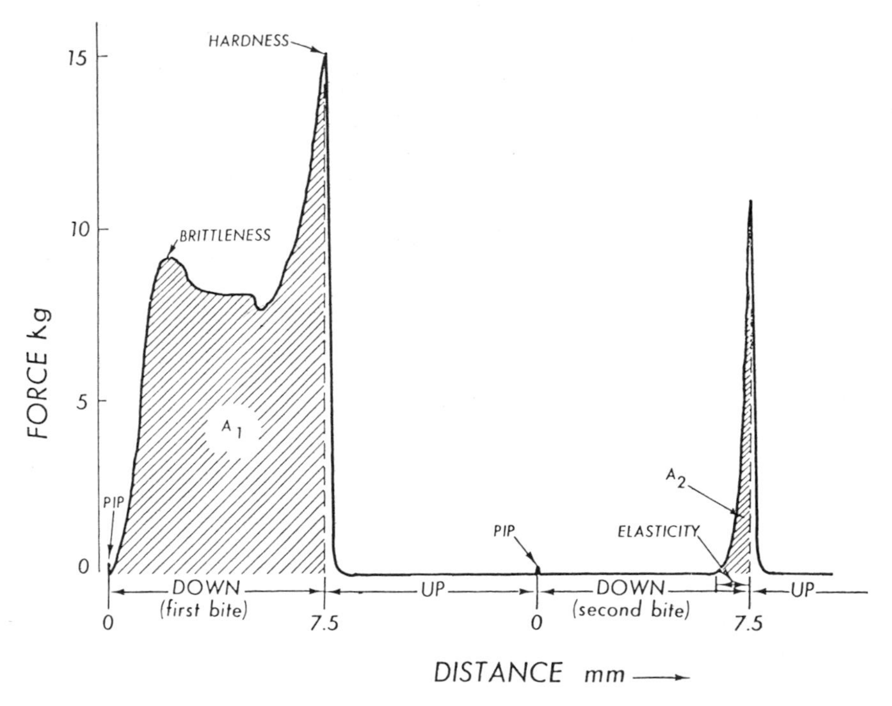

Fig. 1.15. Typical texture profile analysis curve obtained with the Instron on pears.
From Bourne (1968).

involves multiple "chews" (or stress–strain curves) on the same sample and interpretation of the recorded curves in terms of the textural characteristics as defined by Szczesniak (1963A) and summarized in Table 1.9. Typical instrumental profiles for foodstuffs obtained with the GF Texturometer and the Instron are shown in Figs. 1.14 and 1.15, respectively. A good review of the principles and applications of the texture profile analysis has been published by Breene (1975).

CONCLUDING REMARKS

Since the physical properties of foods represent such a vast area, this discussion was limited to the key elements. Certain properties (e.g., enthalpy, surface tension) were, of necessity, omitted, but interested readers may be guided to the proper sources by the extensive bibliography provided. Other chapters in this book will fill some of the gaps present in this treatment of the subject.

BIBLIOGRAPHY

AMERINE, M.A., PANGBORN, R.M., and ROESSLER, E.B. 1965. Principles of Sensory Evaluation of Food, pp. 230–235. Academic Press, New York.

ANON. 1981. The Almanac of Canning, Freezing, Preserving Industries, 66th ed., p. 403. Edward E. Judge & Sons, Westminster, MD.

ARNOLD, P.C., and MOHSENIN, N.N. 1971. Proposed techniques for axial compression tests on intact agricultural products of convex shape. Trans. ASAE *14*, 78–84.

ASAE. 1976. Compression tests of food materials of convex shape. ASAE Recommendation R368. Agric. Eng. Handb. pp. 415–418.

BENGTSSON, N.E., MELIN, J., REMI, K., and SODERLIND, S. 1963. Measurements of the dielectric properties of frozen and defrosted meat and fish in the frequency range 10–200 MHz. J. Sci. Food Agric. *14*, 592–604.

BIRTH, G.S., and OLSEN, K.L. 1964. Nondestructive detection of water core in Delicious apples. Proc. Am. Soc. Hort. Sci. *85*, 74–84.

BREENE, W.M. 1975. Application of texture profile analysis to instrumental food texture evaluation. J. Texture Stud. *6*, 53–82.

BRENNAN, J.G. 1980. Food texture measurement. *In* Developments in Food Analysis Techniques, Vol. 2. R.D. King (Editor). Applied Science Publishers, Barking, Essex, England.

BOURNE, M.C. 1966. A classification of objective methods for measuring texture and consistency of foods. J. Food Sci. *31*, 1011–1015.

BOURNE, M.C. 1968. Texture profile of ripening pears. J. Food Sci. *33*, 223–226.

BOURNE, M.C. 1975. Is rheology enough for food texture measurements? J. Texture Stud. *6*, 259–262.

BOUTON, P.E., HARRIS, P.V., and SHORTHOUSE, W.R. 1975. Possible relationships between shear, tensile and adhesion properties of meat and meat structure. J. Texture Stud. *6*, 297–314.

CALZADA, J.F., and PELEG, M. 1978. Mechanical interpretation of compressive stress-strain relationships of solid foods. J. Food Sci. *43*, 1087–1092.

CHARM, S.E. 1978. The Fundamentals of Food Engineering, 3rd ed. AVI Publishing Co., Westport, CT.

CHEVALLEY, J. 1975. Rheology of chocolate. J. Texture Stud. *6*, 177–196.

CHONG, J.S., CHRISTIANSEN, E.B., and BAER, A.D. 1971. Rheology of concentrated suspensions. J. Appl. Polym. Sci. *15*, 2007–2021.

CHUMA, Y., SHIGA, T., and IWAMOTO, M. 1978. Mechanical properties of Satsuma orange as related to the design of a container for bulk transportation. J. Texture Stud. *9*, 461–479.

COPSON, D.A. 1975. Microwave Heating, 2nd ed. AVI Publishing Co., Westport, CT.

CUEVAS, R., and CHERYAN, M. 1979. Thermal conductivity of liquid foods. J. Food Process Eng. *2*, 283–306.

CULIOLI, J., and SHERMAN, P. 1976. Evaluation of Gouda cheese firmness by compression tests. J. Texture Stud. *7*, 353–372.

CZERKASKYJ, A. 1971. Consumer response to color in canned cling peaches. J. Food Sci. *36*, 671–673.

DEALY, J.M. 1982. Rheometers for Molten Plastics. Van Nostrand-Reinhold Co., New York.

DeLOOR, G.P., and MEIJBOOM, F.W. 1966. The dielectric constant of foods and other materials with high water contents at microwave frequencies. J. Food Technol. *1*, 313–322.

DeMAN, J.M. 1976. Mechanical properties of foods. *In* Rheology and Texture in Food Quality. J.M. deMan, P.W. Voisey, V.F. Rasper and D.W. Stanley (Editors). AVI Publishing Co., Westport, CT.

DICKERSON, R.W., JR. 1965. An apparatus for the measurement of thermal diffusivity of foods. Food Technol. *19*(5) 198–204.

DU BOSE, C.N., CARDELLO, A.V., and MALLER, O. 1980. Effects of colorants and flavorants on identification, perceived flavor intensity, and hedonic quality of fruit-flavored beverages and cakes. J. Food Sci. *45*, 1393–1399.

EDE, A.J., and HADDOW, R.R. 1951. The electrical properties of food at high frequencies. Food Manuf. *26*, 156–160.

EVESON, G.F. 1959. The viscosity of stable suspensions of spheres at low rates of shear. *In* Rheology of Disperse Systems. C.C. Mill (Editor). Pergamon Press, London.

FRANCIS, F.J., and CLYDESDALE, F.M. 1975. Food Colorimetry: Theory and Applications. AVI Publishing Co., Westport, CT.

FRIEDMAN, H.H., WHITNEY, J.E., and SZCZESNIAK, A.S. 1963. The Texturometer—A new instrument for objective texture measurement. J. Food Sci. *28*, 390–396.

GOLDBLITH, S.A. 1967. Basic principles of microwaves and recent developments. Adv. Food Res. *15*, 277–301.

GOTO, H., and KUNO, H. 1982. Flow of suspensions containing particles of two different sizes through a capillary tube. J. Rheol. *26*, 387–398.

HELDMAN, D.R., and SINGH, R.P. 1981. Food Process Engineering, 2nd ed. AVI Publishing Co., Westport, CT.

HIGGS, S.J., and NORRINGTON, R.J. 1971. Rheological properties of selected foodstuffs. Process Biochem. *6*(5) 52–54.

HOLDSWORTH, S.D. 1971. Applicability of rheological models to the interpretation of flow and processing behavior of fluid food products. J. Texture Stud. *2*, 393–418.

HOLT, J.E., and SCHOORL, D. 1977. Bruising and energy dissipation in apples. J. Texture Stud. *7*, 421–432.

HUNTER, R.S. 1973. The measurement of appearance. Hunter Associates Lab., Fairfax, VA.

HUNTER, R.S. 1975. The Measurement of Appearance. John Wiley & Sons, New York.

JASON, A.C., and JOWITT, R. 1968. Physical properties of foodstuffs in relation to engineering design. 3rd Eur. Symp. European Federation of Chemical Engineering, Bristol, England, April 1968. DECHEMA Monogr. 63, Theodor-Heuss-Allee 24, Frankfurt, West Germany.

JEPPSON, M.R. 1964. Consider microwaves. Food Eng. *36*(11) 49–52.

KATTAN, A.A., OGLE, W.G., and KRAMER, A. 1956. Effect of process variables on quality of canned tomato juice. Proc. Am. Soc. Hort. Sci. *68*, 470–481.

KOSTYLA, A.S., and CLYDESDALE, F.M. 1978. The psychophysical relationship between color and flavor. CRC Crit. Rev. Food Sci. Nutr. *10*, 303–321.

KRAMER, A., and SZCZESNIAK, A.S. 1973. Texture Measurements of Foods. D. Reidel Publishing Co., Dordrecht, Holland.

KRAMER, A., and TWIGG, B.A. 1970. Quality Control for the Food Industry, Vol. 1, 3rd Ed. AVI Publishing Co., Westport, CT.

MAGA, J.A. 1973. Influence of freshness and color on potato chip sensory preferences. J. Food Sci. *38*, 1251–1252.

MEDALIA, A.I. 1980. Three-dimentional shape parameters. *In* Testing and Characterization of Powders and Fine Particles. J.K. Beddow and T. Meloy (Editors). Heyden & Son Ltd., London.

MOHSENIN, N.N. (Editor). 1965. Terms, definitions and measurements related to mechanical harvesting of selected fruits and vegetables. Pennsylvania Agric. Exp. Stn. Progr. Rep. 257.

MOHSENIN, N.N. 1980. Physical Properties of Plant and Animal Materials. Gordon & Breach Science Publishers, New York.

MUDGETT, R.E. 1982. Electrical properties of foods in microwave processing. Food Technol. *36*(2) 109–115.

MULLER, H.G. 1973. An Introduction to Food Rheology. Crane, Russak & Co., New York.

NELSON, S.O. 1973A. Electrical properties of agricultural products. A critical review. Trans. ASAE *16*, 384–400.

NELSON, S.O. 1973B. Microwave dielectric properties of grain and seed. Trans. ASAW *16*, 902–905.

NELSON, S.O. 1978. Electrical properties of grain and other food materials. J. Food Proc. & Preserv. *2*, 137–154.

NELSON, S.O. 1979. RF and microwave dielectric properties of shelled, yellow-dent field corn. Trans. ASAW *22*, 1451–1457.

NELSON, S.O. 1980. Microwave dielectric properties of fresh fruits and vegetables. Trans. ASAE *23*, 1314–1317.

NELSON, S.O. 1982. Dielectric properties of some fresh fruits and vegetables at frequencies of 2.45 to 22 GHz. ASAE Paper No. 82-3053, ASAE, St. Joseph, MI 49085.

PACE, W.E., WESPHAL, W.B., GOLDBLITH, S.A., and VAN DYKE, D. 1968. Dielectric properties of potatoes and potato chips. J. Food Sci. *33*, 37–42.

PALL, R., and MOHSENIN, N.N. 1980. Permeability of porous media as a function of porosity and particles size distribution. Trans. ASAE *23*, 742–745.

PAUL, P.C., and PALMER, H.H. (Editors). 1972. Food Theory and Applications. John Wiley & Sons, New York.

PELEG, M. 1976. Considerations of a general rheological model for the mechanical behavior of viscoelastic solid food materials. J. Texture Studies *7*, 243–255.

PELEG, M. 1977A. Contact and fracture elements as components of the rheological memory of solid foods. J. Texture Studies *8*, 67–76.

PELEG, M. 1977B. Operational conditions and the stress-strain relationship of solid foods—Theoretical evaluation. J. Texture Studies *8*, 283–295.

PELEG, M., and CALZADA, J.F. 1976. Stress relaxation of deformed fruits and vegetables. J. Food Sci. *41*, 1325–1329.

POMERANZ, Y., and MELOAN, C.E. 1971. Food Analysis: Theory and Practice, pp. 372–376. AVI Publishing Co., Westport, CT.

POTTER, N.N. 1978. Food Science, 3rd Ed. AVI Publishing Co., Westport, CT.

QASHOU, M.S., VACHON, R.I., and TOULOUKIAN, Y.S. 1972. Thermal conductivity of foods. ASHRAE Trans. *78*(1), 165–182.

RAO, M.A. 1977A. Rheology of liquid foods. J. Texture Stud. *8*, 135–168.

RAO, M.A. 1977B. Measurement of flow properties of liquid foods—Developments, limitations and interpretation of phenomena. J. Texture Stud. *8*, 257–282.

RAO, M.A., and ANANTHESWARAN, R.C. 1982. Rheology of fluids in food processing. Food Technol. *36*(2) 116–126.

SCHOORL, D., and HOLT, J.E. 1977. The effects of storage time and temperature on the bruising of Jonathan, Delicious and Granny Smith apples. J. Texture Stud. *8*, 409–416.

SHAMA, F., and SHERMAN, P. 1966. The texture of ice cream. 2. Rheological properties of frozen ice cream. J. Food Sci. *31*, 699–706.

SHERMAN, P. 1961. Rheological methods for studying the physical properties of emulsifier films at the oil–water interface in ice cream. Food Technol. *15*(9) 394–399.

SHERMAN, P. 1970. Industrial Rheology. Academic Press, New York.

SINGH, R.P. 1982. Thermal diffusivity in food processing. Food Technol. *36*(2) 87–91.

STEVENS, M.A., DICKINSON, G.L., and AQUIRRE, M.F. 1976. UC82, A High-Yielding Processing Tomato. Vegetable Crops Ser. 183, Vegetable Crop Dept., Univ. of California, Davis.

STONE, H., and OLIVER, S. 1966. Effect of viscosity on the detection of relative sweetness intensity of sucrose solutions. J. Food Sci. *31*, 129–134.

SURAK, J.G., MATTHEWS, R.F., WANG, V., PADUA, H.A., and HAMILTON, R.M. 1979. Particle size distribution of commercial tomato juices. Proc. Fla. State Hort. Soc. *92*, 159–163.

SZCZESNIAK, A.S. 1963A. Classification of textural characteristics, J. Food Sci. *28*, 385–389.

SZCZESNIAK, A.S. 1963B. Objective measurement of food texture. J. Food Sci. *28*, 410–420.

SZCZESNIAK, A.S. 1966. Texture measurements. Food Technol. *20*, 1292–1298.

SZCZESNIAK, A.S., and FARKAS, E. 1962. Objective characterization of the mouthfeel of gum solutions. J. Food Sci. *27*, 381–385.

SZCZESNIAK, A.S., HUMBAUGH, P.R., and BLOCK, H.W. 1970. Behavior of different foods in the standard shear compression cell of the shear and the effect of sample weight on peak area and maximum force. J. Texture Stud. *1*, 356–378.

VAN WAZER, J.R., LYONS, J.W., KIM, K.Y., and COLWELL, R.E. 1963. Viscosity and Flow Measurement—A Laboratory Handbook of Rheology. Interscience Publishers, New York.

VOISEY, P.W., and DEMAN, J.M. 1976. Applications of instruments for measuring food texture. *In* Rheology and Texture in Food Quality. J.M. deMan, P.W. Voisey, V.F. Rasper, and D.W. Stanley (Editors). AVI Publishing Co., Westport, CT.

VON ELBE, J.H., and JOHNSON, C.E. 1971. Sales pattern changing—Color may be factor. Canner/Packer Aug., 10–11.

VON HIPPEL, A.R. (Editor). 1954. Dielectric Materials and Applications. MIT Press, Cambridge, MA.

WOODAMS, E.E., and NOWREY, J.E. 1968. Literature values of thermal conductivities of foods. Food Technol. *22*(4) 150–158.

2

Electron Microscopy of Foods

Miloslav Kalab[1]

INTRODUCTION

Interest of food scientists in using electron microscopy has been growing rapidly, particularly in areas related to food analysis (Stasny *et al.* 1981), interactions between food ingredients, and food texture (Stanley and Geissinger 1972; Stanley and Swatland 1976; Stanley and Tung 1976; Lewis 1981).

Although this review can make food scientists aware of some of the electron microscopy techniques useful in food research, comprehensive information on all electron microscopic techniques cannot be provided. This review also points to sources of information. Such information may be found in monographs, some of which have been listed in the Appendix at the end of this review, and which deal with electron microscopy of biological samples in general. A monograph entitled "Food Microscopy" (1979) reviews the results of microscopic studies of a great variety of foods and also mentions sample preparation. Valuable information may also be found in a monograph entitled "Studies of Food Microstructure" (1981), consisting of reviews and tutorials, as well as original research papers published earlier in various volumes of *Scanning Electron Microscopy*. Food science journals and, in particular, the new international journal, *Food Microstructure*, which specializes in microscopy and microanalysis of foods, are additional sources of information as well as a forum for the presentation of results.

[1] Food Research Institute, Research Branch, Agriculture Canada, Ottawa, Ontario, Canada.

PRINCIPLES OF ELECTRON MICROSCOPY

Principles of electron microscopy have been described in many handbooks, some of which are listed in the Appendix. In general, a focused beam of electrons is accelerated by high voltage *in vacuo* and interacts with the specimen. Because the specimen is normally examined *in vacuo*, ideally it must not contain volatile substances. In practice this means that all foods must first be dehydrated or frozen to temperatures that prevent the generation of measurable amounts of vapor.

Two types of electron microscopy are distinguished: scanning (SEM) and transmission electron microscopy (TEM).

In SEM, the sample is scanned by an electron beam focused into a point. On impact, the primary electron beam produces secondary and back-scattered electrons from the specimen surface, which are used for imaging the specimen. X-Rays produced by this interaction are used in X-ray spectrometry (Marshall 1975) to detect the presence, ratios, and quantities of elements in a specimen. X-Ray spectrometry will not be dealt with in this review.

Secondary or reflected electrons from the specimen surface are processed for display and photography on cathode ray tubes (CRT). The visual CRT is used for focusing the beam and selecting areas of interest in the specimen. Another, higher resolution CRT is used for photography.

Scanning electron microscopes operate typically at resolutions on the order of 10 nm. As accelerating voltage determines the kinetic energy of the electrons, the electrons penetrate more deeply into the specimen at high accelerating voltages; damage to the surface of the specimen may be associated with high accelerating voltage (Black 1974). Specimen charging, a local accumulation of electrical charge produced by the incident electron beam striking a non- or poorly conducting specimen, produces a variety of serious problems in imaging the specimen. For this reason, most food samples must be rendered electrically conductive by coating with a thin layer of metal or by chemical treatment of the sample (Murphy 1978, 1980). The former technique is considerably more usual in food microscopy. Gold or a gold-palladium alloy has been used for coating of the samples. Defects in the continuity of the metal coating are sources of specimen charging. Rapid scanning at a low accelerating voltage, the use of backscattered electrons, or additional coating of the sample are some ways to combat the charging defect.

Although SEM micrographs give the impression of three-dimensional appearance, more accurate information concerning the microstructure of the specimen may be obtained by viewing pairs of stereo micrographs. One set of such micrographs taken at different angles of the incident electron beam can provide more information than many individual conventional micrographs. Standard photogrammetric techniques can be used to determine height differences in the micrographs.

In TEM, the electron beam, accelerated by a voltage of 20–100 kV (and for some special purposes up to 1000 kV) penetrates a thin specimen. The enlarged shadow of the specimen is viewed on a fluorescent screen for focusing and selecting the area of interest and is photographed on photographic plates or films.

The specimen may be observed directly if it is sufficiently thin and at least partly translucent to electrons (bacteria, viruses, isolated cell wall fragments, macromolecules). Negative or positive staining may be used to increase contrast in the micrographs of such specimens. Embedding of larger specimens in a resin makes it possible to obtain thin sections (less than 90 nm), stain their characteristic features (structures) with heavy metal salts, and examine the sections in the path of the electron beam.

Electron microscopic methods based on fixation of samples by rapid freezing (cryofixation) are becoming increasingly popular. The sample is frozen, fractured, the fractures are replicated with platinum and carbon, and the replicas are examined by TEM.

Some time ago, a new type of so-called scanning-transmission electron microscope (STEM) was developed, whereby the use of both principal techniques has been combined in a single instrument. There are, however, some special requirements in sample preparation which will be not dealt with in this review.

SCANNING ELECTRON MICROSCOPY

Conventional Scanning Electron Microscopy

Foods exist in various forms as diverse as liquids (beverages), powders (flour, spray-dried products), compact solids (meat, pasta), foams (whipped products and some bakery products). As far as chemical composition is concerned, foods contain a variety of substances with proteins, carbohydrates, fats, and salts forming their basis. In some foods the biological origin is clearly evident (fruits, vegetables, meat, milk), whereas in some others (bakery products, chocolate, snack foods) it is not.

With most foods, SEM has been used to study the internal microstructure of meat products (Jones *et al.* 1976; Józsa *et al.* 1980; Voyle 1981), milk products (Eino *et al.* 1976; Hall and Creamer 1972; Kalab 1977A; Schmidt and Van Hooydonk 1980; Taranto *et al.* 1979), soybean foods

(Saio 1981; Wolf and Baker 1980), vegetables (Davis and Gordon 1980; Fedec *et al.* 1977), doughs (Evans *et al.* 1981), cooked spaghetti (Dexter *et al.* 1978), and others. With foods in the form of powders, the dimensions, shapes, and characteristic surface features were studied in addition to internal microstructure (Buma 1971, 1978; Buma and Henstra 1971; Chabot *et al.* 1976; Dronzek *et al.* 1972; Evers 1971; Evers and Juliano 1976; Hall and Sayre 1969, 1970; Kalab 1979A, B, 1980A; Moss *et al.* 1980; Saito 1973; Smith 1979; Wolf and Baker 1972, 1980).

Many foods contain water. Samples of such foods must be examined either dried or frozen at a temperature below 173K (−100°C) in order to minimize generation of volatile vapors *in vacuo* inside the electron microscope unless special chambers are used. Drying may severely alter the microstructure of the sample because food components dissolved in the aqueous phase such as salts, carbohydrates, peptides, soluble proteins and others are converted into dry materials such as powders, crystals, filaments, or films and usually obscure the subject under study. Special procedures have been developed to minimize unwanted changes in the sample; such procedures include fixation, dehydration, and special drying techniques. The dried sample is fractured and the fragments are mounted on metal stubs, rendered conductive by coating with carbon and gold, and examined in the microscope.

The first step in preparing food for SEM is to obtain an average sample of suitable dimensions. Preparation of the sample depends on a number of variables and was discussed to a great extent by Angold (1979), Chabot (1979), Geissinger and Stanley (1981), Kalab (1981), and others. In view of subsequent preparatory steps, the sample should be small. Samples are usually taken from a depth of 1 to 2 cm below the surface of the food in order to avoid areas affected by external effects.

Powders are examined in the form in which they are produced and used, if the particle surface features are of interest. To examine internal microstructure, the powder particles are fractured by a blade or in a mortar. Preparation of powder particles for SEM has been dealt with in great detail by Johari and DeNee (1972) and DeNee (1978).

Randomly selected samples of solid foods, which do not possess an oriented microstructure such as yogurt, tofu, pasta, and some cheeses are trimmed into sections 0.5 × 5 × 5 mm; they are cut into smaller prisms (0.5 × 1 × 5 mm) before further preparation for SEM. Foods with an oriented microstructure such as spun proteins (Wolf and Baker 1980), meat (Jones *et al.* 1976; Carroll and Jones 1979; Cohen and Trusal 1980; Geissinger and Stanley 1981; Voyle 1981), traditional mozzarella cheese (Kalab 1977A) and others are trimmed into longitudinal and transverse sections 0.5 × 0.5 × 5 mm (Geissinger and Stanley 1981) and the sections are processed for electron microscopy separately.

Carroll and Jones (1979) devised a minitensile stage to stress muscle tissue samples. Raw and heated meat samples were stressed either in parallel or perpendicular directions to the meat fiber axis to elucidate relationships between meat tenderness and microstructure. Preparation of fish and seafood samples was described in detail by Howgate (1979). Schaller and Powrie (1971, 1972) studied beef, chicken, rainbow trout, and turkey muscles. A new preparation technique for SEM of skeletal muscles was suggested by Józsa *et al.* (1980): the samples were embedded in a resin without polymerization and were freeze-fractured. Seeds ingested by birds and partially digested were sampled for SEM by Smith (1981). A description of dough sample preparation was published by Evans *et al.* (1981); stretched samples were obtained by pulling dough strips (2 mm in diameter) until the dough was near the breaking point. Preparation of cereals for SEM was reviewed by Pomeranz (1976).

In some instances, the sample must be fixed before drying. As Chabot (1979) pointed out, however, running every food sample through a routine procedure such as fixation, dehydration, and critical-point drying is not a reasonable practice. Fixation means rendering structures in the sample rigid and resistant to changes during subsequent preparatory steps such as washing and drying. Most fixation techniques are aimed at stabilizing proteins which are part of many biological samples. However, not all foods contain proteins. Fixation is achieved either by crosslinking protein molecules using low-molecular aldehydes such as formaldehyde, glutaraldehyde, or acrolein (Carroll *et al.* 1968; Hayat 1970A; Glauert 1974), or oxidizing fixatives such as osmium tetroxide (OsO_4) or potassium permanganate ($KMnO_4$). Of the aldehydes, glutaraldehyde at various concentrations either buffered or unbuffered, or in combination with formaldehyde (e.g. Karnovsky 1965), is used most frequently. Buchheim (1982) cautioned that the pH value of unbuffered glutaraldehyde dropped considerably during the fixation of a soy protein sol. Hayat (1970A) defined the optimal properties of fixatives for both SEM and TEM. Because SEM is concerned with surfaces, Geissinger and Stanley (1981) recommended the use of fixatives isotonic with the sample. However, because many authors prefer to dry-fracture their samples before SEM, the fixative should be somewhat hypertonic because it will become isotonic or hypotonic as it penetrates into the deeper parts of the sample, which will be subsequently exposed by fracturing. It should be understood that the osmolality of the fixatives depends on the osmolality of the buffer used.

Temperature of fixation may range from 0° to 37°C (Hayat 1970A; Glauert 1974; Geissinger and Stanley 1981) and duration may vary from 10 min to several hours. Sabatini *et al.* (1963A) reported that samples could be stored in glutaraldehyde for several months without change. As a rule, denser samples require a longer fixation. Small dimensions of the sample ($< 5\ mm^3$) are important for rapid fixation as well as other preparatory steps such as dehydration, fat extraction, and freezing.

Both oxidizing fixatives, OsO_4 and $KMnO_4$, react with unsaturated lipids, but OsO_4 also reacts with proteins and preserves a greater variety of food components than does $KMnO_4$. Because OsO_4 fixes lipids (Crozet and Guilbot 1974; Geyer 1977), it should be used only with samples that are free of lipids or if the lipids have to be retained in the sample for SEM. For example, Hall and Creamer (1972) and Taranto *et al.* (1979) etched cheese with trypsin, which partially digested and removed the protein matrix and exposed fat globules, the dimensions and aggregation of which were subsequently studied by SEM. On occasion, however, lipids present at high concentrations interfere during SEM by obscuring the underlying protein matrix or by melting under the electron beam and by leading to charging artifacts (Kalab 1978). If fat must be removed, postfixation with OsO_4 is omitted: the sample fixed in glutaraldehyde is washed with distilled water, dehydrated in a graded ethanol or acetone series, defatted in chloroform, *n*-hexane, or diethyl ether, returned in ethanol or acetone, and critical-point dried. With fat absent (soy protein gels, yogurt), postfixation with OsO_4 may be useful because it contributes to the bulk conductivity of the sample and facilitates SEM.

The postfixed sample is washed with distilled water and is either freeze-dried or dehydrated in a graded ethanol or acetone series or in acidified 2,2-dimethoxypropane (Muller and Jacks 1975) and is critical-point dried.

Water-soluble components (e.g. salts, sugars) present in the sample are washed out of the specimen with distilled water. Incompletely removed water-soluble substances may later appear in the form of efflorescences on the surface of freeze-dried samples. If the specimen is destined for critical-point drying, residual soluble substances are usually washed out during dehydration in a graded ethanol or acetone series.

There are no suitable techniques for fixation of starch. Artifacts formed in bread by the use of various fixation techniques routinely used in studies of foods of animal origin were shown by Chabot (1979). In most foods containing starch, the starch is present in gelatinized form. Gelatinization takes place in starch granules heated with water. The granules absorb water, increase in volume, change shape, and lose birefringence. Gelatinized starch granules are delicate and are easily destroyed, but it is the gelatinized form that controls many properties of the foods containing starch. Chabot (1979) concluded that freeze-fracturing should be preferably used with foods containing gelatinized starch, as other steps used to prepare the samples for SEM examination in the dry state such as dehydration and drying were found to pose additional problems.

Food samples of appropriate dimensions, fixed, and washed free of fixatives and of soluble substances are ready to be dried.

There are several ways to dry the sample. Air-drying, characterized by evaporation of water from the moist sample, is used rarely [for example with bread (Chabot 1979), with acetone-extracted seeds (Allen and Arnott 1981), with soybeans (Wolf and Baker 1972; Wolf *et al.* 1981), or with meat under special circumstances (Geissinger and Stanley 1981)] because it may lead to severe artifacts caused by the collapse of fine structures due to a high surface tension at the water–air interface. Drying is most frequently accomplished by freeze-drying or by critical-point drying (Fig. 2.1).

In freeze-drying, the specimen is rapidly frozen and the ice is subsequently sublimed *in vacuo*. Ideally, the ice in the sample must be in a glassy form free of crystals. Because ice crystals develop during slow freezing of water, as will be discussed in the section on freeze-fracturing and replication, the sample is frozen as rapidly as possible. Ice crystals distort the initial microstructure by pushing the structural components aside (Kalab and Emmons 1974) and by leaving void spaces in the matrix after the ice has sublimed during freeze-drying. Angold (1979) cautioned that samples destined for electron microscopy should be rapidly frozen and then maintained at a temperature below 198K (−75°C) to prevent recrystallization. The effect of slow freezing on ice crystal formation was shown by Lawrence and Jelen (1982) and in fact was utilized to produce characteristic texture in protein concentrates such as *kori-tofu* (Saio 1981). For this reason, experimental procedures in which food samples are frozen at −20°C, in a cryostat, or in dry ice may be suitable for light microscopy, where the minute ice crystals may not be noticeable, but should under no circumstances be used in electron microscopy because they would show the disruptive effects at the higher magnifications.

Even liquid nitrogen is not a suitable cryofixative for freezing samples for electron microscopy, because it does not prevent ice crystal formation. Liquid nitrogen is used at its boiling point and as the warm sample is immersed into it, an insulating layer of gas immediately forms around the sample and reduces the rate of freezing. Melting nitrogen slush, which does not immediately boil around the sample (Angold 1979; Umrath 1974), is a considerably better medium (cryo-

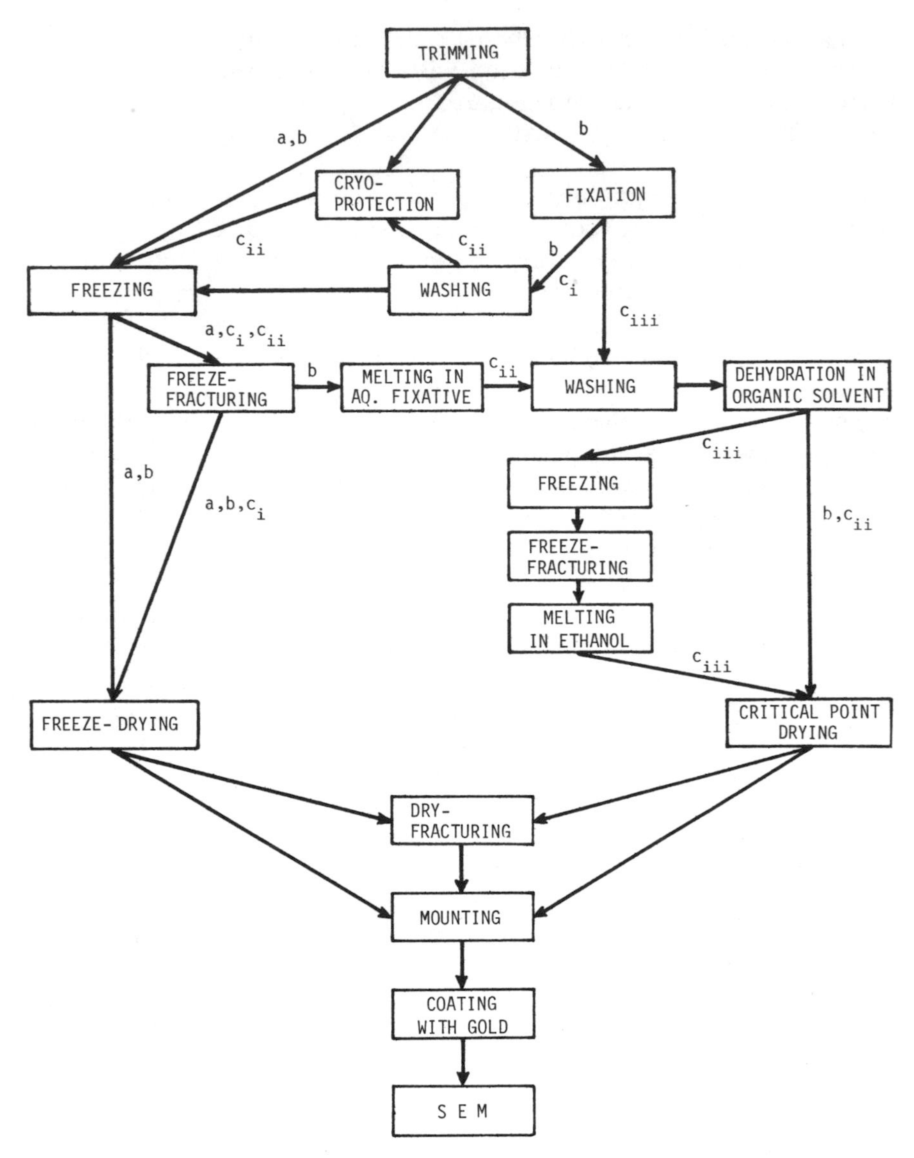

Fig. 2.1. Sample preparation for SEM—a flow diagram. Various pathways may be followed as discussed in the text.

fixative). The slush is prepared by exposing liquid nitrogen in an insulated cup to a reduced pressure of 12 kPa (94 torr) for 1 min. Liquid nitrogen vigorously evaporates and its temperature decreases to the freezing point of nitrogen (63K, i.e. −210°C).

In the most frequently used "standard freezing" technique (Bachmann and Schmitt-Fumian 1973), a small sample (1 mm^3) in a metal holder is dipped into liquid Freon cooled to near its freezing point (approximately −150°C) with liquid nitrogen. To carry out this operation, liquid Freon in a ladle is cooled with liquid nitrogen until it freezes solid; immediately before freezing the sample, a metal rod 8 to 10 mm in diameter is pushed into the solid Freon where it forms a small pool of molten Freon into which the sample is plunged and where it is frozen. Additional methods suitable for rapid freezing of samples for electron microscopy are discussed in the section on freeze-fracturing and replication for TEM.

The frozen sample is transferred onto the freeze-drier cold stage precooled to 193K (−80°C) and vacuum is rapidly applied. Phosphorus pentoxide is used to absorb water vapor from the sample. Any rise in the temperature of the sample to 233K (–40°C) or higher before the sample is completely dry would be harmful as it would lead to the development of large ice crystal artifacts.

Critical-point drying (CPD) (Cohen 1974, 1977, 1979) is another method of drying samples for SEM. The fixed and washed sample is first dehydrated in an organic solvent, such as a graded series of ethanol or acetone (20, 40, 60, 80, 96, and 100% solvent; Eino *et al.* 1976) or in acidified 2,2-dimethoxypropane (Muller and Jacks 1975). Fat may be extracted from the sample at this stage. Acetone, chloroform, *n*-hexane, or a mixture of diethyl ether and petroleum ether may be used for this purpose. The fat-free sample is then returned into absolute alcohol, acetone, or amyl acetate if CPD is to be carried out in carbon dioxide (CO_2), because it is not known whether chloroform or petroleum and diethyl ethers are miscible with liquid CO_2 (Lewis *et al.* 1975).

The principle of CPD is substitution of liquid CO_2 as one of several suitable transitional fluids (Bartlett and Burstyn 1975) for the organic solvent with which the dehydrated sample has been impregnated and the subsequent conversion of liquid CO_2 into gaseous CO_2. CO_2 exists in the liquid form only under pressure at temperatures below 304.3K (31.3°C), and thus this conversion is carried out in a pressurized cell. Above the critical temperature of 304.3K, CO_2 exists as a gas irrespective of the pressure applied. Therefore, by placing the sample into liquid CO_2 at a low temperature (5–10°C) and by substituting the organic solvent, with which the sample had been impregnated, with CO_2 and

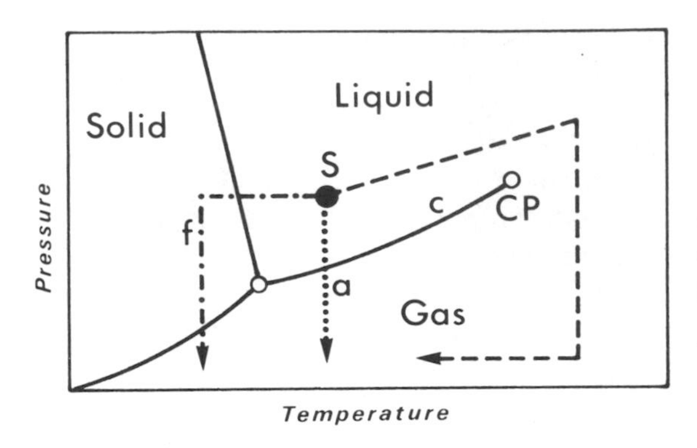

Fig. 2.2. Course of different drying processes indicated in a phase diagram. (a) air drying; (c) critical-point drying; (f) freeze drying; (CP) critical point; (S) sample.
From Moor (1973A).

thereafter raising the temperature above 304.3K, the liquid CO_2 is converted into gas. After gaseous CO_2 is released from the pressurized cell, the sample is dry without having passed through any phase boundary (liquid–gas; solid–gas; Fig. 2.2) unlike during other drying procedures.

In spite of considerable advantages, such as rapidity of the procedure, CPD is not absolutely free of producing artifacts (Cohen 1977; Boyde 1978). CPD is suitable for food products based on protein (meat—Geissinger and Stanley 1981; milk products—Kalab 1981; vegetable proteins—Taranto and Yang 1981) or for plant tissues (Humphreys *et al.* 1974; Davis and Gordon 1980), but leads to artifacts in products consisting of gelatinized starch such as pasta, puddings, and baked products, or in products in which gelatinized starch is used as a thickening agent (salad dressings, some yogurts etc.). Being impossible to fix, gelatinized starch is susceptible to changes in terms of its microstructure during dehydration in organic solvents and during CPD; freeze-drying is a better alternative for products containing gelatinized starch (Chabot 1979), yet not even this procedure prevents artifacts.

The internal microstructure of the sample under study can be revealed by fracturing, which is carried out either with a dried sample (dry-fracturing) or with a frozen sample (freeze-fracturing). Dry-fracturing is done by hand under a low-magnification stereoscopic microscope (Flood 1975). The tip of a scalpel (blade) is briefly pressed into the edge of the particle until the particle ruptures. This procedure is easy to perform with brittle samples such as muscle (Geissinger and Stanley 1981) or cheese (Kalab 1981). Fractures of a dry sample that is relatively porous such as tofu (Saio 1981), yogurt (Kalab *et al.* 1975), gelled milk (Harwalkar and Kalab 1981), or cottage cheese (Glaser *et al.* 1979; Kalab 1981) (Fig. 2.3A) are ragged. Smooth fractures obtained by fracturing the sample while it is frozen are more suitable for the evaluation

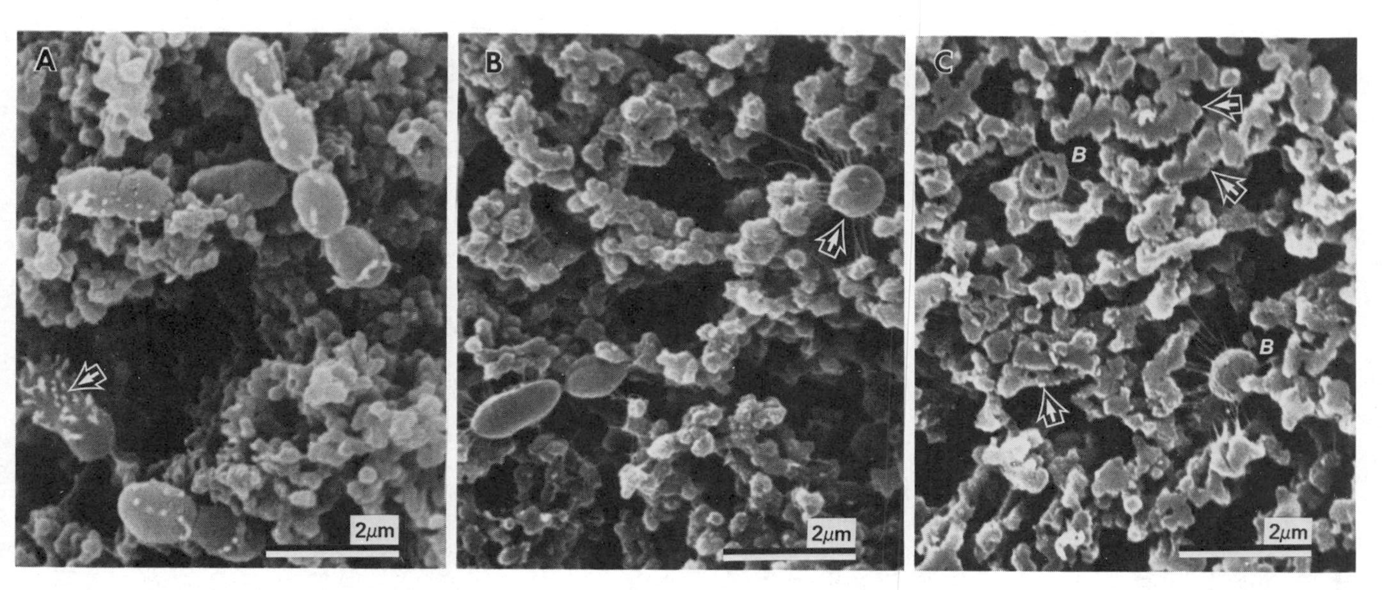

Fig. 2.3. SEM of cottage cheese fractured by different procedures. (A) Sample was fixed in glutaraldehyde, freeze-dried, and dry-fractured. Filaments anchoring streptococci in the protein matrix have been broken (arrow). (B) Sample fixed in glutaraldehyde was dehydrated in a graded ethanol series, impregnated with absolute ethanol, frozen in melting Freon, freeze-fractured under liquid nitrogen, melted in absolute ethanol, and critical-point dried from carbon dioxide. Only streptococci were cross-fractured at weak points (arrow). (C) Sample was impregnated with 30% glycerol, frozen in melting Freon, freeze-fractured under liquid nitrogen, melted and fixed in glutaraldehyde, dehydrated in a graded ethanol series, and critical-point dried from carbon dioxide. Both casein micelles (arrows) and streptococci, B, were cross-fractured.
From Kalab (1981).

of sample porosity and network structure (Fig. 2.3B,C). Freeze-fracturing may be carried out at various stages of sample preparation. The sample is rapidly frozen as in freeze-drying and is immersed in liquid nitrogen; frozen Freon adhering to the sample is removed with a scalpel. Freeze-fracturing is carried out under liquid nitrogen by a procedure similar to dry-fracturing. All the instruments used must be cooled with liquid nitrogen. The fragments are subsequently processed according to one of the several pathways shown in Fig. 2.1. It is evident from this figure that there is a wide variety of combinations in which a sample may be prepared for SEM. These combinations may be divided into three groups.

a. Samples that do not require fixation (for example, cooked spaghetti) are frozen, freeze-fractured, and freeze-dried.
b. Samples that contain proteins and require fixation, may be freeze-fractured either before or after fixation. Freeze-fracturing preceding fixation is aimed at the removal of soluble solids from the fractured matrix in the sample (Haggis *et al.* 1976). The fragments are rapidly plunged in a warm (293–298K, 20–25°C) fixative, fixed, washed with water, dehydrated in a graded ethanol or acetone series, and are critical point dried. Melting must be carried out rapidly (for example, with stirring) to avoid recrystallization of the ice in the sample.
c. Fixed samples may be freeze-fractured in three different ways: (1) the fixed sample is washed with water, frozen, freeze-fractured, and freeze-dried. (2) The fixed and washed sample is impregnated with a cryoprotective agent such as 5–30% glycerol, ethylene glycol, or dimethoxy sulfoxide solutions (Rebhun 1972) and is frozen. Cryoprotective agents retard the development of ice crystals during subsequent freezing. The frozen sample is freeze-fractured and melted in an aqueous medium; the cryoprotective agent is washed out and the sample is dehydrated and critical point dried. (3) The fixed sample is dehydrated in a graded ethanol series, defatted in chloroform if necessary, and returned to absolute ethanol. Then the sample is frozen in liquid nitrogen. Ethanol is not susceptible to ice crystal formation during freezing. After the ethanol-impregnated sample is fractured under liquid nitrogen, the fragments are rapidly melted in ethanol and CPD.

Each procedure has some advantages and some deficiences, but their nature can be revealed only by experiments. In cottage cheese, for example, dry-fracturing led to a ragged surface at which neither casein micelles nor bacteria were cross-fractured (Fig.2.3A). Casein micelles were also left intact in cottage cheese impregnated with absolute ethanol and fractured while frozen, but bacteria were ruptured at weak points (Fig. 2.3B). Both casein micelles and bacteria were cross-fractured in cottage cheese impregnated with 30% glycerol and fractured in liquid nitrogen (Fig.2.3C; Kalab 1981).

Dry fragments are mounted on metal SEM stubs*. In the author's laboratory, new aluminum stubs are first sanded and then painted with a conductive silver cement; in this way bonding of the fragments is stronger than to a polished stub surface. Highly porous dried fragments (milk gels) are lifted one by one using a fine brush in which static electric charge had been generated by several strokes on a plexiglass plate. More compact fragments (cheese, pasta) are positioned with a pair of tweezers. The fractured planes must not be touched. Each fragment is positioned on a droplet of freshly applied silver cement. After the cement hardens, the vertical sides of each particle (less than 1 mm high) are carefully painted leaving only the fracture exposed. A proper consistency of the silver cement is critical, as a cement containing too much solvent penetrates the fragment and causes its matrix to collapse, whereas a cement too thick does not provide a sufficient bond between the metal support and the fragment. Almost any viscous cement is suitable provided that it does not generate bubbles *in vacuo* when dry (Wells 1974; Nickerson *et al.* 1974). Minute dust particles adhering to the fractured surfaces can be removed by a gentle stream of dry air.

To mount powdered foods for SEM, a double-sided sticky tape is first attached to a metal stub and its edges are painted with a conductive silver cement to help provide an uninterrupted conductive path between the particles and the metal stub. The tape is then lightly sprinkled with the powder under study. Loose particles are removed from the tape by knocking them off and/or by gently blowing a stream of dry air or nitrogen. This prevents the particles from overlapping, which could lead to charging artifacts after the particles are coated with carbon, gold, or a gold–palladium alloy (Fig. 2.4).

Stubs with powder particles or with sample fragments firmly attached are coated with a conducting layer. Vacuum evaporation and sputter coating are two common techniques, each with specific advantages and deficiences (Echlin 1978B).

Vacuum evaporation is carried out in the author's laboratory with the sample rotated inside an evacuated bell jar, while the coating material is evaporated consecutively from two to three different angles. Controlled admission of nitrogen to 66 mPa (5×10^{-4} torr) provides a low concentration of nitrogen molecules in the bell jar, with which the

* See Murphy (1982).

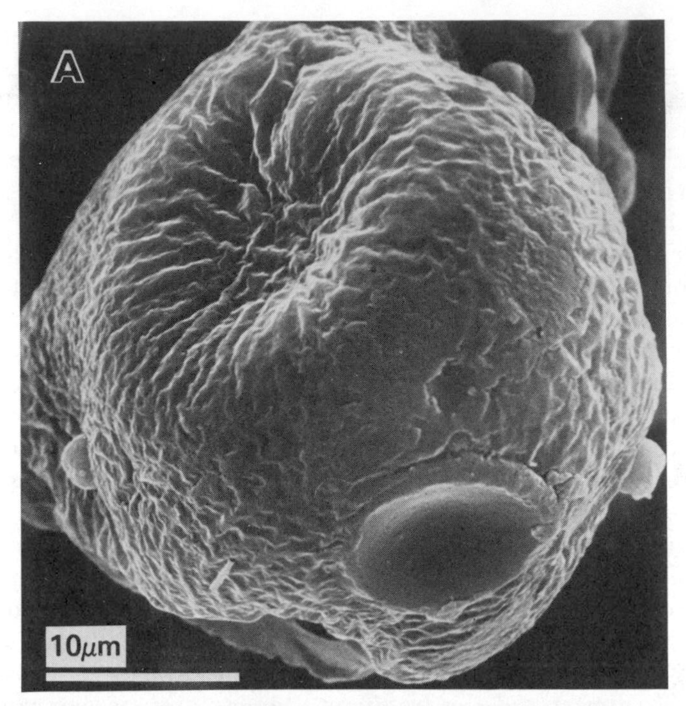

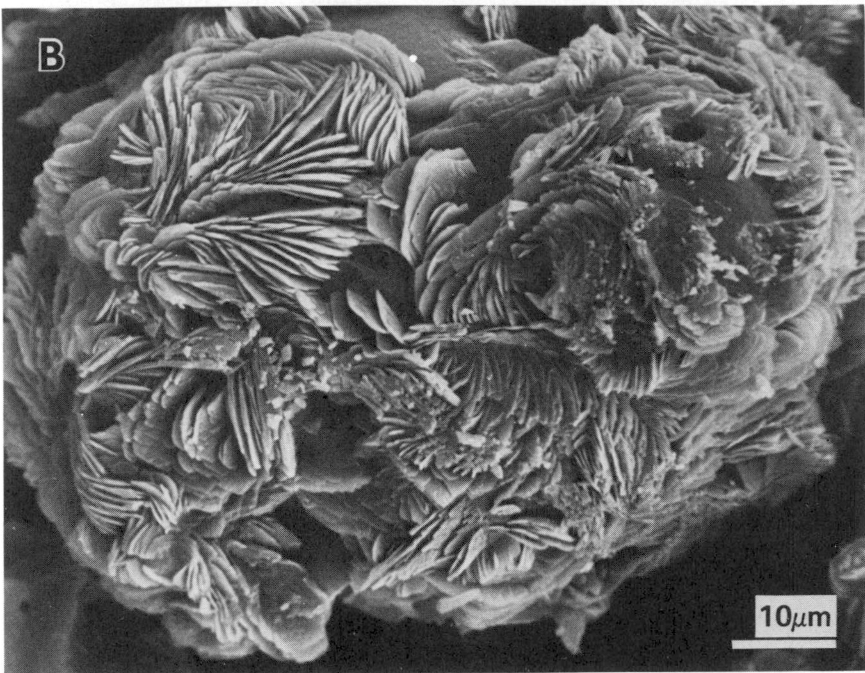

Fig. 2.4. SEM of spray-dried buttermilk. (A) A particle of spray-dried buttermilk stored in a dry atmosphere. (B) A particle of spray-dried buttermilk exposed to a high-humidity atmosphere for 1 hr. Lactose present in the particle crystallized on the particle surface.

evaporated atoms collide and are thus deflected, coating the sample from various directions.

In sputter coating, the coating material (gold) is the cathode and the specimen is placed on the anode. Sputtering is carried out in an argon atmosphere at 1.3–13 Pa (10^{-2}–10^{-1} torr). At 2–3 keV, argon is ionized and the positive ions bombard the cathode (coating material) which is thus eroded, and the particles ejected coat the sample from various directions (Echlin 1975, 1978B).

The amount of gold or of a gold–palladium alloy needed for adequate coating varies, depending on the porosity (i.e., on the specific surface of the sample). The more porous the sample, the greater the amount of metal needed to form a uniform conductive layer on the sample surface. Some authors lightly coat the sample with carbon before coating with gold, particularly if the sample is destined for examination at higher magnifications (e.g., 20,000×); a carbon film on the sample reduces the grain size of the gold coating. Metal-coated samples should be examined by SEM as soon as possible after coating and should be stored in a desiccator over dry silica gel. With very porous samples, only samples that can be examined during a day's session at the microscope should be coated. Compact samples are less susceptible to damage during storage.

There have been several chemical treatments developed to render biological samples conductive (Murphy 1978, 1980) eliminating the need for carbon and gold coating; the use of such techniques, however, has not yet been reported with foods.

For viewing in the electron microscope, the accelerating voltage selected depends on the nature of the sample and the magnification. Most foods are examined at magnifications between 100× and 20,000×. Although, in general, resolution increases with increasing accelerating voltage, most work with food is carried out at 5–20 kV.

In addition to charging artifacts, localized overheating caused by the electron beam may develop during focusing, particularly if the conductivity of the gold coating is insufficient to carry the beam current. Samples containing fat are especially susceptible to this kind of damage because fat may melt and the gold coating on its surface may crack. Lactose α-hydrate crystals (Fig. 2.4) are also extremely susceptible to electron beam damage (Buma and Henstra 1971; Saltmarch and Labuza 1980; Kalab 1980A; Kalab and Emmons 1974). It was hypothesized (Kalab 1981) that the removal of water from the α-hydrate crystals *in vacuo* inside the electron microscope leaves the crystals in an unstable "fluffy" state as the anhydride retains the original crystalline shape of the hydrate. These are only several examples of artifacts that may be encountered during scanning electron microscopy. The objective of each electron microscopist should be to obtain micrographs at the optimal operating conditions of the microscope using samples in which the incidence of artifacts has been reduced to a minimum by selecting appropriate preparatory techniques. Understanding the capabilities and limitations of the equipment and techniques is the prerequisite for achieving this objective.

In addition to examining external appearance and internal microstructure of the foods under study, SEM may be used to evaluate dimensions of particles, pores, cavities, etc., provided that certain precautions are observed.

As magnification is related to the working distance in the microscope, particles viewed at a shorter distance (i.e., particles located higher on the stub) appear to be larger than particles viewed at a longer working distance (i.e., particles located lower on the stub). A single micrograph would be insufficient to distinguish between the distances and provide the information required for accurate calculation of the particle dimensions. For this reason, pairs of stereo micrographs are taken and are evaluated by stereological techniques (Boyde 1974; Clarke 1975;Weibel and Bolender 1973). The angle by which stereo micrographs differ may vary between 6 and 20°. The greater the angle, the shorter the apparent distance of a fixed reference point from the observer's eyes. With the eyes 7.0 cm apart, a fixed point in a pair of stereo micrographs taken with a 6° angular separation would appear at a distance of 67 cm, whereas the same point in a pair of micrographs with a 20° separation would appear 20 cm away. These differences are illustrated in Figs. 2.5 and 2.6 using the same sample. Too great separation produces images in which dimensions may appear altered. As evident from both figures, a pair of stereo micrographs may provide more information than a large number of single micrographs.

In order to obtain a stereo pair at the same magnification, it is necessary to use the same working distance in both instances. The area under study is continuously checked during the tilting of the specimen and the sample is repositioned to keep it in the field of vision. It is useful to mark characteristic features with a felt-tip pen on the protective shield of the viewing screen (CRT) at the magnification used as well as at a considerably lower magnification. This facilitates the proper repositioning of the sample after it has been tilted through the desired angle.

In approximation, a single micrograph may be used to compare dimensions of particles provided either that it is taken at a relatively long working distance or that all the measured particles are positioned in a plane scanned by the electron beam striking the plane at normal incidence. In the case of powders this means that the powder particles must be attached to the metal stub in a single layer. To measure dimensions of particles, pores, cavities (e.g., those initially occupied by fat droplets, Fig. 2.19B), or air cells in solid foods, freeze-fracturing is preferable to dry-fracturing because the former fracturing technique produces smooth fracture planes. Viewing at normal incidence is beneficial even in instances other than those aimed at gaining approximate data on particle dimensions. Although the image may lack contrast compared with micrographs obtained at a 30 or 45° tilt, it provides information on surface or structural features reflecting the actual microstructure of the sample with the entire field of vision in focus (Fig. 2.7).

Cold-Stage SEM

Some foods, particularly those with a high fat content such as butter or mayonnaise, foods containing gelatinized starch (baked products), foods in the form of foam (whipped cream), and foods consumed in the frozen state (ice cream) cannot be dehydrated and dried for SEM without introducing artifactual changes. Most such foods, however, can be examined by SEM after they have been rapidly frozen and freeze-fractured. A cold stage mounted inside the scanning electron microscope is a prerequisite for this technique (Robards and Crosby 1979). The cold stage makes it possible to keep the sample at a low temperature (below 203K, −70°C). Most convenient are so-called biochambers attached to the electron microscope. The rapidly frozen sample is placed in the chamber and vacuum is applied. The sample is freeze-fractured, coated with carbon and/or gold, and positioned in the path of the electron beam while it is maintained at the low temperature *in vacuo*. Carrying out the procedure in the biochamber offers several advantages, such as rapidity and a reduced risk of contaminating the sample or the microscope. Requirements for low-temperature SEM of biological samples were discussed by Robards and Crosby (1979) who also reviewed equipment from commercial sources. Other valuable information on low-temperature SEM can be found in reviews by Echlin (1973, 1978A), Echlin and Moreton (1976) and Davis and Gordon (1978).

Biochambers are relatively expensive and, thus, in spite of their advantages, various authors have devised methods for carrying out some steps of sample preparation outside the microscope. One example is the procedure used by Schmidt and van Hooydonk (1980) in studies of the microstructure of whipped cream. The sample was placed in a 10 mm diameter, aluminum foil tube, the inner wall of which was covered with a layer of a conductive carbon cement. The bottom of the tube was wrapped around a copper disc with a short pin in its center. After the sample had been frozen, the aluminum foil was peeled off; the layer of carbon cement remained attached to the sample and formed a thin-walled tube around it. The sample was fractured in liquid nitrogen, transferred under liquid nitrogen into an apparatus for coating with

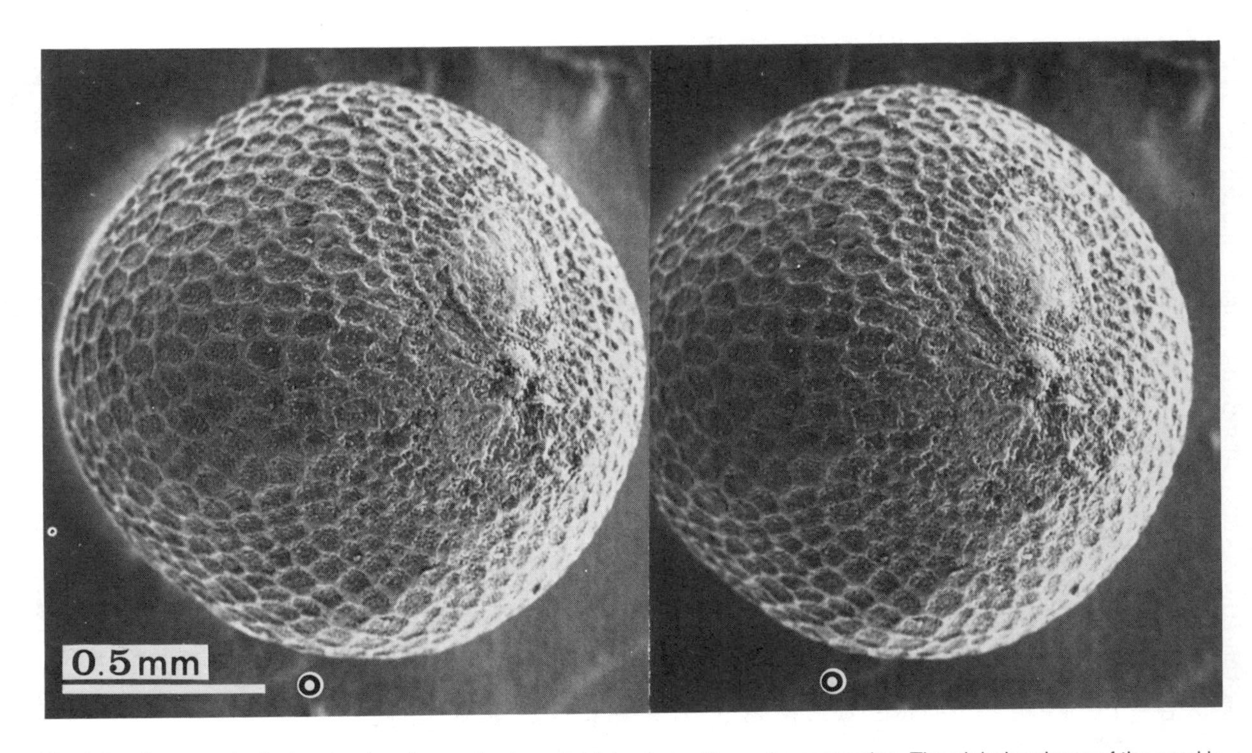

Fig. 2.5. Stereo pair of micrographs of a mustard seed obtained at a 6° angular separation. The globular shape of the seed is clearly evident. A pair of black circles is provided to facilitate focusing of the eyes.

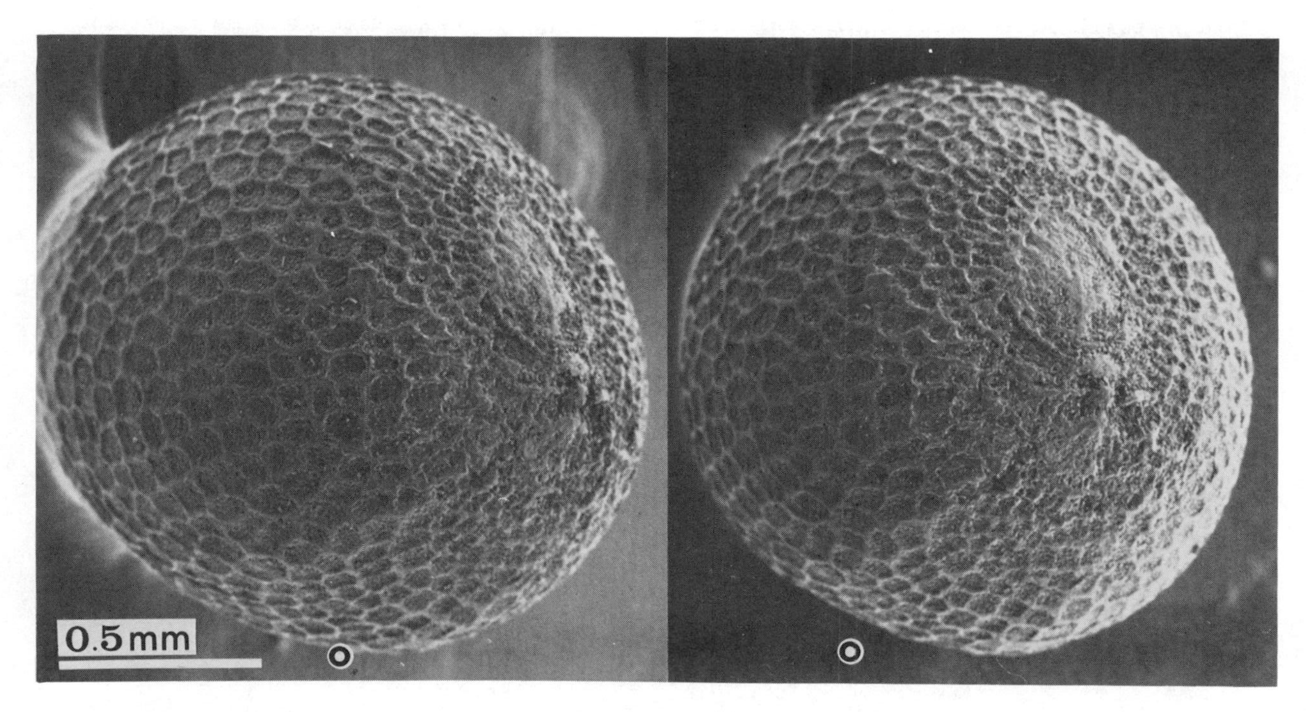

Fig. 2.6. Stereo pair of micrographs of a mustard seed (same as in Fig. 2.5) obtained at a 20° angular separation. When compared with Fig. 2.5, this seed seems to be closer to the observer's eyes; the seed appears to be elongated in the vertical direction. A pair of black circles is provided to facilitate focusing of the eyes.

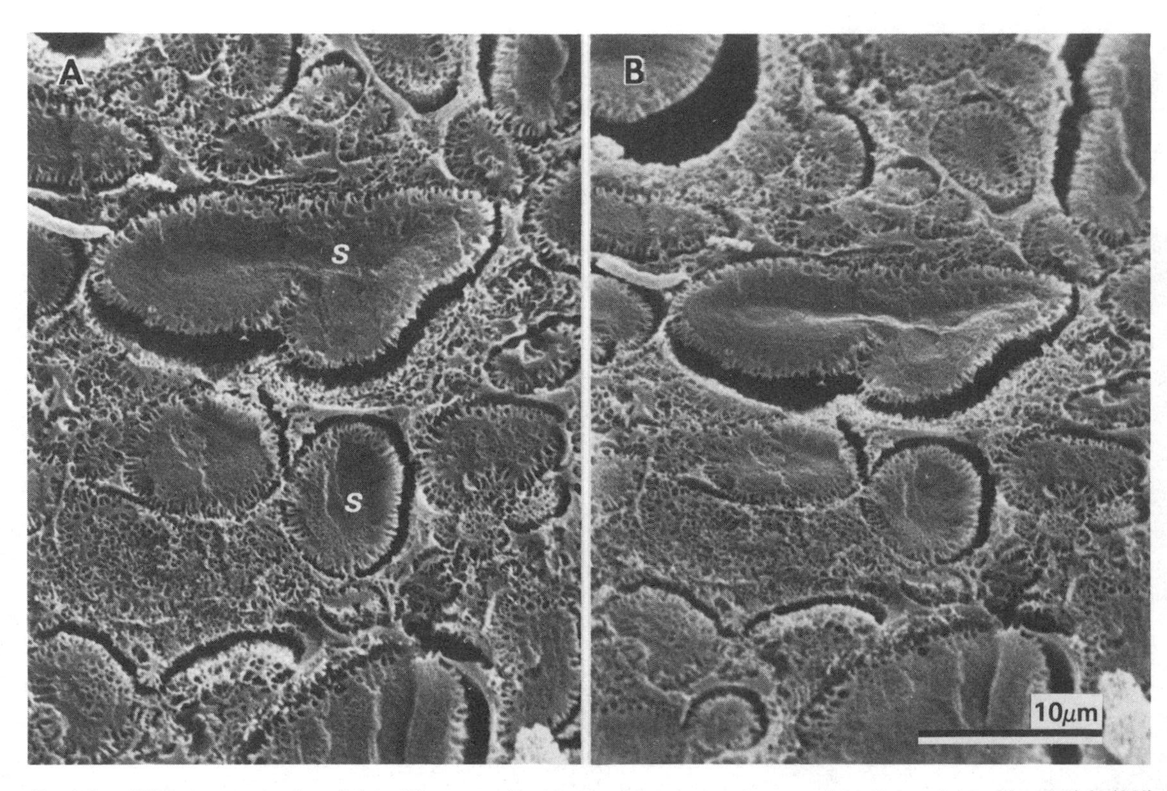

Fig. 2.7. SEM of a spaghetti particle at (A) a normal incidence of the electron beam striking the surface of the particle (90°) and (B) at a 45° tilt. The spaghetti was cooked for 9 min, frozen in melting Freon, freeze-fractured in liquid nitrogen, and freeze-dried. S, starch granule. True dimensions shown in (A) are distorted by the 45° tilt in (B); in addition, the bottom and top parts of the micrograph taken at the 45° tilt (B) are out of focus as they have been moved forward and backward, respectively.

approximately 20 nm carbon, and again transferred under liquid nitrogen into an electron microscope equipped with a cold stage. Because some water vapor condensed on the sample during the last transfer, the frost was removed by sublimation in the microscope. This technique makes it possible to examine the state of fat globules in whipped cream made from unhomogenized and homogenized creams (Fig. 2.8). Differences in the dimensions of the fat globules and in their clustering as well as in the dimensions of the air cells depending on the extent of whipping were well documented. The authors claim that the depth of freeze-etching, i.e., the thickness of the frozen aqueous phase removed by sublimation did not exceed, on an average, the equivalent of one or two fat globules and that there was no risk of the exposed structures having collapsed. This technique is a modification of an earlier method by Schmidt *et al.* (1979) developed for low-temperature SEM of cheese. The modification presented here is suitable for the examination of viscous samples.

In many samples of biological origin, ions present in the aqueous phase may provide sufficient conductivity to permit examination of uncoated specimens if the scanning electron microscope is operated at low accelerating voltage (3–10 kV) and low beam currents (<40 μA). This technique was used by Kalab (1981) in a study of the microstructure of cream cheese. The sample was placed in a hollow grid holder, 3 mm in diameter (Fig. 2.9), a part of the sample (~1 mm) protruding above the holder. The sample was rapidly frozen in melting Freon and transferred into liquid nitrogen, where the protruding part was fractured off with a cooled scalpel. The sample was positioned on the cold stage of the microscope and scanned periodically. Condensed water vapor frost was observed first (Fig. 2.10). The frost vanished within 30 to 40 min, after which time ice present in the aqueous phase of the sample started to sublime, freeze-etching the sample.

In uncoated samples, freeze-etching continues until, finally, the ion-containing aqueous phase recedes deep under the surface of the sample and the exposed structures (fat globules with attached casein micelle clusters in the case of the cream cheese), unable to provide sufficient conductivity, show charging artifacts, making further observation impossible. After losing support, the exposed structures collapse and may suffer from electron beam damage, as well. The quality of micrographs is worse than with carbon- or gold-coated samples.

Freeze-fractured cream cheese was coated with gold in a conventional vacuum evaporator by Kalab (1981) and Kalab *et al.* (1981) using the principle outline by Katoh (1979). Freeze-fracturing was carried out in the same way as described above. The holder with the sample was placed

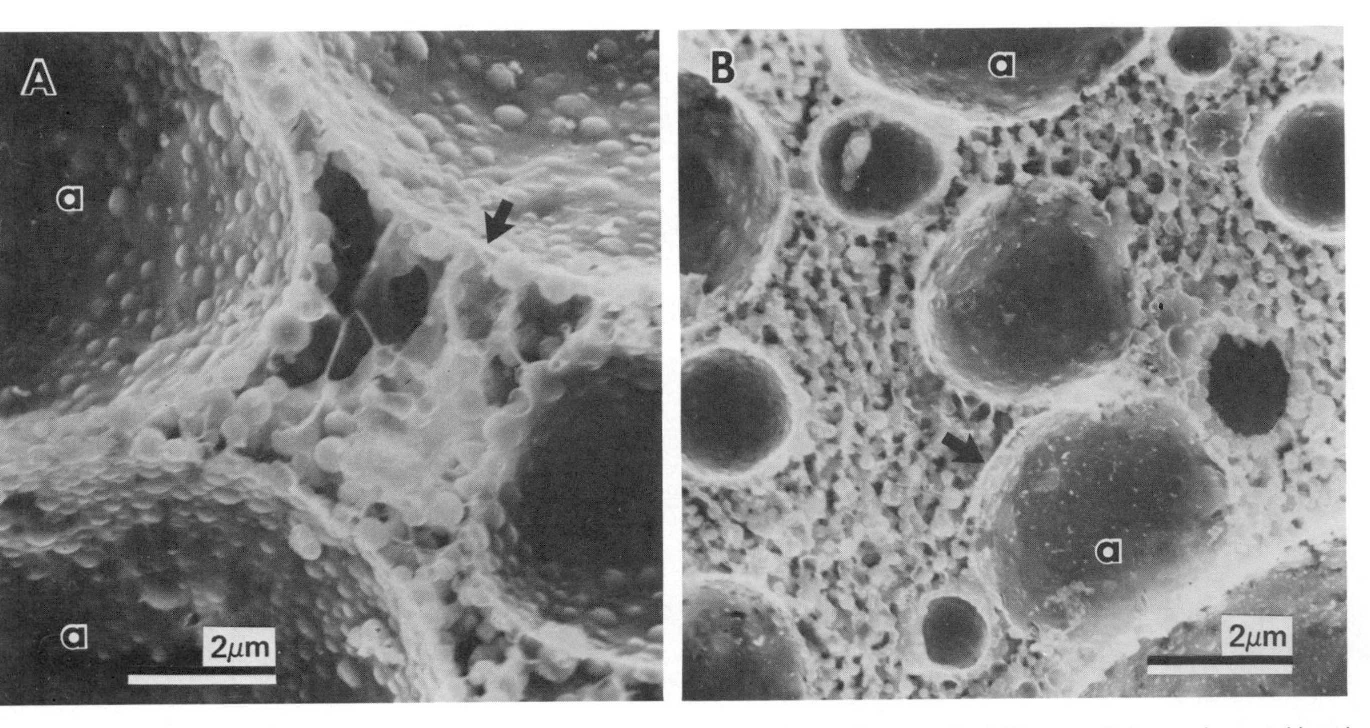

Fig. 2.8. Cold-stage SEM of whipped cream. Whipped cream from unhomogenized (A) and homogenized (B) cream. Both samples are whipped to maximum rigidity and show coalescence of the fat globules, the dimensions of which in both samples are markedly different. a, Air cells; arrows point to the air–cream serum interface.
From Schmidt and van Hooydonk (1980).

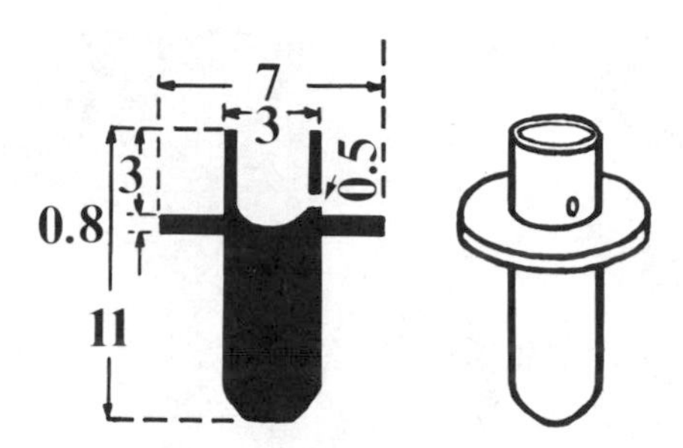

Fig. 2.9. Hollow grid holder for SEM designed by E.F. Bond. Dimensions are given in millimeters. Grids, 3.05 mm in diameter, with the sample attached, are mounted on the top of the holder using a silver cement. Space below the grid is evacuated through an orifice, 0.5 mm in diameter (arrow). The holder is also useful in cold-stage SEM: the sample is placed in the cavity in the holder, fractured under liquid nitrogen, and coated with gold. The holder fits the cold stage in the scanning electron microscope.
From Kalab (1981).

in an aluminum dish filled with liquid nitrogen and the dish positioned on a rotary table in a conventional vacuum evaporator. Vacuum was applied immediately, which led to vigorous boiling of liquid nitrogen resulting in its freezing within several seconds. Solid nitrogen and condensed water vapor were sublimed, the only controls of these processes being the visual observation of the sample and of the dish, and the decrease in the pressure inside the bell jar after the frost had sublimed.

The samples were rotated and coated with gold at two different angles: one portion of gold was evaporated at an angle determined by the height of the wall of the dish and the other portion of gold was evaporated at 90 deg. Four samples were coated simultaneously. After breaking the vacuum, the samples were plunged in liquid nitrogen for transport to the electron microscope. Protective carriers for transport are also available (Robards and Crosby 1979).

Katoh (1979) developed a technique which by-passes the need for a cold-stage attachment in the electron microscope to examine freeze-fractured samples. The technique is based on fracturing the sample under liquid nitrogen, coating it with gold in a conventional vacuum evaporator as described above, and separating the gold coating in the form of a replica: the gold-coated sample is warmed up to ambient temperature, the gold coating is floated on water, and the replica is cleaned using methods described in the section on transmission electron microscopy (freeze-fracturing and replication). Clean replicas are lifted on Formvar and carbon film-coated single slot grids or on bare 150- to 200-mesh (hexagonal) grids (Fig. 2.11). The grids are mounted on hollow grid holders (Fig. 2.9) to provide free space under the replica. With replica positioned directly on a metal support as suggested by Katoh

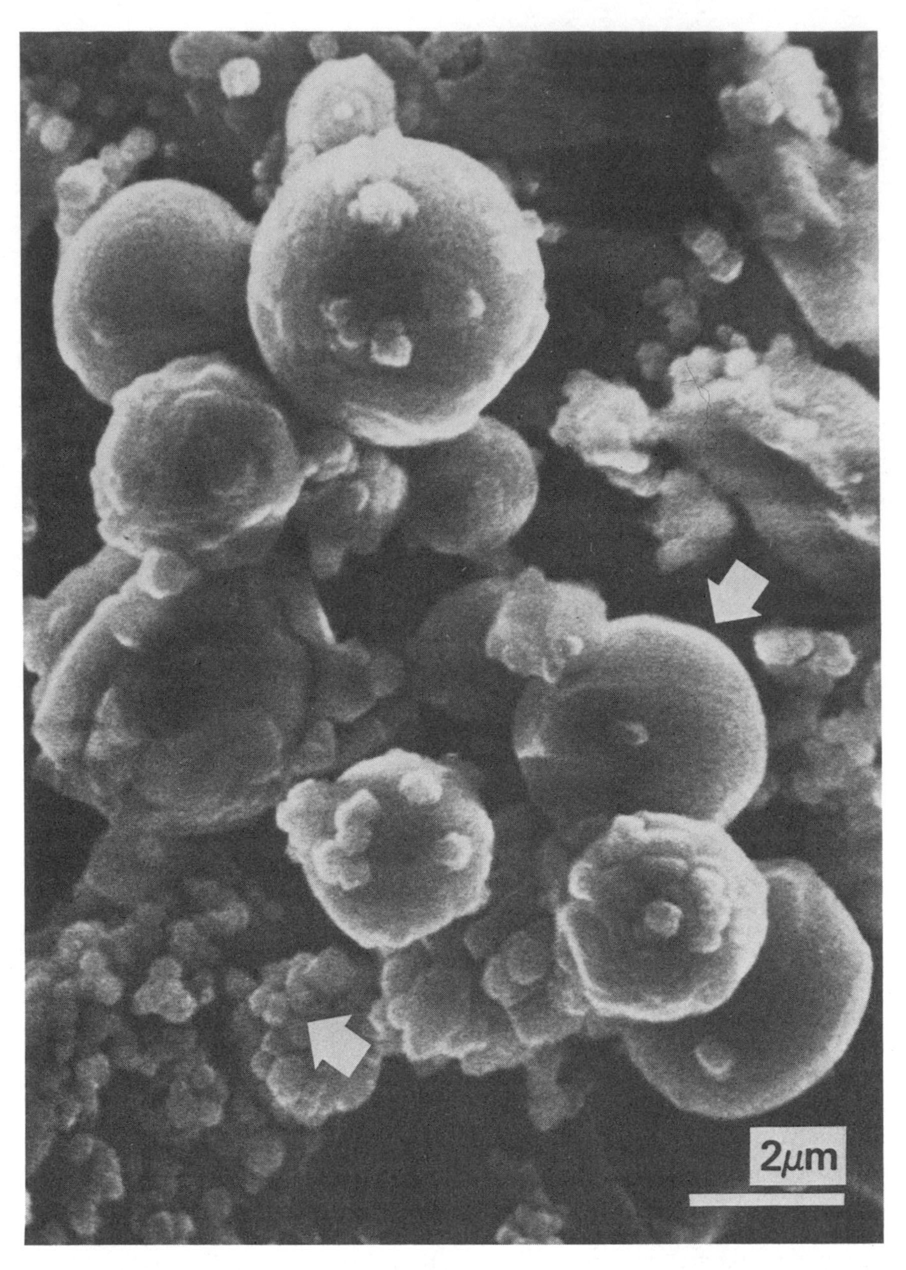

Fig. 2.10. Frost contaminating a frozen particle not coated with gold, examined by SEM at 173K (-100°C). The frost is in the form of globules and minute crystals (arrows). Charging artifacts have been caused by poor conductivity of the frost particles.
From Kalab (1981).

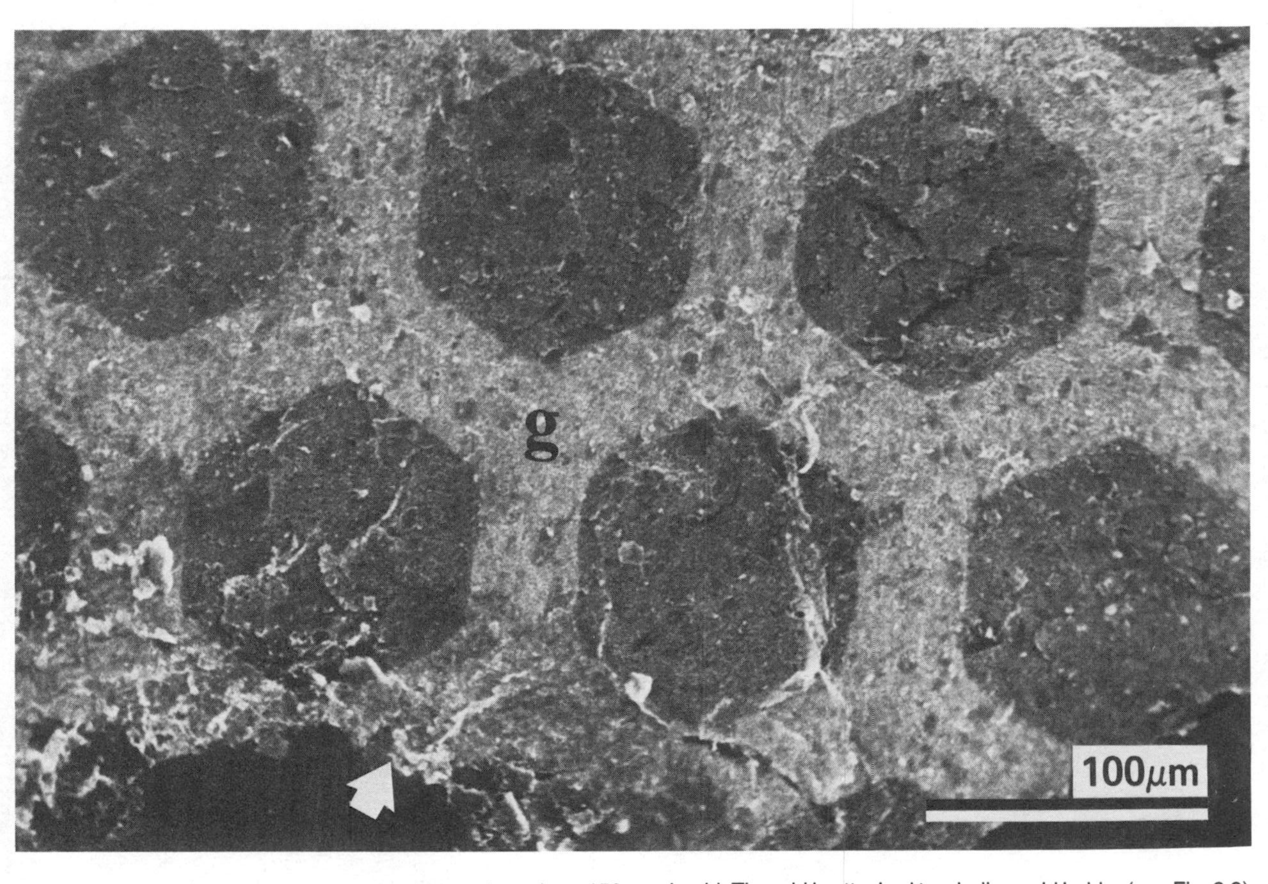

Fig. 2.11. Part of a gold replica (arrow) positioned on a bare 150-mesh grid. The grid is attached to a hollow grid holder (see Fig. 2.9). Interference by backscattered electrons from the metal support makes parts of the replica lying on grid bars (g) lighter.
From Kalab (1981).

(1979), backscattered electrons from the metal support interfere with electrons coming from the replica, particularly at high accelerating voltage. This interference is demonstrated in portions of the replica lying on the grid bars in Fig. 2.11 as markedly lighter areas.

Micrographs of the gold replicas of cream cheese obtained by Katoh's technique resembled those obtained by cold-stage SEM.

Rapid freezing of the sample is essential for a minimal incidence of artifacts caused by the development of ice crystals. In spite of precautions, however, ice crystals occasionally develop in some areas of the sample, particularly at the center of the particle, as shown in Fig. 2.12. It is advisable, therefore, to examine the fractures at a low magnification and to avoid areas affected by ice crystal development.

Most procedures, in which the sample is fractured and metal-coated outside the electron microscope, suffer from disadvantages such as lack of control over the removal of frozen water vapor (frost) contaminating the fractured surface, and over the extent of freeze-etching, i.e. over the partial removal of ice which is part of the genuine structure of the sample.

TRANSMISSION ELECTRON MICROSCOPY

Negative Staining

Negative staining is a rapid microscopic technique (Horne 1965; Hashemeyer and Meyers 1972) suitable for suspensions of submicroscopic particles translucent to the electron beam, such as polysaccharide fibers (Fig. 2.13A), casein micelles (Fig. 2.13B), bacteriophage. Addition of a solution of an electron-dense salt such as sodium phosphotungstate or ammonium molybdate to the suspension and drying of a thin layer of the mixture on an electron-transparent film makes the medium around the particles electron dense. The electron beam passes through the translucent particles under study and is absorbed by the surrounding electron-dense stain. Consequently, the particles appear light in the micrographs against a dark background.

Negative staining has been used with casein micelles (Uusi-Rauva *et al.* 1972; Creamer and Matheson 1980), myofibrillar structures and sarcoplasmic reticulum as reported by Lewis (1979), with pectin (Leeper 1973), carrageenan (Snoeren *et al.* 1976), etc.

Formvar film coated with carbon used as the support film is hydrophobic and must be rendered wettable to accept a uniform layer of the suspension. This can be accomplished by irradiating the Formvar- and carbon-coated grids with ultraviolet light or by their exposure to glow discharge (Milne and Luisoni 1977). Suspensions destined for negative

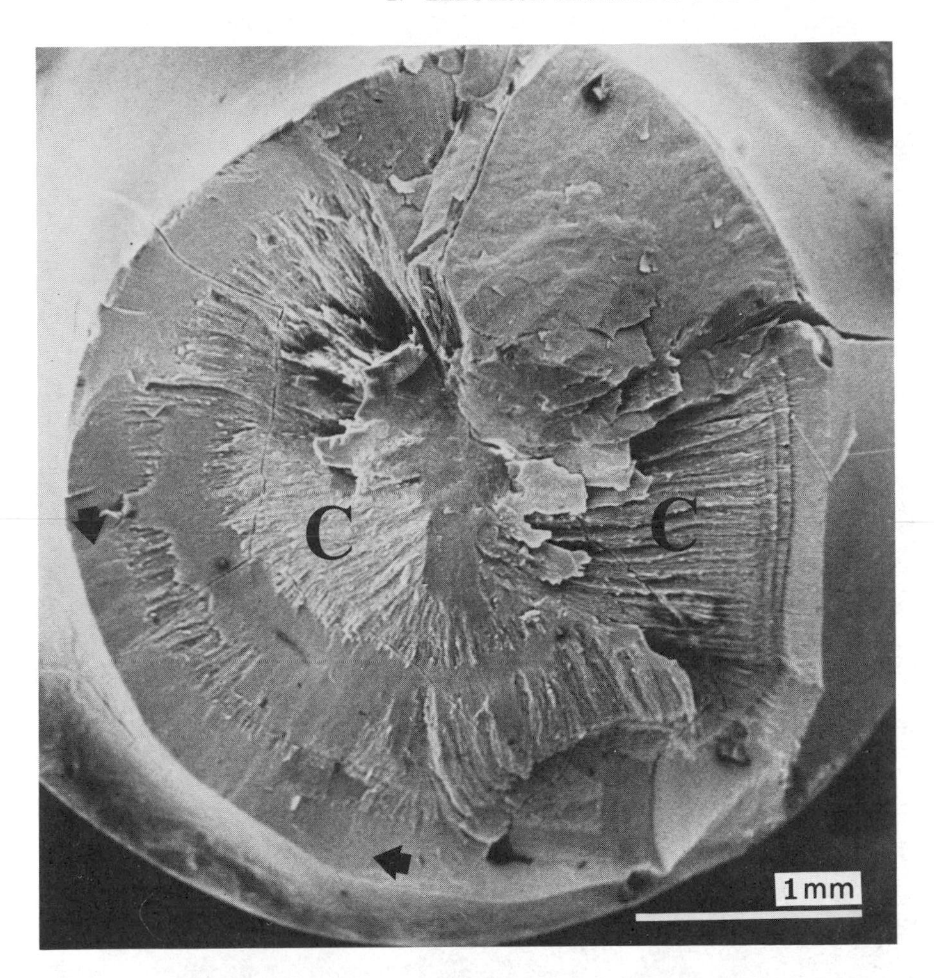

Fig. 2.12. Ice crystal development in cream cheese frozen in melting Freon. Cream cheese in a hollow grid holder for SEM (shown in Fig. 2.9) was frozen in melting Freon, freeze-fractured under liquid nitrogen, coated with gold by the technique of Katoh (1979), and examined by SEM at 173K (−100°C). Areas characterized by lines (C) converging toward the particle center have been affected by ice crystal formation. A thin layer at the particle surface (arrows) has not been affected by ice crystals and is suitable for SEM.
From Kalab (1981).

staining must be fixed, if protein is present, preferably in a glutaraldehyde solution, and adjusted to a proper particle concentration to avoid overlapping of the particles. A proper pH value of the mixture is an important factor to avoid aggregation of the particles. This is particularly important with pectin (Leeper 1973). A small droplet of the mixture is placed on the prepared grid and excessive liquid is removed after several minutes by draining the droplet with a piece of filter paper. The

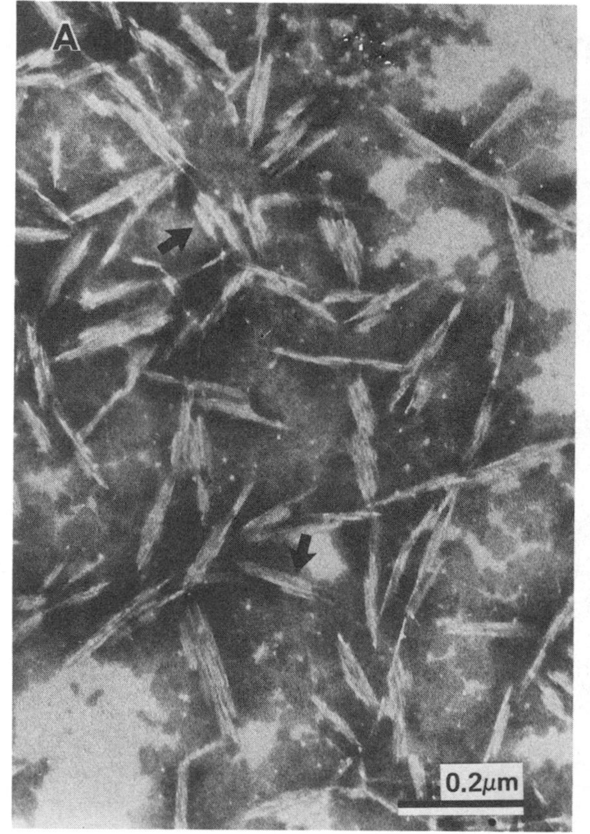

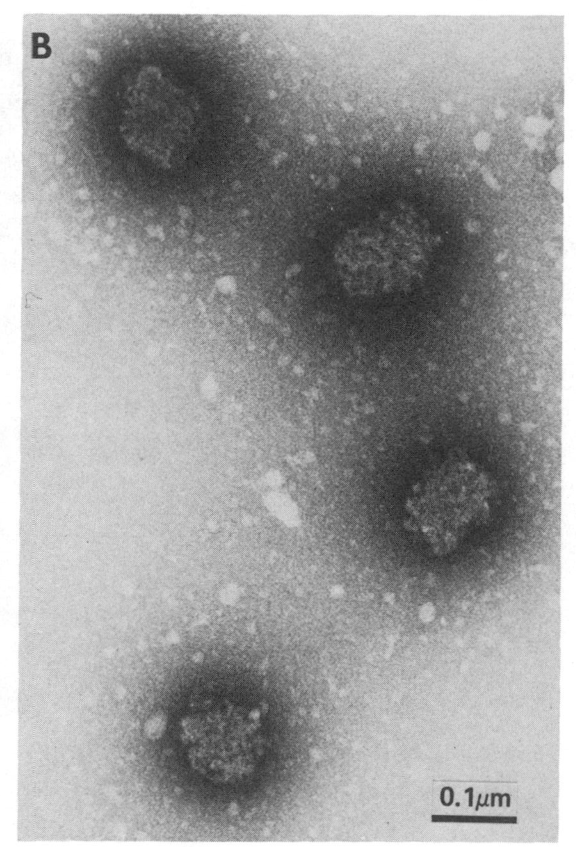

Fig. 2.13. Examples of negative staining. (A) Microcrystalline cellulose used as a food additive (arrows); (B) casein micelles in milk.

remaining thin liquid layer is allowed to dry before electron microscopic examination. In some cases, an aqueous suspension of the particles is dried first and the electron-dense salt solution is added subsequently. It is essential to find the optimum concentrations, pH, and thickness of the layer of the suspension by preliminary experiments.

Nermut (1973) showed that air-drying can lead to distortion of negatively stained particles and that their shapes may be better preserved by freeze-drying. The suspended particles are adsorbed on a Formvar film-coated grid, stained with 2% sodium phosphotungstate, 3% ammonium molybdate, or 1% uranyl acetate, excess liquid is drained by blotting, and the grid is dipped in liquid nitrogen, where the sample is rapidly frozen and then freeze-dried. White powder appearing on the grid consists of dried stain; it is immediately removed by gently blowing a stream of dry air or nitrogen on the sample while it is still on the specimen stage. The sample is examined as soon as possible. If examination must be delayed, the sample is stored *in vacuo* to prevent the particles from rehydrating (Nermut 1973).

Metal Shadowing

Metal shadowing is a broad term for the preparation of samples for electron microscopy by techniques characterized by the deposition of metal as a vapor at a specific angle or angles on the specimen surface, resulting in a three-dimensional appearance of the micrographs (Fig. 2.14). The image is obtained by partial absorption of the electron beam due to the varying thickness of the metal layer on the sample depending, in part, on the angle at which the metal had been deposited at the particular site. During deposition, energetic metal particles migrate over the specimen surface to produce small crystallites that impart granularity to the sample (Moor 1973C.) The high granularity of gold routinely used in SEM not easily recognized at low magnifications used in this type of electron microscopy would be unacceptable in TEM. To obtain low granularity, high-melting point metals such as chromium (Puhan *et al.* 1976), platinum, palladium, iridium, or tungsten are used (Henderson and Griffith 1972). Metal evaporation may be done either by resistance or electron beam heating or by ion beam sputtering *in vacuo* (Echlin 1978B). In resistance heating, a calculated mass of platinum is premelted to form a bead on a carbon electrode; the platinum bead is mounted to face the sample and is evaporated as the carbon electrode is heated (Haggis 1974). In electron beam heating, the metal evaporant is the anode target and is heated from a cathode maintained at 2–3 keV. This is a very efficient method of heating since the region of

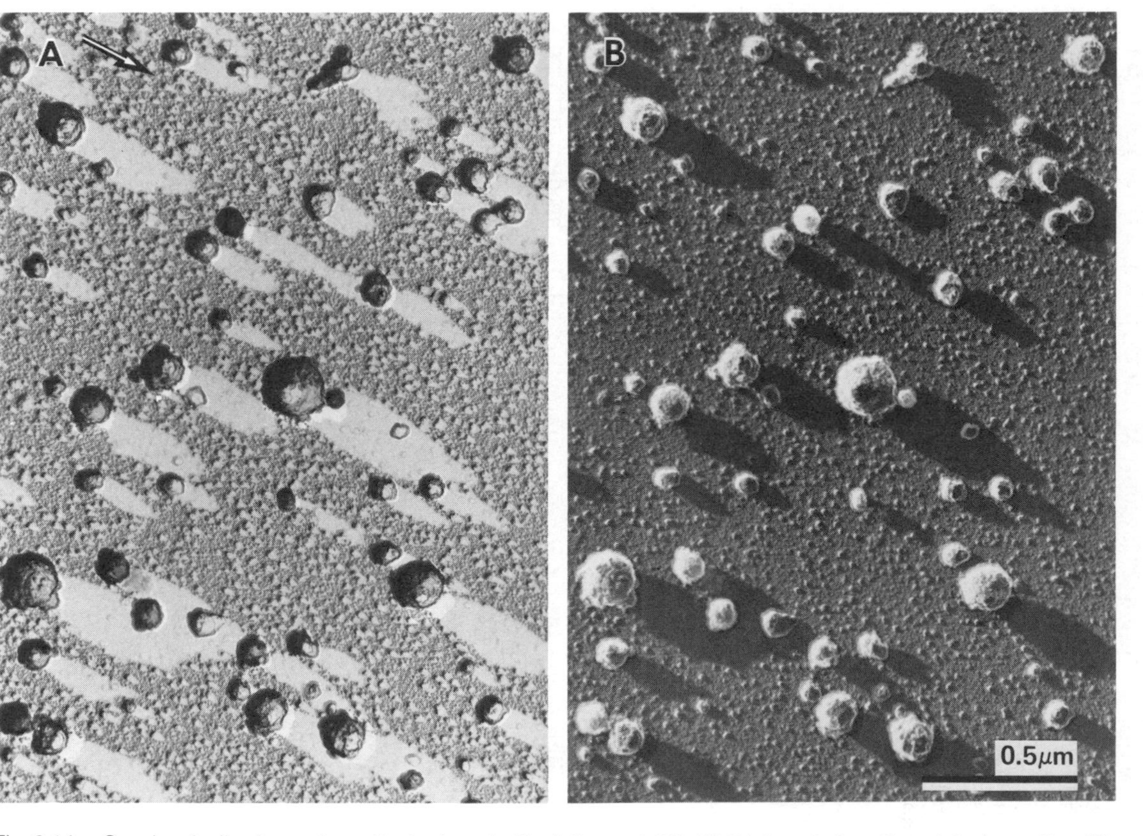

Fig. 2.14. Casein micelles in caprine milk shadowed with platinum at 17°. (A) Print made from the original negative; (B) print made from an intermediate negative; the same area is shown in both A and B. Arrow indicates direction of shadowing. Minute particles of uniform dimensions attached to the background are submicellar casein.

the highest temperature is only the vapor-emitting surface and not the entire evaporant source material (Echlin 1978B). In ion beam sputtering, a beam of argon ions is focused onto a target plate composed of the metal to be deposited; atoms of the metal are ejected from the target plate by the ion beam and are deposited on the sample.

Thickness of the metal deposited on the specimen can be either calculated (Bradley 1965; Echlin 1978B) or measured by commercial film thickness monitors.

To examine a suspension by metal shadowing, the particles are fixed and are subsequently dried on a Formvar film-coated grid, on a glass slide, or on a freshly cleaved mica sheet. Although air-drying has been used most frequently, freeze-drying and CPD produce superior results because they prevent artifacts associated with air-drying. It is possible to attach submicroscopic particles to support surfaces by means of polyelectrolytes. For example, casein micelles were attached to mica treated with poly-L-lysine, where they remained during dehydration and critical point drying (Kalab *et al.* 1982). Dried particles are shadowed in high vacuum of 0.13–1.3 mPa (10^{-6}–10^{-5} torr) on cold surfaces where the scattering of platinum atoms and their deposition on areas shielded by the particles is greatly reduced. The dimensions of the particles under study determine the optimal angle for shadowing: the smaller the particles, the more acute the angle (2–30°). Grids with the shadowed particles are examined directly in the electron microscope, but if shadowing was done on a mica sheet or on a glass slide, carbon is evaporated at 90 deg to reinforce the metal coating by forming a thin continuous film covering the shadowed particles. This film is floated on water, cleaned, and the replica thus formed is lifted on a bare grid for electron microscopic examination.

Because the electron beam passes through the shadowed area, i.e., the area free of the shadowing metal, and exposes photographic material, the shadows appear dark on the negative, whereas areas with a platinum deposit appear light in color. The negative thus gives an impression of a particle lit with light at a low angle. A print made from the negative resembles a negative of the particle lit with light (Fig. 2.14A). In order to obtain a print of the more familiar version, an intermediate negative is made from the initial negative by copying or photographing (Boye and Rønne 1978) it on another photographic film or plate (Fig. 2.14B). Another possibility is developing the film by an image-reversing process similar to slide processing (Schulz and Reynolds 1978). In another technique, the negative is printed on photographic paper and the print is copied on another sheet of photographic paper (Towe 1979). These photographic techniques are useful with all variants of metal shadowing, including replicas of freeze-fractured samples. However, unfamiliar images are encountered only if shadows are cast by the particles. If shadows are not cast on the background or on other particles, the specimen appears as if lit with light from the direction opposite to the metal shadowing.

The shadowing technique offers several advantages. (1) The technique is rapid even if freeze-drying or CPD is used because the thin layer of the sample is freeze-dried within several hours, taking approximately the same time as dehydration either in an alcohol series or in acidified 2,2-dimethoxypropane (Muller and Jacks 1975) followed by CPD. (2) Because three-dimensional objects are examined, tilting the grid with the sample and taking micrographs at two different angles differing, for example, by 12° produces pairs of stereo micrographs that further enhance the three-dimensional images of the samples when viewed under a stereoscope.

However, there are also disadvantages, the greatest one probably being that a part of each particle is obscured by its shadow and is not available for viewing. Mangino and Freeman (1981) raised an objection to using shadowed casein micelles for the evaluation of their dimensions by claiming that between 2 and 20 nm of metal was deposited on different micelles in the same field of vision. This variability in the metal thickness was not negligible at the mean micelle diameter of 80 to 90 nm.

A technique closely related to unidirectional shadowing is rotary shadowing similar to coating samples for SEM. The sample is rapidly rotated (approximately 530 rpm) as platinum deposition takes considerably less time than coating with gold. If rotary shadowing is carried out with the sample attached to a mica sheet, the metal coating is reinforced with carbon evaporated at 90 deg and the metal-and-carbon film is separated from the mica in the form of a replica. It is advisable to score the film on the mica support with a sharp needle because this facilitates fragmentation of the replica on the water surface and makes it easier to lift fragments of appropriate dimensions on the bare copper grid.

Although this technique produces no well-defined shadows, the three-dimensional particles may be examined stereoscopically to enhance the three-dimensional impression; if a thin platinum coating is applied, particles sufficiently translucent to the electron beam stacked one above the other may appear in the form of "ghosts" visible one through the other. The outlines of the particles produce severe contrast, as the electron beam is absorbed by the thick layer of the metal on the vertical particle walls. For this reason, the greatest advantage of rotary coating is experienced with very small horizontal structures, for example, parts of the surface of casein micelles (Fig. 2.15) examined at magnifications in excess of 50,000×.

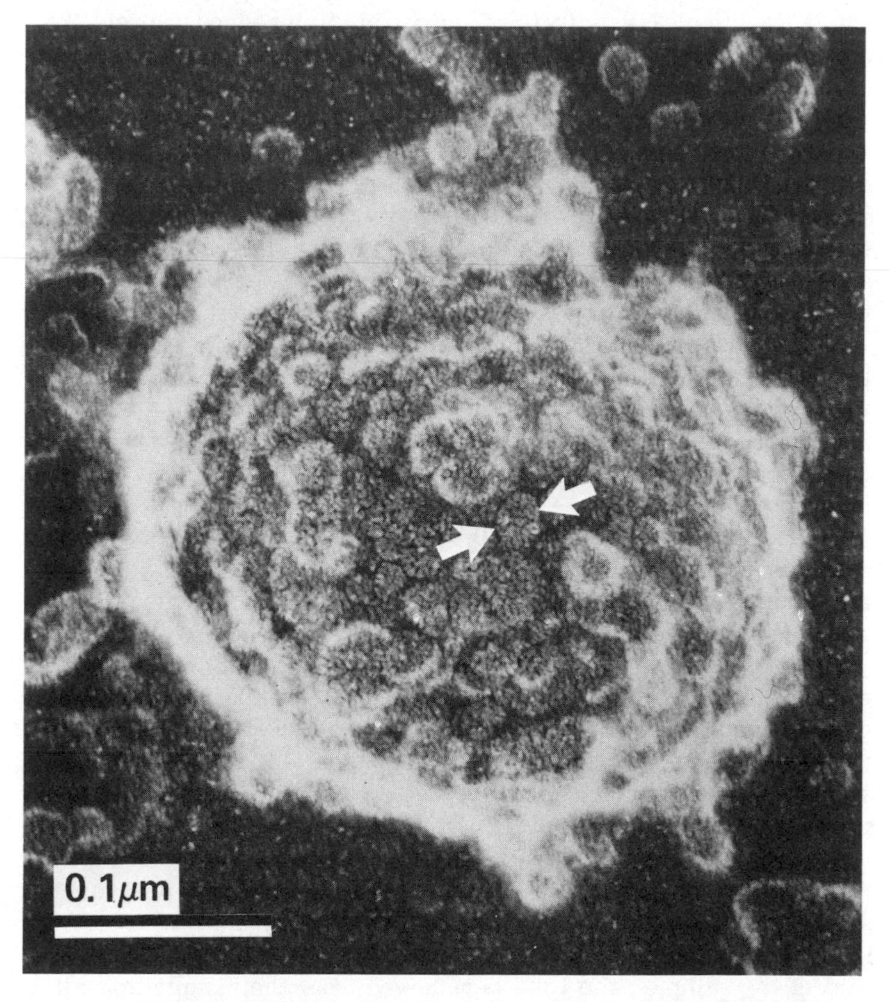

Fig. 2.15. Casein micelle in heated bovine milk coated with platinum while rotated. "Spikes" attached to the micelle and submicellar composition of the micelle proper (arrows) are clearly evident.

Thin-Sectioning

Thin-sectioning is a standard electron microscopic technique for studying the internal microstructure of the sample; it is suitable for solid samples as well as for suspensions. Apart from an unusual direct examination of sections obtained from a frozen hydrated muscle using cryomicrotomy and cold-stage TEM (Varriano-Marston *et al.* 1977, 1978), the sample is usually fixed, dehydrated, embedded in an electron-transparent resin (Hayat 1970B; Luft 1973; Glauert 1974), and sectioned using a glass or diamond knife (Hayat 1970C). The sections (< 100 nm thick) are collected on grids coated with a support film (Formvar, Butvar, Parlodion) on which a thin layer of carbon was subsequently evaporated (Hayat 1970E). To emphasize certain components of the sample, the sections are stained (Hayat 1970D) with electron-dense stains such as uranyl acetate and lead citrate (Reynolds 1963). The image in the electron microscope is formed by the passage of electrons through areas containing varying amounts of the stains.

To prepare fluid samples (emulsions and suspensions) for thin-sectioning, Salyaev (1968) suggested encapsulating them first in a narrow tube (~0.5 mm) made from 2–4% agar gel as shown in Fig. 2.16. The agar gel tube is permeable for fixatives as well as for dehydrating and embedding agents but retains corpuscular components (casein micelles, fat globules, bacteria, etc.) in the sample. The technique was used with milk (Henstra and Schmidt 1970A, B; Harwalkar and Vreeman 1978) and with orange juice (Jewell 1981); its advantage is that is does not affect the original distribution of the corpuscular components in the sample.

The procedure is described step by step in Fig. 2.16. Stainless steel rods used for cleaning syringes have been found superior to glass rods because of their strength and uniformity. The rod is dipped several times in a warm agar sol until a uniform agar gel cylinder is formed, after which it is trimmed at both ends forming a tube approximately 10 mm long. The tube is pushed with fingers about 1 mm down the rod and the sample is slowly aspirated. A small air bubble between the rod functioning as a piston and the sample during aspiration prevents the sample from spreading around the rod by capillary forces. After a sufficient volume of the sample is aspirated (filling the tube with 2 to 3 mm of the sample), the tube is removed from the sample and air is aspirated. The lower end of the tube is blotted with filter paper and a drop of agar sol is used to seal the agar gel tube. In addition, it is advisable to dip the tube containing the sample briefly into agar sol to reinforce the seal. The gel sets within several seconds after which time

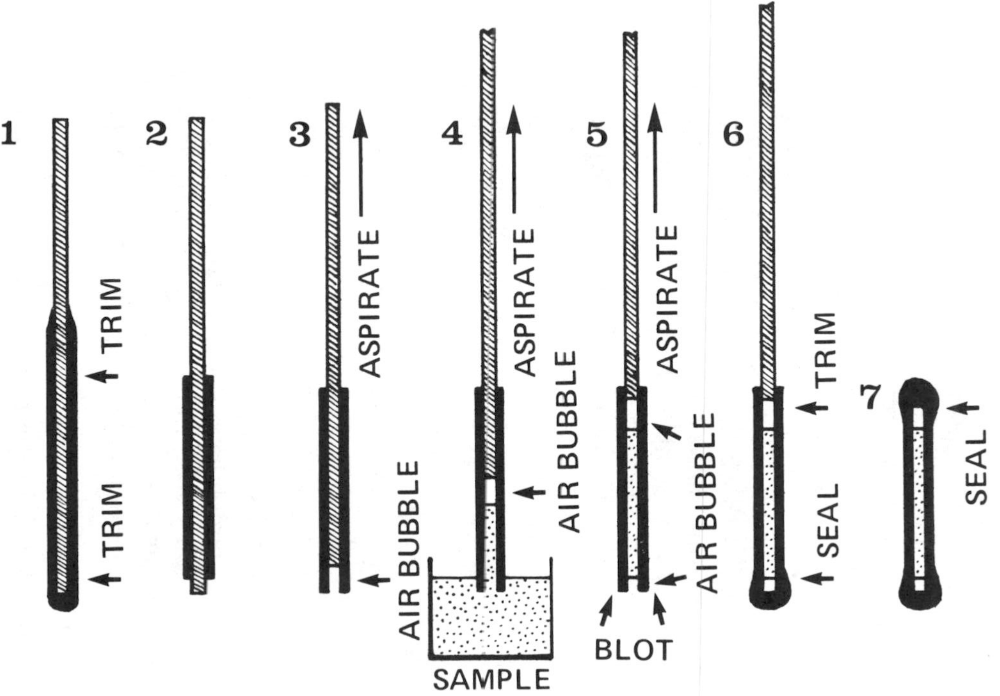

Fig. 2.16. Microcapsulation of liquid samples in agar gel according to Salyaev (1968) for subsequent embedding in a resin. 1. A glass rod or stainless steel wire (shaded) is coated with agar gel (solid black line) and the gel is trimmed; 2. Trimmed agar gel forms a tube; 3. A small air bubble is aspirated by pushing the gel tube down; 4. Sample is aspirated into the agar gel tube by using the rod or wire as a piston; 5. A small air bubble is aspirated following the aspiration of the sample and the lower end of the agar gel tube is blotted with a piece of filter paper; 6. The lower end of the tube is sealed with warm agar sol and the upper end is trimmed leaving a small air bubble above the sample; 7. The upper end of the agar gel tube is sealed with warm agar sol and the sample is immediately fixed (unless a fixed sample has been encapsulated) and dehydrated.

the tube is cut away from the rod leaving a small air bubble at the upper end, and the tube is sealed. The sample is fixed in glutaraldehyde immediately afterward. It is recommended that several tubes be prepared with a single sample because some agar gel tubes may occasionally leak, leading to the loss of the sample. It is imperative that the sample be small, particularly if it contains fat (cream, mayonnaise) or it will be difficult to impregnate the sample with the resin. The agar gel tube walls should also be as thin as possible.

If only the corpuscular components are of interest, for example, if dimensions of fat globules or the ultrastructure of fat globule membranes are studied in mayonnaise (Chang *et al.* 1972; Tung and Jones 1981) or cream, the emulsion or suspension may be mixed with a warm agar sol; the resulting gel, in which the corpuscular components of the sample are immobilized and with which they are diluted, is cut into small blocks, 1–2 mm^3, and the blocks are processed for electron microscopy in the same way as solid samples. In contrast to diluting the sample with agar gel it is possible to concentrate and compact suspended particles for electron microscopy by centrifugation. Depending on their density, the particles either float (fat globules) or sediment (protein aggregates such as casein micelles and bacteria) and form a pellet. The pellet may be processed for electron microscopy (Kalab 1980A), or it may be resuspended and fixed and the particles retrieved by ultracentrifugation. Hobbs (1979) used this technique to isolate fat globules from cream. The fat globules were finally collected on a Nucleopore filter (200 nm pore sizes), and the filter was covered with a thin layer of a 2% agar gel. After fixation, the agar gel was block-stained with a uranyl acetate solution, cut into small blocks, and embedded in Araldite for sectioning.

Some foods (cream cheese or mayonnaise) may disintegrate during fixation in aqueous media in spite of their apparently solid nature. Such samples are dip-coated with a warm agar sol (Kalab 1981) and the resulting minute beads are handled without the risk of disintegration. Mayonnaise examined in the original state using this technique is shown in Fig. 2.17.

In another procedure, agar gel is spread ~ 1 mm thick on a microscope slide and after gelling, small wells (~ 1 mm in diameter) are made with the tip of a scalpel. Approximately 1 mm^3 of the sample is placed in the wells and another thin layer of warm agar sol is spread over the sample with a spatula taking care that the sample is not touched and disturbed. Samples thus sandwiched between two layers of agar gel are trimmed immediately after the agar gel solidifies and are fixed. Incorporation of a glutaraldehyde solution in the warm agar sol accelerates the onset of

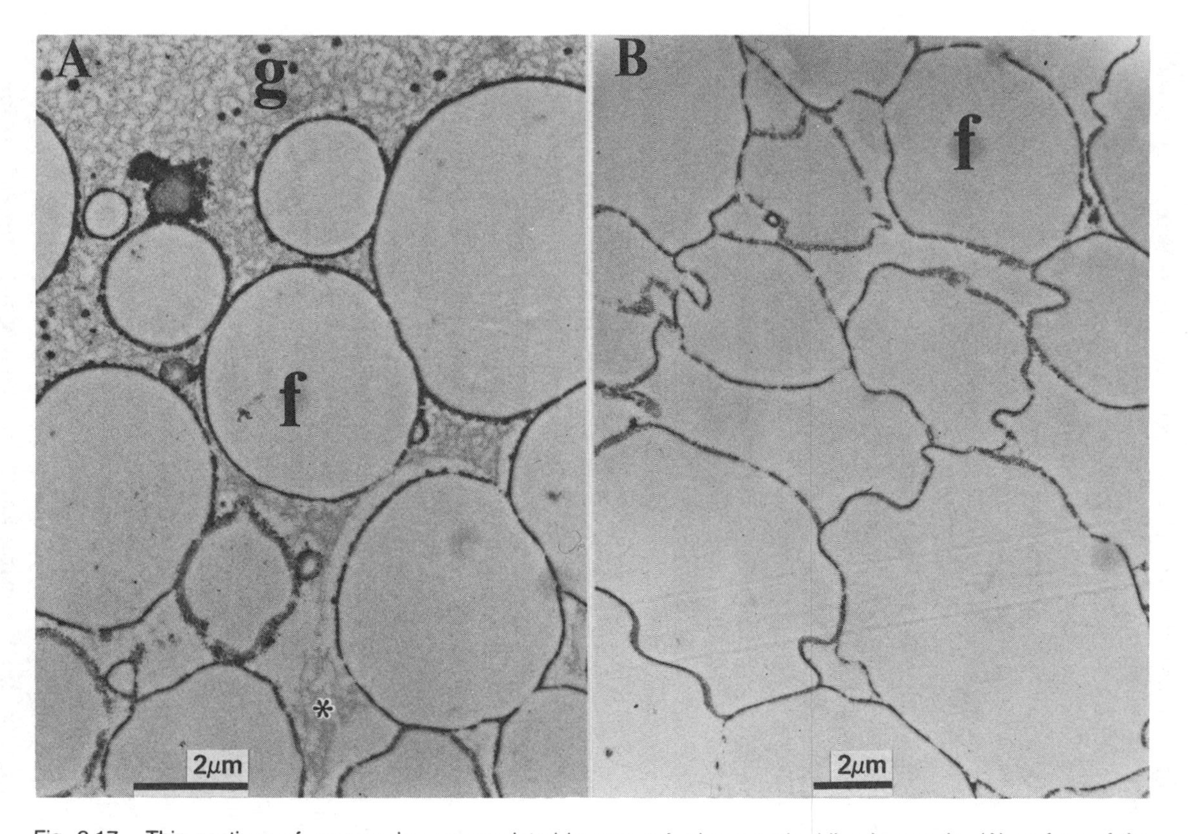

Fig. 2.17. Thin sections of mayonnaise encapsulated in agar gel prior to embedding in a resin. (A) surface of the mayonnaise particle in contact with agar gel (g) which penetrated (asterisk) a layer of several fat globules (f); (B) Fat globules (f) are compacted in the interior of the mayonnaise particle in agreement with results obtained by freeze-fracturing.

fixation. It seems that the effect of the aqueous phase in the agar gel on the encapsulated sample has not yet been assessed although it is probably not negligible.

Pudding composed of low-methoxyl pectin and milk is relatively solid, yet it becomes dispersed rapidly in aqueous fixatives within a short time (Kalab 1977B). Rather than encasing the pudding in agar gel, the author modified the fixative by incorporating calcium ions. Because the pudding matrix was composed of a calcium pectate gel network in which calcium-depleted casein micelles were dispersed (casein micelles were the donors of calcium), additional calcium ions in the fixative kept the pudding matrix insoluble while the casein micelles were fixed; both components were dehydrated for embedding in a resin.

It is evident that there is a variety of techniques suitable for the preparation of fluid and soft samples for thin-sectioning that will meet the objectives of the study.

To prevent artifactual changes from taking place in the sample during dehydration and embedding, the sample is fixed, as was mentioned in the section on SEM. Aldehydes such as glutaraldehyde, formaldehyde, and acrolein (Hayat 1970A; Glauert 1974) react with proteins and fix them by cross-linking. Unbuffered glutaraldehyde or glutaraldehyde in acetate, cacodylate, phosphate and other buffers are used most commonly. The sample is usually postfixed with buffered (pH 6–7) osmium tetroxide. Tung and Jones (1981) reported that OsO_4 buffered to pH 4 to be compatible with the acidity of the sample failed to fix mayonnaise unless the pH of the fixative was increased to 6 to 7. Fixatives containing other heavy metals are also used in specific cases. Lead hydroxide, citrate, and acetate (Fig. 2.18) and potassium permanganate were used to fix membranes, and ruthenium salts present in the fixatives were used to increase the contrast of polysaccharides such as pectin (Hayat 1970A; Glauert 1974).

Hard and compact samples such as cereal and oil seeds, potato chips, or chocolate require special attention because fixation as well as impregnation with resin monomers takes a considerably longer time than with more porous samples.

Peculiarities of preparing cereal seeds for thin sectioning were dealt with by Angold (1979). The fixatives used include both oxidizing (OsO_4 and $KMnO_4$) and cross-linking substances (formaldehyde, glutaraldehyde, and acrolein). The former also react with unsaturated lipids. Because they contain heavy elements, they give higher contrast in the micrographs partially acting as stains. Osmium tetroxide preserves more cell components, whereas $KMnO_4$ emphasizes membrane systems (Angold 1979). Aldehyde fixatives provide no staining but are more

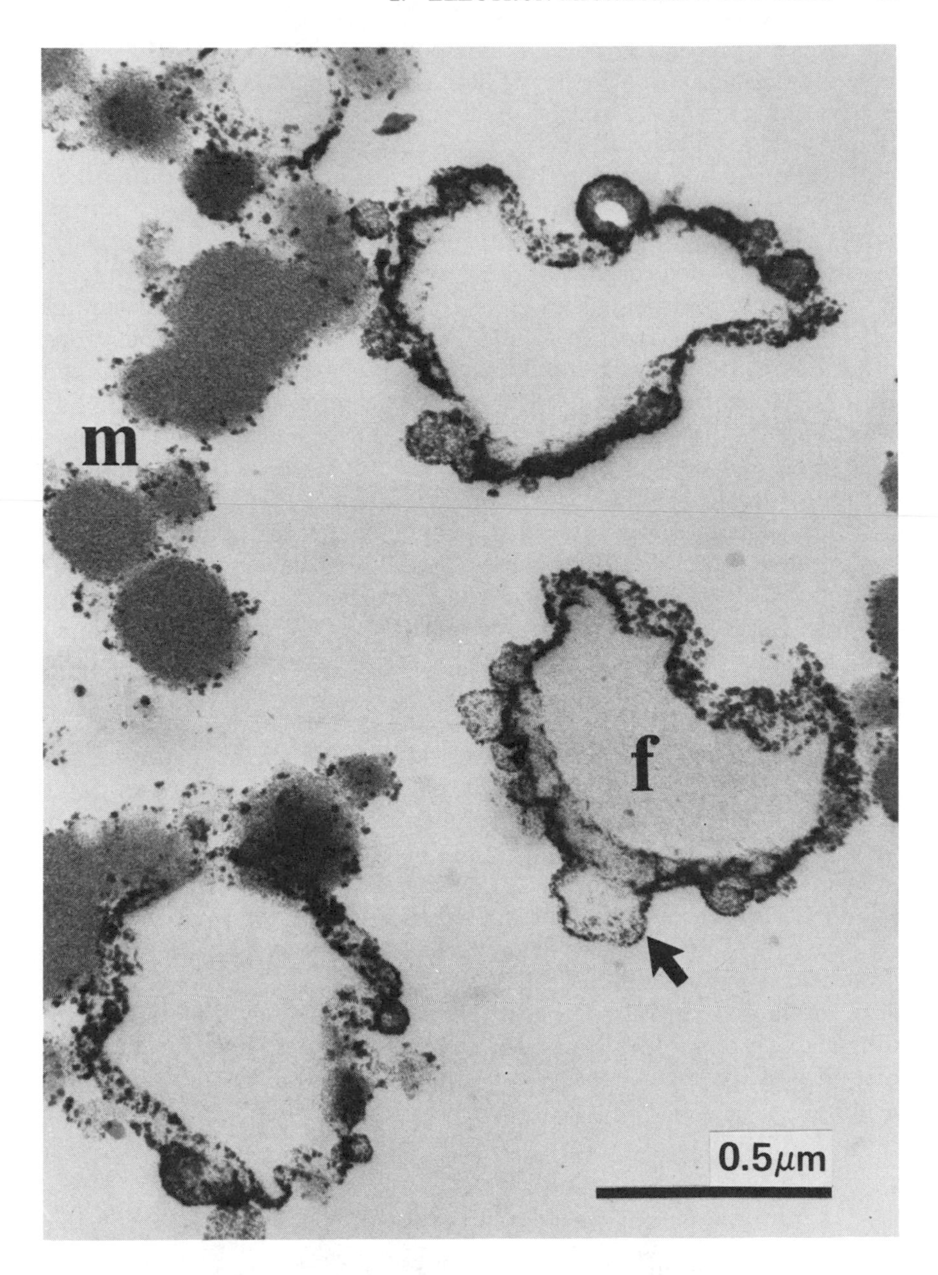

Fig. 2.18. Lead acetate present in the fixative increases the contrast of fat globule membranes in yogurt made from heated milk (Kalab 1977B). (m) Casein micelles; (f) fat globules; arrow points to a bleb in the fat globule membrane; lead has imparted high granularity to the images of the membranes.

effective in preserving cellular components. Sabatini *et al.* (1963B) suggested a double fixation using both the aldehydes and heavy metals.

Fixation leads to the hydration of seeds but must not be accompanied by enzymatic changes in the seed microstructure.

Powders such as powdered milk products and flour are difficult to embed in a resin without changing the nature of the particles. Fixation with OsO_4 vapors was used (Müller 1964; Kalab 1980A) by placing OsO_4 crystals in a sealed container along with the particles under study. A droplet of water was added before sealing the container. Fixation cannot take place in a total absence of water. However, the physical appearance of spray-dried buttermilk and skim milk particles was seriously changed in the presence of the water vapor needed to fix the samples (Fig.2.4). It was found that lactose initially present in the milk powders in the form of an amorphous glassy mixture of α- and β-lactose was converted at a humidity as low as 40% (Saltmarch and Labuza 1980) into α-hydrate crystals, which rapidly altered the particle surfaces (Kalab 1980A).

Concerning high-fat foods, Berger *et al.* (1979) reported that Cruickshank (1976) and Lewis (1975) fixed chocolate and fatty tissues with OsO_4 for a period of 2 to 4 weeks and Wortmann (1965) fixed butter for as long as 6 months because of lack of effective fixatives for lipids. Lewis (1979) recommended keeping fatty tissues of meat in contact with OsO_4 for a minimum of 2 weeks, but he warned that this long-term fixation may render protein more labile and cause its loss during the subsequent treatment. On the other hand, OsO_4 partially fixes only the liquid fat portion but not crystalline fat that shows up in the micrographs as clear areas against the dark background of the liquid fat. This is evidently associated with reactivity of OsO_4 with unsaturated lipids. Cruickshank (1976) warned that the appearance of electron micrographs of cocoa butter differed with various types of embedding procedures.

The fixed sample is dehydrated either in a graded ethanol or acetone series (Glauert 1974; Kalab and Harwalkar 1974) or in acidified 2,2-dimethoxy propane (Muller and Jacks 1975; Kalab 1977A) which in the presence of hydrogen ions reacts with water to form methanol and acetone:

$$CH_3\text{—}\underset{\underset{OCH_3}{|}}{\overset{\overset{OCH_3}{|}}{C}}\text{—}CH_3 + H_2O \xrightarrow{H^+} 2\,CH_3OH + CH_3\text{—}CO\text{—}CH_3$$

Dehydration prepares the sample for embedding in a resin. There is a great variety of resins with different characteristics commercially available (Hayat 1970B; Luft 1973; Glauert 1974; Spurr 1969). The resin is selected so that its characteristics (viscosity of the monomer mixture and hardness of the polymer) suit the sample; hard resins are used to embed hard samples such as cereal grains, whereas meat emulsions or cheese with a high fat content are best embedded in a soft resin to reduce the difference between the soft fat and the resin.

The compact structure of seeds poses difficulty during embedding in resins because infiltration with monomers is a slow process that is difficult to overcome even by the use of low-viscosity media (Spurr 1969). Angold (1979) reported satisfactory results by impregnating seeds in a 25% solution of the Spurr's resin in 1 : 2 epoxy propane. By leaving the vessel uncovered at 30°C overnight, most of epoxy propane evaporated, leaving the seed material in a concentrated resin ready for polymerization.

Blocks containing the embedded sample are trimmed and sectioned (Hayat 1970C). Diamond knives are preferred to glass knives, particularly with samples that will be examined at relatively low magnifications—for example, cheese—because diamond knives produce large areas free from defects such as scratch marks. Although diamond is an extremely hard material, food samples containing crystals—for example, calcium phosphate (Fig. 2.19) or oxalate—should be sectioned with caution and the knife should be checked frequently. As far as the presence of soluble salt crystals in foods is concerned, such crystals are most probably washed out from the sample during preparatory steps leaving behind cavities of the initial crystalline shapes in the surrounding medium (Fig. 2.19) (Kalab 1981).

Thickness of the sections depends on the subject of study and varies from 30 to 100 nm.

Hard and compact samples such as seeds and powders cause various difficulties during sectioning: some foods in the powder form have a tendency to chip out from the resin. The only way to overcome this problem is to select very hard resins to match the hardness of the powder particles. It seems, however, that the technique by Buchheim (1982) of embedding food powder particles in polyethylene glycol with subsequent freeze-fracturing and replication produces better results as will be discussed later.

The major difficulty in preparing high-fat foods such as fatty tissues, chocolate, and meat emulsions for thin-sectioning is the solvent action of the embedding media coupled with the requirement to polymerize the resin at an elevated temperature (Berger *et al.* 1979). A high fat content or extremely large fat globules in the sample may cause difficulties during sectioning as they cause inhomogeneities in the block; being

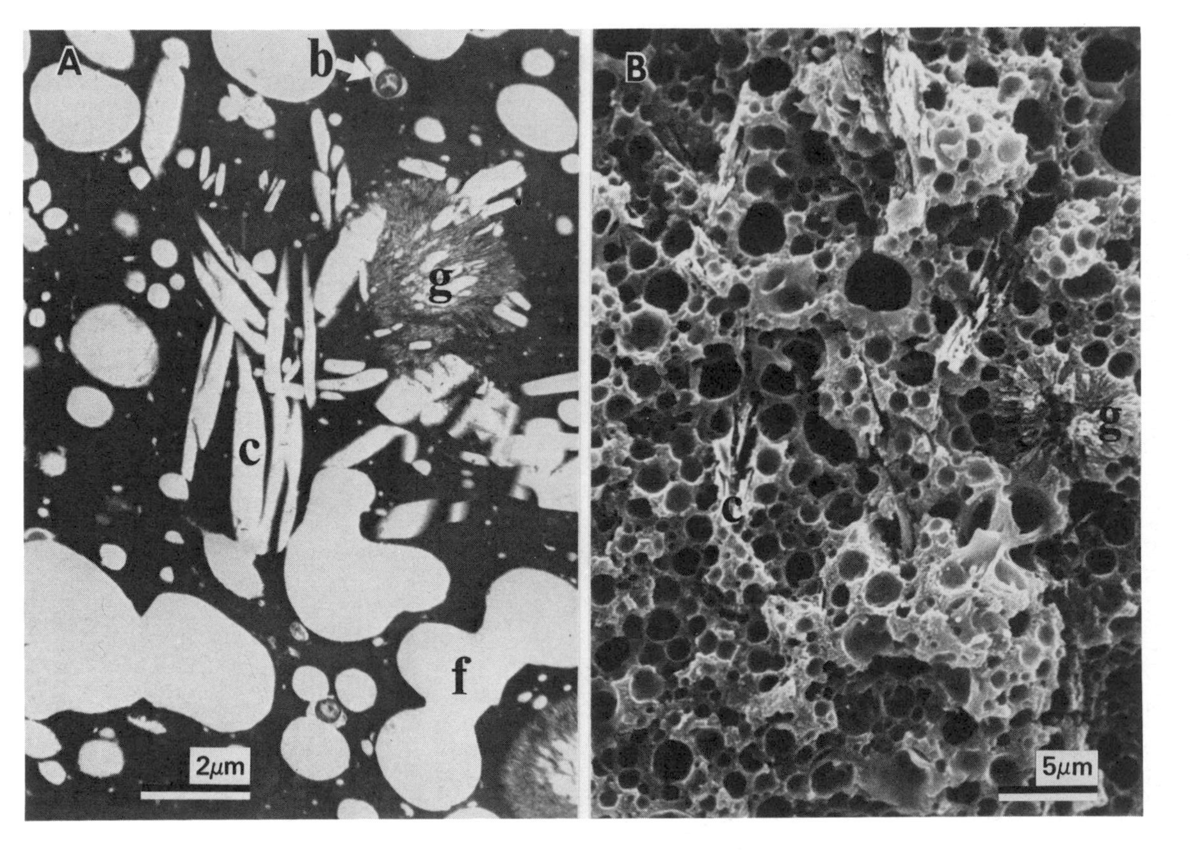

Fig. 2.19. Detection of water-soluble crystals in process cheese. (A) Thin section of process cheese after 10 min in the cooker: (c) Sodium citrate crystals; the salt was probably washed out from the sample during fixation leaving behind cavities in the form of the initial crystals; (f) Fat particles being emulsified into smaller globules; (g) Initially present crystals of insoluble calcium phosphate. (B) SEM of process cheese held in the cooker for 40 min: (c) Cavities left in the cheese by sodium citrate crystals washed out during the preparation of the sample for SEM; (g) initially present crystals of insoluble calcium phosphate.

From Rayan et al. (1980).

softer than the resin, the fat globules become compressed during sectioning (chatter) which may result in a striated image of the fat globules (Fig. 2.20). Starch causes a different problem: during embedding the compact starch granule resists penetration by the resin and remains unimpregnated. While sections are floated on water during sectioning, the starch granule becomes hydrated and the resultant swelling gives rise to radial folds. In the micrographs, the folds appear as if they are part of the microstructure (Fig. 2.21A), although they are artifacts (Fig. 2.21B). Gallant and Guilbot (1971) explained the process by which this artifact is formed (Fig. 2.22).

Staining of samples for electron microscopy is limited compared with staining of samples destined for light microscopy because the images in electron micrographs are in the form of shadows of areas more or less translucent to the electron beam. Some authors prefer block-staining (Knight 1977) with uranyl acetate, which means that either the sample is stained with an aqueous solution of uranyl acetate immediately after fixation, or the fixed and dehydrated sample is stained with uranyl

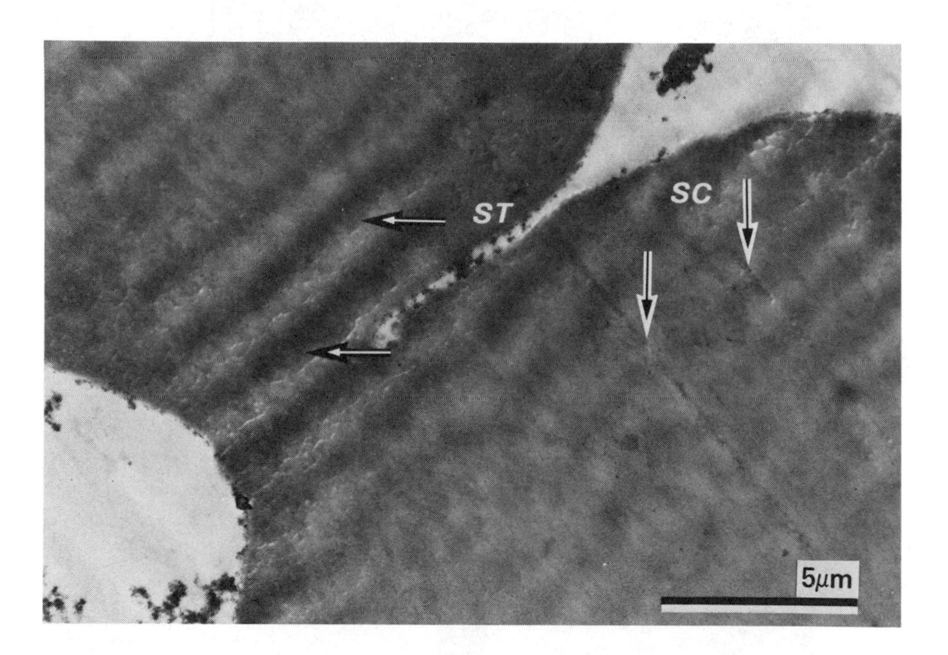

Fig. 2.20. Striation (chatter) of large fat globules due to a defect during sectioning. The striations (ST) caused by chatter during sectioning are perpendicular to the direction of sectioning indicated in this micrograph by scratch marks (SC) caused by a defective knife. *From Kalab (1981).*

Fig. 2.21. Thin sections of starch granules. (A) A starch granule with characteristic radial "shadows." (B) Thin section of another (larger) starch globule was shadowed with gold (direction of shadowing is indicated by an arrow) by Gallant and Guilbot (1971) to explain that the "shadows" in the sections are caused by folds (see also Fig. 2.22).

(B) From Gallant and Guilbot (1971).

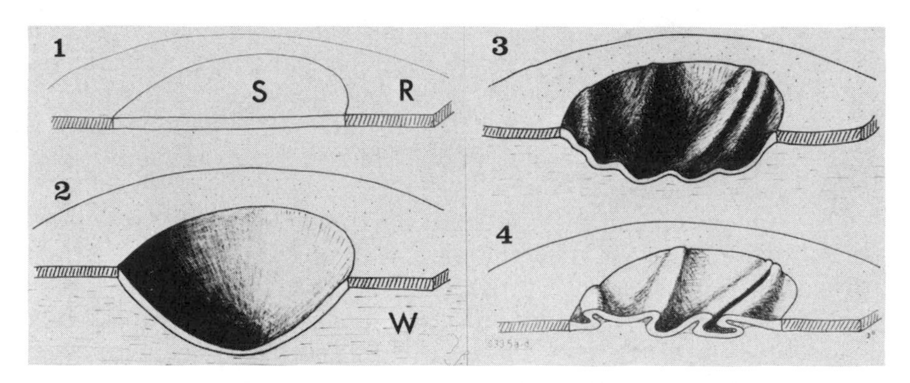

Fig. 2.22. Mechanisms by which folds develop in starch granule sections. (1) Section immediately after separation from the block; (R) resin; (S) starch granule. (2) Starch granule swells when the section is floated on water. (3) Undulations develop in the swollen starch granule section. (4) Undulations form folds after the section is lifted on a grid and dried. In the microscope, the folds cast shadows as shown in Fig. 2.21A.
From Gallant and Guilbot (1971).

acetate dissolved in methanol prior to embedding. Others stain sections with uranyl acetate in methanol followed by staining with aqueous lead citrate solutions (Reynolds 1963). Knoop *et al.* (1973) studied the distribution of calcium phosphate in casein micelles using unstained sections and resorting to the absorption of the electron beam by calcium alone. Kudo *et al.* (1979) used periodic acid–silver methenamine staining for κ-casein in casein micelles. Horisberger and Vonlanthen (1980) studied κ-casein using lectin-labeled gold granules; the use of gold granules as markers has recently been reviewed by Horisberger (1981).

Dimensions of globular particles (fat globules, casein micelles, etc.) may be calculated from the apparent diameter of the sections using mathematical procedures developed by Bach (1964), Goldsmith (1967), Groves and Freshwater (1968) and amended by Rose (1980). These procedures were used to calculate the size distribution of casein micelles in bovine milk (Schmidt *et al.* 1973; Schmidt and Buchheim 1976). Methods for determining fat globule size distribution were published by Walstra *et al.* (1969). Such calculations are facilitated by digital image analyzers that can be programmed to evaluate the diameters, areas, eccentricity, orientation of the particles, and other parameters.

Freeze-Fracturing and Replication

All of the preceding TEM techniques require dehydrating the sample prior to the electron microscopy. Dehydration has been known to change the initial microstructure to a varying extent and may be the source of some severe artifacts. It is assumed that fixation of samples by rapid freezing (cryofixation) best preserves the original microstructure provided that ice crystal formation is prevented. Requirements for the rapidity of freezing are even stricter with TEM than with SEM because of the generally higher resolution of TEM and, thus, frequent use of higher magnifications.

The principle of replication is similar to that described in the section on SEM examination of gold replicas: the sample is frozen, freeze-fractured, and replicated with a high-melting metal and carbon either immediately or after a certain predetermined period of freeze-etching. The metal-and-carbon replica of the fractured surface is cleaned, placed on a grid, and examined by TEM.

All foods that can be examined by other electron microscopic methods may also be examined by freeze-fracturing and replication. For some foods, particularly for suspensions (fruit juices, milk), emulsions (salad dressings, margarine, chocolate), and foams (ice cream, whipped products), freeze-fracturing may be the only reliable electron microscopic technique. Recent developments have rapidly expanded the variety of foods examined by freeze-fracturing. The technique, as it relates to foods, has been recently reviewed in great detail by Buchheim (1982) (Fig. 2.23).

The most critical step in this technique is rapid freezing of the sample. If the temperature of water is lowered below its freezing point, supercooling occurs until impurities start acting as nuclei for crystallization of ice. Pure water crystallizes between 273 and 143K (0° and −130°C). Ideally, the objective of freezing samples for electron microscopy is to transform liquid water into amorphous ice, i.e., to achieve so-called vitrification of the sample. This process is characterized by extracting heat from the sample more rapidly than it can be produced by crystallization. Crystallization may be prevented by supercooling. Critical freezing rate of pure water is supposed to be higher than 10^6 K/sec (Moor 1973A), but the critical freezing rate of water present in live cells is approximately 10^4 K/sec; this is probably also true for water in foods and the value may be even lower in foods containing high concentrations of soluble substances such as sugar. The more rapidly the sample is frozen, the lower the risk of its damage by the development of ice crystals. Ice crystals developing in the samples frozen slowly distort the original microstructure by compressing structural components aside and even cause some individual structural components to fuse (Kalab and Emmons 1974).

In general, the rate of freezing is affected by (1) sample dimensions, (2) composition of the sample, and (3) the nature of the freezing agent (cryofixative).

Fig. 2.23. Microstructure of butter as revealed by freeze-fracturing. (f) Fat globules; asterisk marks impurities in the replica. A similar micrograph has been published by Precht and Buchheim (1979).
Courtesy of Wolfgang Buchheim (1979).

"Standard freezing" is the term used (Bachmann and Schmitt-Fumian 1973) for dipping a small metal holder with approximately 1 mm^3 sample into liquid Freon at 123K (−150°C) as mentioned in the section on SEM. Sample holders are made of silver or gold and are in the forms of minute dishes or thin-walled two-part tubes.

If tubular holders are used, the sample fills both the lower and the upper parts and, thus, a single rigid cylinder is formed by freezing. The holder is mounted under liquid nitrogen onto a special specimen table;

the table is attached to the specimen stage precooled to 173K (−100°C) in a freeze-fracturing apparatus. After appropriate pressure (1.3 mPa, 10^{-6} torr) is achieved, the upper tube of the sample holder is knocked off by a cooled (77K, −196°C) mechanical arm and the fracture is replicated either immediately or after a period of freeze-etching, during which a predetermined thickness of the frozen aqueous phase is sublimed from the surface of the sample. Up to three tubular holders may be accommodated on specimen tables developed in the author's laboratory (Kalab 1981).

Smaller samples may be used with holders in the form of minute dishes and up to four samples may be fractured at the same time. In this case, the mechanical arm consists of a cooled blade that shaves off thin layers of the sample; the mechanisms of the fracturing arm must allow the blade to advance in the vertical direction.

The rate of freezing increases as the dimensions of the sample are reduced. Emulsions, especially if not too viscous, are rapidly frozen by spraying into a suitable coolant (spray-freezing).

A small volume of the sample (~0.7 ml) is sprayed into liquid propane (83K, −190°C) and the vessel is transferred into a cryostat (188K, −85°C), where propane is evaporated under reduced pressure. Frozen droplets of the sample (up to 50 μm in diameter) are mixed with a drop of precooled *n*-butylbenzene and the mixture is placed on precooled sample holders which are then dropped into liquid nitrogen. *n*-Butylbenzene acts as a binder and freezes solid in liquid nitrogen. The sample is then freeze-fractured and, if required, freeze-etched in the regular way (Bachmann and Schmitt-Fumian 1973).

An alternate method of rapid freezing consists of placing a holder with the sample into a fine jet stream of liquid propane. In this procedure, no insulating layer of gas forms around the sample. There is no time for propane to vaporize because the heat from the sample is rapidly carried away by the jet stream. Caution must be exercised when working with liquid propane because there is a great danger of oxygen condensation and the formation of an explosive mixture (Bachmann and Schmitt-Fumian 1973).

Spray-freezing was used to obtain quantitative data such as molecular weights, dimensions, and shapes of molecules of individual milk proteins (Buchheim and Schmidt 1979); Schmidt and Buchheim 1980).

Buchheim (1972) developed a method of forming fine emulsions of milk or fine suspensions of milk gels in paraffin oil, which were then subjected to "standard freezing." Other protein gels were studied by Hermansson and Buchheim (1981). In 1981, Buchheim extended the use of nonaqueous media for the preparation of milk powder suspensions.

The powders are suspended in polyethylene glycol (molecular weight 400) to form highly viscous yet flowing suspensions. Small amounts of such suspensions (1 to 2 μl) are transferred on dish-like sample holders and are frozen by the standard procedure. Freeze-fracturing and replication of the fractured surfaces make it possible to examine the internal microstructure of spray-dried particles. The replicas are floated on pure polyethylene glycol and are cleaned with acetone and with a sodium hypochlorite solution to removed fats and protein. Cleaned replicas are picked up on bare 200-mesh grids previously dipped into a 4% solution of Bedacryl 122X in a 1 : 1 xylene and benzene mixture in order to improve their adhesiveness.

Some substances are capable of reducing or eliminating the development of ice crystals in freezing water. Such substances are called "cryoprotective agents." Glycerol, dimethoxysulfoxide, and polyethylene glycol are used most frequently (Rebhun 1972). It was established that a 20% glycerol solution reduces the critical freezing rate of water in live cells from 10^4 K/sec to approximately 10^2 K/sec and leads to vitrification of samples up to 0.5 μm thick (Moor 1973A). Because the sample must be thoroughly impregnated with the cryoprotective agent, these agents are probably of greater importance to food samples of a strictly biological origin (plant and animal tissues) than to food mixes. Cryoprotective agents are known to introduce artifacts in unfixed samples, and for this reason it is advisable to fix susceptible samples in glutaraldehyde in advance. Only limited freeze-etching is possible if cryoprotective agents are used (Buchheim 1982) as they do not sublime. Some foods already contain natural cryoprotective agents such as sucrose.

A low water content also reduces the risk of ice crystal development. Freeze-fracturing of dehydrated samples was suggested by Buchheim (1976). The author found, however, that whereas in aqueous media casein micelles were fractured in the same plane as the medium, casein micelles suspended in dioxane were fractured along their surfaces and these surfaces were revealed by freeze-etching. Differences in the fracture planes in cottage cheese impregnated with 30% glycerol and with ethanol were shown earlier in the section on SEM (Fig. 2.3).

As was mentioned earlier, Freons cooled with liquid nitrogen to their freezing points and a nitrogen slush are excellent freezing media for standard freezing because these media do not generate gas that would insulate the sample and decelerate freezing. Liquid propane cooled to 83K (−190°C) is used in spray-freezing and jet-freezing. Good coolants are characterized by a low melting point, high thermal conductivity, and high specific heat. The boiling point must be high to avoid formation of an insulating gas layer and the vapor pressure must be high to allow evaporation under reduced pressure within a short time at temperatures lower than 188K (−85°C) to prevent recrystallization of ice in the sample.

Samples that have been freeze-fractured, are shadowed at 45 deg with platinum or another high-melting metal and the metal layer is reinforced with carbon evaporated at 90 deg. If freeze-etching is carried out (Moor 1973B) to expose structures obscured by (buried in) the frozen aqueous phase, it is accomplished by sublimation of ice from the surface of the fracture; a shield, cooled to 77K (−196°C) is placed over the fractured sample to prevent contamination of the fracture.

The replica is separated from the fractured sample by flotation on water, then cleaned and lifted on a grid for electron microscopy. Various cleaning agents such as sodium hydroxide, sodium hypochlorite, and nitric, hydrochloric, and chromic acids have been recommended (Koehler 1972; Steere 1973). Organic solvents such as acetone, petroleum and diethyl ethers, or chloroform are used to remove fat residues from the replicas. Replicas are easily retrieved from organic solvents immiscible with water (petroleum ether) by carefully pipetting water under the organic solvent; the replicas settle at the water–petroleum ether interface.

The replicas are examined and micrographs are printed in the same way as mentioned in the section on metal shadowing.

It is suggested that beginning microscopists consult the literature for examples of the most frequently occurring artifacts (Böhler 1975, 1979). Ice crystal development in the sample, contamination of the fractures with condensed frost, and insufficient cleaning of the replicas are the most common artifacts.

Replication of Dried Specimens

The extent of freeze-etching in freeze-fractured samples is limited to a relatively thin layer of ice that may be removed from the sample by sublimation before the onset of various artifacts (Böhler 1975, 1979). However, examination of three-dimensional structures by high-resolution TEM has been made possible by a technique developed by Haggis and Bond (1979) and applied to milk gels by Kalab (1980B).

The sample to be replicated is freeze-fractured and dried by following appropriate procedures outlined in Fig. 2.1. The fragments are wrapped in an aluminum foil leaving only the fracture exposed and mounted on a glass cover slip using vacuum grease. Platinum is evaporated *in vacuo* (1.3 mPa, 10^{-5} torr) at 45° while the sample is rotated (530 rpm) and carbon is evaporated at 133 mPa (10^{-3} torr) around a circular shield, 13 mm in diameter, protecting the sample from direct sputtering of carbon;

only carbon atoms deflected by N_2 molecules are deposited on the fractured surface of the fragment to reinforce the platinum coating. The coated surface of each fragment is cut away with approximately 0.5 mm of the sample, placed on a microscope slide facing up, and covered with a Perspex film held on a 12.5 mm diameter rubber ring. [The film is made from a 0.6% Perspex solution in chloroform (w/v) similar to the preparation of a Formvar film (Mercer and Birbeck 1972)]. The film is tapped lightly around the inner wall of the rubber ring and attached to the glass slide; the ring is removed and the slide is placed in a glass Petri dish saturated with chloroform vapors. In the Petri dish the film stretches and adheres firmly to the metal- and carbon-coated fracture and attaches the fragment to the glass slide. The Perspex film is scored with a scalpel approximately 1 mm away from the fragment and the cut-out rectangle is floated on water. Material adhering to the replica is removed overnight by digestion with chromic acid. Alkaline solutions are not suitable for this purpose because they cause the milk proteins to swell and damage the replica. Cleaning is completed by several exchanges of distilled water and the replicas are lifted on 400-mesh grids, dried in air, and carefully bathed in chloroform to dissolve the Perspex film. The micrographs resemble those obtained by SEM but show considerably more detail as evident in Fig. 2.24. Cottage cheese presented in this figure was also examined by SEM (Fig. 2.3). The topography of the replica depends on the porosity of the sample; after the Perspex film is dissolved from the replica, the replica is susceptible to mechanical damage.

LABORATORY SAFETY

It has to be recognized that workers in electron microscopy are exposed to a variety of safety hazards. These hazards were recently reviewed by Bance *et al.* (1981). They arise from work with toxic chemicals such as fixatives, critical-point drying fluids, embedding media, staining salts, flammable and combustible liquids used in dehydration and defatting of samples and in cleaning of replicas, and from work with electric equipment such as centrifuges, ovens, disintegrators, etc. X-Ray hazards may arise from X-radiation produced within the electron microscope. Critical-point dryers are potentially very dangerous because of the high pressure generated within the pressurized cell. The danger of liquid propane has already been mentioned in the text.

Every worker intending to use electron microscopy in food science should become acquainted not only with the individual research procedures and techniques but also with the hazards inherent in the work, in order to provide safe working conditions for everyone involved.

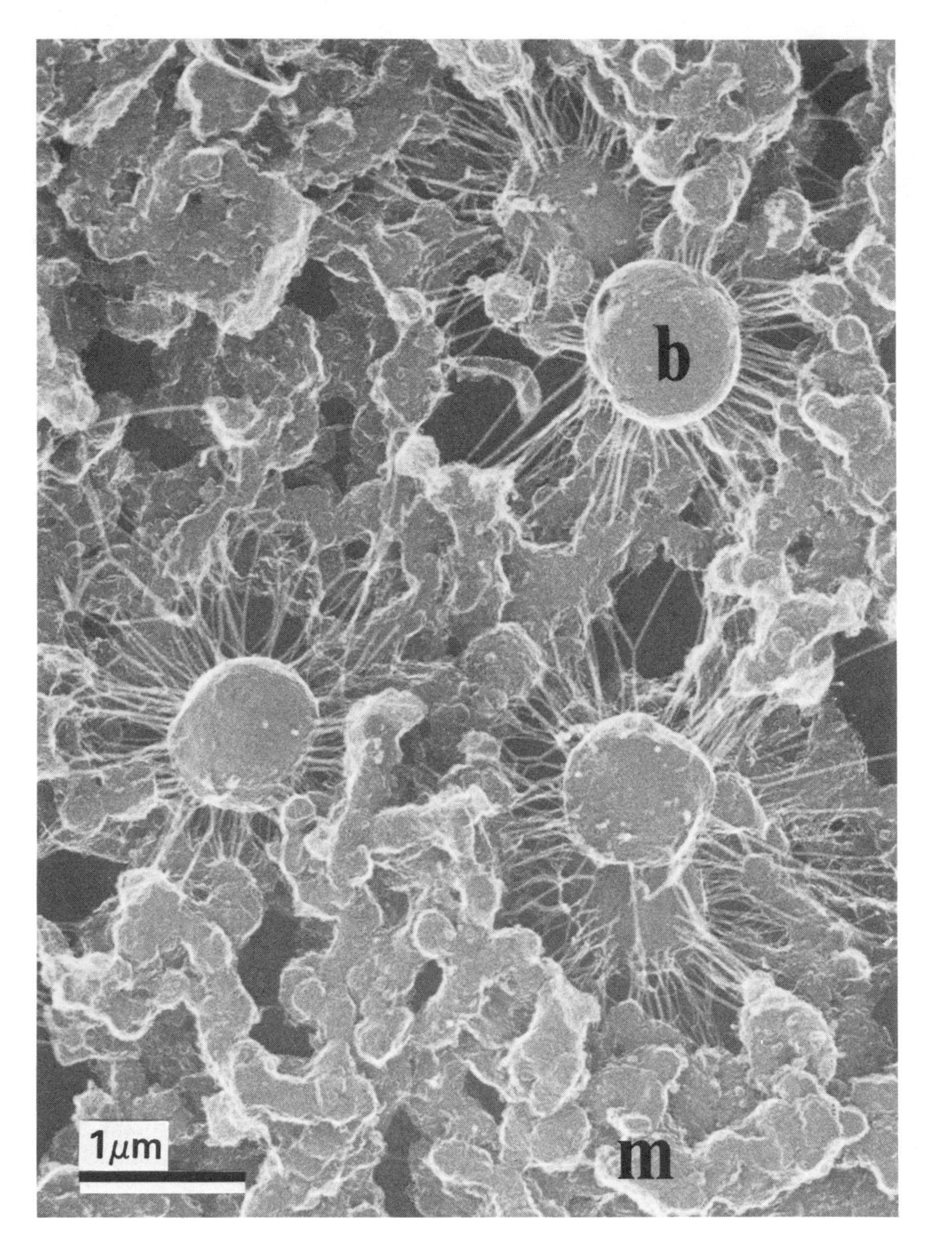

Fig. 2.24. Replica of a dried specimen. Freeze-fractured cottage cheese was critical-point dried and replicated with platinum and carbon by rotary coating (Kalab 1980B). (b) Streptococci attached by filaments to the protein matrix; (m) casein micelle chains forming the protein matrix.

ACKNOWLEDGMENT

This review is Contribution 500 from the Food Research Institute, Agriculture Canada in Ottawa.

The author thanks Mrs. Beverley E. Phipps-Todd, Mrs. Paula Allan-Wojtas, Mr. J.A.G. Larose, and Mr. Sierk Itz for skillful assistance with electron microscopy. The Electron Microscope Centre, Research Branch, Agriculture Canada in Ottawa provided facilities.

The author is grateful to Dr. D.G. Schmidt, Nederlands Instituut voor Zuivelonderzoek, Ede, The Netherlands, for providing Fig. 2.8; to Dr. D. Gallant, Institut National de la Recherche Agronomique, Nantes, France, for Fig. 2.21B and Fig. 2.22; and to Dr. W. Buchheim, Bundesanstalt für Milchforschung, Kiel, Federal Republic of Germany, for Fig. 2.23. Appreciation is expressed to Mr. E.F. Bond for a number of useful suggestions and assistance with the manuscript.

BIBLIOGRAPHY

ANGOLD, R.E. 1979. Cereals and bakery products. *In* Food Microscopy, pp. 75–138. J.G. Vaughan (Editor), Academic Press, New York.

ALLEN, R.D., and ARNOTT, H.J. 1981. Effects of exogenous enzymes on oilseed protein bodies. Scanning Electron Microsc. 1981/III, 561–570.

BACH, G. 1964. Determination of the frequency determination of radii of spherical particles from the frequencies of their cross sections in random sections of the thickness of ∂. z. Wiss. Mikrosk. *66*, 193–200 (German).

BACHMANN, L., and SCHMITT-FUMIAN, W.W. 1973. Spray-freeze-etching of dissolved macromolecules, emulsions and subcellular components. *In* Freeze-Etching Techniques and Applications, pp. 63–72. E.L. Benedetti and P. Favard (Editors). Soc. Franc. Microsc. Electr., Paris, France.

BANCE, G.N., BARBER, V.C., and SHOLDICE, J.A. 1981. Safety in the SEM laboratory—1981 update. Scanning Electron Microsc. 1981/II, 87–94.

BARTLETT, A.A., and BURSTYN, H.P. 1975. A review of the physics of critical point drying. Scanning Electron Microsc. 1975 *1*, 305–316, 368.

BERGER, K.G., JEWELL, G.G., and POLLITT, R.J.M. 1979. Oils and fats. *In* Food Microscopy, pp. 445-497. J.G. Vaughan (Editor). Academic Press, New York.

BLACK, J.T. 1974. The scanning electron microscope: Operating principles. *In* Principles and Techniques of Scanning Electron Microscopy. Biological Applications, Vol. 1, pp. 1–43. M.A. Hayat (Editor), Van Nostrand Reinhold Co., New York.

BÖHLER, S. 1975. Artefacts and Specimen Preparation Faults in Freeze Etch Technology. Balzers Aktiengesellschaft, Balzers, Liechtenstein.

BÖHLER, S. 1979. Artifacts and defects of preparation in freeze-etch technique. *In* Freeze-Fracture: Methods, Artifacts and Interpretations, pp. 19–29. J.E. Rash and C.S. Hudson (Editors). Raven Press, New York.

BOYDE, A. 1974. Three-dimensional aspects of SEM images. *In* Scanning Electron Microscopy (by O.C. Wells), pp. 277–307. McGraw-Hill Book Co., New York.

BOYDE, A. 1978. Pros and cons of critical point drying and freeze drying for SEM. Scanning Electron Microsc. 1978/II, 303–314.

BOYE, H., and RØNNE, M. 1978. Low cost internegatives from electron micrographs of metal shadowed objects. J. Microsc. *112* (3), 353–358.

BRADLEY, D.E. 1965. Replica and shadowing techniques. *In* Techniques for Electron Microscopy, pp. 96–152. D.H. Kay (Editor), Blackwell Scientific Publications, Oxford, England.

BUCHHEIM, W. 1972. A new technique for improved cryofixation of aqueous solutions. Proc. Fifth Eur. Congr. Electron Microsc., pp. 246–247, Manchester, U.K.

BUCHHEIM, W. 1976. Freeze-etching of dehydrated biological material. Proc. Fourth Eur. Congr. Electron Microsc., pp. 122–124. Jerusalem, Israel.

BUCHHEIM, W. 1981. A comparison of the microstructure of dried milk products by freeze-fracturing powder suspensions in non-aqueous media. Scanning Electron Microsc. 1981 *3*, 493–502.

BUCHHEIM, W. 1982. Aspects of sample preparation for freeze-fracture/freeze-etch studies of proteins and lipids in food systems. A review. Food Microstruct. *1*(2), 189–209.

BUCHHEIM, W., and SCHMIDT, D.G. 1979. On the size of monomers and polymers of β-casein. J. Dairy Res. *46*, 277–280.

BUMA, T.J. 1971. Free Fat and Physical Structure of Spray-Dried Whole Milk. Ph.D. Thesis. A collection of ten research papers published in Neth. Milk Dairy J. 1968–1971. Wageningen, The Netherlands.

BUMA, T.J. 1978. Particle porosity of spray-dried milk. Milchwissenschaft *33* (9), 538–540(German).

BUMA, T.J., and HENSTRA, S. 1971. Particle structure of spray-dried milk products as observed by a scanning electron microscope. Neth. Milk Dairy J. *25*, 75–80.

CARROLL, R.J., and JONES, S.B. 1979. Some examples of scanning electron microscopy in food science. Scanning Electron Microsc. 1979/III, 253–259.

CARROLL, R.J., THOMPSON, M.P., and NUTTING, G.C. 1968. Glutaraldehyde fixation of casein micelles for electron microscopy. J. Dairy Sci. *51*, 1903–1908.

CHABOT, J.F. 1979. Preparation of food science samples for SEM. Scanning Electron Microsc. 1979/III, 279–286, 298.

CHABOT, J.F., HOOD, L.F., and ALLEN, J.E. 1976. Effect of chemical modifications on the ultrastructure of corn, waxy maize and tapioca starches. Cereal Chem. *53*, 85–91.

CHANG, C.M., POWRIE, W.D., and FENNEMA, O. 1972. Electron microscopy of mayonnaise. Can. Inst. Food. Sci. Technol. J. *5* (3), 134–137.

CLARKE, I.C. 1975. Stereographic techniques. *In* Principles and Techniques of Scanning Electron Microscopy. Biological Applications, Vol. 3, pp. 154–194. M.A. Hayat (Editor). Van Nostrand Reinhold Co., New York.

COHEN, A.L. 1974. Critical point drying. *In* Principles and Techniques of Scanning Electron Microscopy. Biological Applications. Vol. 1, pp. 44–112. M.A. Hayat (Editor). Van Nostrand Reinhold Co., New York.

COHEN, A.L. 1977. A critical look at critical point drying—theory, practice and artefacts. Scanning Electron Microsc. 1977/I, 525–536.

COHEN, A.L. 1979. Critical point drying—Principles and procedures. Scanning Electron Microsc. 1979/II, 303–323.

COHEN, S.H., and TRUSAL, L.R. 1980. The effect of catheptic enzymes on chilled bovine muscle. Scanning Electron Microsc. 1980/III, 595–600.

CREAMER, L.K., and MATHESON, A.R. 1980. Effect of heat treatment on the proteins of pasteurized skim milk. N.Z.J. Dairy Sci. Technol. *15*, 37–49.

CROZET, N., and GUILBOT, A. 1974. A note on the influence of osmium fixation on wheat flour lipids observation by transmission and scanning electron microscopy. Cereal Chem. *51*, 300–304.

CRUICKSHANK, D.A. 1976. Brit. Fd. Mfg. Ind. Res. Assoc., Tech. Circ. 611. Cited by Berger *et al.* (1979).

DAVIS, E.A., and GORDON, J. 1978. Application of low temperature microscopy to food

systems. J. Microsc. *112*, 205–214.
DAVIS, E.H., and GORDON, J. 1980. Structural studies of carrots by SEM. Scanning Electron Microsc. 1980/III, 601–611.
DENEE, P.B. 1978. Collecting, handling and mounting of particles for SEM. Scanning Electron Microsc. 1978/I, 479–486.
DEXTER, J.E., DRONZEK, B.L., and MATSUO, R.R. 1978. Scanning electron microscopy of cooked spaghetti. Cereal Chem. *55*, 23–30.
DRONZEK, B.L., HWANG, P., and BUSHUK, W. 1972. Scanning electron microscopy of starch from sprouted wheat. Cereal Chem. *49*, 232–239.
ECHLIN, P. 1973. Scanning electron microscopy at low temperature. *In* Freeze-Etching Techniques and Applications, pp.211–222. E.L. Benedetti and P. Favard (Editors). Soc. Franc. Microsc. Electr., Paris, France.
ECHLIN, P. 1975. Sputter coating techniques for scanning electron microscopy. Scanning Electron Microsc. 1975, 217–224, 332.
ECHLIN, P. 1978A. Low temperature electron microscopy: A review. J. Microsc. *112* (1), 47–61.
ECHLIN, P. 1978B. Coating techniques for scanning electron microscopy and x-ray microanalysis. Scanning Electron Microsc. 1978/I, 109–132.
ECHLIN, P., and MORETON, R. 1976. Low temperature techniques for scanning electron microscopy. Scanning Electron Microsc. 1976 *1*, 753–762.
EINO, M.F., BIGGS, D.A., IRVINE, D.M., and STANLEY, D.W. 1976. Microstructure of Cheddar cheese: sample preparation and scanning electron microscopy. J. Dairy Res. *43*, 109–111.
EVANS, L.G., PEARSON, A.M., and HOOPER, G.R. 1981. Scanning electron microscopy of flour-water doughs treated with oxidizing and reducing agents. Scanning Electron Microsc. 1981/III, 583–592.
EVERS, A.D. 1971. Scanning electron microscopy of wheat starch. III. Granule development in the endosperm. Stärke *23*, 157–162.
EVERS, A.D., and JULIANO, B.O. 1976. Varietal differences in surface ultrastructure of endosperm cells and starch granules of rice. Stärke *28*, 160–166.
FEDEC, P., OORAIKUL, B., and HADZIYEV, D. 1977. Microstructure of raw and granulated potatoes. Can. Inst. Food Sci. Technol. J. *10* (4), 295–306.
FLOOD, P.R. 1975. Dry-fracturing techniques for the study of soft internal biological tissues in the scanning electron microscope. Scanning Electron Microsc. 1975, 287–294.
GALLANT, D., and GUILBOT, A. 1971. Artefacts during the preparation of sections of starch granules. Studies under light and electron microscope. Stärke *23* (7), 244–250 (French).
GEISSINGER, H.D., and STANLEY, D.W. 1981. Preparation of muscle samples for electron microscopy. Scanning Electron Microsc. 1981/III, 415–426, 414.
GEYER, G. 1977. Lipid fixation. Acta Histochem., Suppl. *XIX*, 209–222.
GLASER, J., CARROAD, P.A., and DUNKLEY, W.L. 1979. Surface structure of cottage cheese curd by electron microscopy. J. Dairy Sci. *62*, 1058–1068.
GLAUERT, A.M. 1974. Fixation, dehydration and embedding of biological specimens. *In* Practical Methods in Electron Microscopy. A.M. Glauert (Editor). North-Holland Publ. Co., Amsterdam, The Netherlands.
GOLDSMITH, P.L. 1967. The calculation of true particle size distributions from the sizes observed in a thin slice. Br. J. Appl. Phys. *18*, 813–830.
GROVES, M.J., and FRESHWATER, D.C. 1968. Particle-size analysis of emulsion systems. J. Pharm. Sci. *57*, 1273–1291.
HAGGIS, G.H. 1974. Deposition of platinum in freeze-etch work. Microsc. Soc. Can. Bull. *2* (1), 8.

HAGGIS, G.H., and BOND, E.F. 1979. Three-dimensional view of the chromatin in freeze-fractured chicken erythrocyte nuclei, J. Microsc. *115*(3), 225–234.

HAGGIS, G.H., BOND, E.F., and PHIPPS, B. 1976. Visualization of mitochondrial cristae and nuclear chromatin by SEM. Scanning Electron Microsc. 1976/I, 281–286.

HALL, D.M., and CREAMER, L.K. 1972. A study of the submicroscopic structure in Cheddar, Cheshire and Gouda cheese by electron microscopy. N. Z. J. Dairy Sci. Technol. *7*, 95–102.

HALL, D.M., and SAYRE, J.G. 1969. A scanning electron microscope study of starches. I. Root and tuber starches. Textile Res. J. *39*, 1044–1052.

HALL, D.M., and SAYRE, J.G. 1970. A scanning electron microscope study of starches. II. Cereal starches. Textile Res. J. *40*, 257–266.

HARWALKAR, V.R., and KALAB, M. 1981. Effect of acidulants and temperature on microstructure, firmness, and susceptibility to syneresis of skim milk gels. Scanning Electron Microsc. 1981 *3*, 503–513.

HARWALKAR, V.R., and VREEMAN, H.J. 1978. Effect of added phosphates and storage on changes in ultra-high temperature short-time sterilized concentrated skim-milk. 2. Micelle structure. Neth. Milk Dairy J. *32*, 204–216.

HASHEMEYER, R.H., and MEYERS, R.J. 1972. Negative staining. *In* Principles and Techniques of Electron Microscopy, Biological Applications. Vol. 2, pp. 101–147. M.A. Hayat (Editor). Van Nostrand Reinhold Co., New York.

HAYAT, M.A. 1970A. Fixation. *In* Principles and Techniques of Electron Microscopy. Biological Applications. Vol. 1, pp. 5–107. M.A. Hayat (Editor). Van Nostrand Reinhold Co., New York.

HAYAT, M.A. 1970B. Embedding. *In* Principles and Techniques of Electron Microscopy. Biological Applications, Vol. 1, pp. 111–179. M.A. Hayat (Editor). Van Nostrand Reinhold Co., New York.

HAYAT, M.A. 1970C. Sectioning. *In* Principles and Techniques of Electron Microscopy. Biological Applications, Vol. 1, pp. 183–237. M.A. Hayat (Editor). Van Nostrand Reinhold Co., New York.

HAYAT, M.A. 1970D. Staining. *In* Principles and Techniques of Electron Microscopy. Biological Applications, Vol. 1, pp. 241-319. M.A. Hayat (Editor). Van Nostrand Reinhold Co., New York.

HAYAT, M.A. 1970E. Support films. *In* Principles and Techniques of Electron Microscopy. Biological Applications, Vol. 1, pp. 323–332. M.A. Hayat (Editor). Van Nostrand Reinhold Co., New York.

HENDERSON, W.J., and GRIFFITH, K. 1972. Shadow casing and replication. *In* Principles and Techniques of Electron Microscopy. Biological Applications, Vol. 2, pp. 151–193. M.A. Hayat (Editor), Van Nostrand Reinhold Co., New York.

HENSTRA, S., and SCHMIDT, D.G. 1970A. Ultrathin sections of milk by means of the microcapsulation method. Naturwissenschaften *57*(5), 247 (German).

HENSTRA, S., and SCHMIDT, D.G. 1970B. On the structure of the fat–protein complex in homogenized cow's milk. Neth. Milk Dairy J. *24*, 45–51.

HERMANSSON, A.M., and BUCHHEIM, W. 1981. Methods for characterization of protein gel structures by scanning and transmission electron microscopy. J. Colloid Interface Sci. *81*, 519–529.

HOBBS, D.G. 1979. An improved method for preparing bovine milk fat globules for electron microscopy. Milchwissenschaft *34*, 201–202.

HORISBERGER, M. 1981. Colloidal gold: A cytochemical marker for light and fluorescent microscopy and for transmission and scanning electron microscopy. Scanning Electron Microsc. 1981/II, 9–31.

HORISBERGER, M., and VONLANTHEN, M. 1980. Localization of glycosylated κ-casein micelles by lectin-labelled gold granules. J. Dairy Res. *47*, 185–191.

HORNE, R.W. 1965. Negative staining methods. *In* Techniques for Electron Microscopy, pp. 328–355. D. Kay (Editor), Blackwell Scientific Publications, Oxford, England.

HOWGATE, P. 1979. Fish. *In* Food Microscopy, pp. 343–392. J.G. Vaughan (Editor). Academic Press, New York.

HUMPHREYS, W.J., SPURLOCK, B.O., and JOHNSON, J.S. 1974. Critical point drying of ethanol infiltrated cryofractured biological specimens for scanning electron microscopy. Scanning Electron Microsc. 1974, 275–282.

JEWELL, G.G. 1981. The microstructure of orange juice. Scanning Electron Microsc. 1981/III, 593–598.

JOHARI, O., and DENEE, P.B. 1972. Handling, mounting and examination of particles for scanning electron microscopy. Scanning Electron Microsc. 1972, 249–256.

JONES, S.B., CARROLL, R.J., and CAVANAUGH, J.R. 1976. Muscle samples for scanning electron microscopy: preparative techniques and general morphology. J. Food Sci. *41*, 867–873.

JÓZSA, L., JÄRVINEN, M., and REFFY, A. 1980. A new preparation technique for scanning electron microscopy of skeletal muscle. Microsc. Acta *83*, 45–47.

KALAB, M. 1977A. Milk gel structure. VI. Cheese texture and microstructure. Milchwissenschaft *32*, 449–458.

KALAB, M. 1977B. Milk gel structure. VII. Fixation of gels composed of low-methoxyl pectin and milk. Milchwissenschaft *32*, 719–723.

KALAB, M. 1978. Milk gel structure. VIII. Effect of drying on the scanning electron microscopy of some dairy products. Milchwissenschaft *33*, 353–358.

KALAB, M. 1979A. Scanning electron microscopy of dairy products: An overview. Scanning Electron Microsc. 1979/III, 261–272.

KALAB, M. 1979B. Microstructure of dairy foods. 1. Milk products based on protein. J. Dairy Sci. *62*, 1352–1364.

KALAB, M. 1980A. Possibilities of an electron microscopical detection of buttermilk made from sweet cream in adulterated skim milk. Scanning Electron Microsc. 1980/III, 645–652.

KALAB, M. 1980B. Milk gel structure. XII. Replication of freeze-fractured and dried specimens for electron microscopy. Milchwissenschaft *35*, 657–662.

KALAB, M. 1981. Electron microscopy of milk products: A review of techniques. Scanning Electron Microsc. 1981/III, 453–472.

KALAB, M., and EMMONS, D.B. 1974. Milk gel structure. III. Microstructure of skim milk powder and gels as related to the drying procedure. Milchwissenschaft *29*, 585–589.

KALAB, M., and HARWALKAR, V.R. 1974. Milk gel structure. II. Relation between firmness and ultrastructure of heat-induced skim milk gels containing 40-60% total solids. J. Dairy Res. *41*, 131–135.

KALAB, M., EMMONS, D.B., and SARGANT, A.G. 1975. Milk gel structure. IV. Microstructure of yoghurt in relation to the presence of thickening agents. J. Dairy Res. *42*, 453–458.

KALAB, M., SARGANT, A.G., and FROEHLICH, D.A. 1981. Electron microscopy and sensory evaluation of commercial Cream cheese. Scanning Electron Microsc. 1981/III, 473–482, 514.

KALAB, M., PHIPPS-TODD, B.E., and ALLAN-WOJTAS, P. 1982. Milk gel structure. XIII. Rotary coating of casein micelles for electron microscopy. Milchwissenschaft *37*(9), 513–518.

KARNOVSKY, M.J. 1965. A formaldehyde-glutaraldehyde fixative of high osmolality for use in electron microscopy. J. Cell Biol. *27*, 137A–138A.

KATOH, M. 1979. SEM replica technique for butter and cheese. J. Electron Microsc. *28*, 199–200.

KNIGHT, D.P. 1977. Cytochemical staining methods in electron microscopy. *In* Practical Methods in Electron Microscopy, Vol. *5*, pp. 25–76. A.M. Glauert (Editor). North-Holland Publishing Co., Amsterdam.

KNOOP, A.-M., KNOOP, E., and WIECHEN, A. 1973. Electron microscopical investigations on the structure of the casein micelles. Neth. Milk Dairy J. *27*, 121–127.

KOEHLER, J.K. 1972. The freeze-etching technique. *In* Principles and Techniques of Electron Microscopy. Biological Applications, Vol. 2, pp. 53–98. M.A. Hayat (Editor). Van Nostrand Reinhold Co., New York.

KUDO, S., IWATA, S., and MADA, M. 1979. An electron microscopic study of the location of κ-casein in casein micelles by periodic acid-silver methenamine staining. J. Dairy Sci. *62*, 916–920.

LAWRENCE, R.A., and JELEN, P. 1982. Freeze-induced fibre formation in protein extracts from residues of mechanically separated poultry. Food Microstruct. *1*(1), 91–97.

LEEPER, G.F. 1973. Pectin ultrastructure. I. Pectate elementary fibrils. J. Texture Stud. *4*, 248–253.

LEWIS, D.F. 1975. Brit. Fd. Mfg. Ind. Res. Assoc., Res. Rep. 238. Cited by Berger *et al.* (1979).

LEWIS, D.F. 1979. Meat products. *In* Food Microscopy, pp. 231–272. J.G. Vaughan (Editor). Academic Press, New York.

LEWIS, D.F. 1981. The use of microscopy to explain the behaviour of foodstuffs—A review of work carried out at the Leatherhead Food Research Association. Scanning Electron Microsc. 1981/III, 391–404.

LEWIS, E.R., JACKSON, L., AND SCOTT, T. 1975. Comparison of miscibilities and critical-point drying properties of various intermediate and transitional fluids. Scanning Electron Microsc. 1975/I, 317–324.

LUFT, J.H. 1973. Embedding media—old and new. *In* Advanced Techniques in Biological Electron Microscopy, pp. 1–34. J.K. Koehler (Editor). Springer-Verlag, New York.

MANGINO, M.E., and FREEMAN, N.W. 1981. Statistically reproducible evaluation of size of casein micelles in raw and processed milks. J. Dairy Sci. *64*, 2025–2030.

MARSHALL, A.T. 1975. Electron probe x-ray microanalysis. *In* Principles and Techniques of Scanning Electron Microscopy. Biological Applications, Vol. 4, 103–173. M.A. Hayat (Editor). Van Nostrand Reinhold Co., New York.

MERCER, E.H., and BIRBECK, M.S.C. 1972. Electron Microscopy. A Handbook for Biologists, p. 55. Blackwell Scientific Publications, Oxford, England.

MILNE, R.G., and LUISONI, E. 1977. Rapid immune electron microscopy of virus preparations. *In* Methods in Virology, Vol. VI, pp. 265–281. K. Maramorosch and H. Koprowski (Editors). Academic Press, New York.

MOOR, H. 1973A. Cryotechnology for the structural analysis of biological material. *In* Freeze-Etching Techniques and Applications, pp. 11–20. E.L. Benedetti and P. Favard (Editors). Soc. Franc. Microsc. Electr., Paris, France.

MOOR, H. 1973B. Etching and related problems. *In* Freeze-Etching Techniques and Applications, pp.21–26. E.L. Benedetti and P. Favard (Editors), Soc. Franc. Microsc. Electr., Paris, France.

MOOR, H. 1973C. Evaporation and electron guns. *In* Freeze-Etching Techniques and Applications, pp. 27–30. E.L. Benedetti and P. Favard (Editors), Soc. Franc. Microsc. Electr., Paris, France.

MOSS, R., STENVERT, N.L., KINGSWOOD, K., and POINTING, G. 1980. The relationship between wheat microstructure and flourmilling. Scanning Electron Microsc. 1980/III, 613–620.

MÜLLER, H.R. 1964. Electron microscopic investigations of milk and milk products. 1. Explanation of the structure of milk powders. Milchwissenschaft *19*, 345–356 (German).

MULLER, L.L., and JACKS, T.J. 1975. Rapid chemical dehydration of samples for electron microscopic examination. J. Histochem. Cytochem. *23*, 107–110.

MURPHY, J.A. 1978. Non-coating techniques to render biological specimens conductive. Scanning Electron Microsc. 1978/II, 175–194.

MURPHY, J.A. 1980. Non-coating techniques to render biological specimens conductive. Scanning Electron Microsc. 1980/I, 209–220.

MURPHY, J.A. 1982. Considerations, materials, and procedures for specimen mounting prior to scanning electron microscopic examination. Scanning Electron Microsc. 1982/III, 657–696.

NERMUT, M.V.. 1973. Freeze-drying and freeze-etching of viruses. *In* Freeze-Etching Techniques and Applications, pp. 135–150. E.L. Benedetti and P. Favard (Editors). Soc. Franc. Microsc. Electr., Paris, France.

NICKERSON, A.W., BULLA, L.A., and KURTZMAN, C.P. 1974. Spores. *In* Principles and Techniques of Scanning Electron Microscopy. Biological Applications., Vol. 1, pp. 159–180. M.A. Hayat (Editor). Van Nostrand Reinhold Co., New York.

POMERANZ, Y. 1976. Scanning electron microscopic applications to cereal science and technology. *In* Principles and Techniques of Scanning Electron Microscopy. Biological Applications. Vol. 5, pp. 193–234. M.A. Hayat (Editor). Van Nostrand Reinhold Co., New York.

PRECHT, D., and BUCHHEIM, W. 1979. Electron microscopic studies on the physical structure of spreadable fats. I. Microstructure of fat globules in butter. Milchwissenschaft *34*(12), 745–749 (German).

PUHAN, Z., CARIC, M., and DJORDJEVIC, J. 1976. Electron microscope study of casein in model systems. I. Shape, size and size distribution of casein micelles in the presence of Ca^{2+}, in combination with Na^+, K^+, PO_4^{3-} and citrate. Lebensm. Wiss. Technol. *9*, 374–379.

RAYAN, A.A., KALAB, M., and ERNSTROM, C.A. 1980. Microstructure and rheology of process cheese. Scanning Electron Microsc. 1980/III, 635–643.

REBHUN, L.I. 1972. Freeze-substitution and freeze-drying. *In* Principles and Techniques of Electron Microscopy. Biological Applications, Vol. 2, 3-49. M.A. Hayat (Editor), Van Nostrand Reinhold Co., New York.

REYNOLDS, E.S. 1963. The use of lead citrate at high pH as an electron-opaque stain in electron microscopy. J. Cell Biol. *17*, 208–212.

ROBARDS, A.W., and CROSBY, P. 1979. A comprehensive freezing, fracturing and coating system for low temperature scanning electron microscopy. Scanning Electron Microsc. 1979/III, 325–344, 324.

ROSE, P.E. 1980. Improved tables for the evaluation of sphere size distributions including the effect of section thickness. J. Microsc. *118*(2), 135–141.

SABATINI, D.D., BENSCH, K., and BARNETT, R.J. 1963A. Cytochemistry and electron microscopy. The preservation of cellular structure and enzymatic activity by aldehyde fixation. J. Cell Biol. *17*, 19–58.

SABATINI, D.D., MILLER, F., and BARNETT, R.J. 1963B. Aldehyde fixation for morphological and enzyme histochemical studies with the electron microscope. J. Histochem. Cytochem. *12*, 57–71.

SAIO, K. 1981. Microstructure of traditional Japanese soybean foods. Scanning Electron Microsc. 1981/III, 553–559.

SAITO, Z. 1973. Electron microscopic and compositional studies of casein micelles. Neth. Milk Dairy J. *27*, 143–162.

SALTMARCH, M., and LABUZA, T.P. 1980. SEM investigation of the effect of lactose crystallization on the storage properties of spray dried whey. Scanning Electron Microsc. 1980/III, 659-665.

SALYAEV, R.K. 1968. A method of fixation and embedding of liquid and fragile materials in agar microcapsulae. Proc. Fourth Eur. Regional Conf. Electron Microsc. *II*, 37–38. Rome. Italy.

SCHALLER, D.R., and POWRIE, W.D. 1971. Scanning electron microscopy of skeletal muscle from rainbow trout, turkey and beef. J. Food Sci. *36*, 552–559.

SCHALLER, D.R., and POWRIE, W.D. 1972. SEM of heated beef, chicken and rainbow trout muscles. Can. Inst. Food Sci. Technol. J. *5*, 184–190.

SCHMIDT, D.G., and BUCHHEIM, W. 1976. Particle size distribution in casein solutions. Neth. Milk Dairy J. *30*, 17–28.

SCHMIDT, D.G., and BUCHHEIM, W. 1980. On the size of α-lactalbumin and β-lactoglobulin molecules as determined by electron microscopy using the spray freeze-etching technique. Milchwissenschaft *35*, 209–211.

SCHMIDT, D.G., and VAN HOOYDONK, A.C.M. 1980. A scanning electron microscopical investigation of the whipping of cream. Scanning Electron Microsc. 1980/*III*, 653–658, 644.

SCHMIDT, D.G., WALSTRA, P., and BUCHHEIM, W. 1973. The size distribution of casein micelles in cow's milk. Neth. Milk Dairy J. *27*, 128–142.

SCHMIDT, D.G., HENSTRA, S., and THIEL. F. 1979. A simple low-temperature technique for scanning electron microscopy of cheese. Mikroskopie (Vienna) *35*, 50–55.

SCHULZ, W.W., and REYNOLDS, R.C. 1978. Enhancement of three-dimensional appearance of freeze-fracture images by reversal processing of electron microscopy sheet film. J. Microsc. *112*(2), 249–252.

SMITH, C.G. 1979. Oil seeds. *In* Food Microscopy, pp. 35–74. J.G. Vaughan (Editor). Academic Press, New York.

SMITH, L.B. 1981. An SEM study of the effects of avian digestion on the seed coats of three common angiosperms. Scanning Electron Microsc. 1981/III, 545–552.

SNOEREN, T.H.M., BOTH, P., and SCHMIDT, D.G. 1976. An electron microscopic study of carrageenan and its interactions with κ-casein. Neth. Milk Dairy J. *30*, 132–141.

SPURR, A.R. 1969. A low-viscosity epoxy resin embedding medium for electron microscopy. J. Ultrastruct. Res. *26*, 31–43.

STANLEY, D.W., and GEISSINGER, H.D. 1972. Structure of contracted porcine psoas muscle as related to texture. Can. Inst. Food Sci. Technol. J. *5*, 214–216.

STANLEY, D.W., and SWATLAND, H.J. 1976. The microstructure of muscle tissue—a basis for meat texture measurement. J. Texture Stud. *7*, 65–75.

STANLEY, D.W., and TUNG, M.A. 1976. Microstructure of food and its relation to texture. *In* Rheology and Texture in Food Quality, pp.28–78. J.M. deMan, P.W. Voisey, V. Rasper, and D.W. Stanley (Editors). AVI Publ. Co., Westport, Ct.

STASNY, J.T., ALBRIGHT, F.R., and GRAHAM, R. 1981. Identification of foreign matter in foods. Scanning Electron Microsc. 1981/III, 599–610, 560.

STEERE, R.L. 1973. Preparation of high-resolution freeze-etch, freeze-fracture, frozen-surface, and freeze-dried replicas in a single freeze-etch module, and the use of stereo electron microscopy to obtain maximum information from them. *In* Freeze-Etching Techniques and Applications, pp. 223–255. E.L. Benedetti and P. Favard (Editors). Soc. Franc. Microsc. Electr., Paris, France.

TARANTO, M.V., and TOM YANG, C.S. 1981. Morphological and textural characterization of soybean Mozzarella cheese analogs. Scanning Electron Microsc. 1981/III, 483–492.

TARANTO, M.V., WAN, P.J., CHEN, S.L., and RHEE, K.C. 1979. Morphological, ultrastructural and rheological characterization of cheddar and mozzarella cheese. Scanning Electron Microsc. 1979/III 273–278.

TOWE, K.M. 1979. Electron micrographs of metal shadowed materials: A simple technique for making negative prints. J. Microsc. *116*(2), 281–283.

TUNG, M.A., and JONES, L.J. 1981. Microstructure of mayonnaise and salad dressing. Scanning Electron Microsc. 1981/III, 523–530.

UMRATH, W. 1974. Cooling bath for rapid freezing in electron microscopy. J. Microsc. *101*(1), 103–105.

UUSI-RAUVA, E., RAUTAVAARA, J.-A., and ANTILA, M. 1972. Effects of various temperature treatments on casein micelles. An electron microscopic study using negative staining. Meijeritiet. Aikak. *31*, 15–25 (German).

VARRIANO-MARSTON, E., GORDON, J., HUTCHINSON, T.E., and GORDON, J. 1977. Cryomicrotomy applied to the preparation of frozen hydrated muscle tissue for transmission electron microscopy. J. Microsc. *109*, 193–202.

VARRIANO-MARSTON, E., DAVIES, E.A., HUTCHINSON, T.E., and GORDON, J. 1978. Postmortem aging of bovine muscle: A comparison of two preparation techniques for electron microscopy. J. Food Sci. *43*, 680–683.

VOYLE, C.A. 1981. Scanning electron microscopy in meat science. Scanning Electron Microsc. 1981/III, 405–413.

WALSTRA, P., OORTWIJN, H., and DE GRAAF, J.J. 1969. Studies on milk fat dispersion. I. Methods for determining globule-size distribution. Neth. Milk Dairy J. *23*, 12–36.

WEIBEL, E.R., and BOLENDER, R.P. 1973. Stereological techniques for electron microscopic morphometry. *In* Principles and Techniques of Electron Microscopy. Biological Applications, Vol. 3, pp. 237–296. M.A. Hayat (Editor). Van Nostrand Reinhold Co., New York.

WELLS, O.C. 1974. Scanning Electron Microscopy. p. 333. McGraw-Hill Book Co., New York.

WOLF, W.J., and BAKER, F.L. 1972. Scanning electron microscopy of soybeans. Cereal Sci. Today *17*, 124–126, 128–130, 147.

WOLF, W.J., and BAKER, F.L. 1980. Scanning electron microscopy of soybeans and soybean protein products. Scanning Electron Microsc. 1980/III, 621–634.

WOLF, W.J., BAKER, F.L., and BERNARD, R.L. 1981. Soybean seed-coat structural features: pits, deposits and cracks. Scanning Electron Microsc. 1981/III, 531–544.

WORTMANN, A. 1965. Electron microscopic examination of the structures of butter, rendered butter and mayonnaise. Fette, Seifen, Anstr. Mittel *67*(4), 279–285 (German).

APPENDIX

A List of Selected Monographs on Electron Microscopy

ADVANCED TECHNIQUES IN BIOLOGICAL ELECTRON MICROSCOPY. 1973-1978. J.K. Koehler (Editor). 2 volumes. Springer Verlag, New York.

ARTEFACTS AND SPECIMEN PREPARATION FAULTS IN FREEZE ETCH TECHNOLOGY. 1975. S. Böhler. Balzers Aktiengesellschaft, Balzers, Liechtenstein.

BASIC ELECTRON MICROSCOPY TECHNIQUES. 1972. M.A. Hayat. Van Nostrand Reinhold Co., New York.

BIOLOGICAL TECHNIQUES IN ELECTRON MICROSCOPY. 1971 C.J. Dawes. Barnes and Noble Inc., New York.
ELECTRON MICROSCOPES. 1970. J.A. Swift, Barnes and Noble, New York.
ELECTRON MICROSCOPY. A HANDBOOK FOR BIOLOGISTS. 1972. E.H. Mercer and M.S.C. Birbeck. Blackwell Scientific Publications, Oxford, England.
FOOD MICROSCOPY. 1979. J.G. Vaughan. (Editor). Academic Press, New York.
FREEZE-ETCHING TECHNIQUES AND APPLICATIONS. 1973. E.L. Benedetti and P. Favard (Editors). Soc. Franc. Microsc. Electr., Paris, France.
FREEZE-FRACTURE: METHODS, ARTIFACTS AND INTERPRETATIONS. 1979. J.E. Rash and C.S. Hudson (Editors). Raven Press, New York.
FREEZE-FRACTURE REPLICATIONS OF BIOLOGICAL TISSUES. TECHNIQUES, INTERPRETATION AND APPLICATIONS. 1975. C. Stolinski and A.S.Breathnach. Academic Press, New York.
LOW TEMPERATURE BIOLOGICAL MICROSCOPY AND MICROANALYSIS. 1978. The Royal Microsc. Soc., Oxford, England.
POSITIVE STAINING FOR ELECTRON MICROSCOPY. 1975. M.A. Hayat. Van Nostrand Reinhold Co., New York.
PRACTICAL ELECTRON MICROSCOPY FOR BIOLOGISTS. 1970. G.A. Meek. John Wiley and Sons, New York.
PRACTICAL METHODS IN ELECTRON MICROSCOPY. 1972-1977. A.M. Glauert (Editor). 5 volumes. American Elsevier Publishing Co., Inc., New York.
PRINCIPLES AND TECHNIQUES OF ELECTRON MICROSCOPY. 1970-1977. BIOLOGICAL APPLICATIONS. M.A. Hayat (Editor). 9 volumes. Van Nostrand Reinhold Co., New York.
PRINCIPLES AND TECHNIQUES OF SCANNING ELECTRON MICROSCOPY. BIOLOGICAL APPLICATIONS. 1974-1976. M.A. Hayat (Editor). 6 volumes. Van Nostrand Reinhold Co., New York.
SCANNING ELECTRON MICROSCOPY. 1974. O.C. Wells. McGraw-Hill, Inc., New York.
SCANNING ELECTRON MICROSCOPY. AN INTRODUCTION FOR PHYSICIANS AND BIOLOGISTS. 1973. J. Ohnesorge and R. Holm. Georg Thieme Publishers, Stuttgart, Germany.
SCANNING ELECTRON MICROSCOPY. APPLICATIONS TO MATERIALS AND DEVICE SCIENCE. 1968. P.R. Thornton. Chapman & Hall Ltd., London, England.
SCANNING ELECTRON MICROSCOPY. SYSTEMATIC AND EVOLUTIONARY APPLICATIONS. 1971. V.H. Heywood (Editor). Academic Press, New York.
SOME BIOLOGICAL TECHNIQUES IN ELECTRON MICROSCOPY. 1970. D.F. Parsons (Editor). Academic Press, New York.
STUDIES OF FOOD MICROSTRUCTURE. 1981. D.N. Holcomb and M. Kalab (Editors). SEM Inc., Chicago.
TECHNIQUES FOR ELECTRON MICROSCOPY. 1965. D. Kay (Editor). Blackwell Scientific Publications, Oxford, England.

3

Colorimetry of Foods

F.J. Francis[1]

INTRODUCTION

The color of food is not a physical characteristic in the same sense as melting point, particle size, specific gravity, etc. Rather it is one portion of the input signals to the human brain that ultimately results in the human perception of appearance. Appearance may be influenced by a number of physical attributes of a food as well as a series of psychological perceptions. It is possible to define color in a purely physical sense in terms of the physical attributes of the food, but this approach has serious limitations when we try to use color measurement as a quality control tool for food processing and merchandising. A more satisfactory approach is to define color in a physical sense as objectively as possible and interpret the output in terms of how the human eye sees color.

PHYSICAL PROPERTIES OF COLOR

We can define color in a physical sense as the energy distribution of the light reflected by, or transmitted through, a particular food. Energy may be described as a continuous electromagnetic spectrum ranging from gamma rays with a wavelength of 10^{-5} nm to wavelengths of 10^{17}nm for power transmission. The portion of the electromagnetic spectrum to which the eye is sensitive (380–770 nm) is an exceedingly small portion of the total. From a chemical point of view, we are accustomed to thinking of energy absorption in the ultraviolet (100–380 nm)

[1] Department of Food Science and Nutrition, University of Massachusetts, Amherst, MA 01003.

and the infrared (770 – 1,000,000 nm) as well as the visible. Obviously it is only energy absorption in the visible range that contributes to the perception of color.

When we shine a beam of white light through a sample of red wine the anthocyanin pigments in the wine absorb energy in the visible region. The energy that is not absorbed passes through and reaches the eye. In this case, the perception is obviously red, and the energy distribution in the transmitted beam contains wavelengths of 700 – 770 nm. If the beam had wavelengths of 380 – 400 nm, it would appear violet. Wavelengths between 400 and 475 are called blue. Those from 500 to 570 are green and the span from 570 to 590 is yellow. This is the conventional concept of colorimetry from a chemical approach and has its historical roots in the development of chemical colorimetry. This use of the word colorimetry is obsolete and has been replaced by the term "absorptimetry." The term "colorimetry" should be reserved for its physical meaning, namely, the science associated with the measurement of color.

The color of an object will obviously be influenced by the absorption of light by particles in the object. The classic demonstration of this is the appearance of large pieces of glass from a broken green beer bottle. Obviously they appear to be green. However, when the pieces of glass are very small, the glass appears white, rather than green, due to the multiple reflections between the particles (Hunter 1975). In this case, the scattering effect has greatly overshadowed the absorption effect. The color of most foods is a combination of both the absorption and the scattering effect. This makes the measurement of color in food empirical, but fortunately it is reproducible and interpretable.

THE PHYSIOLOGICAL BASIS OF COLOR

The physical stimuli received by the human eye, which eventually are interpreted as color, can be estimated quite rigorously, but unfortunately this is not true for the physiological reactions (Dember and Warm 1979). The initial stimuli by which we perceive color have been described, and the responses of the eye have been standardized (Wright 1969). Briefly, the human eye has two types of sensitive cells in the retina, the rods and the cones (Boynton 1979). The rods are sensitive to lightness and darkness, and the cones to color. There are three sets of cones within the retina, one sensitive to red light, another to green, and the third to blue. The cones send a signal to the brain that sets up a response in terms of opposing pairs. One is red-green and the other is yellow-blue. This is why we have individuals who are red-green or blue-yellow colorblind. There are no individuals who are red-yellow or blue-green colorblind.

The interpretation of the signals from the retina by the human brain is a very complex phenomena and is influenced by a variety of psychological aspects. One is color constancy since a sheet of white paper looks white in bright sunlight and also when under the green leaves of a tree. The physical stimuli is obviously different, but the brain knows that the paper should be white. The color as seen by the eye is influenced by the surrounding color. A gray tape on a yellow background looks darker than the same tape on a blue background. Similarly, a large expanse of color appears brighter than a small area. One needs only to paint a whole wall of a room a particular color and see how different it appears from a small color chip in the paint store. There are many examples of this type of interpretation of color by the human brain. The old adage "I believe what I see" is interesting but unfortunately not always true. It is a relatively simple matter to fool the human eye. A classic demonstration shows a triangle with three right angles. This is obviously impossible. It is only when we see a view from another angle that we realize that the sides of the triangle do not meet in space. In this situation, the human brain was not given sufficient information to make a correct judgment. But the brain will make a judgment based on available information which may, or may not, be correct.

The psychological interpretations of color as described previously are well known to those who do visual judgments of color. Fortunately, when the conditions of visual judgment are controlled, the human eye is a very good instrument. The optical engineers have even complained that it is unfair competition. The eye can determine approximately 10 million colors, which is far more than any instrument can do.

A great deal of research has been done on visual systems of color measurement and a number of excellent three-dimensional color solids for visual measurement have been developed. Perhaps the best known in America is the Munsell system (Anon. 1963), which contains 1225 color chips arranged for convenient visual comparison. Each chip has a numerical designation. When the color of a sample is described by its Munsell designation, the color is unambiguous.

The designation of colors of foods by visual comparison to chips from a visual color solid is very appealing since it is simple, convenient, and easy to understand. Many specialized color standards of paint, plastic, or glass are available for use as food standards, and a number of companies have adopted the approach for food quality purposes. For example, the official USDA grading system for tomato juice employs spinning disks of a specific Munsell designation to describe the color grades (Francis and Clydesdale 1975). Glass color standards are available for sugar products. Plastic color standards are available for a large number of commodities such as peas, lima beans, apple butter, peanut butter, orange juice, canned mushrooms, peaches, sauerkraut, salmon, etc.

The glass and plastic color standards have been very successful, but obviously are available in a limited number of colors. Painted paper chips, as in the Munsell system, are available in a much wider range of colors, but even these are limited. They are also fragile and may change color with use. The visual standards also have another problem. Repeated visual judgments are tiring and sometimes tedious. Colors that fall between existing standards are often difficult to communicate to another individual. These are the main reasons why instrumental methods of color measurement have been so appealing. However, unfortunately for the newcomer, the choice of instrumental methods is bewildering both in instrumentation and approach. Perhaps a short discussion of the development of the various approaches would make the area more understandable.

SPECTROPHOTOMETRIC MEASUREMENT OF COLOR

The earliest instrumental methods for color measurement are based on transmission or reflection spectrophotometry (Judd and Wyzecki 1963). The physiologists developed the response of the cones in the human eye in terms of the visible spectrum. They were able to do this in a manner easily reproducible in a laboratory today. Three projectors are required, each with a red, green, or blue filter in front of the lens (Fig. 3.1). Red, green, or blue light beams are focused on a screen such that they overlap over half a circle. The other half is illuminated by another projector or by spectrally pure light from a prism or grating. The observer can see both halves of the circle on the screen simultaneously. Each projector is equipped with a rheostat to vary the amount of light from each of the red, green, and blue sources. By varying the amount of light, the observer can determine the amounts of red, green, and blue required to match almost any spectral color. Therefore we can define spectral color in terms of the amounts of red, green, and blue (Francis and Clydesdale 1975). We can set up a triangle with the RGB stimuli at each corner (Fig. 3.2). Every point within the triangle represents a color and can be specified mathematically by the

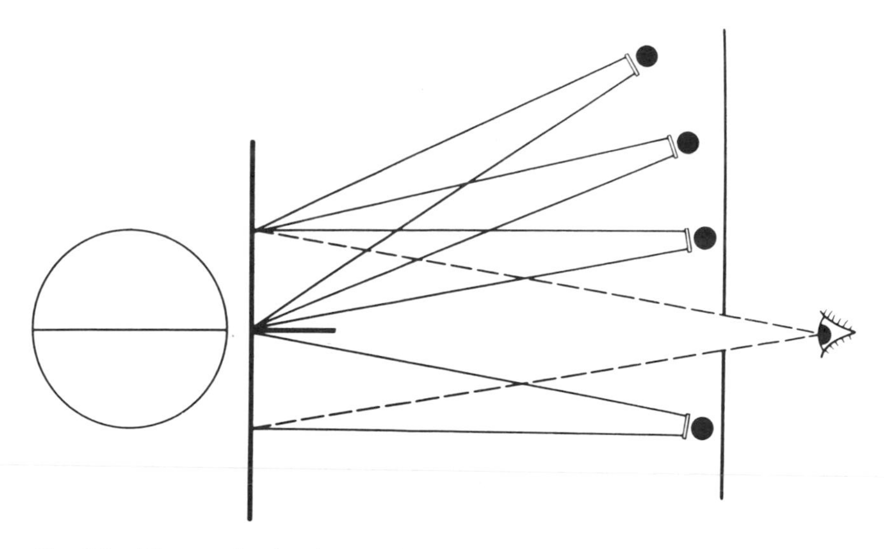

Fig. 3.1. Diagram showing three projectors focused on the upper half of a circle on the screen. The color to be measured is projected on the lower half and the eye can see both halves simultaneously.

amount of red, green, and blue. Unfortunately, red, green, and blue are not particularly good stimuli to use since not all colors can be matched with them. The early researchers chose to use X, Y, and Z as reference stimuli. They cannot be reproduced in the laboratory since they are only mathematical concepts. However, they were chosen for optimum mathematical convenience in constructing a color solid and specifying color coordination. If one wants a crude visual reference, one can think of X as red, Y as green, and Z as blue. The relative positions in space for red, green, and blue and X, Y, and Z are shown in Fig. 3.3.

If we take the red, green, and blue data for the spectral colors, transform them to X, Y, and Z coordinates, and plot the response of the human cones against wavelength (Fig. 3.4) we have the response of the human eye to color. These curves were standardized in 1932, and were called the CIE $\bar{x}$, $\bar{y}$, $\bar{z}$ standard observer curves. The initials stand for "Commission Internationale d'Eclairage."

With the data in Fig. 3.4, it is mathematically simple to calculate the color from a reflection or a transmission spectrum. This is illustrated in Fig. 3.5. The sample spectrum is multiplied by the spectrum of the light source and area under the resultant curve is integrated in terms of the

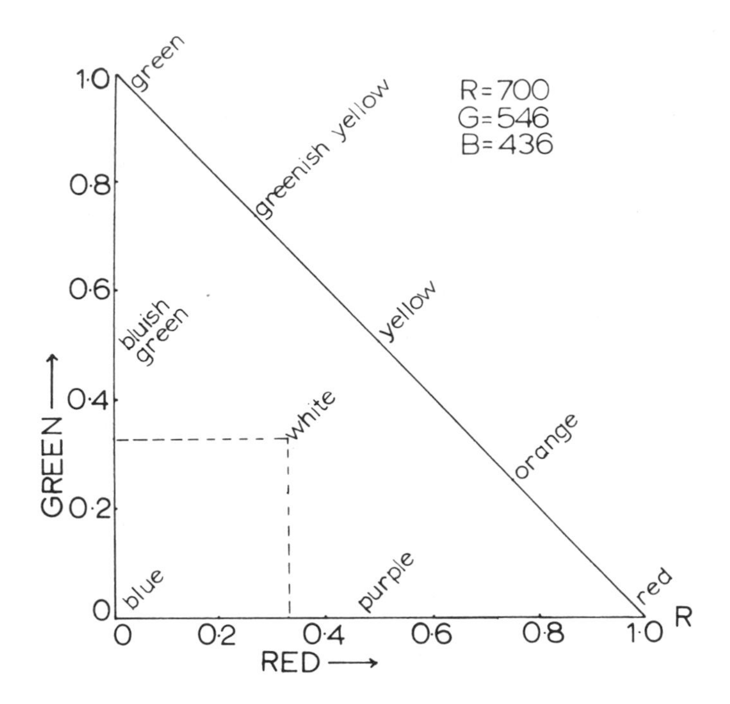

Fig. 3.2. Colors plotted on a red, green, blue color triangle.

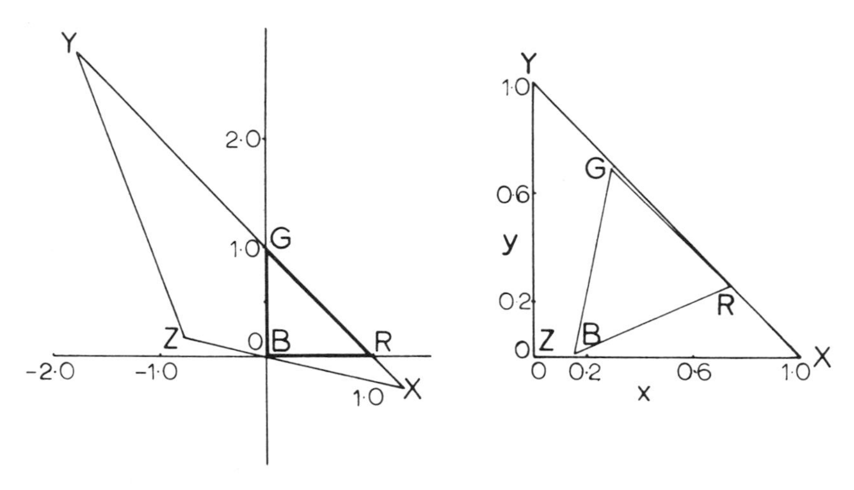

Fig. 3.3. Diagram showing the relative positions in space of the *RGB* and *XYZ* stimuli. (*left*) The *GRB* coordinates for a right-angled triangle and (*right*) the *XYZ* coordinates are plotted as a right-angled triangle.

$\bar{x}, \bar{y}, \bar{z}$ curves of Fig. 3.4. The resulting figures for X, Y, and Z specify the color of the sample. The process can be described mathematically by

$$X = {}^{750}_{380} RE\bar{x}\, d\lambda$$

$$Y = {}^{750}_{380} RE\bar{y}\, d\lambda$$

$$Z = {}^{750}_{380} RE\bar{z}\, d\lambda$$

where R = the sample spectrum; E = the source spectrum; and $\bar{x}$, $\bar{y}$, $\bar{z}$ = standard observer curves.

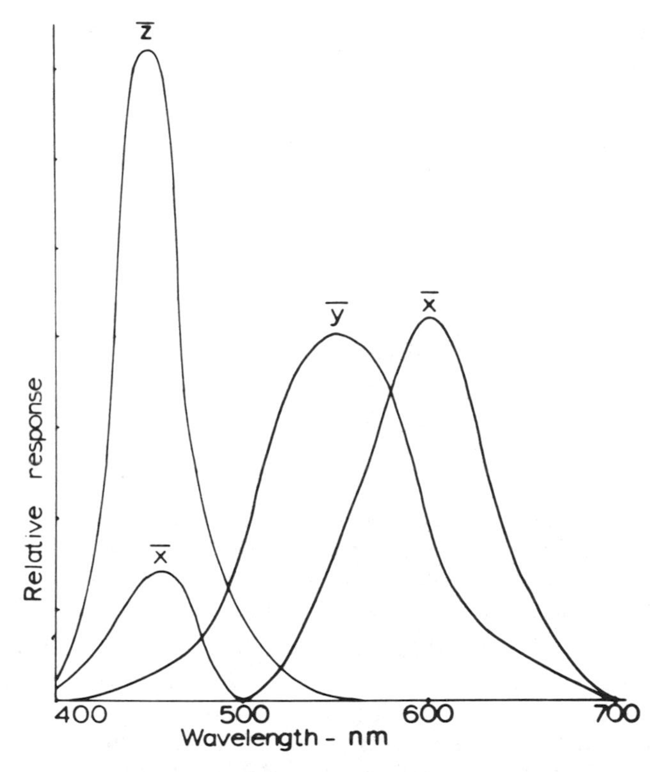

Fig. 3.4. The standard observer curves relationship between the response of the human eye, defined as the standard observer curves ($\bar{x}$, $\bar{y}$, $\bar{z}$), and the visible spectrum.

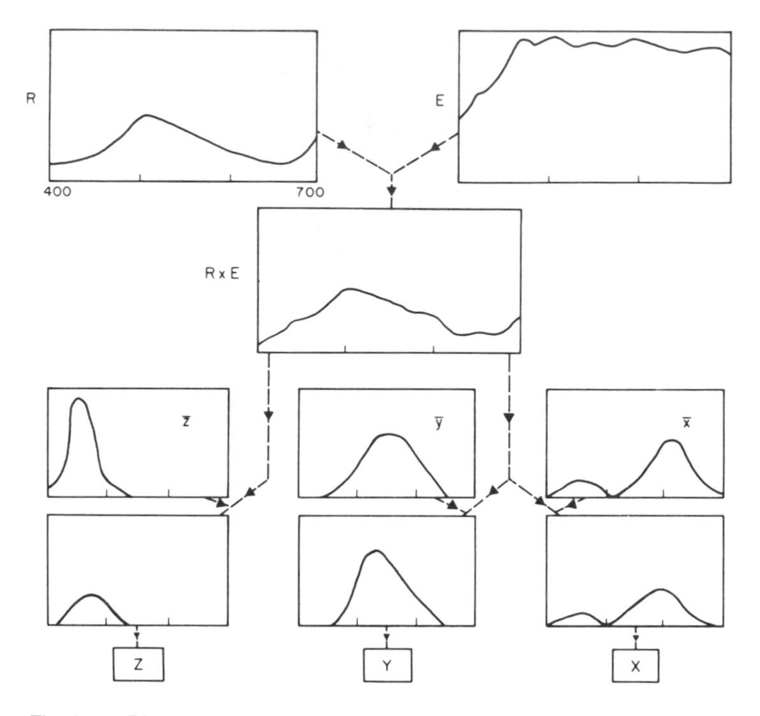

Fig. 3.5. Diagram showing the derivation of *XYZ* data from a spectrophotometric reflection or transmission curve.

The *X, Y, Z* data are usually plotted as *x, y, z* coordinates, where

$$x = \frac{X}{X + Y + Z} \qquad y = \frac{Y}{X + Y + Z} \qquad z = \frac{Z}{X + Y + Z}$$

The spectrum, plotted on an *x, y* diagram is shown in Fig. 3.6. The color solid is actually a solid not a plane with the lightness function perpendicular to the plane of the paper. Figure 3.6 also illustrates another popular way of presenting color data. The point of intersection of a line from the coordinates of white light ($x = 0.333$, $y = 0.333$) through the point to the edge of the solid is the *dominant wavelength* of the point. The relative distance from white light is the *purity* of the color.

The early spectrophotometers provided a reflection or transmission spectrum and the *XYZ* data had to be calculated by hand. This was very tedious, so mechanical integrators were developed that were later replaced by electronic integrators. However, these instruments were complicated and usually expensive. This facilitated the development of tristimulus colorimetry.

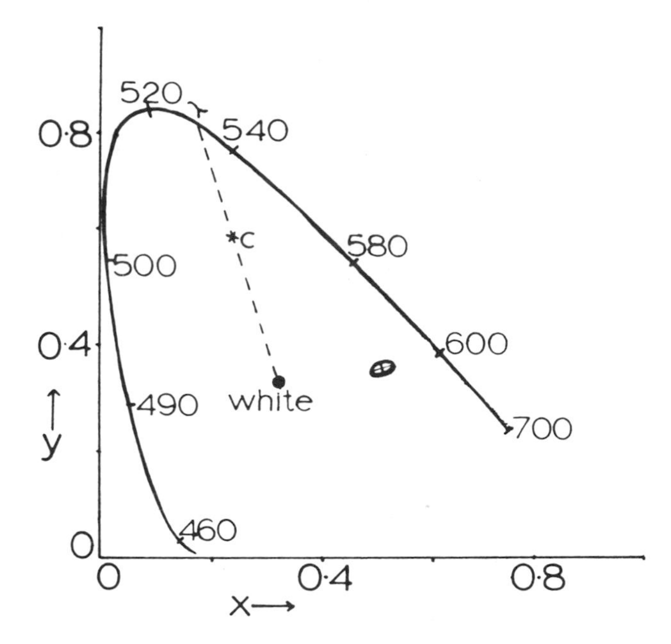

Fig. 3.6. The spectrum colors plotted on x, y coordinates.

TRISTIMULUS COLORIMETRY

The definition of the standard observer curves led to the development of colorimeters designed to duplicate the response of the human eye. The concept is very simple (Fig. 3.7). One needs a light source, three glass filters, with transmittance spectra that duplicate the X, Y, and Z curves, and a photocell. With this arrangement, one can get a XYZ reading that represents the color of the sample. All tristimulus colorimeters available today depend on this principle with individual refinements in photocell response, sensitivity, stability, and reproducibility. However, in spite of the similarity in approach, they do not all use the same units. It is possible to use a number of different filter–photocell combinations as well as different axes in space. In fact, a number of color solids with different axes have been suggested and incorporated into instruments. However, some standardization is developing since two systems seem to be gaining priority. One is the CIE–XYZ system, and the other is the Judd–Hunter $L\ a\ b$ solid. The latter represents a color solid in which L is lightness or darkness, $+a$ is redness, $-a$ is greenness, $+b$ is yellowness and $-b$ is blueness (Fig. 3.8).

Regardless of the particular instrument, or the mathematics of the color solid involved, the limiting factor is the ingenuity of the operator

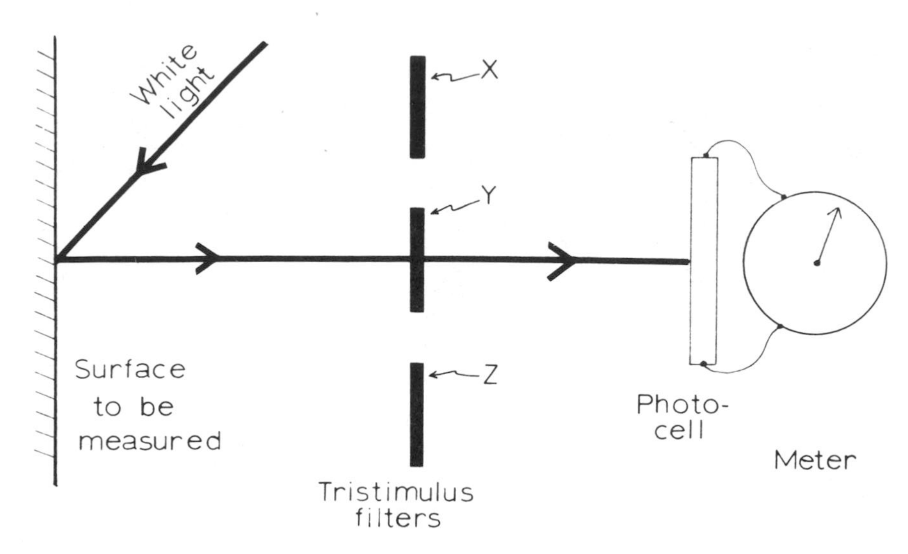

Fig. 3.7. Diagram of a simple tristimulus colorimeter.

in getting a representative signal from the food in question and interpreting it in terms of color. The data are empirical, but nonetheless useful for characterization, quality control, purchase specifications, etc. (Francis and Clydesdale 1975).

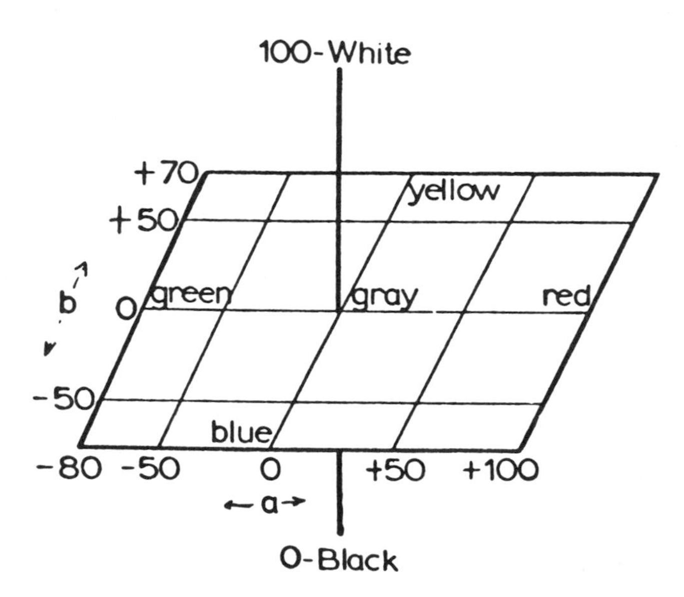

Fig. 3.8. The Judd–Hunter color solid.

SPECIALIZED COLORIMETERS

The tristimulus colorimeters were remarkably successful and led to a great expansion of research in color measurement. Color data were now easily and quickly obtained with relatively inexpensive instrumentation. The time scale coincided with the development of statistical quality control concepts for production control. Unfortunately, SQC charts were two-dimensional and color data came in three dimensions. Requests emerged for color data in one or two dimensions and a series of specialized colorimeters were developed. One of the earliest was the tomato colorimeter (Hunter and Yeatman 1961), designed to measure the color of raw tomato juice. The impetus for the development of this instrument was for incentive payments for growers to deliver more highly colored tomatoes to the processors.

The rationale behind the development of the tomato colorimeter provides an interesting model. Samples of tomatoes representing the range of commercial samples were graded by USDA inspectors into grade A, B, and culls. The juice was then extracted from the tomato samples and measured on a tristimulus colorimeter. A relationship was then established between the graders' decision and the color coordinates. In effect, the equation representing how the graders visualized the color of the tomatoes was established in color space. In this application, tomato color (TC) was represented by

$$\mathrm{TC} = 2000 \cos \theta / L$$

where

$$\cos \theta = \frac{a}{\sqrt{(a^2 + b^2)}}$$

and L, a, and b are Judd–Hunter units. The instrument became known as the USDA Tomato Colorimeter. The Tomato Colorimeter has proved to be a useful instrument for measuring tomato color. It was soon followed (Yeatman 1967) by a modification to read the color of processed tomato juice according to

$$\text{Color Score} = bL/a$$

The same concept was used in developing the citrus colorimeter to measure the color of orange juice (Hunter 1967). This instrument read citrus color (CR) as

$$CR = 200\,(A/Y - 1)$$

where A is the amber filter response representing the long wavelength portion of the CIE X curve. The readout CR actually functions as a new color scale, and three more were added (CY—citrus yellowness, CG—citrus greenness, and CB—citrus blueness) to round out the new color solid. This made the instrument adaptable to a wider range of juices rather than just orange juice.

Specialized instruments were developed for honey (Sechrist 1925), sugar (Bernhardt *et al.* 1962), tea (Hameyer *et al.* 1967), apples (BCRC 1968), cranberries (Francis 1964; Breeze 1972), salmon (Schmidt and Cuthbert 1969), wine (Little and Simms 1971), internal color of pork (MacDougall and Jones 1975), and internal color of beef (Renerre 1981). It may be stretching a point to say that all the instruments above measured color, as such, since they were concerned with the general aspect of quality, but color was a major factor.

The proliferation of specialized instruments led to some dissatisfaction with this approach since, for example, suppliers did not want a roomful of specialized instruments. Fortunately there is another approach. When the original color data from a sample are collected in tristimulus units, it can be read out in any required units by a simple microprocessor in the instrument. For example, the scales for raw and processed tomato juice can be read from the same instrument with an extra circuit. Similarly, the same basic unit could be used for other products by incorporating additional circuits. This trend will discourage the accumulation of color data in other than tristimulus units, but this surely is a progressive step.

The same trend toward simplification of the color data readout was seen with reflectance or transmittance spectrophotometers. Instead of providing a continuous spectrum, a readout at one, two, three, or four discrete wavelengths provides an index of the color of the object. It is possible to design rugged, relatively simple, instruments with these limitations, and they have been very successful in the food industry. However, the "abridged" data obtained from such readouts are not always unambiguous and may require special care in interpretation. Also this type of data cannot be transferred into tristimulus data.

The logic behind the reduction of data from both tristimulus and spectrophotometric instruments led naturally to the investigation of other mathematical transformations, which it is hoped, would either be more accurate or easier to manipulate. One of the more successful approaches, at least for nonfood applications, was the research of Kubelka and Munk.

THE KUBELKA–MUNK CONCEPT

The mathematics behind the calculation of color with a combination of the absorption coefficient (K) and the scattering coefficient (S) are well developed in the paint, plastic, and textile fields and form the basis of computer matching of colors. The concept works well for products that are essentially opaque or transparent. It works less well for the intermediate products, and this is where most foods are found.

The Kubelka–Munk concept (Judd and Wyszecki 1963) has been defined by

$$K/S = kC$$

where K = light absorbed; S = light scattered; C = concentration of colorant; and k = a constant. The value K/S is related to reflectance at a given wavelength by

$$K/S = (1 - R_{\infty})^2 / 2 R_{\infty}$$

where R = the reflectance of an infinitely thick sample. It is possible to refine the relationships between instrument readings as defined by reflectance at a single wavelength, XYZ, or $L\ a\ b$ data and visual responses by transforming the instrument responses into K/S units. This has been done with products such as squash or carrot puree (Huang *et al.* 1970), orange juice (Gullett *et al.* 1972A), Tang (Gullett *et al.* 1972B), applesauce–berry mixtures (Little 1964), powdered milk (MacKinney *et al.* 1966), and meats (Stewart *et al.* 1965; Hunt 1980). The Kubelka–Munk approach is mentioned here because of its success with paints, plastics, and textiles. Variations of the concept have been successful with sugar and tea solutions where their primary consideration was the estimation of both turbidity and color in solutions. The K/S concept has met with limited success in direct food color measurement applications (Francis and Clydesdale 1975).

The Kubelka–Munk concepts, K and S, are useful as definitions of physical constants to describe the optical properties of foods. A knowledge of the K and S values enables one to predict whether the color of a particular sample should be measured by reflectance or transmittance. This approach also shows promise for measuring the internal color and quality characteristics of foods without destroying the sample (Birth 1978, 1979).

HANDLING OF DATA

One of the more difficult concepts to handle in food colorimetry is the handling of data. This is complicated by the diversity of instruments and color solids as well as the variety of approaches to color measurement. Obviously, the handling of data will depend on the reasons for obtaining the data and the uses to which it will be put.

RESEARCH APPROACHES

It may be desirable to follow the changes in color of a given product during storage, processing, maturation, change in physical structure, etc. In such cases it is extremely unlikely that color changes will take place in one or two color axes. It can be taken for granted that the color changes will be three-dimensional. The only way to portray this type of data accurately is to use either a spectrophotometer or a tristimulus colorimeter and record the color change in terms of three coordinates. Again, the *XYZ* or *L a b* systems seem to be gaining over the others for these applications. The data can be tabulated or graphed in three-dimensional models.

Color measurement may be used as an indirect means of analyzing for a colored component of a food since often it is simpler and quicker than chemical analysis. For example, Francis (1962) established the correlation coefficients between total carotenoid content and color for ten varieties of squash. Similar work relating to the color with carotenoid content of sweet potatoes was reported by Ahmed and Scott (1962). If one wanted to use this approach to estimate, for example, β-carotene content as an index of vitamin A activity, one would have to be assured that the predominant carotenoid was, in fact, β-carotene. This approach has some appeal for plant breeding work with orange vegetables since, in addition to being simple and quick, Lauber *et al.* (1967) reported that color measurements of sweet potatoes were more accurate for estimation of carotenoid content than was chemical analyses.

QUALITY CONTROL

Color changes in foods are almost always three-dimensional, but not all three dimensions may be of practical importance. For all quality control situations, it is desirable that the simplest unit of color be used commensurate with the desired accuracy. Statistical quality control charts in three dimensions are difficult to handle, thus the desire to reduce color data from three to two or even one parameter is understandable.

The most accurate way to reduce the number of color parameters is probably via a regression equation. An example of this approach can be taken from the work of Wenzel and Huggert (1969) on color of reconstituted orange juice. The correlation between USDA color score and Hunter color and color-difference readings for R_d alone was −0.815 and for a alone was 0.909. The correlation between the color score and $R_d.a$ was 0.927 and that with $R_d.a.b$ was 0.930. It was obvious that most of the correlation was with the a value and the inclusion of R_d and b in a multiple regression equation would result in some extra accuracy, but the authors concluded that it may not be worth the trouble. Another example was published by Kramer *et al.* (1948). They calculated correlation coefficients between visual scores and Hunter color and color-difference meter readings for canned tomato juice. The correlations between visual score and Rd, a, b, $a.b$, and Rd.$a.b$ were −0.542, +0.635, −0.652, 0.903, and 0.904, respectively. The inclusion of Rd in the multiple regression did not appreciably improve the accuracy of prediction, so obviously the graders were being influenced mainly by the a and b readings. A regression equation from the above data would be color score (PMA grade 1) = $32.6 + 0.553a - 1.478b$.

Kramer (1954) reported examples of regression equations to predict subjective scores from one color attribute, L for lima beans; two attributes, namely a and b for tomato juice; and three attributes, L, a, and b for apple sauce. It is possible to calculate simple and multiple correlation coefficients between visual scores and any number of objective scores, in order to calculate the relative importance of each objective score in the subjective judgment. From these, simple and multiple linear or curvilinear regression equations can be used to construct nomographs for routine calculations (Kramer and Twigg 1973).

There is a tendency, particularly with L a b data, to use the L value (lightness or darkness) as one parameter and reduce the a and b data to some function relating to Munsell hue. The easiest way to do this is to calculate the a/b ratio, and this became well entrenched in the early work with tomatoes. The a/b ratio (Francis 1975; Francis and Clydesdale 1975) is the tangent of the angle that a point in Hunter space makes with the origin. Coincidentally, with the tomato data, the ratios were close to unity, indicating an angle approximating 45°. The a/b ratio has been employed with other products in which the angle is closer to zero or 90°. The tangent function a/b is equal to zero at 0°, unity at 45°, and infinity at 90°. It is difficult to interpret analyses of variance on data that approach infinity, so the use of the a/b function in this instance is absurd. A rule of thumb is that if the a/b ratio is between 0.2 and 2, it may be used as a function of hue. An even better recommendation is to use the angle rather than the tangent of the angle as a function of hue. The angle (θ) is not entirely linear with color change, but it is much better than the tangential function *a*/*b*.

COLOR TOLERANCES

The description of color for purchase specifications of any commodity necessarily involves the concept of color tolerances. The color desired is located in color space and the allowable tolerances are specified in one, two, or three dimensions in color space. Unfortunately, it is not possible to specify a tolerance that is equally acceptable in all portions of the color solid. The reason for this is that the eye is much more sensitive to some colors. This is shown in Fig. 3.9. In the green area, the ellipses are much larger than in the blue or the red areas, indicating a much greater sensitivity. Color tolerances may be plotted in three dimensions as shown in Fig. 3.10. An ellipse is preferable to a rectangle for color tolerances in view of the sensitivity of the eye. In Fig. 3.10, if the point is

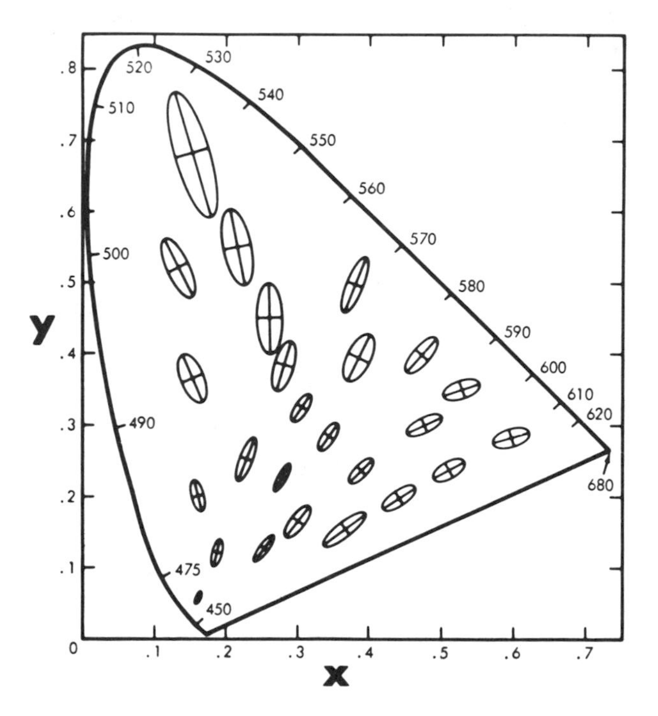

Fig. 3.9. Ellipses in color space representing the sensitivity of the human eye to various colors.

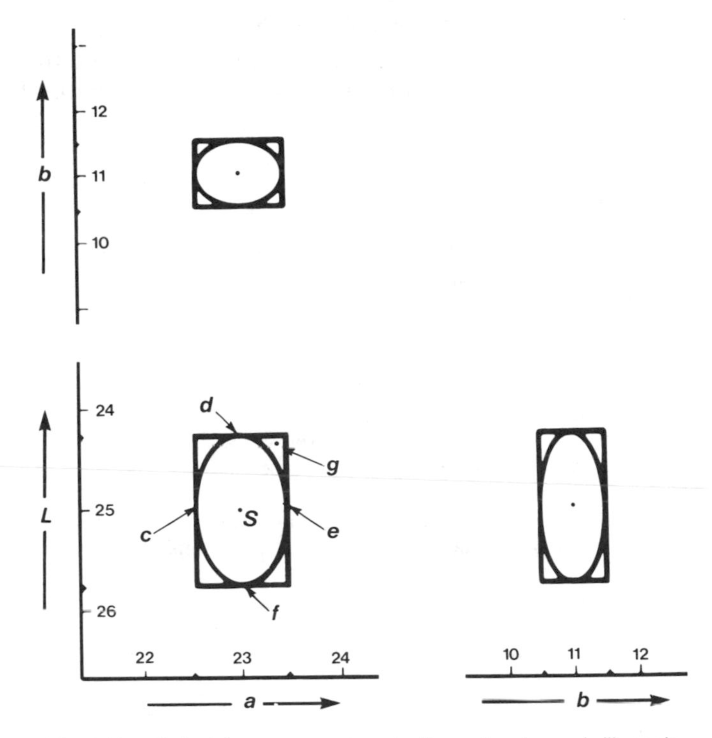

Fig. 3.10. Color tolerances represented by rectangles and ellipses in three color dimensions using *L a b* data.

in the desired color, the points *c, d, e, f* would be within the tolerance when an ellipse is used. The point *g* would be within the tolerance using a rectangle and outside using an ellipse. However, this degree of accuracy may not be necessary. The decision as to whether one, two, or three color parameters are necessary for color specifications is, of course, dependent with use of the product.

CONSIDERATIONS FOR THE FUTURE

The specification of color has come full circle in the development of instrumentation. In the 1920s objective methods of color specification employed spectrophotometers. The labor of computation to get *XYZ* data encouraged the development of mechanical and, later, electronic integrators. However, the expense of the spectrophotometer–integrator combination led to the development, in the 1940s, of the tristimulus colorimeter. The development of microprocessors in the 1970s eliminated the problems of computational labor and spectrophotometers began to come back into favor. The next decades are likely to see spectrophotometers and tristimulus colorimeters equipped with microprocessors to generate and process color data in fundamental units.

BIBLIOGRAPHY

AHMED, E.M., and SCOTT, L.E. 1962. A rapid objective method for estimation of carotenoid content in sweet potato roots. Proc. Am. Soc. Hort. Sci. *80*, 497–506.

ANON. 1963. Munsell Book of Color. 2441 N. Calvert St., Baltimore, MD.

BCRC. 1968. Guidelines to Industrial Progress, No. 7, July. British Columbia Research Council, Vancouver, BC.

BERNHARDT, W.O., EIS, F.G., and MC GINNIS, R.A. 1962. The sphere photometer. J. Am. Soc. Sugar Beet Technol. *12*(2) 106.

BIRTH, G.S. 1978. The light scattering properties of foods. J. Food Sci. *43*, 916–925.

BIRTH, G.S. 1979. Radiometric measurement of food quality: A review. J. Food Sci. *44*, 949–953.

BOYNTON, R.M. 1979. Human Color Vision. Holt, Rinehart and Winston, New York.

BREEZE, J.E. 1972. Personal communication. British Columbia Research Council, Vancouver, BC, Canada.

DEMBER, W.N., and WARM, J.S. 1979. Psychology of Perception. 2nd Ed. Holt, Rinehart and Winston, New York.

FRANCIS, F.J. 1962. Relationship between flesh color and pigment content in squash. Proc. Am. Soc. Hort. Sci. *81*, 408–414.

FRANCIS, F.J. 1964. Cranberry color measurement. Proc. Am. Soc. Hort. Sci. *85*, 312–317.

FRANCIS, F.J. 1975. The origin of $\tan^{-1}a/b$. J. Food Sci. *40*, 412.

FRANCIS, F.J., and CLYDESDALE, F.M. 1975. Food Colorimetry: Theory and Applications. AVI Publishing Co., Westport, CT.

GULLETT, E.M., FRANCIS, F.J., and CLYDESDALE, F.M. 1972A. Colorimetry of foods. 4. Orange juice./J. Food Sci. *37* 389–393.

GULLETT, E.M., FRANCIS, F.J., and CLYDESDALE, F.M. 1972B. Colorimetry of foods. 5. Tang. J. Can. Inst. Food Sci. Technol. *5*, 32–36.

HAMEYER, K., MADLIN, H., and CIACCO, L.L. 1967. A method for measuring turbidity in colored solutions using ratios of scattered to transmitted light. Food Technol. *21*, 199–201.

HUANG, I.L., FRANCIS, F.J., and CLYDESDALE, F.M. 1970. Colorimetry of foods. 3. Pureed carrots. J. Food Sci. *35*, 771–773.

HUNT, M.C. 1980. Meat color measurements. Proc. 33rd Annu. Recip. Meat Conf., pp. 41–46.

HUNTER, R.S. 1967. Development of the citrus colorimeter. Food Technol. *21*, 906–1005.

HUNTER, R.S. 1975. The Measurement of Appearance. John Wiley and Sons, New York.

HUNTER, R.S., and YEATMAN, J.N. 1961. Direct reading tomato colorimeter. J. Opt. Soc. Am. *51*, 552–554.

JUDD, D.B., and WYSZECKI, G. 1963. Color in Business, Science and Industry. John Wiley and Sons, New York.

KRAMER, A. 1954. Color in Foods. Natl. Acad. Sci.–Natl. Res. Council Symp., p. 39–48.

KRAMER, A., and TWIGG, B.A. 1973. Quality Control for the Food Industry. 3rd ed., Vol. 1. AVI Publishing Co., Westport, CT.
KRAMER, A., GUYER, R.B., and SMITH, H.H. 1948. A rapid method of measuring the color of raw and canned tomatoes. Proc. Am. Soc. Hort. Sci. *51*, 381–389.
LAUBER, J.J., TAYLOR, G.A., and DRINKWATER, W.O. 1967. The use of tristimulus colorimetry for the estimation of carotenoid content of raw sweet potato roots. Proc. Am. Soc. Hort. Sci. *80*, 497–506.
LITTLE, A.C. 1964. Color measurement of translucent food samples. J. Food Sci. *29*, 782–789.
LITTLE, A.C., and SIMMS, R.J. 1971. The color of white wine. III. The design, fabrication and testing of a new instrument for evaluating white wine color. Am. J. Enol. Vitic. *22*, 203–209.
MAC DOUGAL, D.B., and JONES, S.J. 1975. The use of a fibre optic probe for the detection of pale pork. Proc. 21st European Meet. Meat Research Workers, Berne, Switzerland, p. 114.
MAC KINNEY, G., LITTLE, A.C., and BRINNER, L. 1966. Visual appearance of foods. Food Technol. *20*, 1300–1308.
RENERRE, M. 1981. Personal communication. Institut National de la Recherche Agronomique, Theix, France.
SCHMIDT, P.J., and CUTHBERT, R.M. 1969. Color sorting of raw salmon. Food Technol. *23*, 32, 98–100.
SECHRIST, E.L. 1925. The Color Grading of Honey. USDA Circular No. 364. U.S. Dept. Agriculture, Washington, DC.
STEWART, M.R., ZIPSER, M.W., and WATTS, B.M. 1965. The use of reflectance spectrophotometry for the assay of raw meat pigments. J. Food Sci. *30*, 464–469.
WENZEL, F.W., and HUGGART, R.L. 1969. Instruments to solve problems with citrus products. Food Technol. *23*, 147–150.
WRIGHT, W.D. 1969. The Measurement of Colour. Van Nostrand-Reinhold Co., New York.
YEATMAN, N.J. 1967. Tomato products: Read tomato red. Food Technol. *21*, 906–1005.

4

Applications of Differential Scanning Calorimetry in Foods

Daryl B. Lund[1]

INTRODUCTION

When a material undergoes a change in physical state such as melting or transition from one crystalline form to another or when a material reacts chemically, heat is either absorbed or liberated. Instrumentation that measures and records the amount of heat involved in the process can provide valuable qualitative and quantitative information. In the food industry, there are many examples in which materials undergo physical changes requiring the addition or removal of heat, including freezing of water, and phase changes of fats and lipids, denaturation of proteins, and gelatinization of starches. The instrumental technique used to study these transition phenomena is referred to as differential scanning calorimetry (DSC). The purpose of this chapter is to review some of the fundamental aspects of differential scanning calorimetry and illustrate its applications in foods.

DEFINITION OF DIFFERENTIAL SCANNING CALORIMETRY

The term "differential scanning calorimetry" was initially a source of some confusion in thermal analysis. The parent technique to DSC was "differential thermal analysis (DTA)" and the indiscriminate use of the terms DSC and DTA resulted in IUPAC proposing definitions of these processes. Several excellent books on thermal analysis that provide detailed information on DTA and DSC include McNaughton and Mortimer (1975), Wendlandt (1974), Pope and Judd (1977), and

[1]University of Wisconsin-Madison, Madison, WI 53706

Mackenzie (1970, 1972). Here it will suffice to point out the differences in instrumentation for DTA and DSC.

The purpose of differential thermal systems is to record the difference between an enthalpy change that occurs in a sample and that in some inert reference material when both are heated. Systems to accomplish this may be classified into three types: (1) classical DTA, (2) "Boersma" DTA, and (3) DSC. The main elements of each type are illustrated in Fig. 4.1.

In the classical and Boersma DTA systems, both sample and reference are heated by a single heat source. Temperatures are measured by

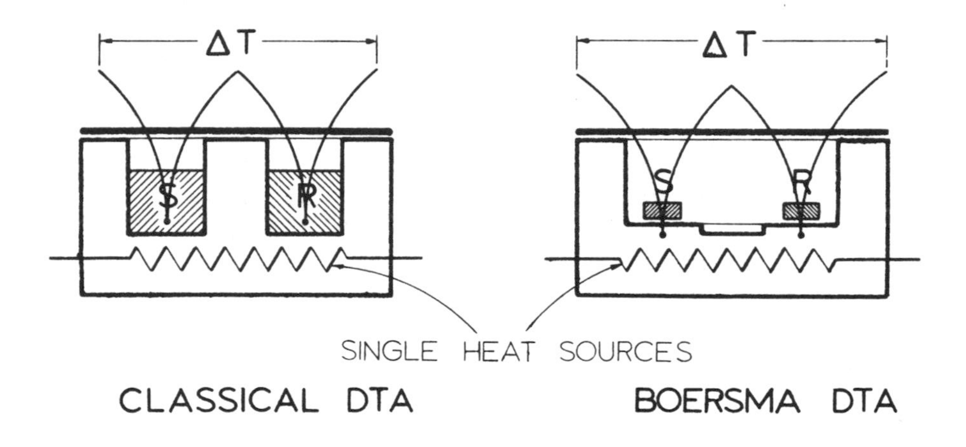

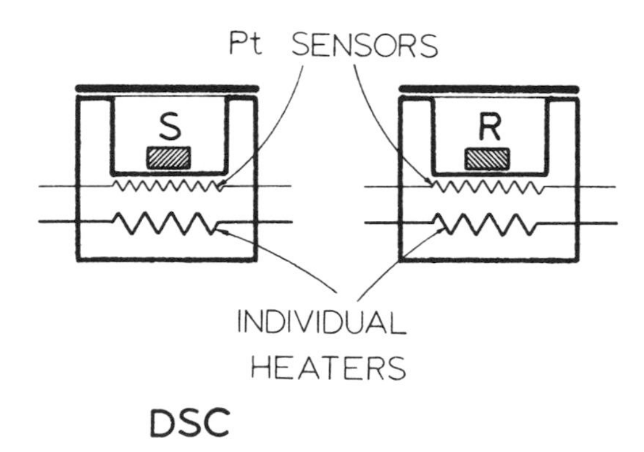

FIG. 4.1 Schematic representation of the three principal thermal analysis systems. *From Perkin-Elmer Corp., Norwalk, CT. Thermal Analysis Newsletter, No. 9, 1970.*

thermocouples embedded in the sample and reference material (classical), or attached to the pans (Boersma). The instrument measures the temperature difference between the sample and reference as a function of temperature and presents the data as a plot of ΔT versus T. The magnitude of ΔT, at any given T (or time since the instrument is programmed to heat at a constant rate of temperature change dT/dt) is a function of (1) the enthalpy change, (2) the heat capacities, and (3) the total thermal resistance to heat flow. The thermal resistance to heat flow is dependent on the nature of the sample, the way it is packed into the sample pan, and the extent of thermal contact between the sample pan and holder. In Boersma DTA the temperature sensors are attached directly to the pans in an attempt to reduce variations in thermal resistance caused by the sample itself. In either classical or Boersma DTA, the thermal resistance, and hence the calibration constant for the instrument, is a function of temperature. Although data from DTA provide quantitative information on temperatures of transition, it is very difficult to transform these data to enthalpy changes. To calculate enthalpy changes from DTA data it is necessary to know heat capacities and the variation of the calibration constant with temperature. Consequently, DTA systems are not very suitable for calorimetric measurements.

The important difference between DTA and DSC is that in DSC the sample and reference are each provided with individual heaters, which allows the determination to be conducted with no temperature difference between sample and reference. Running in this mode allows two important simplifications compared to DTA. First, because the sample and reference pan are maintained at the same temperature, the calibration constant for the instrument is independent of temperature. Obviously this greatly simplifies experimental technique since the calibration constant need only be determined for one standard material. Second, since the sample and reference pan have independent heaters, the difference in heat flow to the sample and reference in order to maintain equal temperature can be measured directly. Thus data are obtained in the form of differential heat input (dH/dt) versus temperature (or time since constant heating rates are used). These data are readily used to obtain temperatures and enthalpy of transitions or reactions.

OPERATION

Like any instrumental technique, there are aspects of the operation that must be carefully attended to in order to obtain data with the desired precision and accuracy. DSC is no exception, especially since a calibration procedure must be used to obtain the calibration constant and fix the temperature scale accurately. Although McNaughton and Mortimer (1975) presented an excellent detailed review of considerations in operation of DSC, a brief summary will be presented here.

Most of the DSC units currently in use have small sample pans which generally hold 10–20 μl of sample. Obviously, for pure materials, obtaining a representative sample of this size does not present any great difficulty. For optimum peak sharpness and resolution, the contact surface between pan and sample should be maximized. This is usually accomplished by having the sample as thin disks, or films, or fine granules. Frequently in applications in foods, the sample will be dispersed or soluble in water, which obviates any problems with contact surface. For heterogeneous materials, sampling can present a problem because of the small quantity required. In these cases, homogenizing the sample may be required with care being exercised not to heat the sample during homogenizing (causing some irreversible transitions which are being looked for in DSC).

Generally the samples are encapsulated in aluminum pans with lids that are crimped into position. For samples containing water, the pan must be hermetically sealed to prevent evaporation of water (with its large enthalpy of vaporization). Generally, the pans and crimping devices currently in use will withstand internal pressures to 2–3 atm. For higher pressures, there are high pressure DSC cells available from instrument manufacturers, and flint glass ampules can be used that withstand internal pressures up to 30 atm.

Calibration of the instrument is generally carried out with a high purity metal with accurately known enthalpy of fusion and melting point. The most commonly used calibrant is indium (ΔH_{fusion} = 6.80 cal/g; m.p. 156.4°C). McNaughton and Mortimer (1975) discussed the procedure that should be used to obtain temperature data from the endo/exotherm to within ± 0.2°C and ± 0.1°C.

Determination of enthalpy for the process under study requires measurement of the area of the endo/exotherm. This can present some difficulty because the baseline for the endo/exotherm may not be horizontal (a function of the match of heat capacities of the sample and reference pans) and the peak is generally not symmetrical. An accepted procedure for determining the baseline for the peak is illustrated in Fig. 4.2. First the baseline on each side of the peak is extrapolated across the peak. The linear portion of each side of the peak is then extrapolated to intercept the extrapolated baseline on its respective side (i.e., right of peak extrapolated to intercept right baseline). The intersections (points

T_0 and T_c on Fig. 4.2) represent the initial and final temperature of the transition. The baseline for determination of the peak area starts at the deviation of the pen from the left-hand baseline and terminates at the return of the pen to the right-hand baseline. The area is indicated by the shaded area in Fig. 4.2.

The three basic approaches to determination of the peak area are "cut and weigh," planimeter, and integrator. Each method has some disadvantages and all require experience by the experimenter. For cut and weigh, it may be necessary to retrace the data onto paper with a uniform density and to allow the paper to reach constant moisture content before weighing. With a planimeter, it may be necessary to enlarge the peak to improve the accuracy of measuring the area.

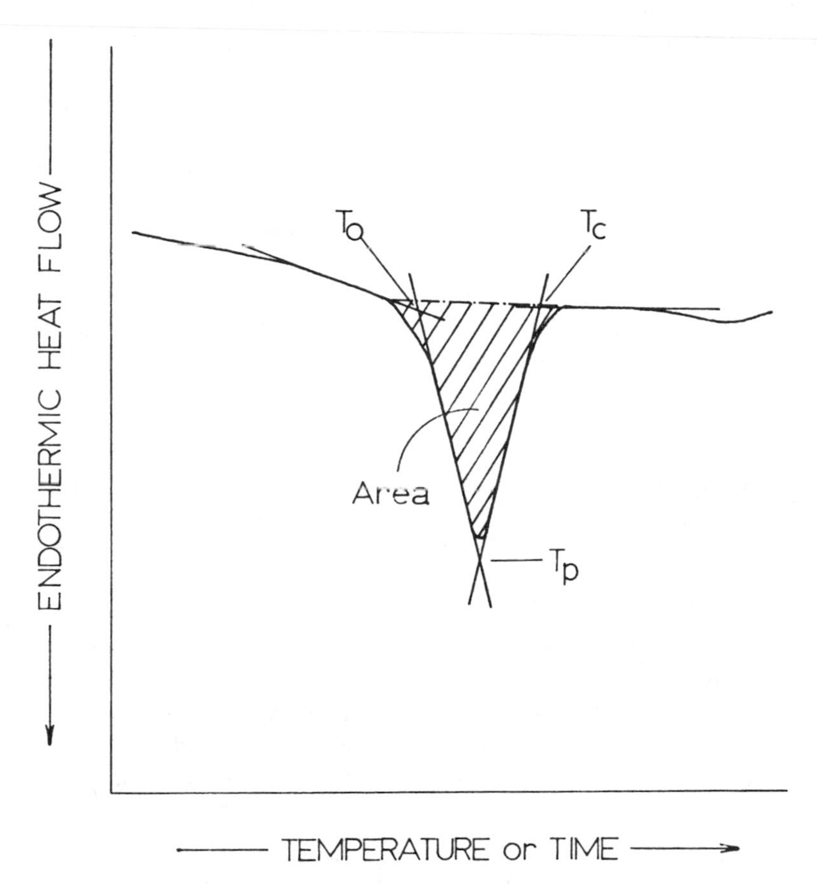

FIG. 4.2 DSC endotherm for measuring area and temperatures.

Enlarging can conveniently be done by photocopying the original tracing onto transparency, showing the transparency on a screen, and tracing the enlarged peak. The magnification factor can be easily determined from the grid on the DSC paper or by marking a known length on the transparency. Electronic integrators would certainly have the required accuracy for use with DSC data; however, it may be necessary to correct the data for a nonhorizontal baseline.

APPLICATIONS OF DSC

In the last decade, DSC has become the most widely used of all the thermal analysis techniques. Qualitative data from DSC include temperatures of glass transitions, crystallization, solid–solid transitions, fusion, dehydration, and denaturation. Quantitative data from DSC include enthalpy of transitions or reactions, heat capacity measurements, purity determinations, degree of crystallinity, and quantitative analyses of multicomponent mixtures.

Lóránt (1972) reviewed the applications of thermal analysis techniques in the food industries, particularly the simultaneous application of thermal gravimetric analyses (TGA) and DTA. TGA measures the change in weight of a sample subjected to a programmed temperature change and has been used to study thermal decomposition of fats and steroids, proteins, carbohydrates, and other minor food constituents. Lóránt also has used TGA to determine moisture and fat content of meats, dairy products, food emulsions, and confections. The method is fast and apparently sufficiently accurate, promoting its use for routine quality control.

Although DSC would appear to have countless applications in foods, its use has been primarily restricted to studies on protein denaturation and starch gelatinization.

Studies with DSC in Foods

Three of the most important phase transitions in foods are (1) water/ice, (2) protein denaturation, and (3) starch gelatinization. DSC has been used extensively to study all three transitions. The purpose here is not to review the current state of the knowledge regarding the physical chemistry of these phenomena but rather to illustrate the use of DSC. To do that, selected studies on protein denaturation and starch gelatinization will be cited.

Protein Denaturation. DSC has been used extensively to study thermal denaturation of proteins. Proteins of egg white have received considerable attention because of their importance in baked goods. Example studies include (1) stability of avidin and avidin/biotin complex (Donovan and Ross 1973), (2) stability of egg white proteins as influenced by pH, aluminum, and sucrose (Donovan *et al.* 1975), (3) conversion of ovalbumin to S-ovalbumin (Donovan and Mapes 1976), (4) effect of pH and neutral salts on thermal aggregation of ovalbumin (Hegg *et al.* 1979), and (5) stability of ovalbumin as influenced by sodium dodecyl sulfate (Hegg *et al.* 1978). Proteins of milk have also been studied: (1) thermal denaturation of whey protein in simulated milk ultrafiltrate (Ruegg *et al.* 1977), (2) β-lactoglobulin denaturation as influenced by sugar, salt and κ-casein (Itoh *et al.* 1976), and (3) thermal stability of β-lactoglobulin as a function of pH and sodium dodecyl sulfate (Hegg 1980). Eliasson and Hegg (1980) studied the thermal stability of wheat glutin.

DSC has been used to provide qualitative information on the effects of various constituents on thermal denaturation/aggregation of proteins and quantitative data on temperatures and enthalpy of denaturation.

Although DSC provides a method of producing a large amount of data on the phase transitions of proteins, it is important to realize its limitations. Proteins that do not heat denature do not appear on the thermogram (e.g., κ-casein), and in a mixture of proteins, those present in very small concentration may not appear on the thermogram (e.g., macroglobulin of egg white).

Starch Gelatinization. Heating starch in the presence of water results in a series of physical changes referred to collectively as "gelatinization." Actually a sequence of events occurs consisting of at least the following: (1) granule swelling, (2) conformational change of the starch accompanied by hydration, (3) translational and rotational diffusion of portions of starch chains within the granule, (4) breaking or disintegration of the granule, and (5) increase in measured viscosity resulting from loss of water to the granules and from loss of hydrated starch polymer from disrupted granules (Donovan 1977). Process 2, which is endothermic, is conveniently measured calorimetrically.

Stevens and Elton (1971) appear to have been the first to apply DSC to the study of gelatinization of starches. Since then many investigators have used DSC to study starch and starch systems. For example, Wootton and Bamunuarachchi (1979A, B) used DSC for native and modified starches and investigated the effect of heating rate and moisture level. Eliasson (1980) studied the influence of water content on the gelatinization of wheat starch using DSC. A semicrystalline polymer model for starch gelatinization was presented by Biliaderis *et al.* (1980) to explain data from DSC experiments. The interaction of amylose and lipids was studied by Kugimiya *et al.* (1980) and was subsequently used as the basis for a calorimetric determination of the amylose content of starches (Kugimiya and Donovan 1981). Wirakartakusumah (1981) applied DSC in the investigation of gelatinization of rice starch. To illustrate the use of DSC to study starch gelatinization, the effect of water-to-starch (W/S) ratio on gelatinization of rice starch as reported by Wirakartakusumah will be presented.

When a rice starch suspension is heated from room temperature to 150°C at 10°C/min, up to three endothermic processes are observed, depending on the water-to-starch ratio (Fig. 4.3). At W/S ratio ⩾2/1, only two endotherms are obtained. The first endotherm is observed at an onset temperature (T_0) of ca. 71°C. This is the single endotherm observed by Stevens and Elton (1971), Heidemann (1978), and Eberstein *et al.* (1980), which has been characterized as the "gelatinization" phase change. A second endotherm is obtained in the temperature range between 90° and 110°C. Although Stevens and Elton (1971) and

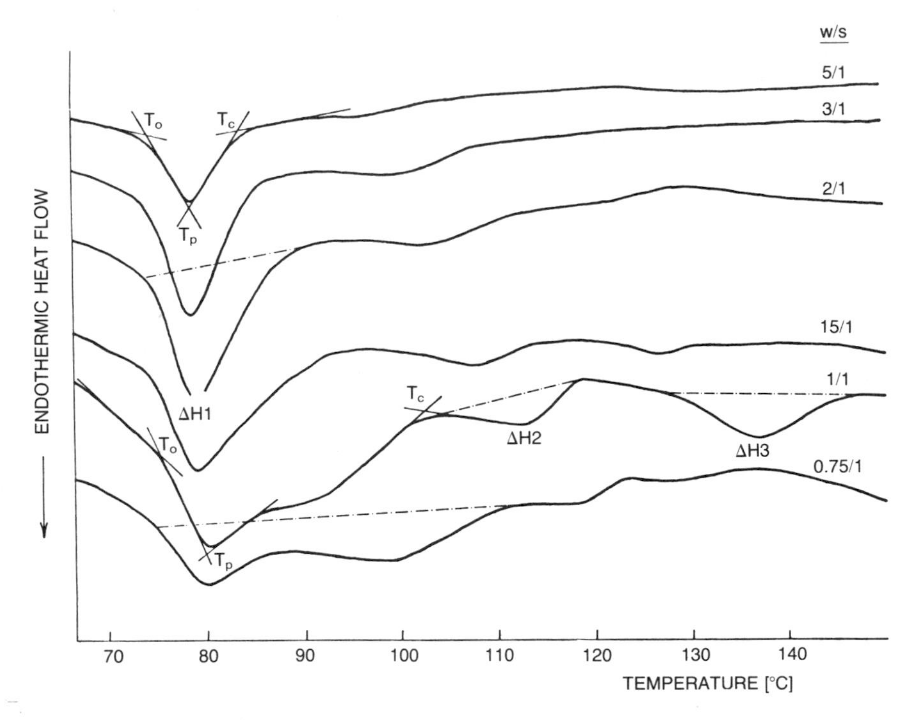

FIG. 4.3 Typical endotherms when rice starch suspensions with various W/S ratios are heated to 150°C at 10°C/min.
From Wirakartakusumah (1981)

Heidemann (1978) did not report this second endotherm, Eberstein *et al.* (1980) observed it when rice starch with a W/S ratio of 4/1 was heated to 120°C. It was suggested that this higher temperature endotherm was associated with melting of an amylose–lipid complex. A similar type of endotherm that occurred at 100°C was also reported by Kugimiya *et al.* (1980) from maize and wheat starches. When lipids of maize and wheat starches were extracted with methanol, the areas of the first endotherm (ΔH_1) were not altered; however, for maize, the second endotherm essentially disappeared after 150 hr of extraction, while for wheat starch the area decreased to about two-fifths of the original area (Kugimiya *et al.* 1980). Czaja (1980) reported that there were several types of lipids in rice starch including lysolecithin, ethanolaminkephalin, and lysoethanolaminkephalin. Generally, the lipid content of starches is of the order of 1% or less. However, if this lipid complexes with starch, this may result in 10–20% starch—lipid complex by weight (Kugimiya *et al.* 1980).

As the water to starch ratio decreases to less than 1.5/1, the gelatinization endotherm (ΔH_1) developed a trailing shoulder with a marked decrease in the area (enthalpy). This shouldering may be due to the destabilizing effect of water and heat treatments on the amorphous and the crystalline regions of starch granules. The onset temperature (T_0) and temperature range of gelatinization shifted to higher temperatures as the water content decreased as shown in Table 4.1. As the water-to-starch ratio decreased to less than 1.5/1, ΔH_1 decreased and the shoulder peak formed its own endotherm, which subsequently shifted to a higher temperature. The second endotherm (ΔH_2) also shifted to a higher temperature. In addition to this shouldering effect with decreasing water-to-starch ratio, a third endotherm (ΔH_3) at temperatures above 130°C was observed. However, at W/S ratio <1/1, the third endotherm was not detected because it occurred above 160°C, which was

TABLE 4.1 EFFECT OF WATER TO STARCH RATIO ON TEMPERATURE OF RICE STARCH GELATINIZATION[1]

Water to Starch Ratio (W/S)	Temperature (°C)[2]		
	Onset (T_o)	Peak (T_p)	Conclusion (T_c)
0.75	74.3	81.5	111.3
1.0	72.3	80.0	103.8
1.2	72.1	77.3	94.3
1.5	71.6	77.3	87.8
2.0	71.3	77.2	85.3

Source: Wirakartakusumah 1981.
[1] Heating rate is 10°C/min.
[2] Average of two replicates with range ± 0.5°C.

beyond the capability of the DSC instrument used. When the heated starch suspension (W/S = 1/1) was cooled from 160° to 10°C immediately, and subsequently reheated to 160°C using the same heating rate (10°C/min), the gelatinization endotherm disappeared, but the second and third endotherms were still present, as shown in Fig. 4.4. Wirakartakusumah (1981) suggested that the second endotherm was the amylose–lipid complex as reported by Kugimiya *et al.* (1980), who obtained similar results when they reheated the previously heated amylose–palmitic acid complex mixture. Furthermore, Kugimiya *et al.* (1980) suggested that the endotherm near 100°C need not necessarily result from melting of a crystalline phase. The occurrence of the third endotherm has not been reported or observed by other investigators (Donovan 1979; Biliaderis *et al.* 1980; Eliasson 1980). This endotherm was probably associated with thermal decomposition of some of the components of starch when heated to these high temperatures. It can be seen from Fig. 4.4 that the area of the second endotherm in a once-heated sample (curve A) and a reheated sample (curve B) was essentially constant. However, the area of the third endotherm for reheated

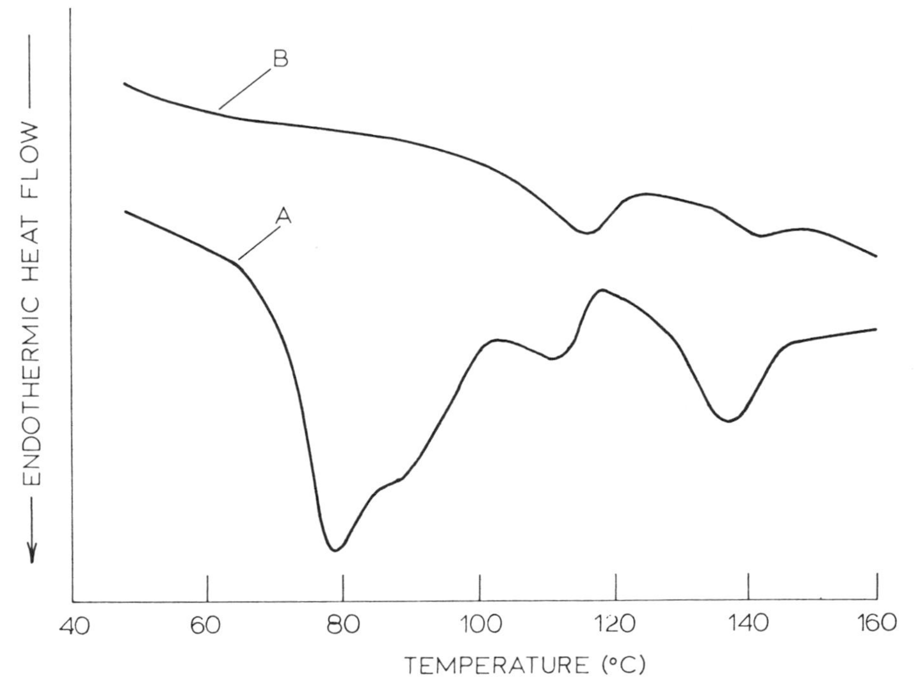

FIG. 4.4 Some typical endotherms after the first and second heat treatments using a W/S of 1/1 (*A* = first heating; *B* = second heating).
From Wirakartakusumah (1981).

starch (curve B) was significantly reduced compared to its original size in curve A. This may indicate that starch granules were irreversibly damaged during the first heating to 150°C.

To examine the nature of this damage, Wirakartakusumah (1981) performed the following. A starch suspension (W/S = 1/1) was heated to 100°C at 10°C/min and subsequently cooled down to 10°C. The pan was then opened and an excess amount of water added. Application of a reheating treatment of 100°C in DSC resulted in an endotherm characterized by T_0, T_p, and T_c of 47°C, 50°C, and 65°C, respectively, with ΔT equal to 18°C. The broader temperature range and lower temperature for the endotherm indicates that the starch granule was damaged in the previous heating, but the crystalline structure has not been disordered completely. Thus, subsequent reheating with excess water resulted in a lower temperature endotherm and wider temperature range due to the lower degree of crystallinity in the granules.

The enthalpies of the phase transitions were determined from the areas of the endotherms for various water-to-starch ratios and are presented in Fig. 4.5. From this figure it can be seen that the "gelatini-

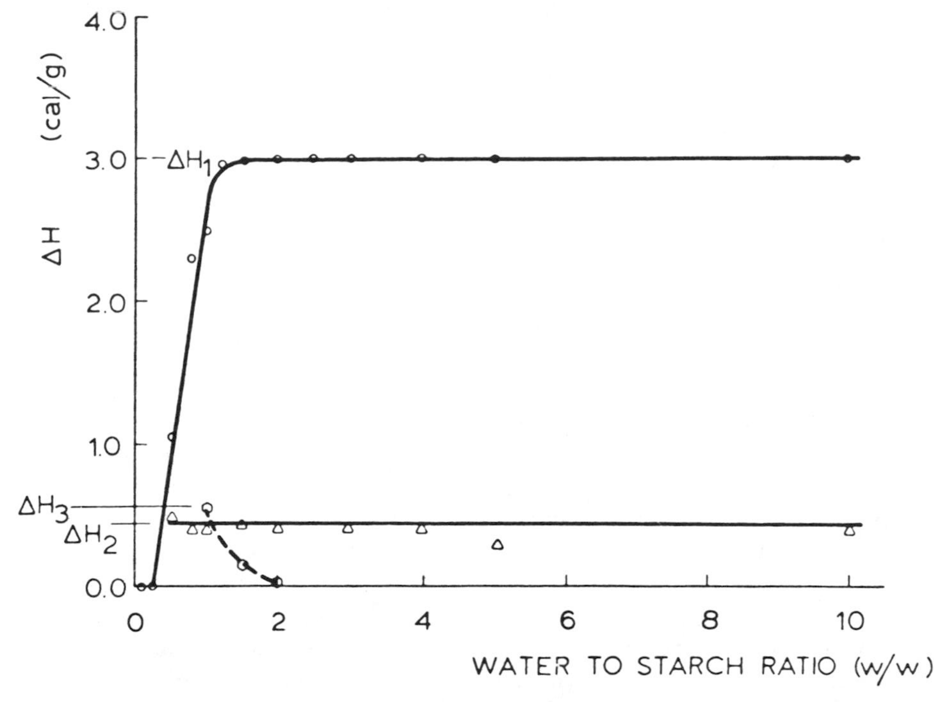

FIG. 4.5 Enthalpies of transition endotherms at various W/S ratios at a heating rate of 10°C/min.
From Wirakartakusumah (1981).

zation" endotherm (ΔH_1) only appeared when the water-to-starch ratio was greater than 0.3/1. This endotherm reached its maximum value at W/S ratio = 1.5/1 and remained constant thereafter, no matter how much water was added. Ultimately with very large water-to-starch ratios, the endotherm was undetectable owing to sensitivity limitations of the instrument. Above W/S ratio of 1.5/1, the enthalpy of ΔH_1 was ca. 3.0 cal/g. Stevens and Elton (1971) reported 3.4 cal/g, Heidemann (1978) 3.6 cal/g, and Eberstein *et al.* (1980) 3.1 cal/g. Slight differences in these values may be due to variability of the rice starch source. Wirakartakusumah reported the enthalpy of ΔH_2 (melting of amylose —lipid complex) as 0.4 cal/g, which is in good agreement with the value reported by Eberstein *et al.* (1980). This second endotherm exhibited a constant enthalpy that was independent of the water-to-starch ratio. This further supported the conclusion by Kugimiya *et al.* (1980) that the amylose–lipid endotherm was not the result of melting of crystals. The enthalpy of the third endotherm (ΔH_3) exhibited a strong dependence on water content and only appeared when the water-to-starch ratio was less than 2/1. Thus, this endotherm only appeared when there was insufficient water to gelatinize all the starch in the lower temperature region, and the enthalpy appeared to increase as the water-to-starch ratio decreased.

Wirakartakusumah (1981) used these data to calculate the molar ratio of water/glucose unit. To start gelatinization or to observe the appearance of the first endotherm (ΔH_1), more than 3 moles of water/hexose unit were required (>0.25 W/S ratio). To obtain complete gelatinization, approximately 13.5 moles of water were required per mole of glucose unit. These values are in good agreement with the data reported by Donovan (1979) on potato starch. About 4 moles water per mole hexose unit and 14 moles water per mole hexose unit were required for initiation of gelatinization and to obtain complete gelatinization, respectively. Collison and Chilton (1974) also presented evidence that a critical amount of water was necessary for initiation of gelatinization. Using a dye binding method (chlorazol violet R) to determine damaged granules due to gelatinization at various moisture contents, Collison and Chilton concluded that there was no gelatinization below 30% moisture, which represented 4 moles of water per mole of glucose unit.

At the conclusion temperature (T_c) of the first endotherm (ΔH_1), all the granules are melted. The progressive shift of this temperature to a higher temperature as the water content decreases was also observed on wheat starch crystals as reported by Lelievre (1973). Lelievre used birefringence technique to determine the melting point of the starch crystals and analyzed the data by employing the Flory equation, which characterizes the relationship between the melting point of a crystalline polymer and diluent concentration (water in this case) (Flory 1953):

$$\left(\frac{1}{T_m} - \frac{1}{T_m^{\circ}}\right) = \left[\frac{R}{H_u}\right]\left[\frac{V_u}{V_1}\right](\nu_1 - X_1\nu_1^2)\cdots \tag{1}$$

where T_m = the melting point of the polymer–diluent mixture (K); T_m° = the true melting point of the undiluted polymer (K); R = the gas constant (kcal/mole K); H_u = the change in enthalpy of fusion per repeating unit (glucose)(kcal/mole); V_u/V_1 = the ratio of the molar volume of the repeating unit (glucose) in that chain to that of the diluent (water); ν_1 = the volume fraction of water; X_1 = the Flory interaction parameter (= BV_1/RT_m); and B = interaction energy density characteristic of the solvent solute pair (kcal/ml). A similar analysis using Eq. (1) has been used on the results from DSC studies on potato starch (Donovan 1979) and various legume starches (Biliaderis *et al.* 1980).

Wirakartakusumah (1981) also applied the Flory equation to his data. To calculate the volume fraction of water, ν_1, the density of water was assumed to be one while that of starch was 1.55 (Lelievre 1973; Donovan 1979). Since ν_1 is less than one while (RT_m) in the Flory interaction parameter is on the order of 1000, $X_1\nu_1^2$ is $\ll\nu_1$ and can be neglected and, from Eq. (1), a linear relationship exists between $1/T_m$ and ν_1. When ν_1 equals zero, the intercept will be $1/T_m^{\circ}$, the reciprocal of the melting point of the undiluted polymer. Figure 4.6 is a plot of $1/T_m$ versus ν_1 for rice starch from Wirakartakusumah (1981). From linear regression analysis (correlation coefficient r = 0.98), extrapolation of the line to ν_1 = 0 resulted in a melting point (T_m°) of 471K (198°C). Donovan (1979) reported the T_m° value of 168°C for potato starch, while Biliaderis *et al.* (1980) reported T_m° in the ranges of 166°–203°C for various legume starches. Furthermore, for wheat starch, Lelievre (1973) reported a T_m° value of 210°C. Since it would be expected that the melting point of starch granules is characteristic of the starch, the value reported by Wirakartakusumah for rice starch is reasonable.

Once T_m° has been determined, the values of ΔH_u (heat of fusion) and B (energy of interaction) can be computed from the intercept and the slope, respectively, of the plot of $(1/T_m - 1/T_m^{\circ})/\nu_1$ versus (ν_1/T_m) as shown in Fig. 4.7. From linear least square analysis, ΔH_u is 13.0 kcal/mole (95% confidence interval is from 11 to 15 kcal/mole), which represents the heat required to melt 1 mole of crystalline unit, while the energy of interaction (B) is 0.7 cal/ml (95% confidence interval is from −8 to 9.5 cal/ml). Donovan (1979) reported values for potato

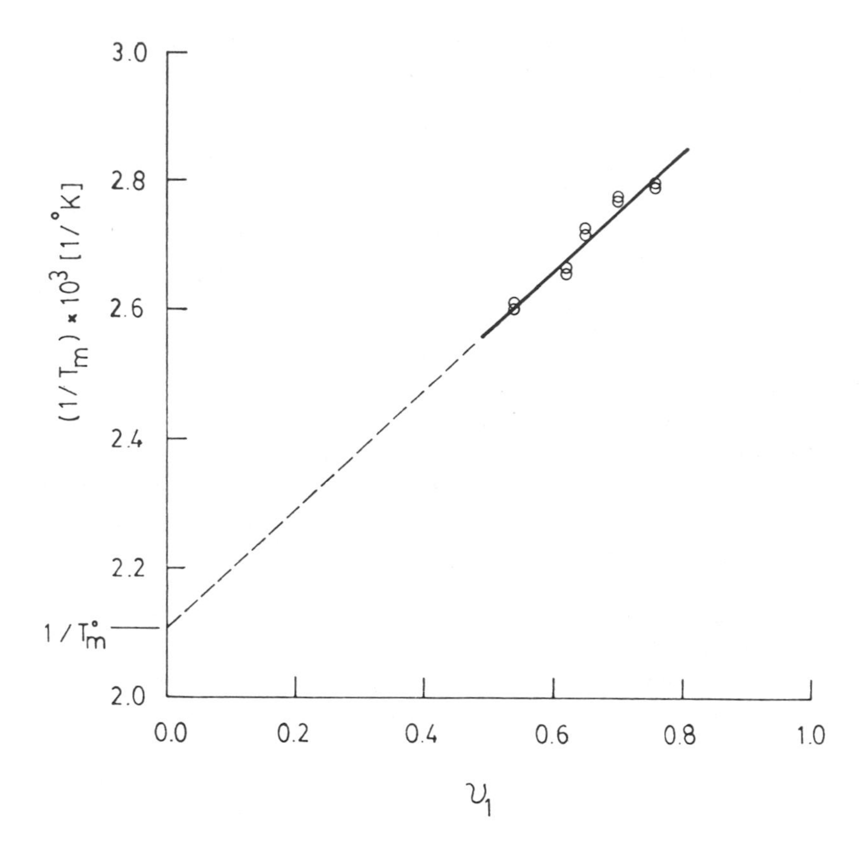

FIG. 4.6 Reciprocal melting point ($1/T_m$) versus the volume fraction of water (ν_1) in water/rice starch mixtures. The line is a linear least-square fit of the data. The equation for the line is $y = 0.00212125 + 0.00090533x$; $r = 0.98$.
From Wirakartakusumah (1981).

starch of 13.5 kcal/mole for ΔH_u and 3.5 cal/ml for B. For various legume starches, Biliaderis *et al.* (1980) reported values between 15.8 and 18.9 kcal/mole for ΔH_u and the B values from 0.3 to 1.03 cal/ml. According to Flory (1953), the positive values of B, which agreed with the reported values by others (Lelievre 1973; Donovan 1979; Biliaderis *et al.* 1980), indicates that water is indeed a poor solvent for starch. The accuracy of these calculated thermodynamic parameters depends on the validity of the extrapolation used in Fig 4.6 and 4.7 and on various experimental limitations. Nevertheless, the analysis does provide at least qualitative evidence that a starch granule may be treated as a spherulite polymer (semicrystalline) and that the Flory equation may be used to explain the dependence of starch gelatinization behavior on moisture content.

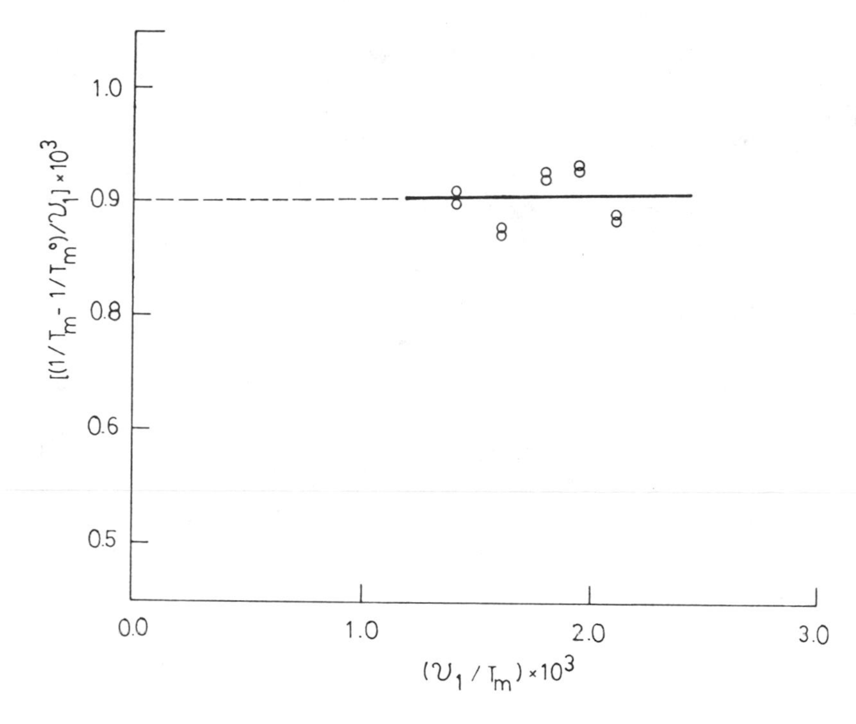

FIG. 4.7 $(1/T_m - 1/T_m^\circ)/\nu_1$ versus ν_1/T_m for rice starch/water mixtures. The line is a linear least square fit of the data. The equation for the line is $y = (8.95 \times 10^{-4}) + (5.98 \times 10^{-3})x$; $r = 0.07$.
From Wirakartakusumah (1981).

DSC as a Method to Obtain Kinetic Parameters

Differential scanning calorimetry provides the opportunity for measuring not only enthalpy change but also rate of reaction. This can lead, in principle, to estimation of kinetic parameters and insight into possible mechanisms of the process. To obtain a measure of reaction rate, DSC may be used in either isothermal or heat evolution (temperature scanning) modes (Sharp 1972; McNaughton and Mortimer 1975).

Isothermal Method. The extent of reaction is measured directly as a function of temperature and time. In separate experiments, several samples are held in the DSC cell, at a given temperature, but for different lengths of time. The decrease in endotherm area per unit weight as a function of previous isothermal exposure time is a measure of reaction rate at the set temperature. Repeating the procedure at various temperatures enables calculation of the kinetic constants at each temperature and the dependence of these constants on temperature is usually expressed by the Arrhenius equation.

Heat Evolution or Temperature Scanning Methods. The method relying on measurement of heat evolution of a temperature-programmed sample in DSC was first described by Borchardt and Daniels (1957) for solutions and later by Uricheck (1966) for solids. The sample was scanned through a temperature range and the thermogram was recorded. The rate of reaction was proportional to the peak height at a given time and the extent of reaction was proportional to the total area of the endotherm to that time. However, Sharp (1972) pointed out that to enable one to use the endotherm data to calculate kinetic parameters, three important assumptions must be met:

1. The temperature within the two holders in the DSC cell must be equal.
2. Heat transfer is by conduction only and the holders are identical, so that the heat transfer coefficient is the same for both.
3. The heat capacities of the liquids in the holders are the same; this will be nearly so if the solution containing the reactant is dilute.

Since DSC measures the energy change per unit time directly by maintaining zero temperature gradient between the two holders, and since the size of the sample is very small (around 10 mg) such that thermal gradients across the sample would be very small (Uricheck 1966), the above assumptions are adequately satisfied. If the reaction is first order, the reaction rate constant can be calculated from the following equation (Sharp 1972):

$$k = \frac{dH/dt}{A - a} \tag{2}$$

where k = first-order reaction rate constant at temperature T; dH/dt = the ordinate displacement from the baseline at temperature T; and $A - a$ = the total area (A) of the reaction peak minus the area (a) up to temperature T.

For simple, first-order reaction of relatively nonvolatile materials, the method works well, is fairly rapid, and results in a precision for activation energy of about 0.5–1.0 kcal/mol (Duswalt 1974). In addition, the method has been applied for studying the heat denaturation of proteins (Beardslee and Zahnley 1973; Donovan and Ross 1973; Donovan and Beardslee 1975). A further advantage of the nonisothermal method is that it covers a wide temperature range continuously and avoids missing regions that can occur in a series of isothermal experiments (Sharp 1972).

CONCLUSIONS

Differential scanning calorimetry offers tremendous potential for studying physicochemical changes that occur in foods. It has been used especially to study the state of water and ice in foods, denaturation of proteins, and gelatinization of starches. It has also been used to study very complex mixtures such as cake mixes (Donovan 1977). DSC will provide valuable qualitative and quantitative data on changes in food as a function of processing. It is a rapid, relatively simple technique. The main limitations with current instruments are (1) cost of the instrument and (2) the necessity of obtaining a very small (order of 20 mg) representative sample.

ACKNOWLEDGMENT

This is a contribution from the College of Agricultural and Life Sciences, University of Wisconsin-Madison.

BIBLIOGRAPHY

BEARDSLEE, R.A., and ZAHNLEY, J.C. 1973. A simple preparation of β-trypsin based on a calorimetric study of the thermal stabilities of α- and β-trypsin. Arch. Biochem. Biophys. *158*, 806–811.

BILIADERIS, C.G., MAURICE, T.J., and VOSE, J.R. 1980. Starch gelatinization phenomena studied by differential scanning calorimetry. J. Food Sci. *45*, 1669–1674, 1680.

BORCHARDT, H.J., and DANIELS, F. 1957. J. Am. Chem. Soc. *79*, 41. Cited by Sharp, J.H. 1972. Reaction kinetics. *In* Differential Thermal Analysis, Vol 2. R.C. Mackenzie (Editor). Academic Press, New York.

COLLISON, R. and CHILTON, W.G. 1974. Starch gelation as a function of water content. J. Food Technol. *9*, 309–315.

CZAJA, A. TH. 1980. Stärke Korner im pflanzenreich. Mikroskupische und submikroskopische morphologie und evolution. Stärke *32*, 253 (German).

DONOVAN, J.W. 1977. A study of the baking process by differential scanning calorimetry. J. Sci. Food Agric. *28*, 571–578.

DONOVAN, J.W. 1979. Phase transitions of the starch–water system. Biopolymers *18*, 263–275.

DONOVAN, J.W., and BEARDSLEE, R.A. 1975. Heat stabilization produced by protein –protein association. A differential scanning calorimetry study of the heat denaturation of the trypsin–soybean trypsin inhibitor and tryspin–ovomucoid complexes. J. Biol. Chem. *250*, 1966.

DONOVAN, J.W., and MAPES, C.J. 1976. A differential scanning calorimetric study of conversion of ovalbumin to S-ovalbumin in eggs. J. Sci. Food Agric. *27*, 197–204.

DONOVAN, J.W., and ROSS, K.D. 1973. Increase in the stability of avidin produced by binding of biotin. A differential scanning calorimetric study of denaturation by heat. Biochemistry *12*, 512–517.

DONOVAN, J.W., MAPES, C.J., DAVIS, J.G., and GARIBALDI, J.A. 1975. A differential scanning calorimetric study of the stability of egg white to heat denaturation. J. Sci. Food Agric. *26*, 73–83.

DUSWALT, A.A. 1974. The practice of obtaining kinetic data by differential scanning calorimetry. Thermochim. Acta *8*, 57–68.

EBERSTEIN, VON K., HOPCKE, R., KONIECZNY-JANDA, G., and STUTE, R. 1980. DSC-untersuchungen an starken. Teil 1. Moglichkeiten thermoanalytischer methoden zur starkecharakterisierung. Stärke *32*, 397–404 (German).

ELIASSON, A.-C. 1980. Effect of water content on the gelatinization of wheat starch. Stärke *32*, 270–272.

ELIASSON, A.-C., and HEGG, P.-O. 1980. Thermal stability of wheat gluten. Cereal Chem. *57*, 436–437.

FLORY, P.J. 1953. Principles of Polymer Chemistry. Cornell University Press, Ithaca, New York.

HEGG, P.-O. 1980. Thermal stability of β-lactoglobulin as a function of pH and the relative concentration of sodium dodecyl sulfate. Acta Agric. Scan. *30*, 401–404.

HEGG, P.-O., MARTENS, H., and LOFQVIST, B. 1978. The protective effect of sodium dodecyl sulfate on the thermal precipitation of conalbumin. A study on thermal aggregation and denaturation. J. Sci. Food Agric. *29*, 245–260.

HEGG, P.-O., MARTENS, H., and LOFQVIST, B. 1979. Effects of pH and neutral salts on the formation and quality of thermal aggregates of ovalbumin. A study on thermal aggregation and denaturation. J. Sci. Food Agric. *30*, 981–993.

HEIDEMANN, R.S. 1978. Studies on respiration in wild rice and characteristics of wild rice starch. M.S. Thesis. Department of Food Science, University of Wisconsin-Madison, Wisconsin.

ITOH, T., WADA, Y., and NAKANISKI, T. 1976. Differential thermal analysis of milk proteins. Agric. Biol. Chem. *40*, 1083–1086.

KUGIMIYA, M., and DONOVAN, J.W. 1981. Calorimetric determination of the amylose content of starches based on formation and melting of the amylose–lysolecithin complex. J. Food Sci. *46*, 765–770, 777.

KUGIMIYA, M., DONOVAN, J.W., and WONG, R.Y. 1980. Phase transitions of amylose–lipid complexes in starches: A calorimetric study. Stärke *32*, 265–270.

LELIEVRE, J. 1974. Starch gelatinization. J. Appl. Polym. Sci. *18*, 293–296.

LÓRÁNT, B. 1972. Food industries. *In* Differential Thermal Analysis, Vol. 2. R.C. Mackenzie (Editor). Academic Press, New York.

MACKENZIE, R.C. (Editor). 1970. Differential Thermal Analysis, Vol. 1. Academic Press, New York.

MACKENZIE, R.C. (Editor). 1972. Differential Thermal Analysis, Vol. 2. Applications. Academic Press, New York.

McNAUGHTON, J.L., and MORTIMER, C.T. 1975. Differential scanning calorimetry. *In* IRS. Physical Chemistry Ser. 2, Vol, 10. Butterworths, London.

POPE, M.I., and JUDD, M.D. 1977. Differential Thermal Analysis. A Guide to the Technique and Its Application. Heydon and Sons Ltd., London.

RUEGG, M., MOOR, U., and BLANC, B. 1977. A calorimetric study of the thermal denaturation of whey proteins in simulated milk ultrafiltrate. J. Dairy Res. *44*, 509–520.

SHARP, J.H. 1972. Reaction kinetics. *In* Differential Thermal Analysis, Vol. 2. R.C. MacKenzie (Editor). Academic Press, New York.

STEVENS, D.J., and ELTON, G.A.H. 1971. Thermal properties of the starch/water system. Part 1. Measurement of heat of gelatinization by differential scanning calorimetry. Stärke *23* 8–11.

URICHECK, M.J. 1966. The determination of kinetic parameters by differential scanning calorimetry. Perkin-Elmer Instrument News *17* (2).

WENDLANDT, W.W. 1974. Differential thermal analysis and differential scanning calorimetry. *In* Chemical Analysis, Vol. 19: Thermal Methods of Analysis. Analytical Chemistry and Its Applications Monogr. Ser. John Wiley and Sons, New York.

WIRAKARTAKUSUMAH, M.A. 1981. Kinetics of starch gelatinization and water absorption in rice. Ph.D. Thesis. Department of Food Science, University of Wisconsin-Madison, Wisconsin.

WOOTTON, M., and BAMUNUARACHCHI, A. 1979A. Application of differential scanning calorimetry to starch gelatinization. I. Commercial native and modified starches. Stärke *31*, 201–204.

WOOTTON, M., and BAMUNUARACHCHI, A. 1979B. Application of differential scanning calorimetry to starch gelatinization. II. Effect of heating rate and moisture level. Stärke *31*, 262–264.

5

Multilayer Emulsions

Stig E. Friberg and Magda El-Nokaly[1]

INTRODUCTION

Emulsions have traditionally been described as two-phase systems, the classical treatment by Becher (1965) and the later book by Sherman (1968) being illustrative examples. However, some characteristics of practical emulsion systems such as sudden changes in stability and viscosity could not be fully explained by the classical two-phase theory as pointed out in 1964 by Sherman (Burt 1965).

This opinion led to rather intensive research during the 1960s in order to analyze the phase conditions in emulsions following Salisbury (Salisbury *et al.* 1954; Lachampt and Vila 1967; Swarbrick 1968), leading to the discovery of the specific stabilization of emulsions by liquid crystals (Friberg *et al.* 1969). Subsequent contributions (Friberg and Mandell 1969, 1970) led to a change of the definition of emulsions. "In an emulsion liquid droplets and/or liquid crystals are dispersed in a liquid" (IUPAC 1972). These emulsions stabilized by multilayers are found in cosmetic preparations (Friberg 1979) and in a wide variety of food products (Larsson, 1968, 1972; Larsson and Krog 1973; Boyd *et al.* 1976; Krog and Jensen 1970; Krog 1973). We found a discussion of the conditions necessary to form such multilayers of interest, and in the following pages, will present a few relevant observations on association structures that lead to such multilayers.

MICELLIZATION OF EMULSIFIERS

An emulsifier is an amphiphilic molecule that will strongly adsorb to an interface between water and air or between water and oil. This

[1] Department of Chemistry, University of Missouri, Rolla, MO 65401.

adsorption leads to a reduction of the surface tension. The relation between the surface tension and the logarithm of the lower range emulsifier concentration in the aqueous phase is well known (Fig. 5.1).

In the narrow concentration range during which the slope of this curve changes from its negative value to zero or almost zero (Gunnarsson *et al* 1980) the emulsifier molecules begin association with normal micelles (Fig. 5.2), creating an interface within the system. This association is governed by the size of the cross-sectional area of the polar group (Israelachvili *et al.* 1976).

MICELLES

Micellization is more directly important to emulsion stability than is generally realized. A schematic picture of the conditions during the coalescence of two emulsion droplets illustrates this.

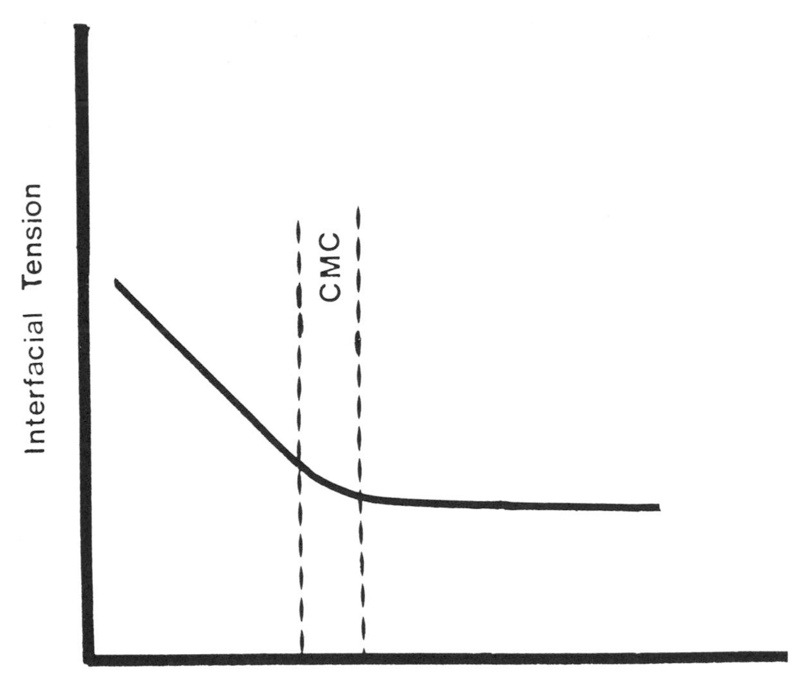

Fig. 5.1. Adsorption of surfactant leads to linear decrease of surface tension when it is plotted against the logarithm of the concentration of the surfactant. In a small concentration range (cmc) of saturated adsorption, micelles begin to form in the system. For concentrations in excess of this value the surface tension remains essentially constant.

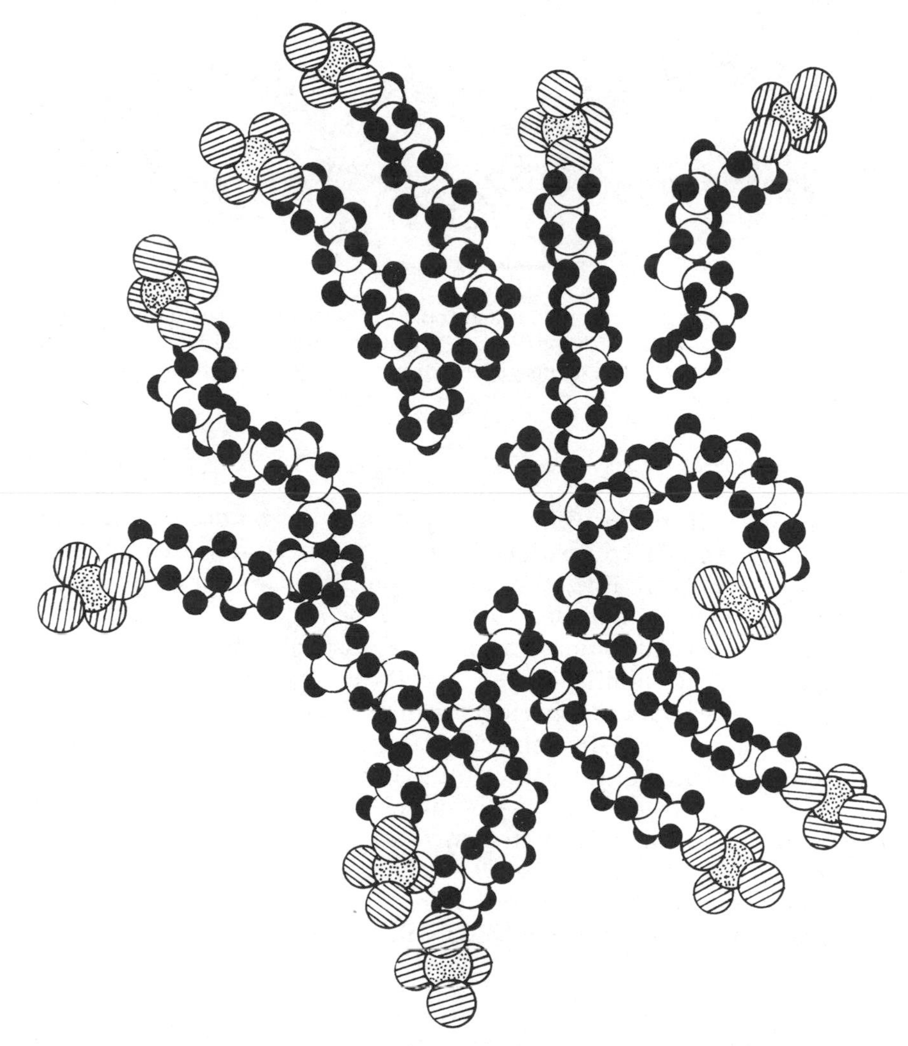

Fig. 5.2. In an aqueous micelle the hydrophobic parts are oriented toward the center and the polar parts toward the surrounding aqueous medium.

The stability against coalescence of the flocculated droplets is determined by the perturbations in the thin film between them. Critical points (Vrij and Overbeek 1968; Vrij 1964; Sheludko 1967) are the ones with minimum thickness (as in Fig. 5.3). For an emulsifier that will primarily associate with micelles it appears obvious that conformations such as the one at M of the upper surface in Fig. 5.3 will lead to disruption of the surface since the local conditions with a high sur-

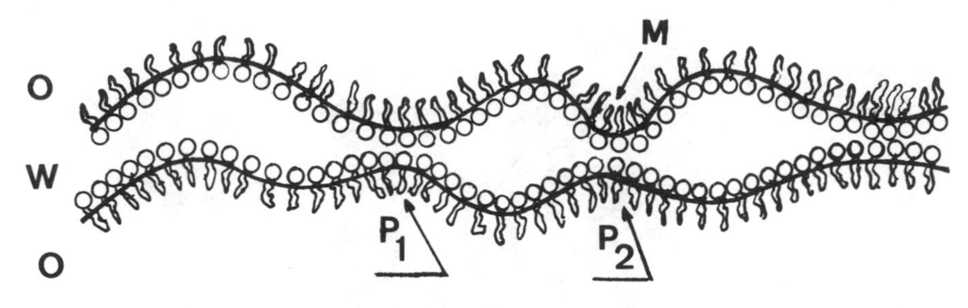

Fig. 5.3. The perturbation pattern in a thin liquid film between the flocculated emulsion droplets may give rise to local thinning (P_1 and P_2) that may reach critical thickness. At points where the radius of curvature is extremely small (M) the tendency to form a normal micelle may lead to destabilization.

factant concentration will favor the formation of a micelle. After the micelle is formed it will be localized in the thin film, causing a new critical point to form if rupture of the thin film does not occur during the formation of the micelle.

With these considerations, it appears reasonable to assume an improved emulsion stability if the surfactant associated to a layered structure rather than to spherical micelles. The factor that determines the shape of the association structure is actually the ratio between the cross-sectional area of the polar group and the hydrocarbon chain volume (Israelachvili *et al.* 1976). A reduction of this ratio favors a layered structure. Since a single chain ionic surfactant necessarily shows a large ratio due to repulsion between charged groups, the properties of surfactant *combinations* are of interest. There are several ways to achieve this reduction; two of them will be discussed in the following section.

Layered Emulsifier Structures

A combination of an ionic surfactant with its large cross-sectional area and a long chain alcohol with its small area should be interesting from the point of view of emulsion stabilization. Lauryl sulfate/lauryl alcohol is a classic example of such a combination; its usefulness as an emulsifier combination has been proved beyond doubt in practical applications.

The combination showed such good properties that the hypothesis of a "complex formation" between the surfactant and the alcohol (Schulman and Cockbain 1940; Davies and Mayers 1960; Blakey and Lawrence

1954) was believed for several years, in spite of strong reservations (Friberg 1971; Vold and Mittal 1972; Friberg *et al* 1976).

There is no evidence for a complex formation between the two species. The following study of the conditions in the water-plus-emulsifier system facilitates the understanding of what happens at the interface (Fig. 5.4).

The alcohol is decanol and the surfactant is an ionic soap, sodium octanoate. The general features are identical for any ionic surfactant/long chain alcohol couple (Ekwall 1974).

The soap is soluble in the water, forming micelles at the critical micellization concentration (cmc) in Fig. 5.4 and in excess of this amount. These micelles have the characteristics of dissolving alcohol molecules in their interior (solubilization) (Fig. 5.2). Alcohol solubility increases at surfactant concentrations in excess of cmc as is seen in Fig. 5.4.

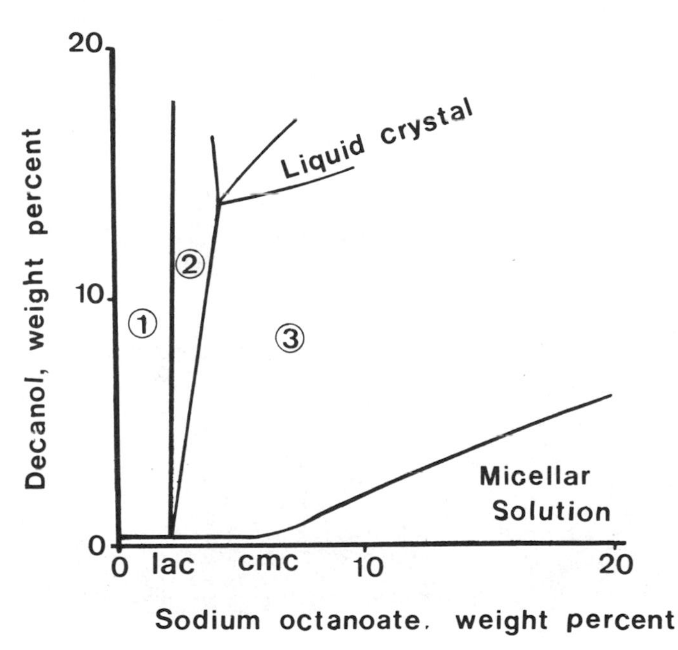

Fig. 5.4. An aqueous solution of a surfactant (sodium octanoate) will solubilize a long-chain alcohol (decanol) at its concentrations in excess of the critical micellization concentration (cmc). The phase in equilibrium with the aqueous phase in area 1 is the liquid alcohol with some water and surfactant dissolved. For concentrations in excess of the lowest association (lac) the phase in equilibrium (area 3) is a lamellar liquid crystal. In area 2 both these phases are in equilibrium with the aqueous phase.

However, the alcohol/surfactant ratio is small, and in order to understand the action of combinations used in practice it is necessary to evaluate the conditions at sufficiently high concentrations to render the alcohol insoluble. The concentration must be focused on the phases appearing above the solubility limit of the alcohol.

At low surfactant concentrations (<1/3 cmc) the phase appearing at alcohol content in excess of the maximum solubility in water is an isotropic liquid of the alcohol with small amounts of water and surfactant dissolved. This changes at the lowest association concentration (lac) in Fig. 5.4. For surfactant concentrations in excess of that value, the phase appearing when the solubility limit of the alcohol is exceeded is an optically anisotropic liquid crystal containing water (82%) in addition to the surfactant–alcohol mixture in a ratio 3:1. This structure is not a liquid but a lamellar liquid crystalline phase (Friberg and Larsson 1976) (Fig. 5.5). The lamellar spacing is exactly identical throughout the structure, giving sharp X-ray diffraction patterns by which the structure may be identified (Fontell 1974).

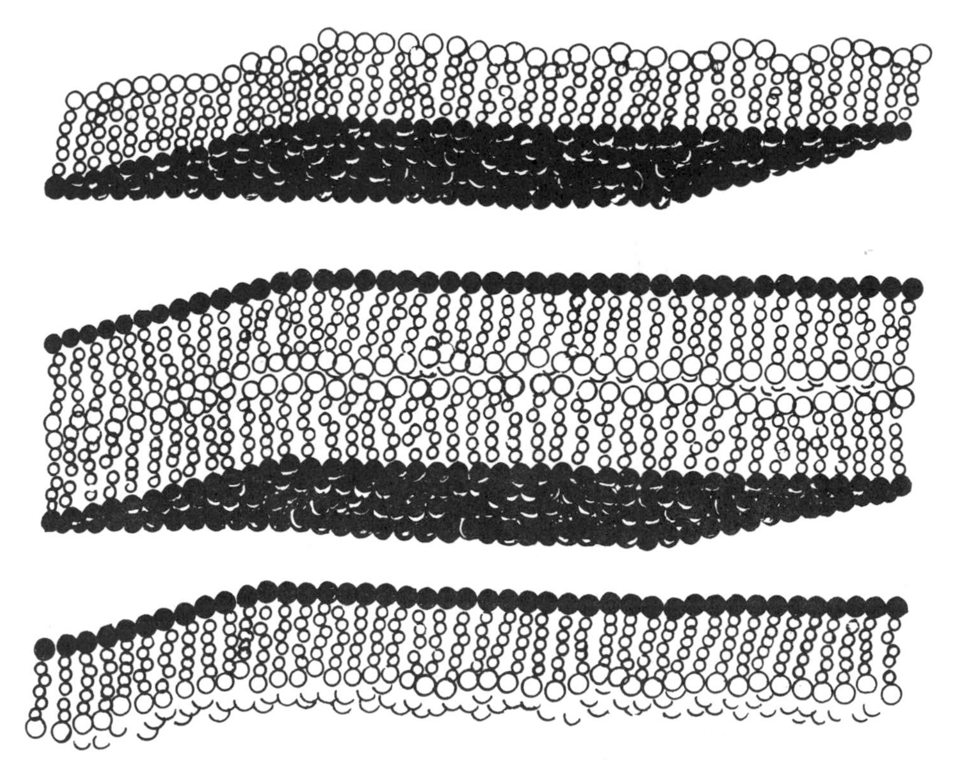

Fig. 5.5. A lamellar liquid crystal consists of aqueous (blank) and amphiphile layers.

The importance for emulsion stability is obvious (Friberg and Larsson 1976; Vold and Mittal 1972) from a comparison with Fig. 5.3. A perturbation of the walls of the thin film between the flocculated droplets will be counteracted by the adsorbed layer of the surfactant–alcohol mixture. The low energy state of the surfactant layer is now a plane, not a sphere. In addition, it should be observed that the saturated layer is attained at surfactant concentrations far below the cmc.

The second way to obtain the correct balance of the polar head cross-sectional area volume of the hydrocarbon is to use glycerol esters. Lecithin is the common material in the central biomolecular layer of biomembranes (Fig. 5.6). In pure form it does not form micelles but associates to bilayers; vesicles (Fendler 1980) are conspicuous examples of this spontaneous bilayer formation.

Monoglycerides also form these layered structures, and the literature on the extensive application of these compounds as emulsifiers and modifiers of emulsion properties has been reviewed by Krog and Lauridsen (1976). In the following section only a brief discussion will be made of the conditions of the structural function of multilayer emulsifiers in emulsions.

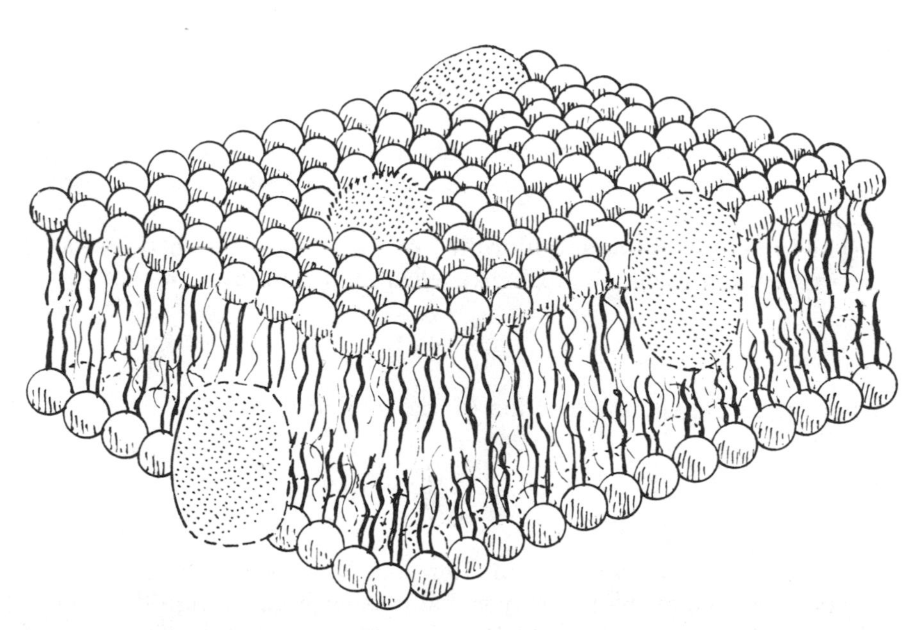

Fig. 5.6. A model of the bimolecular central part of the biomembrane shows similarities with the liquid crystalline structure.

Composition and Structure of the Multilayer

The lamellar structure in the multilayer may be separated from the two remaining phases by centrifugation and the three separated phases form the equilibrium state of the system, so that the structure and composition of the third phase can be conveniently determined in the separated state.

The composition shows that the multilayer contains all three phases (Friberg and Mandell 1969). For lecithin (Friberg and Mandell 1970) the following values were found: water 50 wt %, oil 10 wt %, and lecithin 40 wt % with the result that most of the stabilizing film consists of water and oil. A 2% concentration of emulsifier indicates that the stabilizing film also contained 2.5% water and 0.5% oil.

Nonionic surfactants (Friberg and Mandell 1969) showed an even greater dominance for the two liquids; the stabilizing film contained 80% water, 15% surfactant, and 5% oil. The economy was upset by the fact that the emulsifier was soluble in water to 2% and in *p*-xylene to 3%, meaning a dissolved 2.5% of the total in a 1:1 emulsion with *p*-xylene. In this emulsion the multilayer emulsifier structure acted as an *additional* stabilizer to the initial action of the conventional multilayer.

The structure of the separated multilayer can be determined by X-ray diffraction methods (Fontell 1974). The interlayer spacing is directly given by Bragg's law:

$$n\gamma = 2d \sin \theta$$

where d = interlayer distance perpendicular to the layer; θ = the angle for enhanced intensity; n = the order of diffraction; and γ = wavelength.

The thickness of the amphiphilic layer and the cross-sectional area per molecule (d_a) may be formally calculated (Luzzati *et al.* 1960):

$$d_a = \phi_a d$$

where ϕ_a is the volume fraction of the emulsifier

$$S_a = \frac{2M_a\phi_a}{d_a N} \times 10^{24}$$

M_a = molecular weight of emulsifier; and N = Avogadro's number.

The variation of the interlayer distance with added liquid water or oil may be used to estimate the amount of it that has penetrated between the emulsifier molecules. Different methods of this estimation have been given (Luzzati *et al.* 1960; Moucharafieh *et al.* 1979).

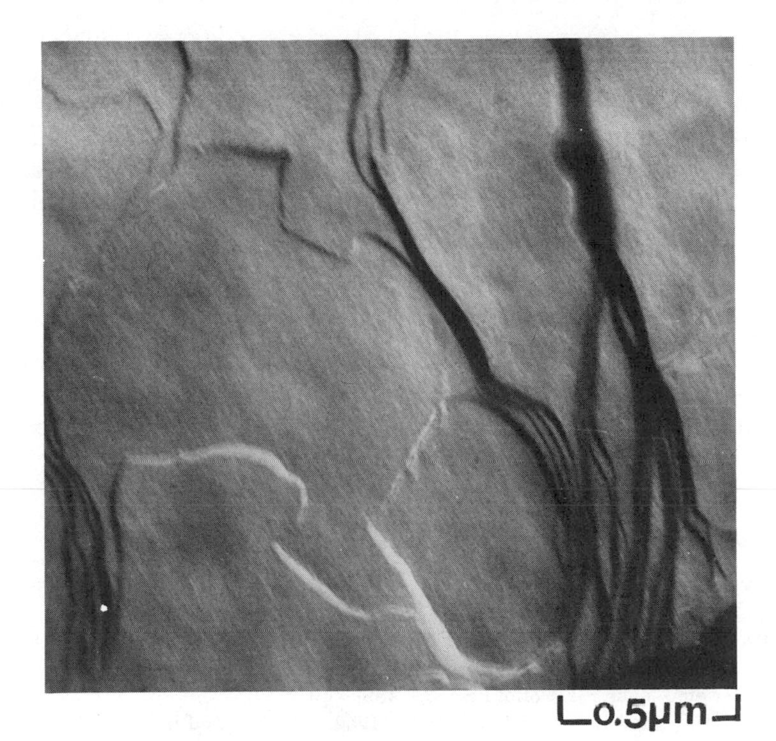

Fig. 5.7. The surface of multilayer particles with no or extremely few dislocations. Magnification 4.8 × 10^4.

Stabilization by the Multilayer

The multilayer organization will stabilize the emulsion with several mechanisms, two of which will be discussed in the following section.

The first has to do with the nonequilibrium state of the emulsion. The formation of the emulsion by the turbulence of vigorous stirring leaves the liquid crystal scattered as bimolecular leaflets through the continuous phase. These leaflets may form bridges between the dispersed droplets, giving a strong increase of the macrorigidity of the system (Fig. 5.5). This phenomenon will enhance the stability against flocculation.

In addition, the multilayer structure has a pronounced influence on the coalescence of the emulsion droplets. The local hydrodynamics (Bernstein *et al.* 1972) justify evaluation of droplet–droplet forces as the decisive component. The calculations (Friberg *et al.* 1976; Jansson and Friberg 1976) show the van der Waals attraction forces to be modified by the presence of the layered structure. They are considerably reduced by it.

In addition, rheological conditions play an important role. The layered structure is highly viscous and the coalescence must take place against considerable energy barriers (de Gennes 1974). In reality, the dislocation pattern probably is decisive. It has not been studied in detail for dispersed systems although preliminary investigations (Moucharafieh and Friberg 1982) (Fig. 5.7) indicate dislocations to be extremely rare in the multilayers oriented around droplets.

CONCLUSIONS

Preliminary investigations into the structural conditions in emulsions stabilized by multimolecular layers have clarified some of the salient factors. Future research giving futher clarification and modification of the concept is needed.

BIBLIOGRAPHY

BECHER, P. 1965. Emulsions: Theory and Practice, 2nd Ed. Reinhold Publ. Corp., New York.

BERNSTEIN, D.F., HIGUCHI, W.I., and HO, N.T.H. 1972. Consideration of the Spielman–Honig theory in the quantitative analysis of coalescence kinetics in dilute oil-in-water emulsions. J. Colloid Sci. *39*, 439–440.

BLAKEY, B.C., and LAWRENCE, A.S.C. 1954. Surface and interfacial viscosity of soap solutions. Discuss. Faraday Soc. *18*, 268–276.

BOYD, J.V., KROG, N., and SHERMAN, P. 1976. Comparison of rheological studies on adsorbed emulsifier films with X-ray studies of the bulk solutions. *In* Theory and Practice of Emulsion Formulation. A.L. Smith (Editor). Academic Press, New York.

BURT, B.W. 1965. An approach to emulsion formulation. J. Soc. Cosmet. Chem. *16*, 465–475.

DAVIES, J.T., and MAYERS, G.R.A. 1960. Interfacial viscosities of monolayers. Trans. Faraday Soc. *56*, 691–696.

DE GENNES, P.J. 1974. The Physics of Liquid Crystals. Oxford Publications, London.

EKWALL, P. (Editor). 1974. Lyotropic Liquid Crystals, Vol. 1. Academic Press, New York.

FENDLER, J. 1980. Surfactant vesicles as membrane mimetic agents: characterisation and utilization. Acc. Chem. Res. *13*, 7–13.

FONTELL, K. 1974. X-ray diffraction by liquid crystal amphiphile system. *In* Liquid Crystals and Plastic Crystals, Vol. 2. G.W. Gray and P.S. Winsor (Editors). Ellis Horwood Ltd., Chichester, England.

FRIBERG, S. 1971. Molecular complexes in liquid crystalline stuctures as a stabilizer at liquid interfaces. Kolloid Z.Z. Poly. *244*, 333–336.

FRIBERG, S. 1979. Three-phase emulsions. J. Soc. Cosmet. Chem. *30*, 309–319.

FRIBERG, S., and LARSSON, K. 1976. Liquid crystals and emulsions. *In* Advances in Liquid Crystals, Vol. 2. G. Brown (Editor). Academic Press, New York.

FRIBERG, S., and MANDELL, L. 1969. Phase equilibria and their influence on the properties of emulsions. J. Am. Oil Chem. Soc. *47*, 149–152.
FRIBERG, S., and MANDELL, L. 1970. Influence of phase equilibria on properties of emulsions. J. Pharm. Sci. *59*, 1001–1004.
FRIBERG, S., MANDELL, L., and LARSSON, K. 1969. Mesomorphous phases, a factor of importance for the properties of emulsions. J. Colloid Interface Sci. *29*, 155–156.
FRIBERG, S., JANSSON, P.O., and CEDERBERG, E. 1976. Surfactant association structure and emulsion stability. J. Colloid Interface Sci. *55*, 614–623.
GUNNARSSON, G., JÖNSSON, B., and WENNERSTRÖM, H. 1980. Surfactant association into micelles. An electrostatic approach. J. Phys. Chem. *84*, 3114–3121.
IUPAC. 1972. Manual on Colloid and Surface Science. Butterworth, London.
ISRAELACHVILI, J.N., MITCHELL, D.J., and NINHAM, B.W. 1976. Theory of self-assembly of hydrocarbon amphiphiles into micelles and bilayers. J. Chem. Soc. Faraday Trans. II *72*, 1525–1568.
JANSSON, P.-O., and FRIBERG, S. 1976. van der Waals potential in coalescing emulsion drops with liquid crystals. Mol. Cryst. Liq. Cryst. *34*, 75.
KROG, N. 1973. Influence of food emulsifiers on pasting temperature and viscosity of various starches. Staerke *25*, 22–30.
KROG, N., and JENSEN, B.N. 1970. Interaction of monoglycerides in different physical states with amylose and their antifirming effects in bread. J. Food Technol. *5*, 77–87.
KROG, N., and LAURIDSEN, J.B. 1976. Food emulsifiers and their associations with water. *In* Food Emulsions. S. Friberg (Editor). Marcel Dekker, New York.
LACHAMPT, F., and VILA, R.M. 1967. A contribution to the study of emulsions. Am. Perfum. Cosmet. *82*, 29–36.
LARSSON, K. 1968. Surface active lipids in foods. SCI Monograph No. 32. Soc. Chem. Ind., London.
LARSSON, K. 1972. Structure of isotropic phases in lipid-water systems. Chem. Phys. Lipids *9*, 181–193.
LARSSON, K., and KROG, N. 1973. Structural properties of the lipid-water gel phase. Chem. Phys. Lipids *10*, 177–181.
LUZZATI, V., MUSTACCHI, H., SKOULIOS, A., and HUSSON, F. 1960. La structure des colloides d'association. I. Les phases liquidecrystallines des systemes amphiphile-caneau. Acta Crystallogr. *13*, 660–677.
MOUCHARAFIEH, N., and FRIBERG, S.E. 1982. Unpublished material.
MOUCHARAFIEH, N., FRIBERG, S.E., and LARSEN, D.W. 1979. Effects of solubilized hydrocarbons on the structure and the dislocation pattern in a lamellar liquid crystal. Mol. Cryst. Liq. Cryst. *53*, 189–206.
SALISBURY, R., LEUALLEN, E.E., and CHAVKIN, L.T. 1954. The effect of phase-volume ratio on emulsion type. I. Beeswax-borax ointments. J. Am. Pharm. Assoc. *43*, 117–119.
SCHULMAN, J.H., and COCKBAIN, E.G. 1940. Molecular interactions at oil in water interfaces. II. Phase inversion and stability of water in oil emulsion. Trans. Faraday Soc. *36*, 661–668.
SHELUDKO, A. 1967. Thin liquid films. Adv. Colloid Interface Sci. *1*, 391–464.
SHERMAN, P. (Editor). 1968. Emulsion Science. Academic Press, New York.
SWARBRICK, J. 1968. Phase equilibrium diagrams: an approach to the formulation of solubilized and emulsified systems. J. Soc. Cosmet. Chem. *19*, 187–209.

VOLD, R.D., and MITTAL, K.D. 1972. Effect of lauryl alcohol on the stability of oil in water emulsions. J. Colloid Interface Sci. *38*, 451–459.

VRIJ, A. 1964. Light scattering by soap films. J. Colloid Sci. *19*, 1–27.

VRIJ, A., and OVERBEEK, J.TH.G. 1968. Rupture of thin liquid films due to spontaneous fluctuations in thickness. J. Am Chem. Soc. *90*, 3074–3078.

6

Relation of Structure to Physical Properties of Animal Material

D. W. Stanley[1]

INTRODUCTION

It is now accepted that physical properties of food are a direct consequence of structural organization. Enough data have been collected from both model systems and actual foods to establish this concept. Elsewhere in this volume other authors have provided evidence linking structure to physical properties for various types of food; this chapter is devoted to a similar approach for animal tissue. Although a large number of food products derived from animal origin are consumed, the present discussion must be limited to voluntary striated muscle, i.e., what the consumer normally calls "meat" and the food scientist often terms "muscle tissue." Also, it will be noticed that most of the data and examples provided herein are from bovine animals. As will be addressed subsequently, several factors, including animal age and related events occurring in the connective tissue, dictate that physical properties are of greater importance in this species than in others consumed by man.

MEAT AS FOOD

Meat is an important, some would say too important, part of the North American diet. The per capita consumption of meat in the United States was about 115 kg in 1974, which translates to around 25% of the total food budget (Milner *et al.* 1978). Meat is the largest supplier of protein for the average North American and also contributes significantly to the dietary intake of vitamins and minerals. Aside from nutrition-

[1]Department of Food Science, University of Guelph, Guelph, Ontario, Canada.

al considerations, meat has certain socioeconomic implications; it is known that meat consumption increases in proportion to the affluence of a given society. It was estimated by Cunha (1978) that a 43% increase above the 1974 level in the production of beef and veal will be needed in the United States to satisfy requirements expected in the year 2000.

What makes meat so attractive to the consumer? Certainly there is no strict nutritional necessity for eating meat; the nutrients provided by animal products can all be obtained from other foods, usually at lower cost. One obvious answer is that meat is an integral component of our cultural heritage (Rubin 1981). But another, and perhaps more important, reason lies in the quality characteristics of meat as a food. Kramer and Twigg (1970) classified sensory quality attributes as appearance, flavor, and kinesthesis—the latter being defined as those characteristics dealing with the sense of feel or, as the consumer would express it, texture.

Texture is the most important quality attribute of meat. When consumers were asked to name the eating characteristic of greatest consequence in meat, tenderness was the factor receiving the largest number of responses (57%), followed by flavor (30%), juiciness (8%), and fat (5%). Perhaps more revealing was the reply to the question, "which element produced the greatest dissatisfaction with meat?" Here tenderness was clearly dominant (72%), again followed by flavor (18%), and juiciness (10%) (Brady 1954). Rhodes *et al.* (1955) also found from consumer surveys that lack of tenderness was the largest single cause of complaints about meat quality, amounting to 55.2% of all dissatisfactions for roasts and 61.9% for steaks.

Given that texture is of great importance to consumers, it is of interest to know what factors are employed in selecting meat cuts at the retail level. Morris *et al.* (1974) found that color, leanness, apparent freshness and, of course, price dictate consumer selection. It seems, then, that texture cannot be accurately discriminated by consumers at the point of purchase.

As a final point on this subject, it will be remembered that during the last decade, many industrial, academic, and governmental scientists predicted major domestic consumption of soy protein in the form of meat analogs. This has not occurred, even considering repeated warnings by nutritionists about cholesterol and saturated fats, and in the face of rapidly escalating meat prices. It would seem that this reflects the failure of the soy products to duplicate adequately, not color or flavor, but meat texture, as well as significant adverse psychological and sociological consumer reactions and an inadequate price differential.

RELATION OF STRUCTURE TO PHYSICAL PROPERTIES AND TEXTURE OF FOOD

Thus far, several terms have been used rather imprecisely—physical properties, structure, and texture. Physical properties of matter are those that depend on the kinetic energy of, and the interactions between, its component molecules. It follows that these properties may be measured; they are usually evaluated by mechanical operations in which force, deformation, and time are measured. In the present sense, structure refers to the often complex organization and interactions of components under the influence of external and internal physical forces. It is when one comes to the final word, texture, that much difficulty is encountered. Although ubiquitously used, it is nearly impossible to arrive at an acceptable meaning for this term. The major reason for this is the uncertainty as to the role of sensory apparatus in texture. Should texture be defined as "that one of the three primary sensory properties of foods which relate entirely to the sense of touch or feel" (Kramer 1973) or "as a manifestation of the rheological properties (rheology can in turn be defined as the study of deformation and flow of materials) of food" (Szczesniak 1966)? Perhaps a definition that incorporates most of the terms to be dealt with is that given by deMan (1976) "texture is the way in which the structural components of a food are arranged in a micro- and macrostructure and the external manifestation of this structure." This takes into account the concept of food as an organized structure of components and its physical and sensory properties. Simply put, food texture is a result of its underlying microstructure and physical properties (Stanley and Tung, 1976); Fig. 6.1 shows how these factors interact.

At this point, it might be thought that the relation of structure to physical properties may be easily determined. It cannot be, for the reasons that foods in general and meat in particular exhibit a complex structure and, while its associated physical properties can be ascertained, the measurements are often of an empirical nature and difficult to interpret. Yet, it is these same physical properties and structure that establish important parameters in such areas as design of processing equipment, packaging requirements, storage conditions and, ultimately, textural quality.

STRUCTURE OF MUSCLE TISSUE

Structure implies not only the size and shape of the components but also how they interact to form an organization. Meat represents an

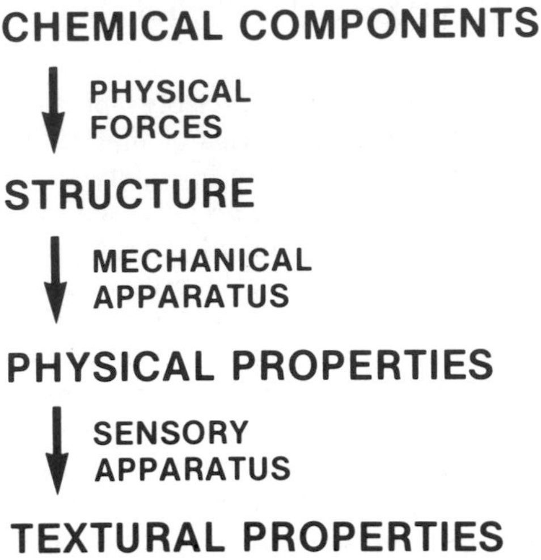

CHEMICAL COMPONENTS

↓ PHYSICAL FORCES

STRUCTURE

↓ MECHANICAL APPARATUS

PHYSICAL PROPERTIES

↓ SENSORY APPARATUS

TEXTURAL PROPERTIES

Fig. 6.1. Dependence of texture upon physical properties and structure. Physical properties are measurable with mechanical apparatus; textural properties require sensory evaluation.

extremely complicated system since it is formed from many individual elements that are arranged in a strikingly elegant but intricate way. Table 6.1 gives a relative structural hierarchy of muscle tissue that emphasizes the wide spread in size of its components. It has been suggested (Raeuber and Nikolaus, 1980) that when dealing with such disparate element sizes, it is useful to break structure down into three levels: macrostructure, observable with the unaided eye; microstructure, cells and groups of cells that may be seen under a light microscope or by scanning electron microscopy; and ultrastructure, cell parts that can only be viewed by the use of a high resolution transmission electron microscope.

In most cases, the organization of muscle structure that is important to food scientists is only visible through some type of enlarging instrument and through the study of magnified images. Previously, instruments designed to provide structural information were used mainly to obtain qualitative data about the morphology of sample features. With the advent of instrumentation described later in this section, it is becoming possible to extract quantitative information about the size, shape, and composition of images and have these results processed, interpreted, and stored by computer. The following sections attempt to sketch the important features of muscle structure as revealed by various instruments and, while not comprehensively, also demonstrate the wide range of techniques available to those studying this subject.

TABLE 6.1. STRUCTURAL HIERARCHY OF MUSCLE TISSUE

Component	Level	Instrument	Approximate size	Relative order of magnitude of diameter
Beef carcass	Macro-structure	Human eye	2 m × 1 m diam	10^9
L. dorsi muscle			1 m × 10 cm diam	10^7
Muscle fiber (cell)	Micro-structure	TLM, SEM	1–40 mm × 10–100 μm diam	10^3
Myofiber		TEM	1–40 mm × 1–2 μm diam	10^2
Myosin filament	Ultra-structure		1.5 μm × 15 nm diam	1

Transmission Light Microscopy (TLM)

Although muscle tissue is composed of many elements including connective tissue, adipose cells, blood vessels, and nerves, food scientists have shown more interest in muscle cells than in any other constituent. Examination of these muscle fibers, so-called because of their thread-like appearance, by TLM (Fig. 6.2) reveals they are multinucleated and that these organelles are located in the peripheral cytoplasm. Each fiber is enclosed by a thin, at this magnification structureless, membrane called the sarcolemma, which is part of the connective tissue surrounding the fiber. The most striking feature of muscle fibers is their cross-striations, from whence the term "striated fibers" originates. Under polarized light the bands that seem darker with ordinary illumination are anisotropic or birefringent, while those that appear light are isotropic. Thus, the dark areas are called A bands and the light ones I bands. Although these bands seem continuous across the fiber, this is only because the underlying elements, myofibrils, are usually in register with one another. For further information on the appearance

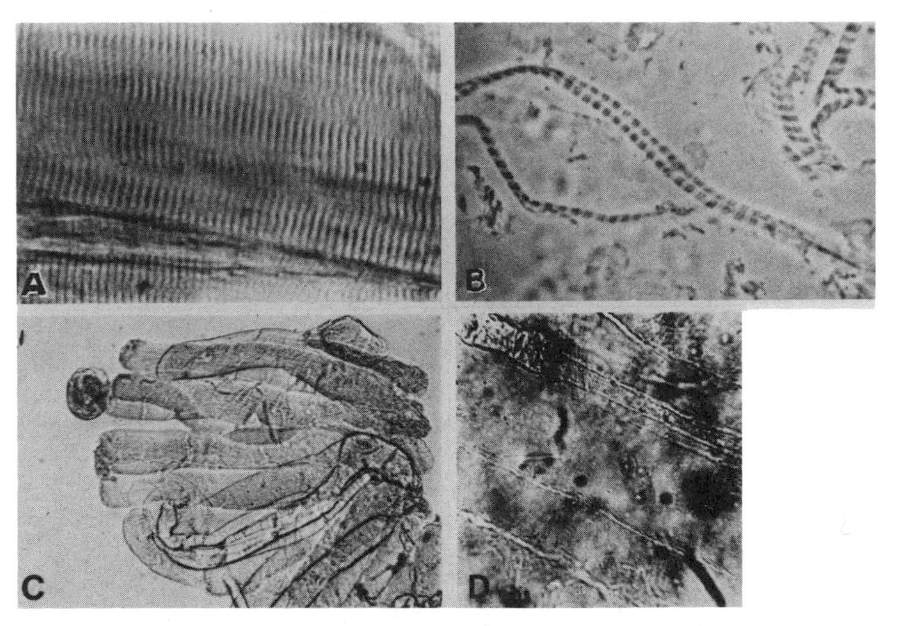

Fig. 6.2. Light microscopy of muscle tissue. (A) Rabbit psoas muscle (phase contrast), several nuclei are visible. (*Courtesy of H.D. Geissinger*). (B) Rabbit psoas myofibrils (polarized light). (C) Trout muscle minced and "teased" onto slide (normal illumination). (D) Sarcolemmae prepared from tissue in (C) by method cited in Stanley and Hultin (1968). Some segments not totally emptied (normal illumination).

of muscle tissue in the TLM, one of the many histology texts available may be consulted. It should be pointed out that augmentative devices such as fiber optics and fluorescence detectors greatly expand the usefulness of TLM.

Scanning Electron Microscopy (SEM)

The advent of the scanning electron microscope has given food scientists a powerful tool for investigating surface microstructure. The major advantages of the SEM that are responsible for its widespread use include depth of focus and convenience; also, sample preparation is much less complex and tedious as compared to that for the earlier transmission electron microscope (TEM), since the material does not have to be thin-sectioned. The basis of SEM operation involves scanning a high energy beam of primary electrons across the surface of a bulk sample, which excites the release of secondary electrons. These are captured and electronically formed into an image of surface topography which can be displayed via a cathode ray tube in raster form. Several reviews of SEM as applied to food materials provide a more in-depth coverage of the operating principles and use of this instrument (Pomeranz 1976; Carroll and Jones 1979; Chabot 1981; Lee and Rha 1979; Lewis 1981). (See also Chapter 2 of this book.) Voyle (1979, 1981A) has reviewed the use of SEM in meat science and the structural features of muscle tissue. Methodology for the preparation of muscle tissue for SEM may be obtained from the work of Jones *et al.* (1976) and Geissinger and Stanley (1981).

When muscle fibers are observed under SEM (Fig. 6.3), several surface features not obvious with the TLM become apparent. Connective tissue including perimysium and endomysium (described in a subsequent section) are visible depending on how much of this material has been exposed during sample preparation. The fiber surface shows transverse banded contours superimposed on the subjacent myofibrils. Part of the sarcolemma may also be removed to reveal these smaller elements. At this point, the repeating nature of muscle structure, the sarcomere, becomes evident as well as parts of the intrafiber membrane system and mitochondria. Using the SEM, sarcomeres appear to be delineated by prominent bands identified as either part of the Z-discs or overlying membrane; there is some question about this, however, see Vriend and Geissinger (1982).

Transmission Electron Microscopy (TEM)

The basis of the transmission electron microscope is to reveal internal ultrastructure by transmitting an electron beam through thin

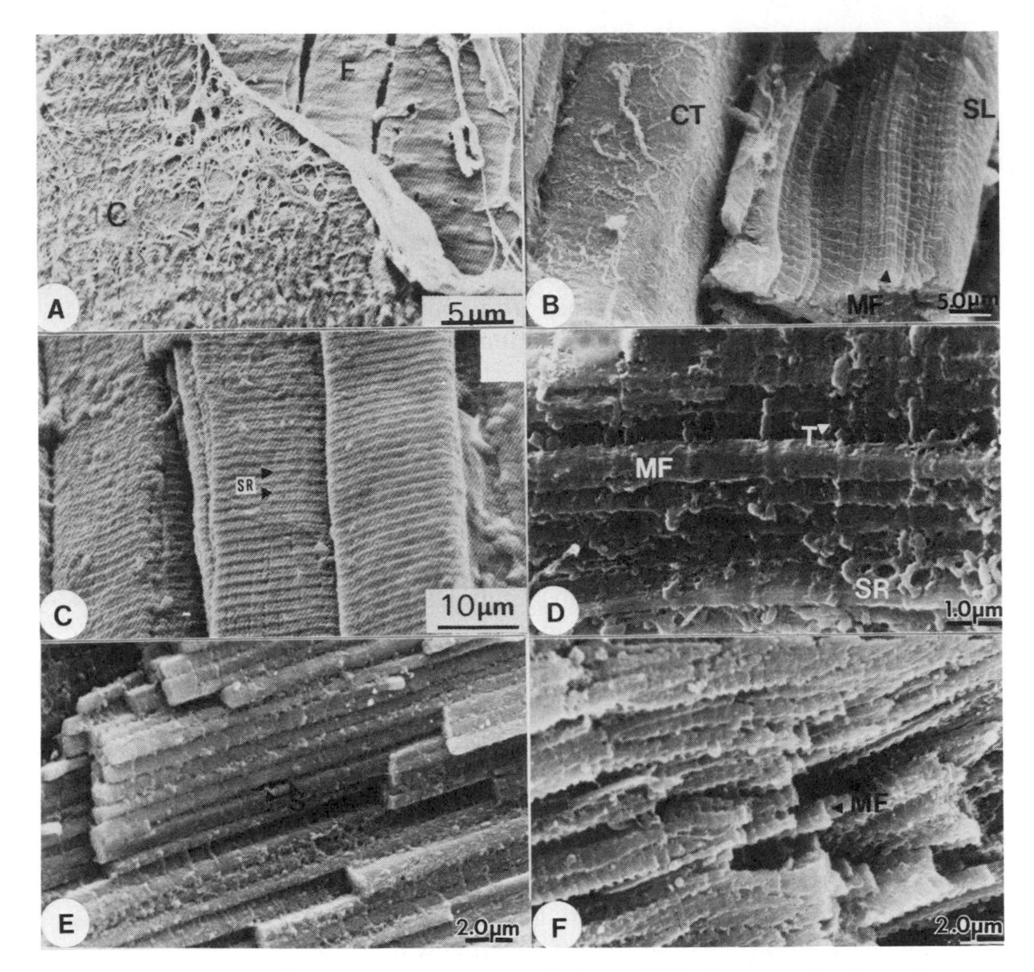

Fig. 6.3. Scanning electron microscopy of muscle tissue. (A) Bovine psoas showing perimysium (C) and muscle fibers (F). (B) Porcine gracilis showing endomysium (CT), sarcolemma (SL), and myofibrils (MF). (C) Bovine psoas showing surface structure of fibers and supposed transverse elements of sarcoplasmic reticulum (SR). (D) Freeze-fracture of mouse gastrocnemius showing myofibrils (MF), T tubules (T), and sarcoplasmic reticulum (SR). (E) Porcine semitendinosus restrained prior to fixation showing sarcomeres at rest length (S); (F) Porcine semitendinosus unrestrained prior to fixation showing extremely contracted myofibrils (MF). (A, C) *from Voyle (1981A)*; (B, D, E, F) *from Geissinger and Stanley (1981), used with permission.*

sections of the specimen. Sectioning muscle tissue longitudinally exposes the architecture of the myofibril and its recurring unit, the sarcomere. This remarkable structure with its associated complex biochemical reactions is the basis for muscle contraction (see Bendall 1969; Huxley 1972; and Harrington 1979 for reviews of this subject). Figure 6.4 shows the classic features of the sarcomere including Z-discs, which limits the contractile unit, and A and I bands. These latter two structures interdigitate during contraction, drawing the Z-discs together and resulting in shortening. Electron microscopy combined with biochemical studies have established that the thin filaments comprising the I band consist mainly of the protein actin with two other proteins, tropomyosin and troponin, being present as well. The A band, consisting of thick filaments containing the protein myosin, remains constant in length during contraction (approximately 1.6 μm in beef) while the thin filaments are drawn along the length of this rod-shaped protein through the action of crossbridge formation. A diagram showing the relation of these structural elements is given in Fig. 6.5.

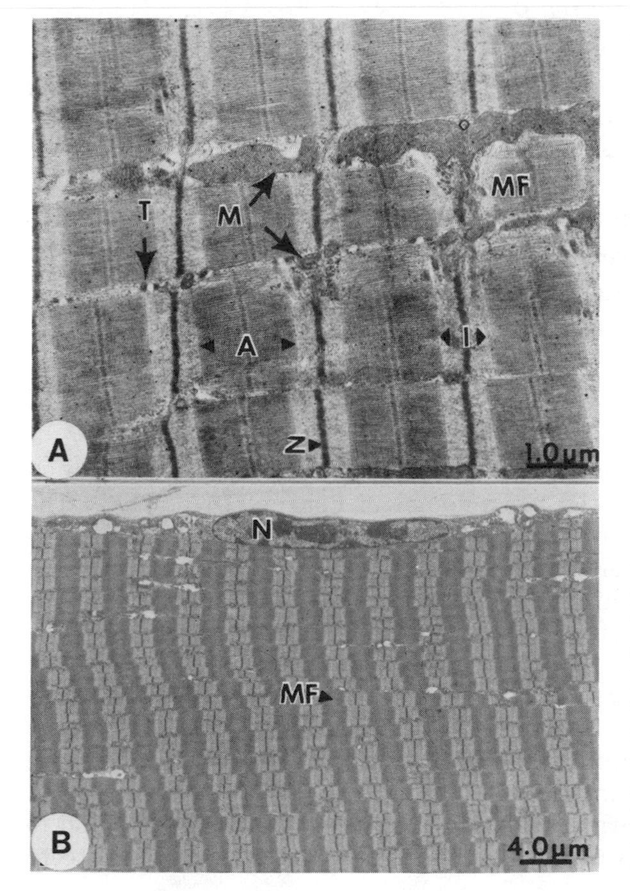

Fig. 6.4. Transmission electron microscopy of muscle tissue. (A) Porcine gracilis showing myofibrils (MF), mitochondria (M), Z-discs (Z), A and I bands (A) (I), and sarcoplasmic triads at A-I level (T). (B) Porcine gracilis showing myofibrils (MF) and nucleus (N). *From Geissinger and Stanley (1981), used with permission.*

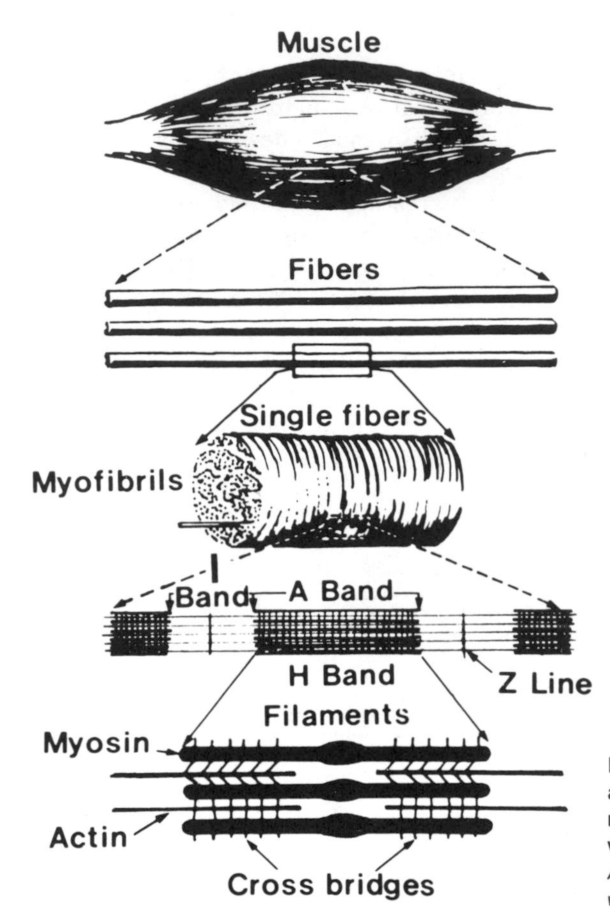

Fig. 6.5. Diagram of the arrangement of protein filaments in a myofibril relative to whole muscle.
Adapted from Huxley (1972), used with permission.

This short description of muscle structure may be profitably augmented by consulting the two volumes of Bourne (1972, 1973). The structural elements of muscle delineated thus far can be modeled in the form of a series of parallel threads of a repeating structure encased in a tube, many of which are interconnected by a fibrous meshwork. This concept will prove useful when physical properties are discussed.

Other Techniques

The previous descriptions have been based on traditional qualitative methods, which are compared in Table 6.2. Scientists are attempting to extract the most information possible from structural studies and to quantitate these data using advanced electronics and computer tech-

TABLE 6.2. COMPARISON OF MICROSCOPES

Criterion	TLM	SEM	TEM
Resolution	200 nm	4 nm	0.2 nm
Depth of field	Poor	High	Moderate
Modes	Many	Many	Few
Specimen			
Preparation	Usually easy	Easy	Difficult
Relative thickness	Thick	Medium	Very thin
Environment	Versatile	Vacuum	Vacuum
Available space	Small	Large	Small
Relative size of field of view	Large	Large	Limited
Signal	Only as image	Available for processing	Only as image
Cost	Low	High	High

Source: Adapted from Hearle *et al.* (1972), used with permission.

nology. The next sections outline some of this newer instrumentation for examining muscle structure.

Scanning Transmission Electron Microscopy. This instrument (STEM), as the name implies, combines characteristics of both SEM and TEM. As in the TEM, electrons are transmitted through a thin-sectioned specimen. However, this instrument also incorporates the concept of scanning the beam of electrons across the section. This equipment provides high resolution (2 nm for a conventional tungsten filament, 0.5 nm with a field emission gun), improved contrast and can show details not visible in either of the earlier electron microscopes. The signal can be stored and processed electronically, thus facilitating image analysis; the capability also exists for the simultaneous display of several detected signals such as backscattered electrons and X-rays. Fig. 6.6 presents the results of Somlyo *et al.* (1979) who demonstrated the close correlation between the periodicity of isolated striated muscle tropomyosin paracrystals stained with mercuric dye as viewed by X-ray mapping and STEM.

Laser Diffraction. Several workers have demonstrated a correlation between sarcomere length of muscle tissue and meat tenderness. Measurement of this characteristic has been most often made by oil immersion or phase contrast in conjunction with an eyepiece reticular, but mechanical disruption of the fibers, in order to obtain myofibrils, tends to produce artifactual shortening. However, more recently a method has been developed to quantitate sarcomere length based on the finding that striated muscle acts as a transmission diffraction grating when impinged upon by a beam of light. The spacing of the diffraction pattern is dictated by the state of contraction. A convenient and simple procedure to measure sarcomere length using a laser as a source of coherent monochromatic light has been described by Voyle (1971). Ruddick and

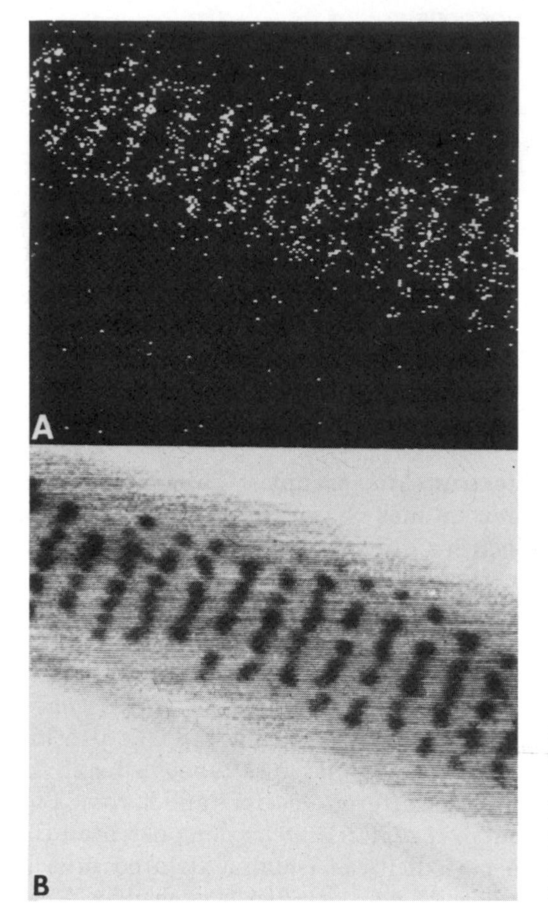

Fig. 6.6. STEM and X-ray mapping analysis of tropomyosin paracrystal with Hg labeled thiol residues. (A) X-Ray map. (B) STEM image. Each repeat = 40 nm.
From Somlyo et al. (1979), used with permission.

Richards (1975), among others, have reported a high correlation between laser diffraction and TLM methods.

X-Ray Microanalysis. When matter is struck by high energy electrons from an external source, X-rays with a characteristic spectrum of the sample are liberated. These X-rays can be used to identify and quantify elements present in the material. Analytical systems are available to resolve the emitted X-ray spectrum through the identification and quantification of specific X-ray wavelengths. Those elements above sodium in the periodic table are resoluble. This type of analysis may be incorporated with SEM or TEM equipment to allow the examination of biological material; it is now practical to detect 10^{-19} g of an element within an ultrathin section with a spatial resolution of

20–30 nm. Specimen preparation, however, still remain a problem due to the dual requirements of tissue integrity and element immobilization [soluble electrolytes (e.g., Na^+, K^+, Cl^-) are unlikely to remain *in situ* when tissues are immersed in solvents]. Cyropreparation, in which tissue is frozen, sectioned, and freeze-dried, presently gives the best results for biological material. Further information on the application of X-ray microanalysis is available (Hall 1979; Chandler 1977, 1979).

As an example of how this technique is useful in the determination of muscle tissue structure, two studies will be cited. Sjostrom and Thornell (1975) examined ultrathin sections of frozen, freeze-dried skeletal muscle in a TEM equipped with an X-ray microanalysis detector and observed the occurrence of a Ca^{2+} peak when the 0.5 μm beam spot was focused on the Z-disc area. It was speculated that this signal may have resulted from the proximity of membraneous material capable of accumulating Ca^{2+}, sarcoplasmic reticulum (the membrane systems associated with muscle tissue are discussed in a subsequent portion of this chapter). Further work by Somlyo *et al.* (1977) used similar procedures but the electron probe measured only 50 nm. Their results (Fig. 6.7) allow a comparison of the ionic composition of the terminal cisternae of

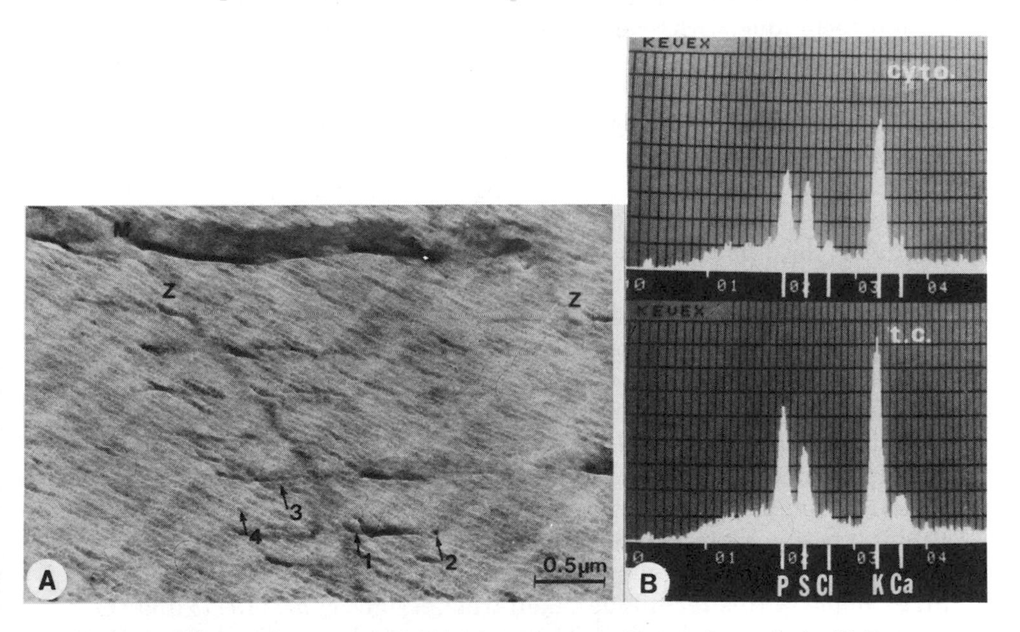

Fig. 6.7. Differential elemental distribution in muscle by X-ray microanalysis. (A) Frozen-dried section of frog toe muscle. Z-discs (Z); areas for X-ray analysis: 1,4, cytoplasm, 2,3-terminal cisternae. (B) X-ray spectra of cytoplasm and terminal cisternae.
From Somlyo et al. (1977), used with permission.

the sarcoplasmic reticulum to undifferentiated cytoplasm. Table 6.3 gives these data from which it may be seen that the Ca^{2+} level is indeed higher in the target area than the surrounding tissue; phosphorus levels are also elevated, which may be a result of membrane phospholipids. Studies of this type can go far in explaining the fine structure of muscle tissue since quantitative elemental analysis is obtained as well as structural data.

TABLE 6.3. ELEMENTAL CONCENTRATIONS IN TERMINAL CISTERNAE AND CYTOPLASM OF NORMAL FROG TOE MUSCLE

Element	Cytoplasm[1]	Terminal cisternae[1]
Na	29 ± 12.6	44 ± 14.1
Mg	39 ± 5.2	40 ± 5.9
P	317 ± 12.3	449 ± 16.0
S	216 ± 8.8	208 ± 9.1
Cl	43 ± 3.8	54 ± 4.5
K	488 ± 15.5	587 − 18.7
Ca	1 ± 2.3	66 ± 4.6

Source: Somlyo *et al.* (1977), used with permission.
[1]mmole/kg dry weight ± standard deviation; $n = 30$.

Differential Scanning Calorimetry (DSC). Although not a magnification technique, DSC can be a useful secondary method to obtain information on structure of muscle tissue. DSC is used to procure calorimetric data under dynamic conditions since the sample is heated at a precise rate (see Chapter 4). The extent of heat flow is compared between the sample and an inert reference heated at the same speed. Events in the sample such as phase transitions, chemical reactions, or any process that involves absorption or evolution of heat causes a change in the differential heat flow, which is recorded as a peak in a plot of differential heat flow versus temperature. The area under the curve gives the heat change in energy units while the direction of the peak indicates whether the transition is exothermic (energy is released) or endothermic (energy is absorbed). Small samples of biological material (1−15 mg) are used and this means high amplification of the resulting signal is required, especially for proteins.

This technique has been used by Quinn *et al.* (1980) to investigate meat proteins as influenced by processing and by Wright *et al.* (1977) on rabbit muscle. DSC is useful for monitoring changes in structural stability of muscle proteins as they are exposed to heat. Figure 6.8 gives an example of this technique taken from Wright *et al.* (1977) and the results that may be obtained. The curve for whole muscle shows endothermic peaks at 60°, 67°, and 80°C that, through the use of purified proteins, were found to correspond to thermal denaturation of myosin,

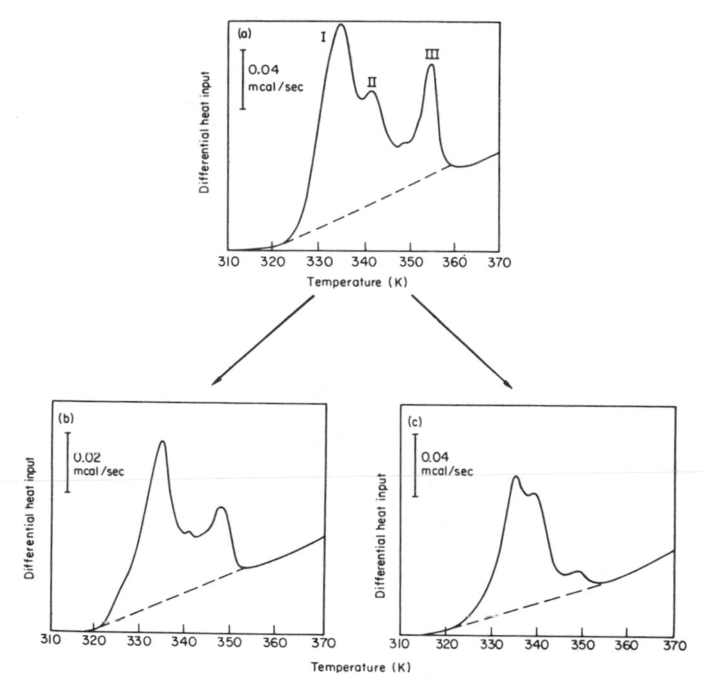

Fig. 6.8. DSC thermograms. (a) Whole rabbit muscle. (b) Myofibrils. (c) Sarcoplasmic proteins.
From Wright et al. (1971), used with permission.

sarcoplasmic proteins, and actin, respectively. It is possible that the second peak also includes the contribution of the thermal transition of collagen. Data of this type are useful when examining the influence of temperature on muscle structure.

Quantification of microscopic data

The preceding discussion demonstrates that extremely sophisticated instrumentation exists for investigating the structure of muscle tissue. Magnification may be used to reveal the structural organization of muscle, but microscopy invariably involves human interpretation of images, and this factor combined with systematic artifacts resulting from sample preparation makes structural investigations fraught with pitfalls. One way to overcome this problem is to remember that the various microscopes are augmentative, not competitive. Synergistic results are often obtained when more than one approach is used and correlative microscopy is a worthwhile approach. Thus, Geissinger *et al.* (1978) have developed methodology to prepare muscle for correlative TLM, SEM, and TEM examination. Figure 6.9 shows correlated SEM and TEM images of skeletal muscle; the advantages of such an approach are obvious.

Almost all published work on muscle structure relies on description of "representative" micrographs to convey results. More quantitative approaches to this problem now exist, however. These range from count or measurement data taken from micrographs to estimates of cell and

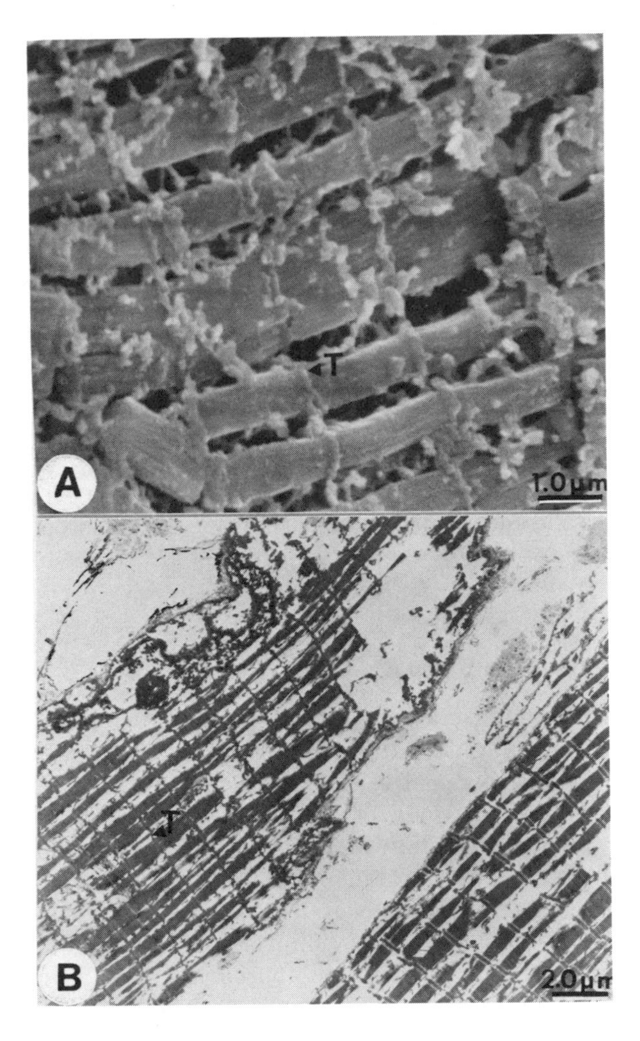

Fig. 6.9. Correlative microscopy of mouse muscle. (A) SEM image, T tubules (T) are visible. (B) Correlated TEM image, the same T tubule (T) is seen.
From Geissinger et al. (1978), used with permission.

organelle dimensions by stereological techniques to autoradiography to immunocytochemistry. Williams (1977A, B) provides a description of how these quantitative methods can be used to place descriptions of cellular structure on a numerical basis.

Another method that is developing for removing subjectivity in structural studies is image analysis, the technique of extracting quantitative geometric and densitometric information from images. Image analyzers can be categorized as either assisted or unassisted by the human eye, designed for use on microscopes or for the analysis of micrographs and as having pattern recognition capability or not. Basically, the apparatus involves some form of feature identification coupled with microprocessor capability allowing the maximization of information obtained while minimizing the time and complexity required for analysis. Some applications of image analysis to biological systems include measurement of organelle content and distribution within cells or of molecules within organelles, particle size distribution, diagnosis of disease and immunological testing. An overview of this field may be obtained in Braggins *et al.* (1971) and Attle *et al.* (1980). This procedure could prove extremely valuable in the analysis of muscle structure.

STRUCTURAL COMPONENTS OF MUSCLE TISSUE RESPONSIBLE FOR PHYSICAL PROPERTIES

Muscle tissue contains numerous structural components, all necessary to fulfill the function of motion. Not all of these, however, are significant contributors to the physical properties of meat. Many researchers now believe that the major structural factors affecting meat texture are associated with connective tissue and myofibrillar proteins (see, e.g., Harris 1976). Accordingly, these structures merit particular interest. Two other components, muscle membranes and water, also deserve attention—not because of their inherent physical properties, but rather as a result of the indirect influence they have on the expression of physical properties. It should be noted that sarcoplasmic proteins may be important for the same reason, although little information on their role in structure is available.

Connective Tissue

Connective tissue is a ubiquitous feature of the animal body. With regards to muscle, this material is present in several forms and serves to provide macro- and microstructural foundations as well as a framework for binding of skin and bone. Many different components are included as connective tissue, but of all these collagen is the most important in terms of muscle physical properties. Another connective tissue protein, elastin, possesses great material strength and is unaffected by temperatures used in cooking. Fortunately, this polymer is found in only very small amounts in muscle, mainly as a component of blood vessels.

Collagen, the most abundant protein in land animals, is a rod-shaped molecule consisting of three subunits that interact to form a compact triple helix. Collagen fibrils are formed by the association of a number of these helices in such a way that leads to a characteristic banding with a 68 nm repeat pattern (Fig. 6.10); bundles of fibrils form a collagen fiber. Also shown in this figure are the three levels of organization in muscle tissue at which these fibers are found: the

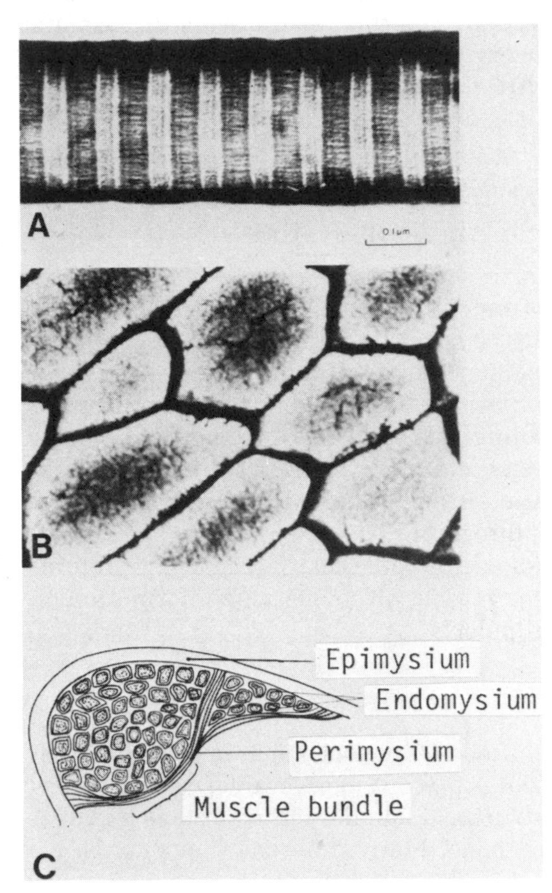

Fig. 6.10. Muscle connective tissue. (A) Transmission electron micrograph of collagen fibril showing 68 nm periodicity. (B) Cross section of porcine muscle stained with silver to demonstrate reticular fibers of the endomysium. (C) Diagram of muscle cross section showing major connective tissue components.

(A) from Bailey (1972); (B) from Swatland (1975); (C) Adapted from Ham (1969), used with permission.

epimysium surrounds the entire muscle and is a continuation of tendon, the perimysium (see also Fig. 6.3) separates bundles of muscle fibers, and the endomysium which insheathes muscle fibers and is part of the sarcolemma. The endomysium forms a continuous intrafascicular meshwork of small diameter collagen fibers, often termed reticular fibers, that are highly branched and stain black with silver. Although individual muscle fibers sometimes appear to be bounded by their own endomysial sheath, these reticular fibers cannot be differentiated from those of adjacent muscle cells.

The nature of the collagen molecule, being a tight helix with a large axial ratio, would suggest it has a significant role in determining physical properties of muscle tissue. In fact, collagen molecules also possess the ability to form heat stable intermolecular crosslinks which is accompanied by a significant change in their physical properties. This is an age-dependent phenomenon. Older animals exhibit increased crosslinks and, hence, toughness, but no increase in total connective tissue (Shimokomaki *et al.* 1972). Thus, measuring total collagen content, usually done by assaying hydroxyproline, the unique constituent amino acid, does not provide an explanation for differences in toughness in animals of different ages since the chemical procedure used does not reflect the degree of crosslinking. Connective tissue content measured in this way may be, however, a useful predictor of meat tenderness when comparing different muscles from animals of the same age (see, e.g., Dutson *et al.* 1976). Recently, Sims and Bailey (1981) have aptly reviewed structural aspects of connective tissue and how they relate to meat texture. Bornstein and Traub (1979) may be consulted for a discussion of the chemistry and biology of collagen.

Although compositional analysis shows that connective tissue is a minor constituent of muscle tissue (approximately 0.5% collagen and elastin in steer L. dorsi, Lawrie 1966), its great strength makes it an important factor when considering physical properties. For example, Abrahams (1967) reported a tensile strength of about 500 kg/cm^2 for tendon (almost pure collagen), and Sims and Bailey (1981) cite studies showing that bovine intramuscular collagen is capable of exhibiting greater tension than tendon. In contrast, the data of Bouton and Harris (1972b) indicate muscle fiber tensile strengths of around 2–3 kg/cm^2 but lower values (approximately 0.5 kg/cm^2) for fiber adhesion in which force is applied transversely to the fiber axis.

In addition to this inherent physical strength, consideration must also be given to the orientation of collagen fibrils. State of contraction influences the arrangement of perimysial connective tissue by increasing the angles of its criss-cross lattice and also its crimp length (Rowe

1974). It was shown by Stanley and Swatland (1976) that if muscle is stretched prerigor, the endomysial collagen fibers are pulled parallel to the longitudinal axis of the muscle fiber, but when unrestrained muscle is allowed to contract, the connective tissue arranges itself so that most of the fibers are perpendicular to the long axis (Fig. 6.11); unrestrained muscle tissue required almost twice the elongation prior to rupture as did restrained tissue.

Connective tissue would, then, appear to influence meat texture in several ways: amount (differences between muscle from animals of the same age), degree of crosslinking (differences between muscles from animals of different ages), and spatial orientation (differences between muscles varying in state of contraction). To these must also be added heat treatment, since collagen is prone to thermal denaturation, which affects its physical properties. This reaction is complicated by an age effect—collagen from younger animals reacts somewhat differently to increasing temperature than does similar tissue from older animals. Postmortem aging may have an influence on the contribution of connec-

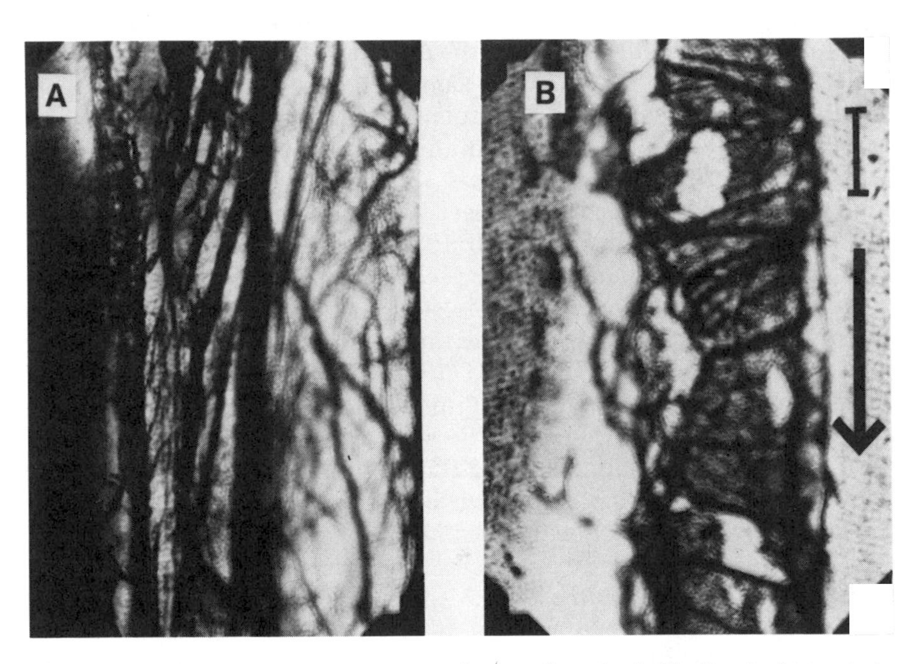

Fig. 6.11. Light microscopy of porcine sartorius muscle stained with silver to demonstrate reticular fibers of endomysium. (A) Restrained during rigor at approximately 150% rest length. (B) Unrestrained during rigor. Bar = 20 μm, arrow indicates orientation of fiber axis. *From Stanley and Swatland (1976), used with permission.*

tive tissue to meat texture as well, since some evidence exists for enzymatic weakening of this material. These points will be taken up in subsequent sections.

Myofibrillar Proteins

The muscle cell has a protein complement of over 200 distinguishable molecular species (Goll *et al.* 1974, 1977). Thankfully, only a few appear to be directly involved in dictating the physical properties of meat. Although knowledge in this area is far from complete, the following section examines what seems to be those components of major importance.

Contractile Proteins. As with any complex biological tissue, the muscle myofibril has been found to be composed of a number of distinct elements. The two major protein components, actin and myosin (which together account for about 70% of the weight of the myofibril), are those involved in the physical act of contraction while seven or so associated proteins are necessary for the regulation of this process. As a consequence of postmortem depletion of adenosine triphosphate (ATP), actin (located in the I band and the major constituent of the thin filament) combines through crossbridges with myosin (located in the A band and the major constituent of the thick filament) to form actomyosin, a protein entity. The degree of overlap of actin and myosin determines sarcomere length, and this measurement is often used as an index of myofibrillar contribution to toughness. It is predictable that, other factors being constant, the more a postrigor sarcomere is contracted, i.e., the greater the number of actin–myosin crossbridges, the tougher the muscle becomes as a result of inextensibility, rigidity, and filament packing density.

Many workers have reported experimental results that show a strong correlation between sarcomere length and physical properties. For example, Stanley and Geissinger (1972) examined muscle structure by SEM and found shorter sarcomeres reflected in increased toughness. This relationship, however, is not a simple one. Muscle that is severely contracted, for example by cold shortening, exhibits increased tenderness at very low sarcomere lengths (< 1.3 μm), shorter than the measured length of the A band. This is thought to be due to a structural disruption of the myofibril with thick filaments penetrating into Z-disc areas (Marsh *et al.* 1974). There is structural evidence for this theory (Fig. 6.12). Differences in tension applied to postmortem muscle can influence sarcomere shortening and, using this technique, it has been

Fig. 6.12. Structure of severely contracted beef sternomandibularis muscle induced by cold-shortening. (A) SEM of passively shortened fibers showing typical wavy structure. (B) TEM of myofibrils in fiber exhibiting supercontraction, arrows indicate points where myosin filaments have penetrated Z discs.
From Voyle (1969), used with permission.

shown that shear force values increase exponentially as sarcomere length decreases from around 3.0 to 1.3 µm (e.g., Herring *et al.* 1967; Bouton *et al.* 1973A); toughening becomes increasingly apparent at sarcomere lengths less than 2.0 µm.

It has not yet been possible to differentiate the effects of connective tissue and contractile proteins experimentally. For example, Szczesniak and Torgeson (1965) provide a wide range of correlations taken from various published reports for connective tissue and meat tenderness. This is not surprising since structural studies, mentioned previously, have shown that these two factors are not independent; contraction affects both sarcomere length and connective tissue configuration. Also, muscles low in connective tissue would be expected to be more prone to changes in contractile proteins than the reverse case. More work will be needed to resolve these factors; one approach to this problem is mentioned subsequently.

Connectin. It has proved difficult to determine the structural elements responsible for the elastic nature of muscle. After numerous experiments designed to measure this property, many workers concluded that it cannot be fully explained by the presence of only actin and myosin filaments. Ultrastructural evidence for an elastic protein component, the so-called third filament, has been offered (McNeill and Hoyle 1967; Locker and Leet 1975; dos Remedios and Gilmour 1978; Takahashi and Saito 1979) and may be summarized as the appearance of a network of thin (around 2 nm), insoluble, and extensible protein fibrils. These fine filaments are seen when muscle is stretched to point where a gap develops between the A and I bands and, hence, were termed "gap filaments." Using chemical methods, Maruyama *et al.* (1976) isolated an elastic protein referred to as connectin from myofibrils that was distinguishable from elastin and collagen and that comprised about 5% of the total myofibrils. Further reports cited the presence of this protein in several different species located as a network between the Z-discs of the sarcomere and described a role for this element in muscle elasticity, tensile strength and postmortem tenderness of meat (Maruyama *et al.* 1977; Takahashi and Saito, 1979). Although there is not total agreement as to the exact location of the connectin (sometimes called titan) containing gap filaments, it has been suggested (Locker and Leet, 1976; Locker *et al.* 1977) that each gap filament forms a core to a thick filament, emerges at one end of the A band, passes between the I band thin filaments, through the Z-disc, continues through the thin filaments of the next sarcomere and into a second thick filament where it ends. Further work will be needed to illustrate the importance of connectin in meat relative to other structural elements. Very recently, several investigators have introduced evidence for components of a muscle cytoskeleton that theoretically are capable of influencing the physical properties of meat (IFT 1982). Proteins such as desmin as well as connectin, may prove of importance as researchers elucidate their structural roles.

Muscle Membranes

The lipoprotein nature of cell and organelle membranes precludes a primary function for this material in the physical properties of muscle, however, these membranes, as a result of the biochemical reactions they control, have an important secondary role in this area. The general structure and function of the membranous systems in muscle fibers is now known (Porter and Franzini-Armstrong, 1965; Franzini-Armstrong, 1973). Briefly, the entire surface of the muscle fiber is covered by a unit membrane called the plasmalemma. This, the distal basement membrane, and the attached endomysial collagen fibrils are collectively termed the sarcolemma. A system of transverse tubules arise at the outer edge of the muscle fiber as invaginations of the plasma membrane. This network, named the T system, enters the fiber at the level of the A–I band junction in vertebrate muscle and penetrates primarily transversely, surrounding the myofibrils. Another membrane system, the sarcoplasmic reticulum (SR) is found in the sarcoplasm that separates myofibrils. The SR is segmented and these repeating units are in register with the sarcomere. The three major portions of the SR are the centrally located fenestrated collar, longitudinal tubules running from either direction of this structure, and the terminal lateral sacs or cisternae. The lateral sacs of the SR abut the T system. The complex formed by two SR sacs and one T system element is termed the triad. The structure and interaction of these membranes are shown in Fig. 6.13.

The importance of these membranes to muscle function was realized when it became known that nerve impulses to the cell membrane produce a depolarization of the connecting T system and allow the impulse to be conducted to the interior of the fiber. At the triad, current excites the lateral sacs of the SR, which then release preferentially accumulated calcium ions that in turn remove an inhibitory effect in the

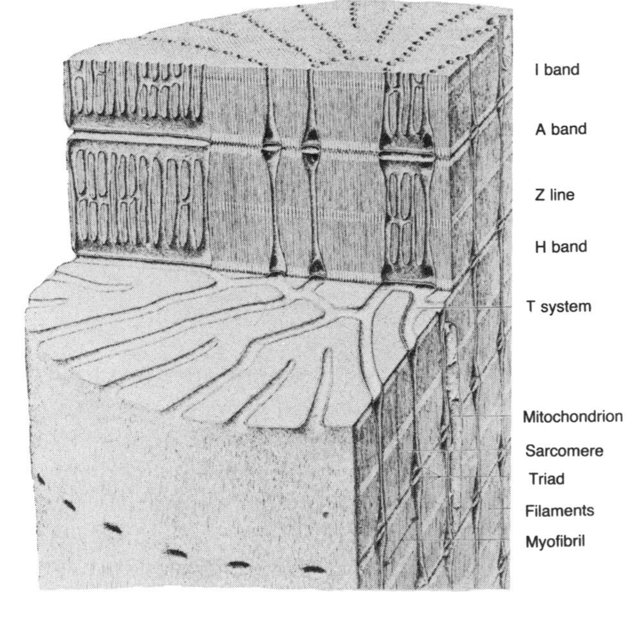

Fig. 6.13. Diagram of membrane systems in fish muscle. Note T-system invaginations and triads at Z-disc level; in mammalian fibers this occurs at the A–I junction.
From Porter and Franzini-Armstrong (1965), used with permission.

contractile proteins and allow all the myofibrils in a fiber to contract simultaneously. It is now evident that the regulatory protein in striated muscle responsible for mediation of calcium, troponin c, is, in fact, a specialized form of calmodulin, a ubiquitous intermediary protein that acts to regulate various enzymes (Cheung 1982). These mechanisms are important to a discussion of physical properties since it is accepted (Goll *et al.* 1972) that in some species postmortem muscle SR may quickly lose its ability to sequester calcium ions and the freed cation then becomes a cofactor in the enzymatic breakdown of Z-discs, as will be described subsequently.

Water

By far the major constituent of muscle is water; it accounts for approximately 75% by weight of fresh tissue. Of more importance than the total amount of water present is the water holding capacity (WHC) of the tissue or the ability to retain its own or added water during application of force. The water bound into muscle structure is primarily a result of its association with myofibrillar proteins. Protein–water interactions significantly affect the physical properties of meat. Changes in WHC are closely related to pH and to variations in muscle proteins. Swelling of muscle protein matrix is thought to improve WHC since more water is trapped within the matrix. The isoelectric point (IEP) of a protein is defined as that pH at which the net charge is zero. Since protein–protein ionic interactions are promoted at this point, it would be expected that the protein matrix would shrink and WHC would be at a minimum. This is shown to be the case in Fig. 6.14, taken from the comprehensive review of the subject by Hamm (1960). It follows that increasing the pH away from the IEP would also result in a higher WHC since protein–water interactions are favored. Bouton *et al.* (1971) were able to increase the ultimate pH and WHC of meat by pre-slaughter injections of epinephrine and found tenderness to parallel increasing pH values (Fig. 6.15). Further work by these authors (Bouton *et al.* 1972, 1973b) showed that as pH (and WHC) increased from normal values of 5.5 toward 7.0, tenderness of the tissue increased as well and became independent of fiber contraction state.

Recent work by Honikel *et al.* (1981) led these authors to conclude that postmortem metabolic processes (ATP hydrolysis and glycolysis) resulting in decreased pH levels has more influence on WHC through alterations in protein structure than does degree of contraction. Currie and Wolfe (1980) measured tensile and adhesive properties of beef muscle strips at various times postmortem. Changes in these physical properties correlated well with determinations of extracellular space, which in

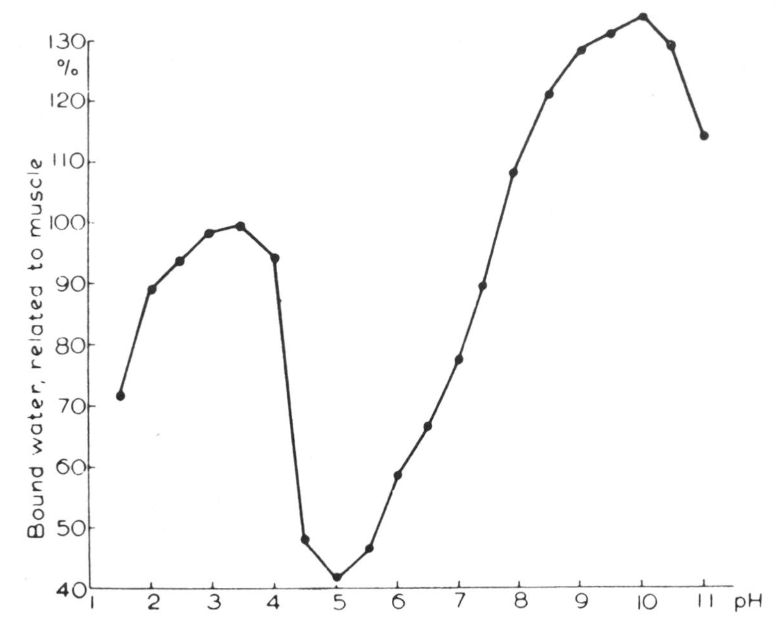

Fig. 6.14. Influence of pH on WHC of beef muscle homogenate.
From Hamm (1960), used with permission.

turn, was thought to be related to the degree of immobilized water in the muscle fiber since low extracellular space is indicative of high intrafiber water and vice versa. These authors postulate that intrafiber water, with its ability to reduce adhesive forces between the myofibrils, is an important third factor—in addition to myofibrillar contraction state and connective tissue—in determining physical properties of muscle tissue.

MEASUREMENT OF PHYSICAL PROPERTIES OF MUSCLE TISSUE

Interaction of the structural components of muscle tissue previously mentioned leads to an organization of elements possessing characteristic physical properties. In order to evaluate the influence of various parameters on these properties and on texture, the quality factor they dictate, it becomes important to quantify them. This may be done by instrumental analysis and/or sensory techniques. Although sensory evaluation (human response is implied) is the ultimate method for

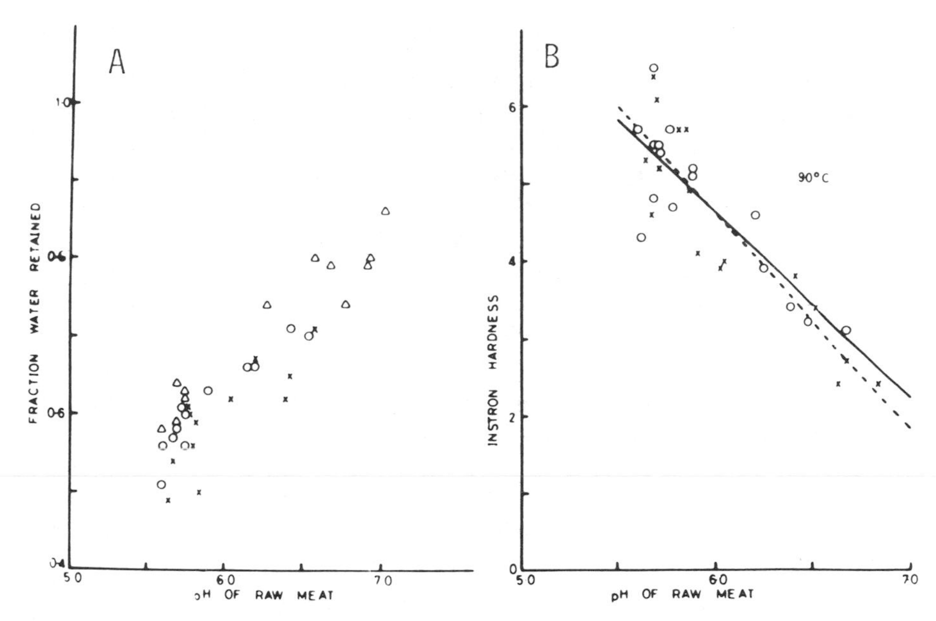

Fig. 6.15. Relation of pH and WHC to physical properties of ovine muscle. (A) WHC as a function of ultimate pH. (B) Instron hardness (penetration force) of cooked muscle as a function of ultimate pH. △, Semitendinosus; ○, semimembranosus; ×, biceps femoris.
From Bouton et al. (1971), used with permission.

measuring meat texture, it will be excluded from the present discussion, because physical properties can be assessed more objectively and reproducibly through the use of instruments. However, the important area of sensory evaluation of meat texture has been reviewed recently by Larmond (1976).

Mechanical properties of food are usually measured by applying a force as either tension, compression, or shear and recording the resulting deformation or vice versa. Muscle tissue possesses a basically fibrous structure and classical rheology has established that the most productive methods for assessing mechanical properties of fibrous materials are (1) shearing, in which the applied force acts at right angles to the fiber axis and (2) tensile testing, wherein the applied force is parallel to the fibers. Since the instrumentation used for evaluating meat texture has been recently surveyed (Voisey 1976), the present discussion will omit most details of instrument construction and test consideration, but attempt to focus more upon the relation to structure of the two major methods in use.

Shear Measurement

By far the most widespread instrumental measurement of muscle mechanical properties is based on so-called shearing forces. It is now known that, in fact, this term is most often used incorrectly. By strict rheological definition, shear force refers only to a force that is parallel to the area on which it acts thus producing deformation in the same plane. For example, with the widely used Warner–Bratzler (WB) shear, in which force is usually applied across muscle fibers via a blunt edge, compressive and tensile forces, as well as shear forces, have been found to be important since sample deformation leads to a complex stress geometry (Voisey and Larmond 1974). This type of test then becomes highly empirical in nature, since several properties are being measured simultaneously including the compression force required to change the cross-sectional dimensions of the sample, the tensile force required to rupture the stretched fibers, and the shear force required to cut those fibers trapped between the blade and the slot. Figure 6.16 shows these actions diagrammatically.

The WB apparatus is an example of an empirical test that has been mistreated by many researchers. It is obvious that any method used to measure physical properties must be strictly controlled and properly interpreted if the resulting data are to be meaningful. A partial list of important factors often not given sufficient attention include (1) deformation rate—both speed of travel and consistency are often not controlled; (2) sample dimensions—size, shape, and particularly orientation are factors too often treated arbitrarily; (3) friction—clearance

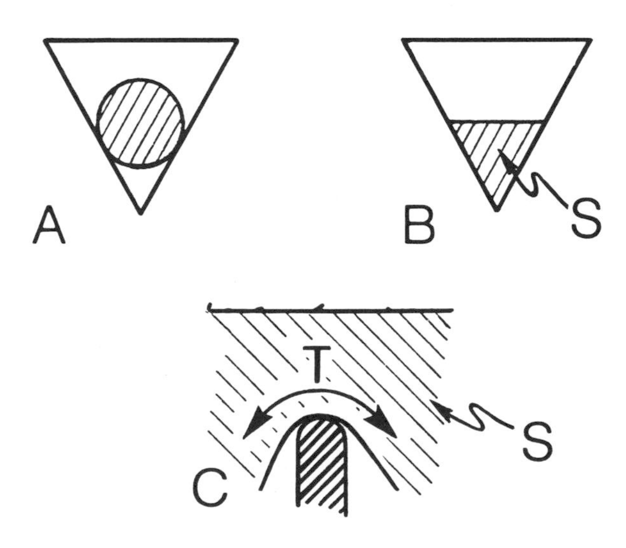

Fig. 6.16. Sample deformation in a Warner–Bratzler attachment. (A) Cylindrical sample in hole of blade. (B) Sample deformation as blade moves through slot. (C) View of blade as sample (S) is bent over edge, creating tensile stress (T).
From Voisey and Larmond (1974), used with permission.

between blade and slot must be standardized and the parts kept clean during testing; (4) recording devices—if the response time is too slow, rapid force fluctuations will not be captured; (5) test cell dimensions—these must be standardized and the influence of wear must be considered; (6) calculating and expressing data—units that have been reported include force, force/sample area, slope of the force curve, etc. As a result of these and other factors, it is virtually impossible to compare WB data found in the literature.

Equally, if not more, important is an understanding of what property of the sample is measured by the WB test. As mentioned previously, it has become a useful simplification to view physical characteristics of muscle tissue as arising from two independent components—connective tissue and myofibrillar proteins. Since the most common parameter recorded from the WB apparatus is peak or maximum force, it would seem impossible to differentiate the influence of these two components using only this technique. Recent work, however, indicates that it may be possible to relate WB data back to their structural origin. It has been indicated that peak shear force values reflect mainly the myofibrillar protein component, but that when the entire force deformation curve is examined, the influence of this structure can be differentiated from connective tissue (Bouton *et al.* 1975; Moller 1981). Initial yield was found to reflect the myofibrillar component while final yield was attributed to connective tissue (Fig. 6.17). As cooking temperature increased, the WB pattern shifted, with the connective tissue peak decreasing while the myofibrillar protein peak increased. Further work (Moller *et al.* 1981) extended this approach; thus far the data are consistent with a model of cooked meat structure in which applied force would be borne first by heat denatured, hardened myofibrillar proteins and then by the elastic connective tissue. Thus, progress is being made toward relating WB back to muscle structure, and it now would appear possible to obtain data resolving both the principal contributors to muscle strength.

Tensile Measurement

The fibrous structure of muscle tissue lends itself to tension testing. This approach provides several advantages over shearing. (1) Defined mechanical parameters can be measured and reported in fundamental units. (2) Problems associated with empirical tests are eliminated. (3) Variations such as cycling can be used in applying tensile force to provide more information about the sample. (4) Tensile testing is widely known and used to determine the strength of engineering materials, and the theoretical basis of this classical method is understood, making it easier to relate results to structural components.

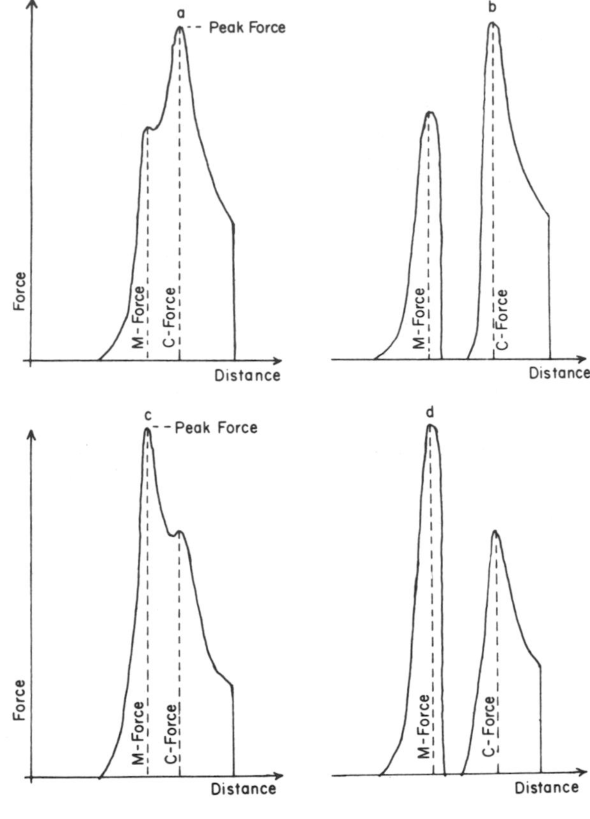

Fig. 6.17. Typical Warner–Bratzler force-deformation curves of bovine semitendinosus muscle. Curve a heated to 60°C; c heated to 80°C. Curves b and d represent the myofibrillar component (M) and connective tissue component (C) force obtained by interrupted crosshead movement.
From Moller (1981), used with permission.

Tensile force measurements have been used by a number of workers to investigate raw (Stanley *et al.* 1971, 1972; Eino and Stanley 1973A, B; Stanley 1976) and cooked meat (Bouton and Harris 1972A; Bouton *et al.* 1973A, B 1975). Two approaches have been taken; strips of muscle fibers can be prepared and the force required to rupture the sample can be determined, or force can be applied perpendicular to the fibers (adhesion measurements). The latter measurement has been regarded as a measure of connective tissue strength, since intact fibers are pulled apart against the collagen network. Both methods have been reported to be sensitive to structural changes. Thus, Bouton *et al.* (1975) found tensile properties of raw and cooked samples and adhesion values of cooked samples to reflect sarcomere length in beef muscle varying only in state of contraction. The interpretation became more complicated when comparisons were made of several muscles varying in both contraction and connective tissue content, but it was suggested that the initial effect of applying tensile force was, as previously shown for WB, to produce a yield in myofibrillar structure after which the connective tissue structure became important. This situation may perhaps be likened to a number of parallel tubes composed of repeating series units and covered by an interconnected, flexible meshwork that possesses greater material strength than the tubes. In either cutting (a term for the action occurring in the WB apparatus and preferred to shearing) or tensile testing the tubes yield first, leading—in the case of cutting—to compression and failure of tubes followed in time by shearing of the meshwork, or in the case of tensile testing to longitudinal separation of tubes at some point in the series unit and straightening and elongation of the meshwork until it fails. In both cases, the weakest element would yield first and a time separation of the two events might be expected. Thus, it may also be possible to divide the force–deformation curves resulting from tensile testing into areas reflecting individual structural elements.

Relationship of Methods Used to Measure Physical Properties

Investigations of the structural basis for measuring physical properties and how these measurements are related have only recently begun. An example of this approach may be found in the work cited earlier of Bouton *et al.* (1975) who measured shear, tensile, and adhesive properties in raw and cooked tissue varying in contraction state and connective tissue content. Figure 6.18 shows typical force–deformation curves obtained with the three methods for tissue of different sarcomere length but similar connective tissue content. Tensile curves reflect increases in

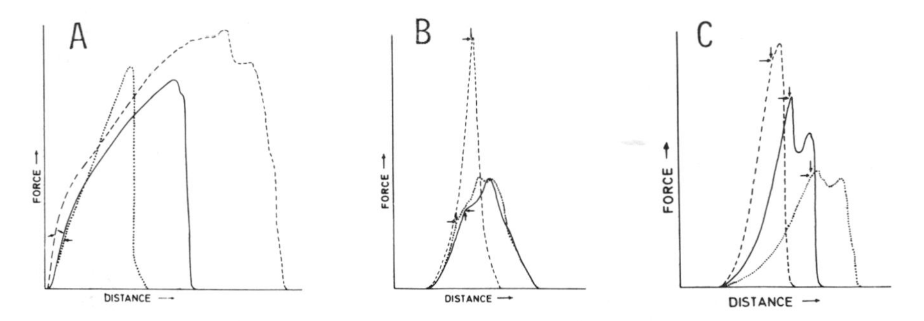

Fig. 6.18. Force–deformation curves of cooked beef deep pectoral muscle as a function of contraction state. Sarcomere lengths: cold-shortened (---), 1.28 μm; control(—),2.24 μm; stretched (····), 3.34 μm. (A) Tensile test. (B) Warner–Bratzler shearing test. (C) Adhesion test. Arrows indicate initial yield points.

From Bouton et al. (1975), used with permission.

yield elongation as a result of muscle shortening. This has also been reported by other workers (Stanley *et al.* 1972; Stanley and Swatland 1976) and probably is due to the interaction of two structural factors, the extension characteristics of the myofibril and the orientation of the connective tissue network. In contracted muscle, tensile forces would be expected to first engage in the myofibrils and only after this will tension be borne by connective tissue that initially lies perpendicular to the fiber axis (Fig. 6.11). With stretched muscle, however, the two components would be encountered simultaneously when tension is applied. More force is required to rupture stretched than contracted tissue, which again is a reflection of the structure of contracted tissue where the strength of myofibers and connective tissue can be overcome separately and, thus, more easily. Cooking produces a large increase in peak force required to rupture contracted muscle but not stretched muscle. Since collagen structure is reduced by heating, this may be a result of the greater interaction of contractile proteins in shortened muscle and their heat coagulation. An increase in initial yield force (the force at the first major inflexion of the curve) for contracted, but not stretched, muscle as a result of cooking supports this, since, as mentioned previously, this parameter may be interpreted as a function of the myofibrillar protein contribution. Peak shear force and initial yield shear force decreases as sarcomere length increases; these data may mean that the myofibrillar contribution increases with degree of contraction and, conversely, that connective tissue plays a greater role in toughening of stretched muscle. Adhesion measurements are also affected by contraction and, generally, these values increased as sarcomere length decreased.

It is interesting to note in this work that the magnitude of tensile forces were generally influenced less by state of contraction than shear or adhesive forces. It may be that, as well as responding to connective tissue and contractile proteins, these parameters are also influenced by an elastic component of muscle—the role predicted for connectin [see Curie and Wolfe (1980) and prior discussion]. Clearly, further studies of this nature are required if a sound structural interpretation of physical property data is to be obtained.

FACTORS INFLUENCING THE RELATION OF STRUCTURE TO PHYSICAL PROPERTIES IN MUSCLE TISSUE

The study of texture–structure interactions in meat would be made much easier if any degree of constancy was seen in the relationship. As it is, virtually every process involved in the conversion of muscle to meat alters the expression of structural elements. Overlying this are variations in composition due to animal factors. It is not surprising, then, that meat texture continues to be a challenge to food scientists. In this section, some of the conditions influencing physical properties will be examined.

Animal Effects

Animal age has a significant influence on physical properties of muscle. It will be remembered that collagen crosslinking increases as a function of age, and it is to be expected that a direct relationship between toughness and age would be found. On the other hand, it would seem unlikely, *a priori*, that myofibrillar structure or strength varies with age. While it is often the case that older animals produce tougher meat, it is not always so, and there are a considerable number of conflicting reports in the literature. For example, Bouton *et al.* (1978) found that all measures of tenderness (instrumental and sensory) decreased for stretched beef semimembranosus muscles as animal age increased from 2 to 120 months. In contrast, WB values decreased for beef longissimus dorsi muscle from Achilles tendon hung sides as age increased from 9 to 42 months. This was thought to be a reflection of less postmortem shortening in tissue from the heavier carcasses of older animals. Table 6.4 provides the data from the former muscle. If initial yield is a reflection of the myofibrillar protein contribution, then subtracting this from peak force may result in a value related to the influence of connective tissue. When this calculation is made, it may be seen that whereas initial yield increases only fractionally as a result of animal age, the derived quantity jumps by over tenfold in the same period. Beef animals are the only source of meat consumed at this state

TABLE 6.4. INFLUENCE OF ANIMAL AGE ON PHYSICAL AND SENSORY PROPERTIES OF COOKED RESTRAINED SEMIMEMBRANOSUS BEEF MUSCLE

	Animal age (mo)					
Parameter	2	9	16	27	42	120
(A) Warner–Bratzler peak force (kg)	3.64	4.75	4.52	4.98	5.16	6.58
(B) Warner–Bratzler initial yield force (kg)	3.45	3.89	3.27	3.38	3.29	4.57
A – B (kg)	0.19	0.86	1.19	1.60	1.88	2.02
Sensory tenderness	6.4	11.7	13.1	13.1	14.5	19.1
Sensory juiciness	16.2	15.2	14.3	14.2	13.7	11.2

Source: Bouton *et al.* (1978), used with permission.

of maturity in North America with the minor exception of mutton. Thus, collagen crosslinking becomes of importance in this species.

Another animal-related factor that may influence physical properties is fiber type. It is now known that commercially important muscle tissue consists of two major types of fibers as well as an intermediate form. Type 1 fibers show weak myofibrillar ATPase reactivity at pH 9.4, are sometimes termed β or slow twitch or red fibers (the latter because of their myoglobin content), have many intermyofibrillar mitochondria, metabolize and store more lipid, are oxidative in nature, and possess wide, distinct and dense Z-discs compared to type 2 fibers. Calkins *et al.* (1981) examined beef longissimus dorsi muscle and found that white fibers exhibited larger diameters (42 μm) than red fibers (34 μm); traditionally, larger fibers have been associated with tougher meat (Szczesniak and Torgeson, 1965). Also, a significant positive correlation was found between shear force and the percentage of type 2 or white fibers. Marbling (visible intramuscular fat located in the perimysial connective tissue between muscle fiber bundles), however, was significantly negatively correlated to this fiber type and, although this quality attribute was more highly related to fiber type than shear force, it may be that genetic selection for fiber type can be used to obtain more tender carcasses.

Processing Effects

When an animal is slaughtered and further processed, changes occur in both structure and physical properties of the muscle tissue. The following discussion does not deal with all possible postmortem events but is aimed at demonstrating the scope of these alterations.

Postmortem events. The cessation of circulation has a number of immediate and long-term consequences to the structure and physical properties of muscle tissue, chiefly as a result of a complex series of chemical reactions. Bendall (1973) provides a comprehensive review of these postmortem changes, many of which are beyond the scope of the present work. The most obvious physical postmortem change in muscle is stiffening and loss of extensibility as a result of rigor mortis (Fig. 6.19). This event occurs between 6 and 24 hr postmortem in beef and is marked by loss of ATP and a decline in pH resulting from the production of lactic acid, the final product of glycogen metabolism. If the muscle is restrained (for example, if it is left on the carcass) shortening does not occur, but if it is unrestrained, severe contraction will be observed. During rigor, which is independent of contraction state, multiple attachments are formed between actin and myosin which lock the inter-

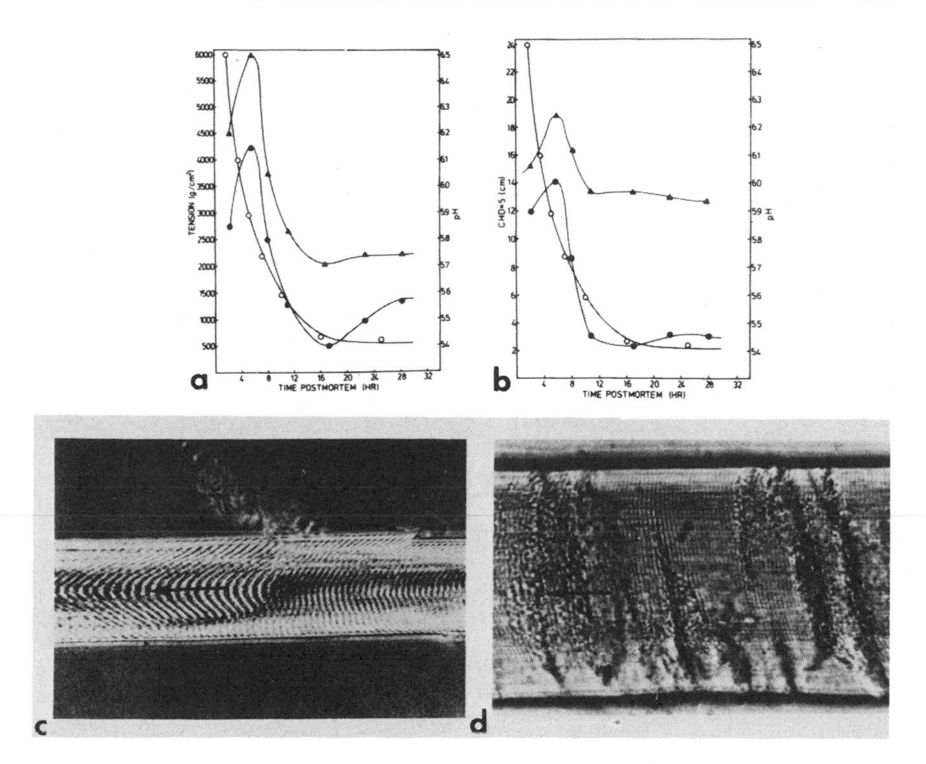

Fig. 6.19. Effect of rigor mortis on physical and structural properties of beef muscle. (a) Tension at initial (●) and final (▲) yield due to longitudinal stretch. (b) Extension at initial and final yield due to longitudinal stretch; (o), pH. (c) TLM of muscle fiber (pH 5.95 – close to tensile peak) fixed at initial yield. Regions of the fiber are extended (sarcomere length > 4.0 μm), myofibrils tend to remain in register (arrow). (d) TLM of muscle fiber (pH 5.65—approaching maximum rigor) fixed at initial yield. Fewer sarcomeres yield under tension and breaks can be seen (arrow).

From Currie and Wolf (1980), used with permission.

digitating mechanism into a rigid, inextensible structure. The number of crosslinkages formed depends on the degree of contraction, and muscles allowed to shorten before the onset of rigor are tougher than if they had been restrained. Fortunately for meat consumers, the textural manifestations of muscle in rigor do not persist. Although there is no resolution of rigor in terms of resynthesis of ATP to allow separation of linked actin and myosin filaments, after a period of time muscle begins a tenderizing process that, in beef, can extend over a period of several days (Fig. 6.20). In terms of physical properties, a decrease is noted in the modulus of elasticity (slope of the stress – strain curve), but extensibility is not returned (Bendall 1973).

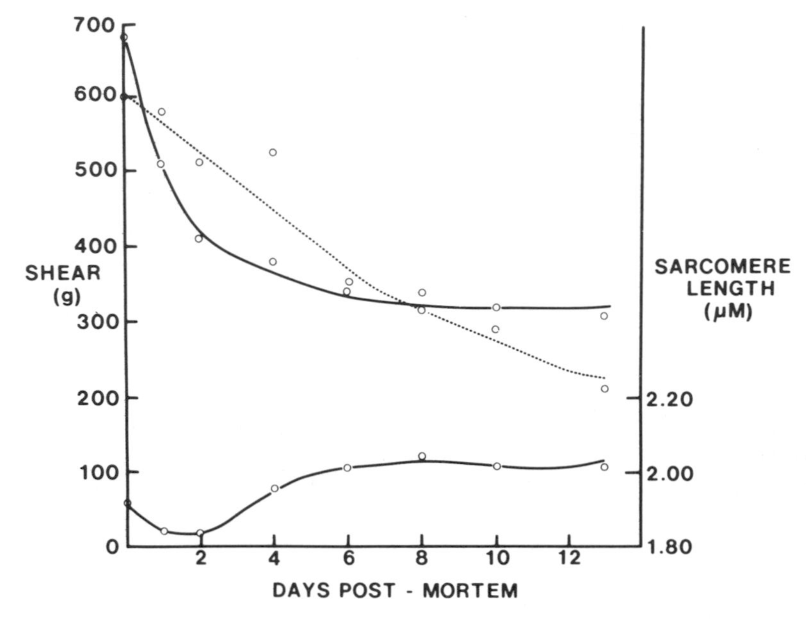

Fig. 6.20. Physical properties of aged beef psoas muscle. (—) Uncooked; (····) cooked.

From Stanley and Tung (1976), used with permission.

Rigor mortis has been shown to have a great influence on texture of the resulting meat. Khan (1977) cites three major variables that influence the course of rigor and, thereby, meat texture: muscle ATP content at the time of death; rate and extent of postmortem glycolysis; prior nutritional status, stress levels, and struggling during slaughter. Variations in these conditions can lead to several different patterns of pH drop. In general, normal tissue exhibits a high pH (6.9–7.1) within 1 hr postmortem and a low ultimate pH (5.7–5.9) and yields tender meat. If a low pH value (6.2 or lower) is attained within 1 hr postmortem as well as a low ultimate pH (5.7–5.9), rapid rigor is indicated and toughness can result as well as the condition known as PSE (pale, soft, and exudative) in pig muscle. Conversely, high initial pH values (6.9–7.1) that do not drop result from factors such as starvation, fatigue, and stress and lead to little or no postmortem glycolysis, high WHC, and the dark-cutting beef phenomenon.

Postmortem events influence the physical properties of meat not only through rigor mortis but also as a result of the action of numerous endogenous enzymes on myofibrillar structure and perhaps connective tissue as well. A major structural alteration that has been observed in postmortem muscle is Z-disc degradation. Several groups (see Goll *et al.* 1974) have identified a calcium activated factor (CAF) located in the sarcoplasmic protein fraction of muscle that is effective in removing Z-discs and releasing α-actinin from isolated myofibrils. In addition, this enzyme hydrolyzes many muscle proteins but does not affect actin, myosin, and several others. It is thought that this enzyme is responsible for structural alterations in myofibrils leading to their fragmentation into lengths of one to four sarcomeres on homogenization. Disruption of myofibrils by this procedure has been termed the fragmentation index and has been shown to have potential as a rapid and accurate method of predicting cooked meat tenderness from measurements on raw muscle (Cole and Davis 1981). Hattori and Takahashi (1979) reported that the degree of fragmentation was strongly related to sarcomere length—maximum disruption occurred in tissue of rest length and decreases were noted in both longer and shorter sarcomeres. The reaction was found to be pH- and temperature-dependent and required calcium.

Not all evidence supports the direct relationship between Z-disc degradation and myofibrillar fragmentation. Gann and Merkel (1978) studied structural changes in beef muscle during postmortem aging by TEM and reported that these two types of alterations occurred independently. Fiber type was found to have an influence since Z-discs from type 1 fibers were unaltered as a result of time postmortem while type 2 fibers showed this reaction. Both fiber types underwent myofibrillar fragmentation that was independent of Z-disc degradation (Fig. 6.21). Other workers have also observed changes in the morphology of muscle as a result of postmortem aging, and SEM has proved to be a valuable tool in this regard. The influence of aging on beef muscle was examined by Stanley (1974) who reported that, as storage time progressed, myofibrils demonstrated an increased propensity to break cleanly at the level of the Z-disc; a collapse of previously swollen transverse SR elements was also noted. Similar results by other workers are cited by Voyle (1981A). Penny (1980) has reviewed the enzymology and structural changes associated with aging.

Several workers have used various enzymes as *in vitro* probes to study their effect on structure and physical properties of postmortem muscle. Thus, Eino and Stanley (1973A, B), Robbins and Cohen (1976), Robbins *et al.* (1979), Cohen and Trusal (1980) and others have demonstrated alterations similar to those produced by aging from the application of either muscle cathepsins or enzymes isolated from organs and, in some work, concomitant changes in physical properties.

While the previous discussion has centered on postmortem aging changes in the proteins of the myofibril, it is also important to determine the fate of connective tissue during this period. For many years, it

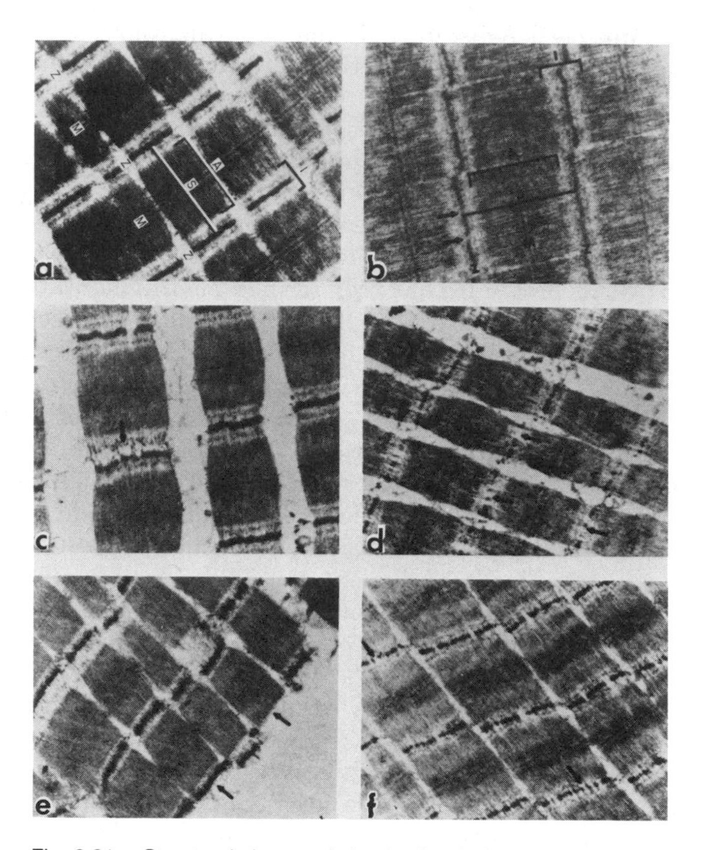

Fig. 6.21. Structural changes in bovine longissimus muscle during postmortem aging. (a) Type 1 myofibers at 1 hr postmortem showing normal A band (A), I band (I), Z-discs (Z), M line (M), and sarcomeres (S). (b) Type 2 myofibers at 1 hr postmortem showing slight Z-disc degradation. Note narrower and less distinct Z-discs. (c) Type 1 myofibers at 48 hr postmortem showing fragmentation at the Z-disc–I band junction (arrow). (d) Type 2 myofibers at 48 hr postmortem showing considerable Z-disc degradation (arrows). (e) Type 1 myofibers at 216 hr postmortem aging showing myofibril fragmentation at the Z-disc–I band junction (arrows). (f) Type 2 myofibers at 216 hr postmortem showing Z-disc degradation (arrows).
From Gann and Merkel (1978), used with permission.

was believed that the unreactive chemical nature of collagen precluded any attack by endogenous muscle enzymes. However, an increase in extractable collagen and the appearance of altered separation patterns for this material was reported by Stanley and Brown (1973); Dutson (1974) reported similar results. Recently, Wu *et al.* (1981) found that certain lysosomal enzymes (β-galactosidase and β-glucuronidase) increased the dissolution of collagen fibers by collagenase. Although other authors have not been successful in demonstrating postmortem changes in connective tissue (e.g., Chizzolini *et al.* 1977), this is a research subject of great importance.

Chilling and freezing. Temperature has a major influence on rigor mortis, and several abnormalities in texture can result as a consequence of failure to properly cool fresh meat (Bendall 1973). Thaw rigor occurs on thawing of muscle frozen in the prerigor state and leads to violent shortening (40–50% of initial length) as the unused ATP triggers contraction. A related but less rapid phenomenon is called cold-shortening (Fig. 6.12). Rapidly chilled prerigor muscles undergo localized shortening in some fibers that then induces passive shortening in those remaining, leading to a characteristic wavy configuration that can be seen by LM, SEM, and TEM (Voyle 1981A). Cold-shortening, perhaps better termed cold-toughening, appears to be characteristic of red fiber types, and it is thought that the less effective SR of red muscle allows the leakage of calcium ions at low temperatures, which stimulates the contractile system. Both of these situations lead to tougher meat but can be prevented by appropriate processing steps. It may be possible that connective tissue plays a role in the toughness resulting from shortening since Rowe (1974) has shown an effect of cold-shortening on the relative number and angle of collagen fibers. Shortening of muscle has also been observed at higher (> 20°C) temperatures; minimum contraction occurs at about 15°C, which would seem to be the optimal point, from textural considerations, for muscle to enter rigor.

Freezing of meat is a common preservation technique. It has been demonstrated that when whole muscle tissue is frozen and stored under accepted processing and packaging conditions, little if any alterations in structural or physical properties are produced during normal storage periods. A recent study (Carrol *et al.* 1981) reinforces this opinion. These authors also demonstrated that repeated freeze–thaw cycles did not alter the structure if thawing occurred at low (4°C) temperatures.

Electrical Stimulation. It follows from the previous discussion that one way to combat the cold-shortening defect and concomitant toughness is to accelerate postmortem breakdown of ATP and pH decline. A process to accomplish this has been developed through electrical stimulation of carcasses soon after slaughter. This reduces the time required for the onset of rigor and increases the rate of glycolysis (Bendall 1976) so that the tissue will no longer contract at lower temperatures. Electrically stimulated meat demonstrates increased tenderness over nonstimulated controls even though the application of current produces zones of severe contraction (Voyle 1981B). Initially, the tenderizing effect of electrical stimulation was ascribed solely to prevention of cold-shortening, but more recent research (Savell *et al.* 1977; Will *et al.* 1980; Voyle 1981B) tends to emphasize the role of tissue disruption in this process. Structural alterations observed as well as contraction bands include intracellular edema, swollen membranous organelles (SR, T tubules, and mitochondria), myofibrillar disruption, and other changes consistent with accelerated autolysis of muscle resulting from tissue disruption. Sarcomere lengths usually do not differ between control and stimulated muscle, further discounting the influence of cold-shortening. The data of Dutson *et al.* (1980) indicate that electrical stimulation also promotes increased lysosomal activity that is capable of producing the autolytic degradation seen microscopically.

Cooking. Apart from health considerations, textural change is a major reason for heating muscle tissue. At least four separate mechanisms important to texture operate during cooking: endogenous proteolytic enzymes are thermally activated; thermal denaturation of connective tissue occurs, resulting in a softening effect; thermal denaturation of contractile proteins occurs, resulting in a hardening effect; WHC is decreased, liquid components such as water and lipid exit, shrinkage in fiber diameter and length occurs and the apparent density increases. Other changes such as denaturation of sarcoplasmic proteins are also observed, but their effect on texture, if any, is not fully understood.

Because of the diverse nature of these reactions, it would not be expected that changes in physical properties would be linear with respect to temperature. This has been demonstrated for beef by Davey and Niederer (1977), who found that when differential in shear force per unit time was plotted against temperature, a discontinuous curve resulted (Fig. 6.22). These authors attribute the initial increase to enzymatic proteolysis and the second to destruction of interstitial collagen and myofibrillar denaturation. A distinct minimum occurred at about 67°C. It is important to attempt to compare these results to the information obtained by DSC. As mentioned previously, Wright *et al.* (1977) observed four major thermal effects using muscle and isolated proteins; an exotherm at 54°C only in pre-rigor muscle that was thought to be associated with contraction, three endotherms at 60°, 67°, and 80°C—the first attributed to myosin, the second to sarcoplasmic proteins (and, from cited work, collagen), and the third to actin. At present, it is difficult to relate these data, performed with milligram quantities of material at precise heating rates, to the former experiment, but it

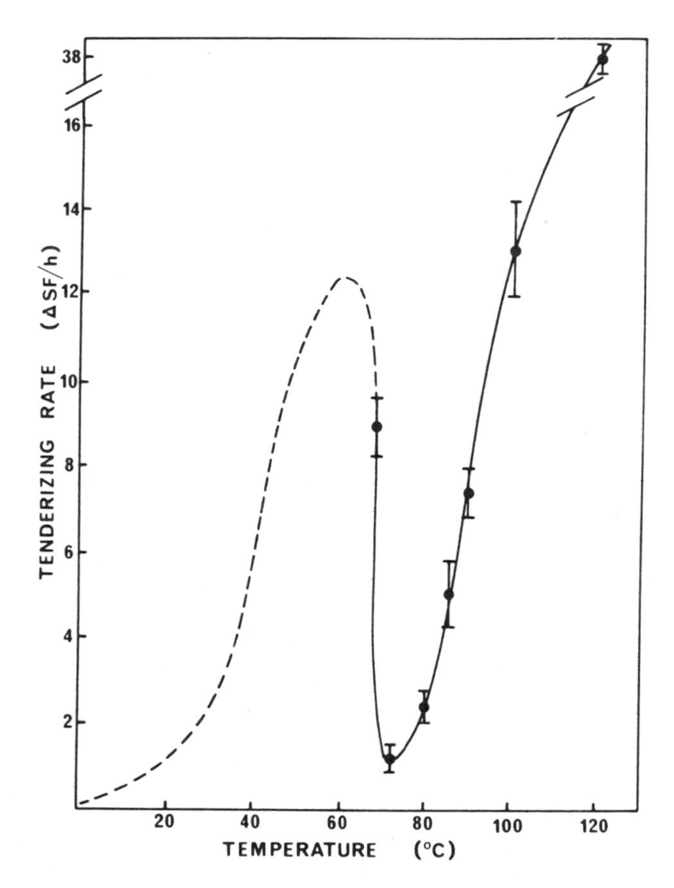

Fig. 6.22. Effect of cooking temperature on tenderization (change in shear force/hr of bovine sternomandibularis muscle. *From Davey and Niederer (1977), used with permission.*

would seem to emphasize the importance of factors other than connective tissue and contractile protein denaturation in the development of tenderness during cooking.

Another approach was taken by Locker and Carse (1976), who measured tensile properties as a function of heating. These workers observed elastic behaviour in cooked muscle and an unchanging yield point, which they take as evidence of a major contribution by gap filaments to tensile strength. TEM work (Davey and Graafhuis 1976) tends to support this contention since fine filaments are seen that connect heat coagulated A and I bands.

Heating produces major changes in muscle structure. Voyle (1981A) has recently reviewed modifications in cooked tissue observable with SEM. Several authors have reported alterations including coagulation of perimysial and endomysial connective tissue, sarcomere shortening, myofibrillar fragmentation, and coagulation of sarcoplasmic proteins, although none of the papers include all these observations. One parameter that is often not standardized in studies of this type is sarcomere length, although Voyle mentions that heat shortening is more evident in stretched muscle since there is less overlap and, thus, less protection against thermal damage. Figure 6.23 presents SEM photomicrographs of raw and cooked muscle in cross- and longitudinal section showing several effects of heating.

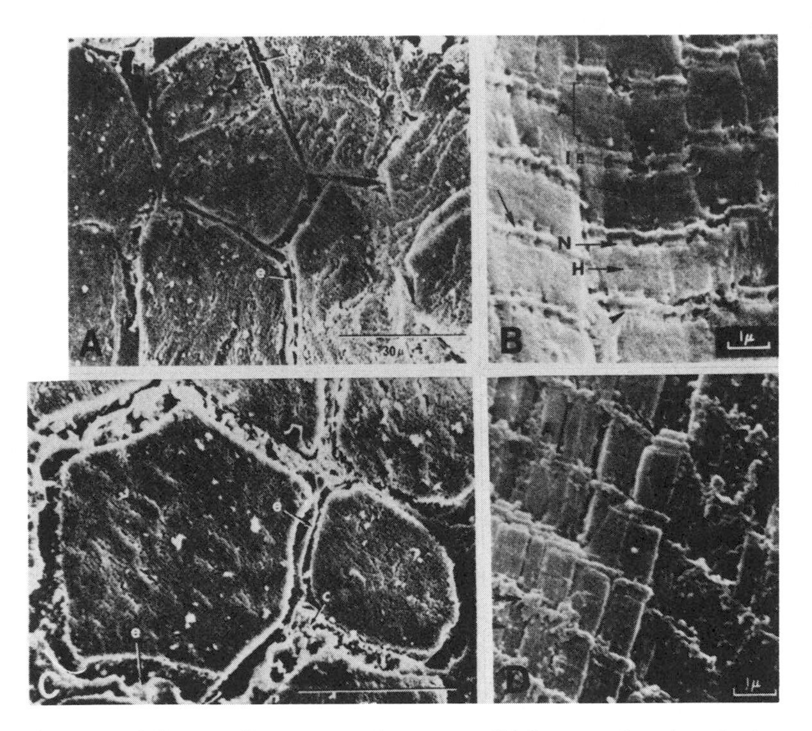

Fig. 6.23. Influence of heat on muscle structure. (A) Cross section of raw bovine longissimus muscle, endomysium (e). (B) Longitudinal section of raw bovine semitendinosus muscle, Z-disc (N), A band (A), I band (I), pseudo H zone (H). (C) Cross section of bovine longissimus muscle heated to 70°C, granular material (C), endomysium (e). (D) Longitudinal section of bovine semitendinosus muscle heated to 60°C, A band (A), Z-disc (arrow), agglomerated material along Z-disc (double arrow), pseudo H zone (arrowhead), splitting of protein layers (double arrowhead).

(A, C) *from Cheng and Parrish (1976)*; (B, D) *Jones et al. (1977), used with permission.*

FINAL REMARKS

It is to be hoped that the broad dependence of meat texture and physical properties of muscle tissue on structure has been amply demonstrated. Means are now available to measure certain physical properties of muscle, but they are by no means standardized and comparison of data from different laboratories is often impossible. Conflicting reports of the degree of correlation between sensory and instrumental tests are frequently a consequence of this. Of greater importance is the present inability to relate properly physical properties back to the specific structures from which they originate. Inspection of force–deformation curves in conjunction with parallel structural studies seems to be a promising area of investigation in this regard.

The structural organization of muscle has been studied extensively and new devices for this purpose continue to become available, yet too little of this work has been done by those scientists investigating meat as a food. Often, structural studies are not done thoroughly or the correct conclusions are not drawn relative to physical properties. As Marsh *et al.* (1981) has pointed out, a single broken sarcomere in every 1000 would rupture the fiber at 2 mm intervals and undoubtedly have an effect on physical properties. The technique of correlative microscopy decreases the incidence of drawing conclusions based on artifacts while quantitative methodology has been developed that helps to remove the subjective factor in evaluating micrographs.

It seems appropriate to end this discussion by stating some pertinent questions, the answers to which will dictate the direction of future research.

(1) Does the model of meat tenderness currently in vogue require more components to be complete? Evidence is building for assigning significant roles to both connectin/gap filaments and other cytoskeletin components and interfiber water. Possibly the use of probes such as DSC, nuclear magnetic resonance (NMR) and X-ray microanalysis can help provide direct evidence for their contribution.

(2) What further control is possible over pre- and postmortem events that can favorably modify physical properties of animal tissue? The genetic manipulation of fiber types may be possible if this factor proves important. Electrical stimulation seems one way to influence ATP and pH levels, but other techniques may be applicable as well. Collagen obviously plays a major role in meat texture and if it can be broken down by endogenous enzymes, it may be possible to accelerate this reaction through alteration of aging conditions.

(3) Considering the importance of physical properties to consumer acceptance of meat, is there a method for determining quickly and accurately a parameter or parameters of raw muscle that will predict the texture of cooked meat? Myofibrillar fragmentation is currently being investigated for this purpose, but other avenues should be pursued as well, particularly those that lend themselves to modern instrumentation and data processing equipment.

(4) A fruitful approach to complex biological problems in the past has been the construction of models capable of explaining physical properties through suitable theoretical and experimental considerations. Would it be worthwhile to attempt an integration of a mechanical and engineering approach to modeling postmortem muscle such as that of Stoner *et al.* (1974) with the biologically and chemically oriented work of Comissiong and Hultin (1978) based on glycerinated muscle at different states of contraction? An imaginative and coordinative approach will be required to provide the complete answer to this and the prior questions which together form only one part of an extraordinarily complex problem.

ACKNOWLEDGMENTS

Some of the original work described herein was supported by the Natural Sciences and Engineering Research Council and the Ontario Ministry of Agriculture and Food.

BIBLIOGRAPHY

ABRAHAMS, M. 1967. Mechanical behaviour of tendon in vitro. Med. Mol. Eng. *5*, 433–443.

ATTLE, J.R., ONEY, D., and SWENSON, R.A. 1980. Applications of image analysis. Am. Lab. *12*(4), 85–99.

BAILEY, A.J. 1972. The basis of meat texture. J. Sci. Food Agric. *23*, 995–1007.

BENDALL, J.R. 1969. Muscles, molecules and movement. Heinemann, London.

BENDALL, J.R. 1973. Post-mortem changes in muscle. *In* The Structure and Function of Muscle, 2nd Ed., Vol. 2, Structure, Part 2. G.H. Bourne (Editor). Academic Press, New York.

BENDALL, J.R. 1976. Electrical stimulation of rabbit and lamb carcasses. J. Sci. Food Agric. *27*, 819–826.

BORNSTEIN, P., and TRAUB, W. 1979. The chemistry and biology of collagen. *In* The Proteins, 3rd Ed., Vol. IV. H. Neurath and R.C. Hill (Editors). Academic Press. New York.

BOURNE, G.H. 1972. The Structure and Function of Muscle, 2nd Ed., Vol. 1, Structure. Academic Press, New York.

BOURNE, G.H. 1973. The Structure and Function of Muscle, 2nd Ed., Vol. 2, Structure, Part 2. Academic Press, New York.

BOUTON, P.E., and HARRIS, P.V. 1972A. A comparison of some objective methods used to assess meat tenderness. J. Food Sci. *37*, 218–221.

BOUTON, P.E., and HARRIS, P.V. 1972B. The effects of some post-slaughter treatments on the mechanical properties of bovine and ovine muscle. J. Food Sci. *37*, 539–543.
BOUTON, P.E., HARRIS, P.V., and SHORTHOSE, W.R. 1971. Effect of ultimate pH upon the water-holding capacity and tenderness of mutton. J. Food Sci. *36*, 435–439.
BOUTON, P.E., HARRIS, P.V., and SHORTHOSE, W.R. 1972. The effects of ultimate pH on ovine muscle: water-holding capacity. J. Food Sci. *37*, 351–355.
BOUTON, P.E., CARROLL, F.D., HARRIS, P.V., and SHORTHOSE, W.R. 1973A. Influence of pH and fiber contraction upon factors affecting the tenderness of bovine muscle. J. Food Sci. *38*, 404–407.
BOUTON, P.E., CARROLL, F.D., FISHER, A.L., HARRIS, P.V., and SHORTHOSE, W.R. 1973B. Effect of altering ultimate pH on bovine muscle tenderness. J. Food Sci. *38*, 816–820.
BOUTON, P.E., HARRIS, P.V., and SHORTHOSE, W.R. 1975. Possible relationships between shear, tensile and adhesion properties of meat and meat structure. J. Texture Stud. *6*, 297–314.
BOUTON, P.E., FORD, A.L., HARRIS, P.V., SHORTHOSE, W.R., RATCLIFF, D., and MORGAN, J.H.L. 1978. Influence of animal age on the tenderness of beef: muscle differences. Meat Sci. *2*, 301–311.
BRADY, D.E. 1954. The consumer's definition of quality in meat. Proc. 7th Annu. Recip. Meat Conf., pp. 111–115.
BRAGGINS, D.W., GARDNER, G.M., and GIBBARD, D.W. 1971. The applications of image analysis techniques to scanning electron microscopy. Scanning Electron Microsc. 1971/I, 393–400.
CALKINS, C.R., DUTSON, T.R., SMITH, G.C., CARPENTER, Z.L., and DAVIS, G.W. 1981. Relationship of fiber type composition to marbling and tenderness of bovine muscle. J. Food Sci. *46*, 708–710, 715.
CARROLL, R.J., and JONES, S.B. 1979. Some examples of scanning electron microscopy in food science. Scanning Electron Microsc. 1979/III, 253–259.
CARROLL, R.J., CAVANAUGH, J.R., and RORER, F.P. 1981. Effects of frozen storage on the ultrastructure of bovine muscle. J. Food Sci. *46*, 1091–1094, 1102.
CHABOT, J.F. 1979. Preparation of food science samples for SEM. Scanning Electron Micros. 1979/III, 279–286, 298.
CHANDLER, J.A. 1977. X-ray Microanalysis in the Electron Microscope. North-Holland Publ. Co., Amsterdam.
CHANDLER, J.A. 1979. Principles of X-ray microanalysis in biology. Scanning Electron Microsc. 1979/II, 595–602.
CHENG, C.S., and PARRISH, Jr., F.C. 1976. Scanning electron microscopy of bovine muscle: effect of heating on ultrastructure. J. Food Sci. *41*, 1449–1454.
CHEUNG, W.Y. 1982. Calmodulin. Sci. Am. *246*(6), 62–70.
CHIZZOLINI, R., LEDWARD, D.A., and LAWRIE, R.A. 1977. Note on the effect of aging on the neutral salt and acid soluble collagen from the intramuscular connective tissue of various species. Meat Sci. *1*, 111–117.
COHEN, S.H., and TRUSAL, L.R. 1980. The effect of catheptic enzymes on chilled bovine muscle. Scanning Electron Microsc.1980/III, 595–600.
COLE, A.B. Jr., and DAVIS, G.W. 1981. Influence of postmortem aging period on the fragmentation index of bovine longissimus muscle. J. Food Sci. *46*, 644–645.
COMISSIONG, E.A., and HULTIN, H.O. 1978. Development of a model system using glycerinated muscle for the study of physical properties. J. Food Biochem. *2*, 305–325.
CUNHA, T.J. 1978. The animal industries are here to stay. II. Food Nutr. News *49*(5), 1–4.
CURRIE, R.W., and WOLFE, F.H. 1980. Rigor related changes in mechanical properties (tensile and adhesive) and extracellular space in beef muscle. Meat Sci. *4*, 123–143.

DAVEY, C.L., and GRAAFHUIS, A.E. 1976. Structural changes in beef muscle during aging. J. Sci. Food Agric. *27*, 301–306.
DAVEY, C.L., and NIEDERER, A.F. 1977. Cooking tenderizing in beef. Meat Sci. *1*, 271–276.
DEMAN, J.M. 1976. Mechanical properties of foods. *In* Rheology and Texture in Food Quality. J.M. deMan, P. Voisey, V. Rasper and D.W. Stanley (Editors). AVI Publishing Co., Westport, CT.
DOSREMEDIOS, C.G., and GILMOUR, D. 1978. Is there a third type of filament in striated muscle? J. Biochem. *84*, 235–238.
DUTSON, T.R. 1974. Connective tissue. Proc. Meat Ind. Res. Conf., pp. 99–107.
DUTSON, T.R., HOSTETLER, R.L., and CARPENTER, Z.L. 1976. Effect of collagen levels and sarcomere shortening on muscle tenderness. J. Food Sci. *41*, 863–866.
DUTSON, T.R., SMITH, G.C., and CARPENTER, Z.L. 1980. Lysosomal enzyme distribution in electrically stimulated ovine muscle. J. Food Sci. *45*, 1097–1098.
EINO, M.F., and STANLEY, D.W. 1973A. Catheptic activity, textural properties and surface ultrastructure of post-mortem beef muscle. J. Food Sci. *38*, 45–50.
EINO, M.F., and STANLEY, D.W. 1973B. Surface ultrastructure and tensile properties of cathepsin and collagenase treated muscle fibers. J. Food Sci. *38*, 51–55.
FRANZINI-ARMSTRONG, C. 1973. Membranous systems in muscle fibers. *In* The Structure and Function of Muscle, 2nd Ed., Vol. II, Structure, Part 2. G.H. Bourne (Editor). Academic Press, New York.
GANN, G.L., and MERKEL, R.A. 1978. Ultrastructural changes in bovine longissimus muscle during postmortem aging. Meat Sci. *2*, 129–144.
GEISSINGER, H.D., and STANLEY, D.W. 1981. Preparation of muscle samples for electron microscopy. Scanning Electron Microsc. 1981/III, 415–426.
GEISSINGER, H.D., YAMASHIRO, S., and ACKERLY, C.A. 1978. Preparation of skeletal muscle for intermicroscopic (LM, SEM, TEM) correlation. Scanning Electron Microsc. 1978/II, 267–273.
GOLL, D.E., STROMER, M.H., SUZUKI, A., BUSCH, W.A., REDDY, M.K., and ROBSON, R.M. 1972. Chemical changes in postmortem muscle. Proc. 18th Meet. Meat Res. Workers *2*, 693–705.
GOLL, D.E., STROMER, M.H., OLSON, D.G., DAYTON, W.R., SUZUKI, A., and ROBSON, R.M. 1974. The role of myofibrillar proteins in meat tenderness. Proc. Meat In. Res. Conf., Am. Meat Inst. Found., p. 75–98.
GOLL, D.E., ROBSON, R.M., and STROMER, M.H. 1977. Muscle proteins. *In* Food Proteins. J.R. Whitaker and S.R. Tannenbaum (Editors). AVI Publishing Co., Westport, CT.
HALL, T.A. 1979. Biological X-ray microanalysis. J. Microsc. *117*, 145–163.
HAM, A.W. 1969. Histology, 6th Ed. Lippincott, Philadelphia, PA.
HAMM, R. 1960. Biochemistry of meat hydration. Adv. Food Res. *10*, 355–463.
HARRINGTON, W.F. 1979. Contractile proteins of muscle. *In* The Proteins, 3rd Ed., Vol. IV. H. Neurath and R.L. Hill (Editors). Academic Press, New York.
HARRIS, P.V. 1976. Structural and other aspects of meat tenderness. J. Texture Stud. *7*, 49–63.
HATTORI, A., and TAKAHASHI, K. 1979. Studies on the postmortem fragmentation of myofibrils. J. Biochem. *85*, 47–56.
HEARLE, J.W.S., SPARROW, J.T., and CROSS, P.M. 1972. The Use of the Scanning Electron Microscope. Pergamon Press, New York.
HERRING, H.K., CASSENS, R.G., SUESS, G.G., BRUNGARDT, V.H., and BRISKEY, E.J. 1967. Tenderness and associated characteristics of stretched and contracted bovine muscles. J. Food Sci. *32*, 317–323.

HONIKEL, K.O., HAMID, A., FISCHER, C., and HAMM, R. 1981. Influence of postmortem changes in bovine muscle on the water-holding capacity of beef. Postmortem storage of muscle at various temperatures between 0° and 30°C. J. Food Sci. *46*, 23–25, 31.

HUXLEY, H.E. 1972. Molecular basis of contraction in cross-striated muscles. *In* The Structure and Function of Muscle, 2nd Ed., Vol. 1, Structure. G.H. Bourne (Editor). Academic Press, New York.

IFT. 1982. Symposium: Fundamental Properties of Muscle Proteins Important in Meat Science. 42nd Annu. Meet. of the Institute of Food Technologists, Las Vegas, NV, June 22–25. Papers 113–120. J. Food Biochem. (To be published).

JONES, S.B., CARROLL, R.J., and CAVANAUGH, J.R. 1976. Muscle samples for scanning electron microscopy: preparative techniques and general morphology. J. Food Sci. *41*, 867–873.

JONES, S.B., CARROLL, R.J., and CAVANAUGH, J.R. 1977. Structural changes in heated bovine muscle: a scanning electron microscope study. J. Food Sci. *42*, 125–131.

KHAN, A.W. 1977. Some factors causing variation in the texture of similar muscles from comparable animals. Meat Sci. *1*, 169–176.

KRAMER, A. 1973. Food texture-definition, measurement and relation to other food quality attributes. *In* Texture Measurements of Foods. A. Kramer and A.S. Szczesniak (Editors). D. Reidel Publishing Co., Dordrecht, Holland.

KRAMER, A., and TWIGG, B.A. 1970. Quality Control for the Food Industry. 3rd Ed., Vol. 1, Fundamentals. AVI Publishing Co., Westport, CT.

LARMOND, E. 1976. Texture measurement in meat by sensory evaluation. J. Texture Stud. *7*, 87–93.

LAWRIE, R.A. 1966. Meat Science. Pergamon Press, Oxford, U.K.

LEE, C.H., and RHA, C.K. 1979. Application of scanning electron microscopy for the development of materials for food. Scanning Electron Microsc. 1979/III, 465–471.

LEWIS, D.F. 1981. The use of microscopy to explain the behaviour of foodstuffs. Scanning Electron Microsc./1981 *3*, 391–404.

LOCKER, R.H., and CARSE, W.A. 1976. Extensibility, strength and tenderness of beef cooked to various degrees. J. Sci. Food Agric. *27*, 891–901.

LOCKER, R.H., and LEET, N.G. 1975. Histology of highly-stretched beef muscle. I. The fine structure of grossly stretched single fibers. J. Ultrastruct. Res. *52*, 64–75.

LOCKER, R.H., and LEET, N.G. 1976. Histology of highly-stretched beef muscle. II. Further evidence for the location and nature of gap filaments. J. Ultrastruct. Res. *55*, 157–172.

LOCKER, R.H., DAINES, G.J., CARSE, W.A., and LEET, N.G. 1977. Meat tenderness and the gap filaments. Meat Sci. *1*, 87–104.

McNEILL, P.A., and HOYLE, G. 1967. Evidence for superthin filaments. Am. Zool. *7*, 483–498.

MARSH, B.B., LEET, N.G., and DICKSON, M.R. 1974. The ultrastructure and tenderness of highly cold-shortened muscle. J. Food Technol. *9*, 141–147.

MARSH, B.B., LOCHNER, J.V., TAKAHASHI, G., and KRAGNESS, D.D. 1981. Effect of early post-mortem pH and temperature on beef tenderness. Meat Sci. *5*, 479–483.

MARUYAMA, K., NATORI, R., and NONOMURA, Y. 1976. New elastic protein from muscle. Nature *262*, 58–60.

MARUYAMA, K., MATSUBARA, S., NATORI, R., NONOMURA, Y., KIMURA, S., OHASHI, K., MURAKAMI, E., HANDA, S., and EGUCHI, G. 1977. Connectin, an elastic protein of muscle. Characterization and function. J. Biochem. *82*, 317–337.

MILNER, M., SCRIMSHAW, N.S., and WANG, D.I.C. 1978. Protein Resources and Technology. AVI Publishing Co., Westport, CT.

MOLLER, A.J. 1981. Analysis of Warner–Bratzler shear pattern with regard to myofibrillar and connective tissue components of tenderness. Meat Sci. *5*, 247–260.

MOLLER, A.J., SORENSEN, S.E., and LARSEN, M. 1981. Differentiation of myofibrillar and connective tissue strength in beef muscles by Warner–Bratzler shear parameters. J. Texture Stud. *12*, 71–83.

MORRIS, J.L., FUNK, T.F., and COURTENAY, H.V. 1974. Consumer attitudes to meat and meat products. Working paper AE 74/4. OAC, Univ. of Guelph, Guelph, Ontario, Canada.

PENNY, I.F. 1980. The enzymology of conditioning. *In* Developments in Meat Science. I.R. Lawrie (Editor). Applied Science Publishers, London.

POMERANZ, Y. 1976. Scanning electron microscopy in food science and technology. Adv. Food Res. *22*, 205–307.

PORTER, K.R., and FRANZINI-ARMSTRONG, C. 1965. The sarcoplasmic reticulum. Sci. Am. *212*(3), 72–80.

QUINN, J.R., RAYMOND, D.P., and HARWALKER, V.R. 1980. Differential scanning calorimetry of meat proteins as affected by processing treatment. J. Food Sci. *45*, 1146–1149.

RAEUBER, H.J., and NIKOLAUS, H. 1980. Structure of foods. J. Texture Stud. *11*, 187–198.

RHODES, V.J., KIEHL, F.R., and BRADY, D.E. 1955. Visual preferences for grades of retail beef cuts. University of Missouri College of Agriculture, Research Bull. 583.

ROBBINS, F.M., and COHEN, S.H. 1976. Effects of catheptic enzymes from spleen on the microstructure of bovine semimembranosus muscle. J. Texture Stud. *7*, 137–142.

ROBBINS, F.M., WALKER, J.E., COHEN, S.H., and CHATTERJEE, S. 1979. Action of proteolytic enzymes on bovine myofibrils. J. Food Sci. *44*, 1672–1677, 1680.

ROWE, R.W.D. 1974. Collagen fiber arrangement in intramuscular connective tissue. Changes associated with muscle shortening and their possible relevance to raw meat toughness measurements. J. Food Technol. *9*, 501–508.

RUBIN, L.J. 1981. Utilization of animal proteins. *In* Utilization of Protein Resources. D.W. Stanley, E.D. Murray, and D.H. Lees (Editors). Food and Nutrition Press, Inc., Westport, CT.

RUDDICK, J.E., and RICHARDS, J.F. 1975. Comparison of sarcomere length measurement of cooked chicken pectoralis muscle by laser diffraction and oil immersion microscopy. J. Food Sci. *40*, 500–501.

SAVELL, J.W., SMITH, G.C., DUTSON, T.R., CARPENTER, Z.L., and SUTER, D.A. 1977. Effects of electrical stimulation on palatability of beef, lamb and goat meat. J. Food Sci. *42*, 702–706.

SHIMOKOMAKI, M., ELSDEN, D.F., and BAILEY, A.J. 1972. Meat tenderness: age related changes in bovine intramuscular collagen. J. Food Sci. *37*, 892–896.

SIMS, T.J., and BAILEY, A.J. 1981. Connective tissue. *In* Developments in Meat Science—2. R. Lawrie (Editor). Applied Science Publishers, London.

SJOSTROM, M., and THORNELL, L.E. 1975. Preparing sections of skeletal muscle for transmission electron analytical microscopy (TEAM) of diffusible elements. J. Microsc. *103*, 101–112.

SOMLYO, A.V., SHUMAN, H., and SOMLYO, A.P. 1977. Elemental distribution in striated muscle and the effects of hypertonicity. Electron probe analysis of cryo sections. J. Cell Biol. *74*, 828–857.

SOMLYO, A.P., SOMLYO, A.V., SHUMAN, H., and STEWART, M. 1979. Electron probe analysis of muscle and X-ray mapping of biological specimens with a field emission gun. Scanning Electron Microsc./1979 *2*, 711–722.

STANLEY, D.W. 1974. The influence of aging on the texture and structure of beef muscle. Proc. Meat Ind. Res. Conf., pp. 109–116.

STANLEY, D.W. 1976. The texture of meat and its measurement. *In* Rheology and Texture in Food Quality. J.M. deMan, P. Voisey, V. Rasper, and D.W. Stanley (Editors). AVI Publishing Co., Westport, CT.

STANLEY, D.W., and BROWN, R.G. 1973. The fate of intramuscular connective tissue in aged beef. Proc. 19th Meet. Meat Res. Workers, pp. 231–248.

STANLEY, D.W., and GEISSINGER, H.D. 1972. Structure of contracted porcine psoas muscle as related to texture. Can. Inst. Food Sci. Technol. J. *5*, 214–216.

STANLEY, D.W., and HULTIN, H.O. 1968. Mechanism of emptying of skeletal muscle cell segments. J. Cell Biol. *39*, C1–C4.

STANLEY, D.W., and SWATLAND, H.J. 1976. The microstructure of muscle tissue—A basis for meat texture measurement. J. Texture Stud. *7*, 65–75.

STANLEY, D.W., and TUNG, M.A. 1976. Microstrostructure of food and its relation to texture. *In* Rheology and Texture in Food Quality. J.M. deMan, P. Voisey, V. Rasper, and D.W. Stanley (Editors). AVI Publishing Co., Westport, CT.

STANLEY, D.W., PEARSON, G.P., and COXWORTH, V.E. 1971. Evaluation of certain physical properties of meat using a universal testing machine. J. Food Sci. *36*, 256–260.

STANLEY, D.W., McKNIGHT, L.M., HINES, W.G.S., USBORNE, W.R., and DEMAN, J.M. 1972. Predicting meat tenderness from muscle tensile properties. J. Texture Stud. *3*, 51–68.

STONER, D.L., HAUGH, C.G., FORREST, J.C., and SWEAT, V.E. 1974. A mechanical model for postmortem striated muscle. J. Texture Stud. *4*, 483–493.

SWATLAND, H.J. 1975. Morphology and development of endomysial connective tissue in porcine and bovine muscle. J. Animal Sci. *41*, 78–86.

SZCZESNIAK, A.S. 1966. Texture measurements. Food Technol. *20*, 1292–1298.

SZCZESNIAK, A.S., and TORGESON, K.W. 1965. Methods of meat texture measurement viewed from the background of factors affecting tenderness. Adv. Food Res. *14*, 33–165.

TAKAHASHI, K., and SAITO, H. 1979. Post-mortem changes in skeletal muscle connectin. J. Biochem. *85*, 1539–1542.

VOISEY, P.W. 1976. Engineering assessment and critique of instruments used for meat tenderness evaluation. J. Texture Stud. 7, 11–48.

VOISEY, P.W., and LARMOND, E. 1974. Examination of factors affecting performance of the Warner–Bratzler meat shear test. Can. Inst. Food Sci. Technol. J. 7, 243–249.

VOYLE, C.A. 1969. Some observations on the histology of cold-shortened muscle. J. Food Technol. *4*, 275–281.

VOYLE, C.A. 1971. Assessment of meat texture by optical diffraction. Proc. 17th Meet. Meat Res. Workers, pp. 95–97.

VOYLE, C.A. 1979. Meat. *In* Food Microscopy. J.G. Vaughn (Editor). Academic Press, New York.

VOYLE, C.A. 1981A. Scanning electron microscopy in meat science. Scanning Electron Microsc. 1981/III, 405–413.

VOYLE, C.A. 1981B. Microscopical observations on electrically stimulated bovine muscle. Scanning Electron Microsc./1981 *3*, 427–434.

VRIEND, R.A., and GEISSINGER, H.D. 1982. Mammalian skeletal muscle: an interpretation of the scanning electron microscopic image. Trans. Am. Microsc. Soc. *101*, 117–125.

WILL, P.A., OWNBY, C.L., and HENRICKSON, R.L. 1980. Ultrastructural postmortem changes in electrically stimulated bovine muscle. J. Food Sci. *45*, 21–25, 34.

WILLIAMS, M.A. 1977A. Autoradiography and Immunocytochemistry. North-Holland Publ. Co., Amsterdam.

WILLIAMS, M.A. 1977B. Quantitative Methods in Biology. North-Holland Publ. Co., Amsterdam.

WRIGHT, D.J., LEACH, I.B., and WILDING, P. 1977. Differential scanning calorimetric studies of muscle and its constituent parts. J. Sci. Food Agric. *28*, 557–564.

WU, J.J., DUTSON, T.R., and CARPENTER, Z.L. 1981. Effect of postmortem time and temperature on the release of lysosomal enzymes and their possible effect on bovine connective tissue components of muscle. J. Food Sci. *46*, 1132–1135.

7

Physical Properties and Structure of Horticultural Crops

Malcolm C. Bourne[1]

In the food industry the words "fruits" and "vegetables" mean those edible fleshy structures of plant origin that are commonly used as foods. Their predominant distinguishing features are that they are of plant origin, cellular in nature and have a high moisture content, usually 75–95% H_2O. They are usually low in protein and fat although there are some exceptions such as avocado and olive.

Fresh fruits and vegetables are living entities after harvest. They actively metabolize organic compounds present in their tissues (principally sugars) generating heat and CO_2, and consuming oxygen as they respire. Few other foods are still alive when brought into the kitchen.

CLASSIFICATION

Most fruits consist of the reproductive organs (ovary) containing the seeds of perennial plants, and tissues closely associated with the ovary. Fruits may be classified botanically as follows.

Aggregate Fruits

Raspberries, blackberries, boysenberries, dewberries, none of which are berries in the botanical sense consist of a collection of small drupes that are loosely held together.

[1] New York State Agricultural Experiment Station and Institute of Food Science, Cornell University, Geneva, NY 14456.

Berry Fruits

Cranberries, gooseberries, blueberries consist of a thick outer wall surrounding a fleshy space in which a number of small seeds are found.

Fleshy Receptacles

Strawberries consist of a large fleshy stem on the surface of which the true fruits or achenes (seeds) are found.

Multiple Fruits

Figs and mulberries are derived from several flowers.

Pome Fruits

Apples, pears, and quinces have a hollow core containing seeds surrounded by fleshy edible stem tissue.

Stone Fruits

Peaches, plums, apricots, cherries, olives are drupes and contain one large hard-shelled pit with a seed in the center.

Vine Fruits

Grapes are true berries and contain from one to several small seeds embedded directly in the flesh.

Citrus Fruits

A type of berry with a leathery rind.

It is necessary to distinguish between the meaning of the word "fruit" to a food technologist and to a botanist. To a food technologist fruits are the fleshy tissues of plant origin that are eaten. To a botanist the word fruit means the reproductive organs or seed-bearing structures of flowering plants. The botanist's definition includes grains, nuts, and legumes as well as a great number of inedible seeds. Therefore, in reading the literature about fruit it is necessary to distinguish in what sense the word is being used.

Vegetables

Almost any part of a growing herbaceous plant may be used as a vegetable, including bulbs (onion), immature flowers (cauliflower, globe artichoke), mature fruits (tomatoes, watermelon, pumpkin), immature fruits (cucumbers, green beans), leaves (celery, lettuce, cabbage, parsley), roots (carrot, radish, yam), mature seeds (dry beans), immature seeds (peas, sweet corn), stems (asparagus), and tubers (potato).

The distinction between fruits and vegetables is not clear cut. Webster's dictionary states "there is no well-drawn distinction between vegetables and fruits in the popular sense, but it has been held by the courts that all those which, like potatoes, carrots, peas, celery, lettuce, tomatoes, etc. are eaten (whether cooked or raw) during the principal part of the meal are to be regarded as vegetables, while those used only for dessert are fruits."

Even this definition does not completely resolve the problem of distinguishing between fruits and vegetables. Applesauce is served with roast pork; pineapple, maraschino cherry, and raisins are served with baked ham; cranberries with turkey; orange with roast duck; and lemon with fish. Do these fruits temporarily become vegetables when served hot with meat dishes? On the other hand, watermelon, cantaloupe and similar melons are considered to be vegetable crops although they are customarily eaten for dessert and botanically would be classified as fruits.

In general, fruits have a fairly high sugar content, sufficient acidity to have a pH below 4.5, and characteristic aromatic flavors while vegetables are lower in sugar content, low in acidity with a pH above 4.5 and do not have the sweet aromatic flavors of fruits. Exceptions can be found to every one of the preceding statements. For example, the tomato has a pH below 4.5 while the ripe fig, avocado, mango and papaya have a pH above 4.5.

STRUCTURE

Rather than strain for a precise distinction between fruits and vegetables we should consider their common features—they are of plant origin, cellular, have a high moisture content and are alive when fresh.

Parenchyma Cells

The edible parts of fruits and vegetables are composed predominantly of fleshy parenchyma cells. Mature parenchyma cells are generally 50–500 micrometer (μm) across. Sometimes they are much larger, as in citrus fruits, where the individual cells can be plainly seen. The typical parenchyma cell is approximately isodiametric and polyhedral. The individual cells are cemented together by an amorphous layer external to the cell wall called the middle lamella, or sometimes the interlamella layer. Parenchyma cells contain the cytoplasm, nucleus, and generally a single large vacuole that accounts for most of the cell volume. The vacuole is surrounded by a membrane and is filled with a watery solution of sugars, acids, and salts that is usually called the "cell sap." Figure 7.1 shows schematically the parenchyma cell of carrot and potato.

Intercellular air space is common in parenchymous tissue and has been estimated as 20–25% in the apple, about 15% in the peach, and 1% in the Irish potato (Sterling 1963). Reeve (1953) measured the intercellular air spaces in several varieties of apples and found they ranged from 210–350 μm wide and from 438–665 μm long, but some individual spaces up to 2000 μm long were observed. Reeve (1970) pointed out that those tissues consisting of small cells with little intercellular space form a compact texture while large cells with considerable intercellular space (e.g., apple) form a coarse or spongy texture.

The living cytoplasmic and vacuolar membranes have the property of differential permeability allowing small molecules such as water to pass through but restricting the passage of larger molecules such as sugars. The physiological processes within the living parenchyma cell enable it to absorb water, thus generating hydrostatic pressure called turgor pressure. It causes the vacuoles to enlarge and press tightly against one another imparting turgidity, rigidity, and crispness to the

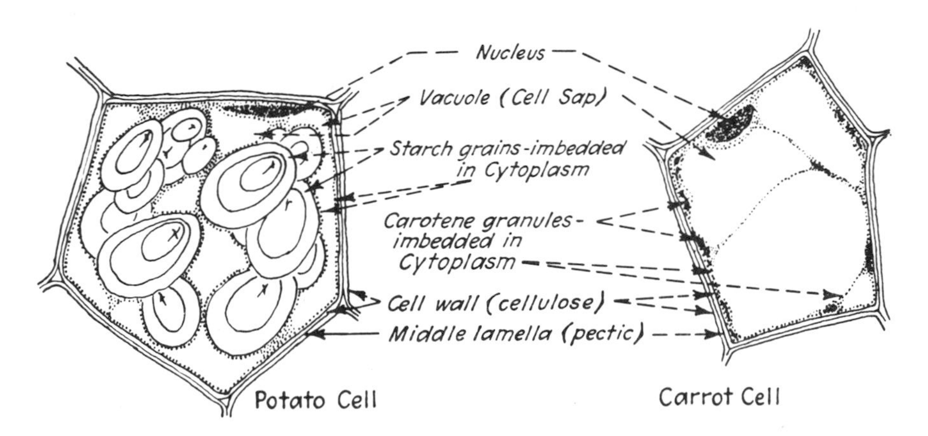

Fig. 7.1. Structure of parenchyma cell of carrot and potato.
From Reeve (1942).

plant tissues. Turgor is lost when the fruit or vegetable is deprived of water or when it dies.

Processing techniques, such as heating or freezing, kill the cell with the result that turgor is absent in processed fruits and vegetables. The differential permeability of the membranes makes it extremely difficult to infuse solutes into the living cell. The differential permeability is lost when the cell dies, and solutes can then be infused into the cell more readily.

The outside or primary cell wall consists of cellulose microfibrils that are loosely woven together in an irregular pattern and embedded in an amorphous matrix composed mainly of pectic substances and hemicelluloses (Esau 1965). Its structure has been likened to that of reinforced concrete. The secondary wall, which lies immediately inside the primary wall, also consists of cellulose microfibrils embedded in an amorphous matrix of hemicellulose and lignin. Figure 7.2 shows the postulated structure of the plant cell wall.

It is remarkable that the cell wall and interlamellar layer, which usually constitute only about 1–3% of the weight of the fresh fruit or vegetable can impart a solid structure to a mixture that is mostly water. Milk and carrot are approximately equal in moisture content (87–88%), and yet milk is a mobile fluid while carrot is a hard solid. The small quantity of polymeric material in or near the cell wall of the carrot is responsible for its hard and rigid texture.

Other Cells

Although parenchyma cells predominate in the edible portion of fruits and vegetables, other types of cells are also present. The conducting cells are long tube-shaped structures that transport food (phloem cells) and water (xylem cells) throughout the tissues. Supporting cells (collenchyma cells) are thick-walled cells that usually take the form of fibers, but may also appear in other shapes such as the rounded lignified sclereids (stone cells) that are conspicuous in the flesh of pears, quinces, and nispero. These supporting and conducting cells form an interconnected branching network throughout the fleshy parenchyma tissues. They are frequently heavily lignified. Generally their physical properties and relative scarcity in the edible tissue are such that their texture is not intrusive on the textural sensations imparted by the parenchyma cells, although fibers (bundles of conducting cells) and grittiness (sclereids) are noticed in some commodities. The texture of the conducting tissues stands out in fibrous celery, asparagus, and mango.

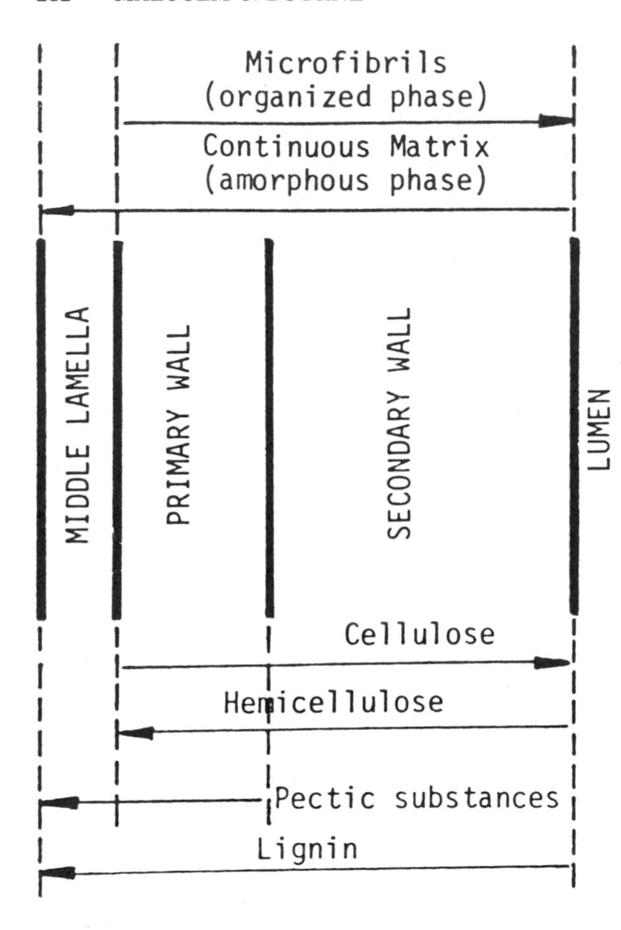

Fig. 7.2. Generalized structure of the plant cell wall. Arrows indicate direction of increasing concentration of each component. *Redrawn from Northcote (1958).*

Fruits and vegetables are covered with one layer of epidermal cells and often several layers of hypodermal cells that constitute the skin and protect the underlying tissues. The epidermal layer contains a number of stoma, which are tiny openings through which gaseous exchange takes place; carbon dioxide and water vapor can be given off to the atmosphere and oxygen enters the tissue through the stoma. The skin provides a natural barrier against invasion of the flesh by rot and mold microorganisms. Fruits and vegetables with damaged skin have a much higher incidence of spoilage than equivalent produce with undamaged skin as they pass through the postharvest system.

The epidermal layer may be very thin (lettuce, strawberry), moderately thick (cherry, potato), or very thick (grapefruit, watermelon); it may be tender (nectarine, green pea), tough (grapes, tomato), fuzzy (peach), hard (pumpkin), or prickly (pineapple). The skin is usually eaten with the fleshy portion when it is tender or thin and removed before eating if it is texturally objectionable. The skins of fruits and vegetables that are of intermediate toughness or thickness are eaten by some people and not by others. Some people peel their apples, potatoes, and tomatoes before eating and some do not.

The epidermal cells excrete a wax (cuticular wax) that helps protect the surface and retard water loss. This wax coating may be so thin that it is unnoticeable (strawberry) or it may be quite prominent (some apples and pears).

The skin of some fruits and vegetables is deliberately waxed after harvest in addition to the naturally present cuticular wax. The waxing process imparts a shiny gloss to the skin giving it more attractive appearance, and it retards moisture loss in storage thus maintaining turgidity. Sometimes preservatives are added to the wax to inhibit the growth of spoilage organisms. Oranges from Florida often have red dye added to the wax to give the fruits a better-looking orange color.

Middle Lamella

The middle lamella is an amorphous layer external to the primary cell wall that cements the individual cells together. Recent evidence indicates that the interlamella layer is not entirely amorphous but contains some elementary pectate fibrils (Leeper 1973). The interlamella layer consists principally of the calcium salts of polymers of galacturonic acid that have been partially esterified with methyl alcohol and is known as the pectic material. Figure 7.3 shows the structural formula for D-galacturonic acid, its methyl ester, and part of a pectin molecule.

The degree of polymerization and esterification of the polygalacturonide chains, and the amount of crosslinking of adjacent pectin molecules by salt bridge formation have profound effects on the physical properties of the middle lamella and the overall textural properties of fruits and vegetables. The changes in the chemical nature of the pectic materials are the primary cause of changes that occur in the textural properties of horticultural products (Kertesz 1951; Doesburg 1965; Van Buren 1979).

In green fruits the pectic material is principally in the form of partially esterified polygalacturonic acid of very high molecular weight called protopectin. This is the generic name given to the water-insoluble pectic substances (Joslyn 1962). It imparts great strength to the tissue.

As the fruit ripens, the chain length of the pectin polymer decreases forming water-soluble pectin, which is not as strong as the protopectin,

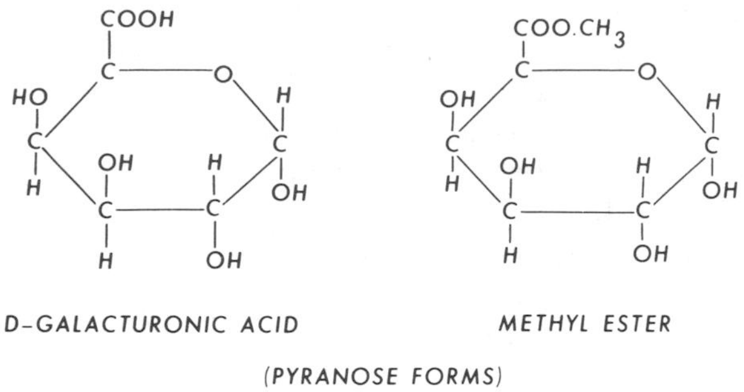

D-GALACTURONIC ACID METHYL ESTER

(PYRANOSE FORMS)

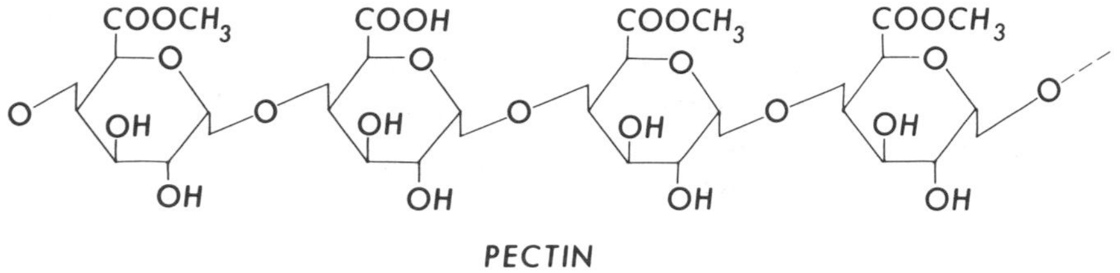

PECTIN

Fig. 7.3. (Top) the monomers D-galacturonic acid and its methyl ester which polymerize to form pectin. (Bottom) portion of a pectin molecule.

and the structure becomes increasingly soft and eventually mushy. Table 7.1 shows the decrease in protopectin and increase in pectin content of four varieties of banana as they ripen. The changes in pectin are caused by two groups of enzymes: (a) pectin methylesterases which catalyze the deesterification of pectin, yielding free polygalacturonic acid and methanol; and (b) polygalacturonases which catalyze the splitting of the 1,4 glycosidic bonds of the pectin molecule.

Heat also promotes the depolymerization of the protopectin and the pectin resulting in the great degree of softening that accompanies the cooking of fruits and vegetables. Table 7.2. shows the changes that occur in the pectic substances of carrot and parsnip as a result of steaming.

Lignin is another amorphous polymeric material that sometimes becomes encrusted in the interlamella layer and cell wall as secondary material to pectin. Lignin renders the tissue hard and woody and provides mechanical strength for the plant cell walls and for the plant. Mature wood may contain as much as 30–40% lignin. The stone cells and fiber cells in fruits and vegetables are highly lignified. Lignin is a heterogeneous highly polymeric material built up largely from *p*-hydroxyphenyl-propane monomers, but the degree of substitution and

TABLE 7.1. PECTIN AND PROTOPECTIN IN BANANA PULP[1]

	Protopectin		Pectin	
Variety	Unripe	Ripe	Unripe	Ripe
Gros Michael	0.53	0.22	0.27	0.40
Lady Finger	0.50	0.35	0.21	0.68
Lacatan	0.59	0.34	trace	0.46
Red banana	0.87	0.48	0.04	0.63

Source: Von Loesecke (1950).
[1]Expressed as % calcium pectate.

TABLE 7.2. EFFECT OF STEAMING ON PECTIC SUBSTANCES IN CARROT AND PARSNIP[1]

	Uncooked		Steamed 20 min		Steamed 45 min	
Pectin Fraction	Carrot	Parsnip	Carrot	Parsnip	Carrot	Parsnip
Total pectic substance	18.6	16.4	16.1	15.8	13.7	15.7
Protopectin	14.1	10.2	9.0	7.7	3.6	5.7
Pectin	3.7	4.7	6.0	6.1	8.8	7.9
Pectic Acid	0.8	1.6	1.0	2.0	1.3	2.1

Source: Simpson and Halliday (1941).
[1]Expressed as % of total dry matter.

type of linkage between the monomers is highly variable (Isherwood 1963; Yan 1982). Unlike pectin, lignin tends to increase in quantity and degree of polymerization as the plant tissue becomes older. Also, unlike pectin, lignin is not depolymerized, solubilized, or removed by naturally occurring enzymes or by heating. Hence, lignification may be considered as irreversible toughening.

The woodiness or stringiness found in old vegetables is largely caused by the deposition of lignin in the walls of plant cells, especially in the walls of the fiber cells. Since there is no acceptable way of removing it, the only practical means of controlling lignification is by prevention. Woody and fibrous tissues that are associated with lignin are avoided by harvesting lignin-prone commodities when they are young before any extensive lignification of the cells can develop.

A successful case of control of lignification is found in green beans, which traditionally had a tough string along the edge of the pod. Plant breeders have been able to postpone lignification of green beans by selection, with the result that modern varieties of green beans no longer have the objectionable string.

The relative strengths of the middle lamella and cell wall determines the locus of the fracture that occurs when edible plant tissue is stressed to its breaking point. When the middle lamella is stronger, the cell walls break releasing the cell contents. Cell sap then pours into the mouth giving the sensation of juiciness. In potatoes, starch grains are released giving rise to gumminess and pastiness in the cooked product.

When the cell wall is stronger, the middle lamella breaks giving intact cells. The cell contents are still confined within the cells. Cell sap is not released and the product feels dry in the mouth. In potatoes, starch grains are not released and the cooked potato has a fluffy, nongummy texture. Apples freshly harvested in autumn have a strong middle lamella; when bitten into the cell walls break and the apple feels juicy. The middle lamella weakens during storage due to degradation of the pectic substances; an apple that has been stored for 6 months breaks through the middle lamella when bitten giving the sensation of dry, slightly abrasive cells and little juiciness even though the moisture content, as determined by chemical analysis, is the same as the juicy apple eaten in autumn.

Figure 7.4 shows sections of the fractured surfaces of hydrated mature lima beans. In the raw bean (Fig. 7.4A) the cell walls have broken and the starch grains can be clearly seen inside the broken cells. Figure 7.4B shows a bean that has been cooked for 20 min. The heat has weakened the middle lamella so that it fractures instead of the cell walls and intact cells are obtained.

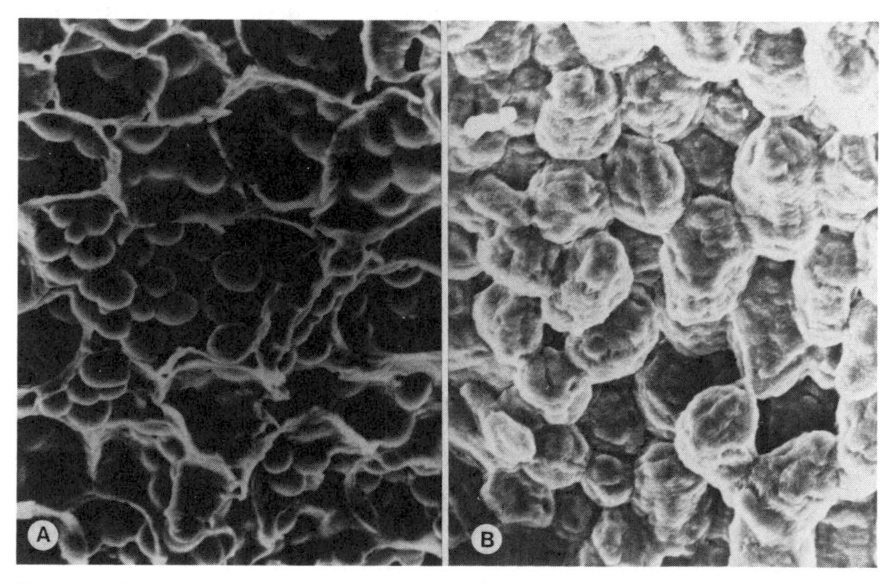

Fig. 7.4. Scanning electron microscope photograph of the fractured surface of water-soaked cotyledons of lima bean. (A) uncooked, (B) cooked 20 min.
From Rockland and Jones (1974).

Turgor Pressure

The living plant tissue has the ability to absorb water through the cell walls and to generate hydrostatic pressure within the living cells. This is called turgor or turgor pressure. It can reach as high as 200 pounds per square inch. It causes the vacuoles to enlarge and press against the partially elastic cell walls causing them to stretch and press tightly against one another imparting turgidity, rigidity, and crispness to the plant tissue. Turgor is largely responsible for the maintenance of young plant parts in a rigid upright condition until the structural support cells are sufficiently developed to provide the mechanical strength necessary to maintain the plant parts in a rigid condition. The desirable property of crispness that is found in many fresh fruits and vegetables is largely due to turgor pressure.

Turgor is lost when the plant is deprived of water. Growing herbaceous plants wilt and become flaccid when the soil in which they grow dries out, causing water deficit in the plant. They soon become rigid again when the soil is moistened and the cells regain their turgor. Fruits and vegetables are cut off from their supply of water when they are separated from the plant and they wilt quickly if allowed to lose moisture. For this reason it is customary to store fruits and vegetables under high humidity conditions after harvest to retard water loss and maintain turgor and crispness.

Vegetables that maintain good vascular tissue can absorb moisture after harvest. Lettuce, parsley, celery and carrots can absorb water and regain their crispness if immersed in water after being allowed to wilt.

The physiological processes of the living plant cell are responsible for producing and maintaining turgor pressure. These processes cease when the cell dies with the result that the turgor pressure is dissipated. Since heating, freezing, and dehydration kill the cell, turgor is absent in processed fruits and vegetables. Hence, processed commodities never have the crispness and succulence of the raw material.

The loss of moisture from transpiration and respiration can result in great loss of quality in those fruits and vegetables that rely heavily on turgor for crispness. Most commodities should be stored in a 90% relative humidity atmosphere (Lutz and Hardenburg 1968). Commercial storages usually maintain high humidity, whereas supermarkets and home refrigerators do not. Consequently, a fruit or vegetable that has successfully withstood a long period of commercial storage with little loss of quality frequently shows extreme wilting within a few days in the store or home. The wrapping of produce in polyethylene film bags reduces the rate of moisture loss and gives better retention of turgor.

The removal of inedible leafy material that drains moisture from the edible portion helps to maintain turgor. For example, carrots stored with the leafy tops removed wilt much more slowly than those that have the tops attached. Hardenburg (1966) found that unpackaged carrots stored at 21°C (70°F) and 50% relative humidity lost 48% of their weight with the tops attached and only 29% with the tops removed. Topless carrots stored in perforated polyethylene bags lost 4% weight.

FACTORS AFFECTING PHYSICAL PROPERTIES

Many factors affect the physical properties of horticultural crops. The major ones are discussed below.

Nature of the Commodity

There are well-recognized differences between commodities. Apples and carrots are firm while apricots and tomatoes are soft.

Cultivar (Variety)

Different varieties of the same fruit or vegetable may have substantial differences in their textural properties. Table 7.3 shows the wide differences in textural properties found between different apple varieties.

Maturity

The maturity of the plant at time of harvest may have a slight, moderate, or profound effect on the textural properties depending on the commodity.

Some plant parts change rather slowly in quality, and the time of harvest has little effect on their textural properties. Carrots, onions, cabbage, and turnips grow larger with time but do not change greatly in texture.

In contrast, some plants are in a phase of rapid change at harvest time. Green peas and sweet corn are immature seeds that are harvested during a time of very rapid physiological change. Every day gives a large increase in yield per acre while quality rapidly rises to a peak and then rapidly declines again. Selecting the optimum time of harvest for these vegetable crops is a subject that has been given great attention by the research community and by vegetable processors. Figure 7.5 shows the changes in total yield and yield of overmature, mature, and imma-

TABLE 7.3. TEXTURAL PROPERTIES OF DIFFERENT APPLE VARIETIES

Variety	Instrumental Texture Profile Analysis Parameters							
	Fracturability (kg)	Hardness (kg)	Area 1	Area 2	Cohesiveness (A_2A_1)	Springiness (mm)	Gumminess	Chewiness
American Forester	10.8	11.0	10.26	0.82	0.079	2.2	0.878	2.06
Ben Davis	15.1	40.4	38.8	3.6	0.094	2.8	3.78	10.55
Crow Egg	11.3	13.5	15.9	1.21	0.076	2.2	0.14	0.25
Cox's Orange Pippin	9.4	10.5	11.9	0.59	0.049	1.8	0.52	0.99
Delcon	6.0	12.0	11.3	0.60	0.052	1.9	0.64	1.31
Delawine	9.3	14.6	18.5	1.50	0.082	2.5	1.19	3.03
Democrat	16.7	39.8	40.4	3.2	0.081	2.2	3.21	6.94
Fyan	15.6	19.2	17.5	2.1	0.120	2.3	2.26	5.15
Bates Giant Lobo	7.5	17.3	16.2	1.27	0.078	2.0	1.36	2.73
Golden Delicious	12.1	13.0	15.8	1.06	0.068	2.0	0.88	1.77
Golden Russet	34.6	79.7	39.0	3.0	0.078	2.2	6.23	13.75
Jonwin	11.8	13.6	16.8	1.28	0.076	2.8	1.04	2.90
Jonathan	9.8	19.1	22.4	1.45	0.065	2.4	1.24	3.02
Kyokko	8.5	7.1	9.39	0.42	0.044	1.4	0.33	0.49
Lyons	3.7	8.9	8.3	0.45	0.054	1.9	4.84	9.52
Limbertwig	14.6	33.8	28.8	1.98	0.068	2.2	2.32	5.14
Melrose	8.8	30.1	29.9	2.40	0.080	2.6	2.43	6.29
Rival	11.6	11.4	14.9	0.74	0.047	1.7	0.58	1.08
Scott Winter	13.0	14.7	17.0	1.16	0.068	2.5	1.01	2.59
Yorking	15.1	22.7	24.2	2.0	0.085	2.4	1.90	4.52

Source: Bourne (1979).

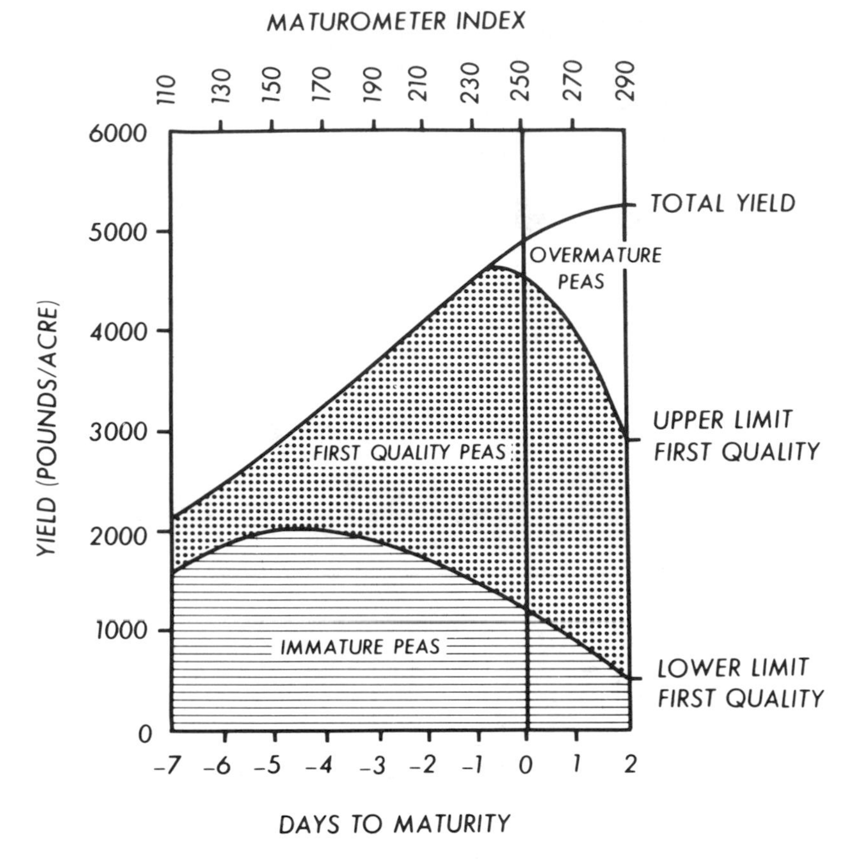

Fig. 7.5. Yield and quality of green peas as affected by days growing in the field. The optimal harvest time (marked 0 on the abscissa) for peas for canning is that day that gives the highest yield of first quality peas.
Redrawn from data of Lynch and Mitchell (1953).

ture peas over a 10-day period using the maturometer (a puncture tester) to measure maturity. If the harvest time is off by even 1 day, the yield of first quality peas is reduced.

Ripeness

Fruits may be divided into two groups (Bourne 1979). (1) Fruits that soften greatly as they ripen (apricots, blackberries, strawberries, cherries, and peaches) in which the firmness in the ripe state is less than half of the firmness in the unripe state. (2) Fruits that soften moderately (apples, cranberries, quinces). The firmness in the ripe condition is more than half of that in the unripe condition (see Table 7.4).

Fruits that soften greatly as they ripen generally have a shorter storage life than fruits that soften moderately. They also require greater care in packaging and transportation because soft fruits bruise more easily than firm fruits. Fruits that soften greatly are picked at a mature but unripe stage while they are still firm enough to withstand bruising damage during their progress through marketing channels. When picked at the correct stage of early maturity they ripen and soften to give the typical texture, flavor, and color of the ripe fruit after reaching their destination.

Size

Small fruits tend to be slightly firmer than larger fruits at an equivalent stage of development. Small apples are slightly firmer than larger apples (Table 7.5). Darrow (1931) found that small strawberries are firmer than medium-size strawberries.

For those vegetables that are harvested as immature seeds the size is highly correlated with textural properties. Green peas and sweet corn increase in size, hardness, and skin toughness as they mature. Figure 7.5 shows the yield of immature, mature, and overmature green peas as a function of harvest day and firmness as measured by the maturometer.

Cultural Factors

Some cultural factors affect the physical properties of horticultural crops. Luh and Liao (1969) found that clingstone peaches harvested from potassium-fertilized trees were slightly softer after canning than peaches growing on trees that had not been fertilized. Williams *et al.* (1964) showed that spraying apple trees in the field with the hormone *N*-dimethylaminosuccinamic acid (Alar, B-9) results in a firmer fruit at the time of harvest and throughout storage (see Table 7.6).

Blanpied *et al.* (1978) showed that nitrogen fertilization of apple trees decreased the firmness of apples at harvest and after 4 months storage. These authors also show that the position of the fruit on the tree affects the firmness. Apples grown in exposed positions that receive much sunlight are generally firmer than well-shaded apples grown on the same tree (Fig. 7.6).

The firmness of canned snap beans can be affected by the nature and amount of mineral fertilizers available to the growing bean plant. Van

TABLE 7.4. FIRMNESS OF FRUITS IN NEW YORK STATE

Fruit	Unripe		Ripe		Firmness Ratio
	Date of Harvest	Firmness	Date of Harvest	Firmness	Ripe/Unripe
A. Fruits That Soften Greatly During Ripening					
Marigold peaches[1]	July 20	113	August 4	24	0.21
Halehaven peaches[1]	August 3	166	August 24	17	0.10
Elberta peaches[1]	August 8	138	September 8	26	0.19
Bartlett pears[1]	August 8	187	September 1	6	0.03
Quackenboss plums[2]	August 17	11.9	September 7	2.6	0.22
Middleburg plums[2]	September 3	22.7	September 24	4.8	0.21
B. Fruits That Soften Moderately During Ripening					
Cortland apples[1]	August 28	108	September 24	91	0.84
McIntosh apples[1]	August 28	91	September 15	75	0.82

[1] Tenderometer test. Data from Lee and Oberle (1948). Firmness in Tenderometer units.
[2] Magness–Taylor pressure test with 7/16 in. tip. Data from Robinson and Holgate (1949). Firmness in pounds.

TABLE 7.5. RELATION BETWEEN FRUIT SIZE AND FIRMNESS[1]

	Fruit Size (diameter in in.)			
Variety	2¼	2½	2¾	3
McIntosh	12.8	12.5	12.5	11.3
Empire	15.6	14.6	14.6	14.3
Delicious	17.0	16.0	15.4	12.2
Stayman	—	15.7	14.6	—
Winesap	—	19.5	18.5	—

[1] Magness–Taylor test with 7/16 in. tip. Data from Blanpied *et al.* (1978).

TABLE 7.6. EFFECT OF HORMONE ORCHARD SPRAY[1] ON APPLE FIRMNESS[2]

	Stored 6 months at 0°C	Stored 10 days at 20°C following 6 months at 0°C
Orchard A		
No spray	14.8	11.7
Sprayed	18.6	15.1
Orchard B		
No spray	15.9	12.4
Sprayed	17.9	15.2

[1] *N*-dimethylaminosuccinamic acid spray applied to Red Delicious apple trees.
[2] Magness–Taylor test with 7/16 in. tip. Data from Williams *et al.* (1964).

Buren and Peck (1963) showed that bean plants grown in sand culture with high calcium levels in the nutrient solution yielded pods with increased calcium content and greater firmness than those grown with low calcium levels. The same authors showed that the firmness of canned green beans was affected by the potassium concentration in the pods, which could be controlled by the level of potassium fertilizer added to the soil (Van Buren and Peck 1982). In this case, increasing potassium concentration reduced the firmness (see Fig. 7.7). They attributed the softening effect of K^+ as being due to a combination of Ca^{2+} displacement and enhancement of pectin degradation.

Storage Conditions

Since fresh horticultural crops are living tissue, the physiological life processes continue during storage. Hence, storage conditions and the length of storage affect the physical properties of fruits and vegetables. In general, fruits that soften during ripening continue to soften during storage although at a slower rate in cool storage than at ambient temperature. Vegetables that toughen because of lignification will continue to toughen during storage. Fruits and vegetables lose moisture

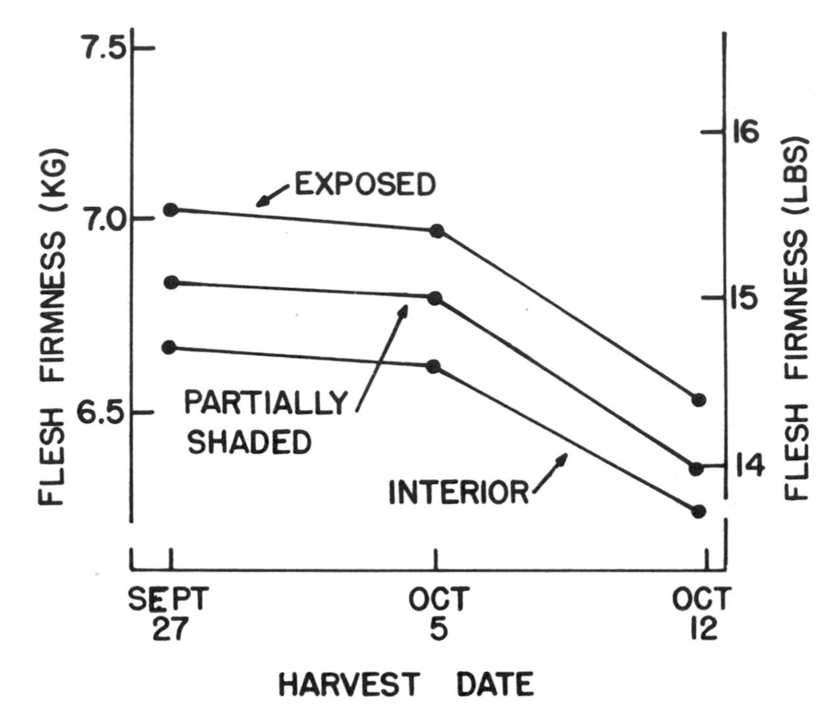

Fig. 7.6. Firmness of Jonathan apples picked from exposed, partially shaded and interior zones of large mature trees. Magness–Taylor test.
From Blanpied et al. (1978).

during storage and become wilted and flaccid unless stored under high humidity conditions. Lutz and Hardenburg (1968) list the optimum conditions for maintenance of quality in stored temperate fruits and vegetables.

Temperature

The temperature at which fruits and vegetables are tested can affect the physical properties. This topic was reviewed by Bourne (1982) who showed that the firmness of most fruits and vegetables showed a decrease in firmness with increasing temperature over the range 0–45°C. Figure 7.8 shows typical data for the effect of temperature on firmness measurements on some fruits and vegetables.

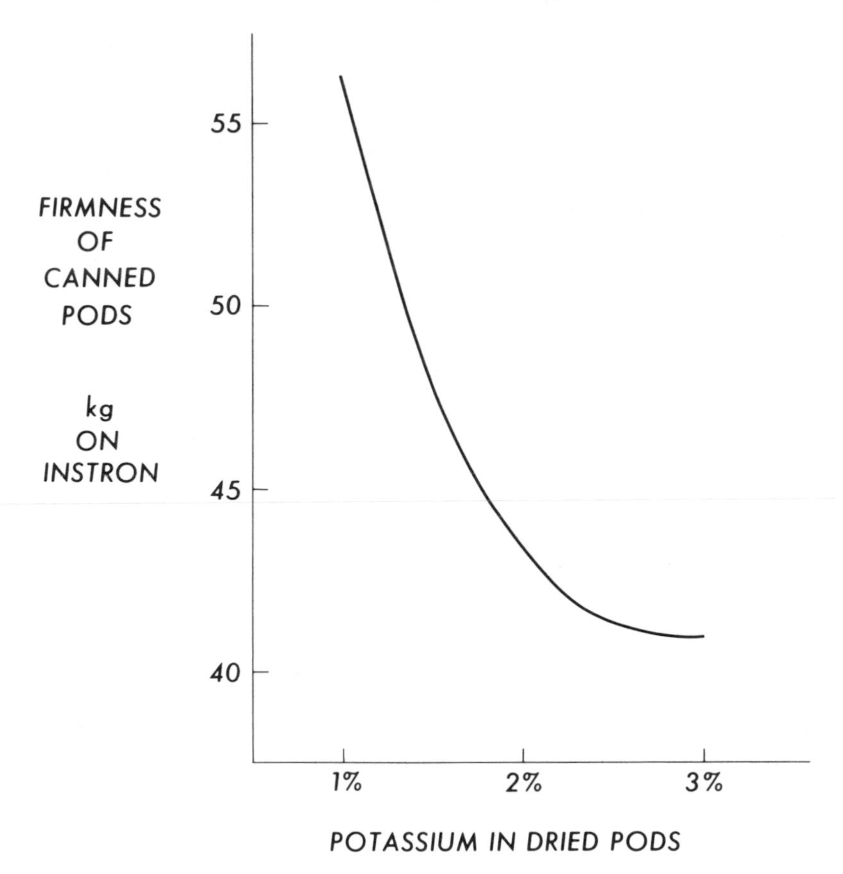

Fig. 7.7. Relation between pod potassium content as influenced by fertilization and firmness of green beans.
From Van Buren and Peck (1982).

Bourne defined a firmness-temperature coefficient as:

$$\%\text{ change in firmness per degree temperature} = \frac{\text{firmness at } T_2 - \text{firmness at } T_1}{\text{firmness at } T_1\,(T_2 - T_1)} \times 100$$

where T_1 = lowest temperature and T_2 = highest temperature at which firmness is measured.

The firmness–temperature coefficient ranged from -1.65 for apricot to $+0.12$ for carrot using the puncture method, and from -0.04 for

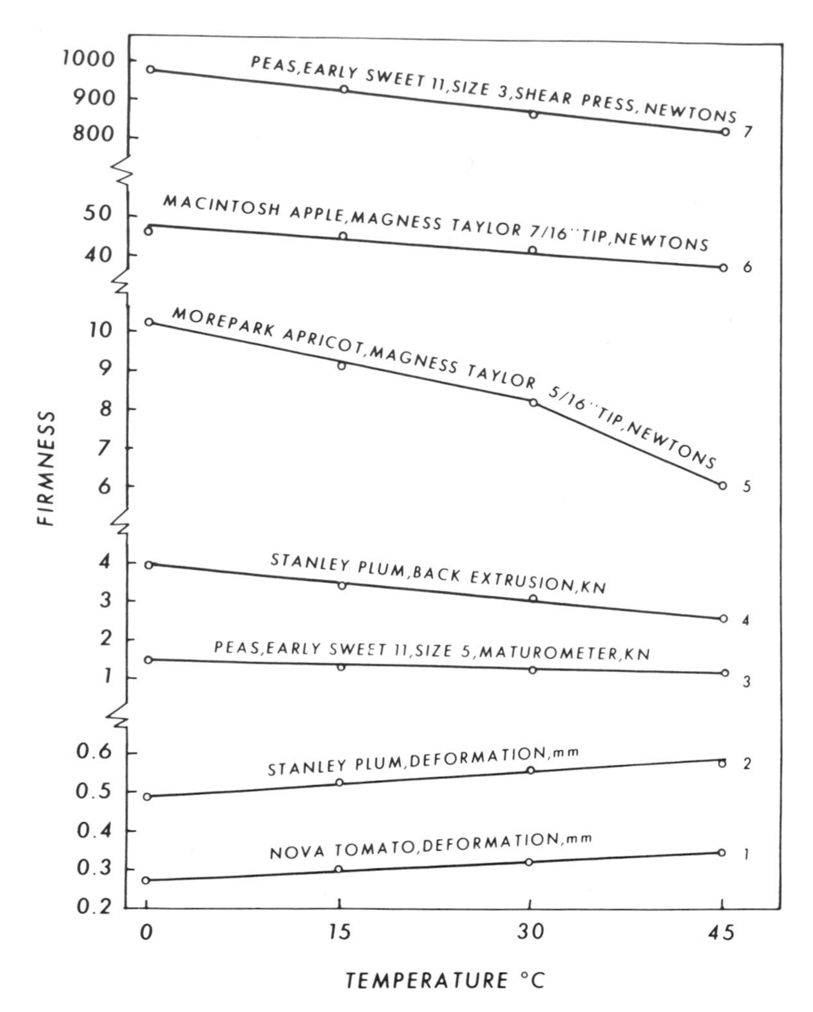

Fig. 7.8. Effect of temperature on firmness of some fruits and vegetables. The units in which each firmness was made are shown alongside the commodity description. The deformation test performed on plum and tomato is a measure of softness, i.e., a high reading indicates low firmness.
From Bourne (1982).

Golden Delicious apples stored for 7 months to −1.34 for NK199 sweet corn using the extrusion method. With a few exceptions which showed a slight firming effect, most fruits and vegetables tested showed decreasing firmness with increasing temperature over the range 0−45°C. The firmness−temperature relationship was approximately linear for most commodities.

The firmness–temperature coefficient is highly variable; it varies from commodity to commodity, from cultivar to cultivar within the same commodity, from one test principle to another on the same commodity and on the same cultivar during storage and from year to year. For those commodities that are graded on a firmness measurement, it is possible to change the apparent grade by adjusting the temperature by a sufficient amount in the right direction. Researchers should endeavor to make firmness measurements within a narrow temperature range unless preliminary studies have shown a small firmness–temperature coefficient for that commodity.

BIBLIOGRAPHY

BLANPIED, G.D., BRAMLAGE, W.J., DEWEY, D.H., LABELLE, R.L., MASSEY, L.M. JR., MATTUS, G.E., STILES, W.C., and WATADA, A.E. 1978. A standardized method for collecting apple pressure test data. N.Y. Food Life Sci. Bull. Pomology No. 1.

BOURNE, M.C. 1979. Texture of temperate fruit. J. Texture Stud. *10*, 24–44.

BOURNE, M.C. 1982. Effect of temperature on firmness of raw fruits and vegetables. J. Food Sci. *47*, 440–444.

BROWN, M.S. 1979. Frozen fruits and vegetables. Their chemistry, physics and cryobiology. Adv. Food Res. 25, 181–235.

CZYHRINCIW, N. 1969. Tropical fruit technology. Adv. Food Res. *17*, 153–207.

DARROW, G.M. 1931. Effect of fertilizers on firmness and flavor of strawberries in North Carolina. Proc. Am. Soc. Hort. Sci. *28*, 231–235.

DOESBURG, J.J. 1965. Pectic substances in fresh and preserved fruits and vegetables. Ins. Res. Storage Processing Hort. Produce Commun. No. 25. Wageningen, The Netherlands.

ESAU, K. 1965. Plant Anatomy. John Wiley and Sons, New York.

HARDENBURG, R.E. 1966. Packaging and protection. *In* Protecting Our Food, pp. 102–117. U.S. Dept of Agriculture, Yearbook of Agriculture. U.S. Govt. Printing Office, Washington, D.C.

ISHERWOOD, F.A. 1963. Lignin. *In* Recent Advances in Food Science, Vol. 3, pp. 300–309. J.M. Leitch and D.N. Rhodes (Editors). Butterworths, London.

JOSLYN, M.A. 1962. The chemistry of protopectin: A critical review of historical data and recent developments. Adv. Food Res. *11*, 1–107.

KERTESZ, Z.I. 1951. The Pectic Substances. Wiley (Interscience), New York.

LEE, F.A., and OBERLE, G.D. 1948. The use of the Tenderometer for the determination of the firmness of apples, peaches and pears. Fruit Products J. *27*(8), 244–6.

LEEPER, D.N. 1973. Pectin ultrastructure I Pectate elementary fibrils. J. Texture Stud. *4*, 248–253.

LUH, B.S., and LIAO, F.W. 1969. Texture, color and chemical composition of cling peaches as affected by potassium fertilization. Fruchtsaft Ind. *14*, 2–8.

LUTZ, J.M., and HARDENBURG, R.E. 1968. The Commercial Storage of Fruits, Vegetables, and Florist and Nursery Stock. U.S. Dept of Agriculture Handbook No. 66. Washington, D.C.

LYNCH, L.J., and MITCHELL, R.S. 1953. The definition and prediction of the optimal harvest time of pea canning crops. Australia CSIRO Bull. No. 273. Melbourne, Australia.

NORTHCOTE, D.H. 1958. The cell walls of higher plants: their composition, structure and growth. Biol. Rev *33*, 53–102.

REEVE, R.M. 1942. Facts of vegetable dehydration revealed by microscrope. Food Ind. *14* (12), 51–54, 107, 108.

REEVE, R.M. 1953. Histological investigations of texture in apples. II. Structure and intercellular spaces. Food Res. *18*, 604–617.

REEVE, R.M. 1970. Relationships of histological structure to texture of fresh and processed fruits and vegetables. J. Texture Stud. *1*, 247–284.

ROBINSON, W.B., and HOLGATE, D.C. 1949. How ripe is a "ripe" plum? Farm Res. *15*(3), 7.

ROCKLAND, L.B., and JONES, F.T. 1974. Scanning electron microscope studies on dry beans. Effect of cooking on the cellular structure of cotyledons in rehydrated large lima beans. J. Food Sci. *39*, 342–346.

SIMPSON, J.I., and HALLIDAY, E.G. 1941. Chemical and histological studies of the disintegration of cell-membrane materials in vegetables during cooking. Food Res. *6*, 189–206.

STERLING, C. 1963. Texture and cell-wall polysaccharides in foods. *In* Recent Advances in Food Science, Vol. 3, p. 259. J.M. Leitch and D.N. Rhodes (Editors). Butterworths, London.

VAN BUREN, J.P. 1979. The chemistry of texture in fruits and vegetables. J. Texture Stud. *10*, 1–23.

VAN BUREN, J.P., and PECK, N.H. 1963. Effect of calcium level in nutrient solution on quality of snap bean pods. Proc. Am. Soc. Hort. Sci. *82*, 316–321.

VAN BUREN, J.P., and PECK, N.H. 1982. Effect of K fertilization and addition of salts on the texture of canned snap bean pods. J. Food Sci. *47*, 311–313.

VON LOESECKE, H.W. 1950. Bananas. Chemistry, Physiology, Technology. Wiley (Interscience), New York.

WILLIAMS, M.W., BATJER, L.P., and MARTIN, G.C. 1964. Effects of *N*-dimethylaminosuccinamic acid (B-9) on apple quality. Proc. Am. Soc. Hort. Sci. *85*, 17–19.

YAN, J.F. 1982. Kinetics of delignification: a molecular approach. Science *215*, 1390–1392.

8

Structural and Textural Characteristics of Baked Goods

M.V. Taranto[1]

INTRODUCTION

Foods in which small bubbles of gas constitute a major portion of the volume are classified as foams. These products consist of a discontinuous phase of gases (e.g., air) and a continuous liquid or solid phase that supports and maintains the structure. Sponges are a type of foam distinguished by having a greater firmness. One of the more economically important examples of foam/sponge foods is leavened bakery products.

Baked products are inherently complex systems because they involve a large number of ingredients. During the cooking (baking) of cake batters and bread doughs, heat induces physical and chemical changes in the components of the batter or dough system. These heat-induced changes yield a stable structure with subjectively desirable flavor, aroma, and texture characteristics.

Texture in breads and cake is, to a large extent, a function of the rheological properties of the material making up the gas cell walls. Some of the other factors affecting the texture of breads and cakes, as perceived in the mouth, are the shape, size, and size distribution of the gas cells.

The gas in the cells is always virtually at atmospheric pressure when the finished product is at rest. However, the gas cell walls may restrict the escape of gas sufficiently to cause a positive pressure that can be detected when biting and chewing the product. This occurs when bread is eaten, but is of no significance in layer cakes.

[1]ITT Continental Baking Company, Research & Development Laboratories, Rye, NY 10580.

A true picture of the structure of the films that form the gas cell walls in bread and cake would be a valuable contribution to many baking studies, including the role of starch in baking and bread staling and the effects various treatments with enzymes, oxidizing agents, and reducing agents have on the structure of both bread and cake products.

Microscopic techniques have proved to be valuable research tools for studying leavened baked products. These techniques yield information that is used to characterize the structure of individual components and also demonstrate the relationships among components in bread doughs and cake batters both before and after heating.

BREAD STRUCTURE

Bread and other wheat products are important components in our diet and, therefore, have been the subject of numerous chemical and microstructural studies. Sheffer (1916) was one of the pioneers in the use of transmitted light microscopy (TLM) to study dough and bread structure. Some of the earliest micrographs of bread and dough were published by Butterworth and Colbeck (1938), Baker (1941), and Burhans and Clapp (1942). Wheat kernel structure, flour components, dough formation, and bread structure have been studied more recently using TLM, transmission electron microscopy (TEM), and scanning electron microscopy (SEM) (Sandstedt *et al.* 1954; Khoo *et al.* 1975; Simmonds, 1972; Bechtel *et al.* 1978).

The type of white bread loaf the U.S. consumer now expects can only be produced from wheat flour. Starch granule size and morphology and gluten-film-forming proteins are important components of bread (Chabot 1979). Substitution of other nonwheat starches or proteins in bread drastically alters crumb structure (Hyder *et al.* 1974; Fleming and Sosulski 1978). Therefore, an understanding of the precise functional relationships between all the components in a bread loaf is critical. This information would be useful in aiding attempts to add other nutrients to the product (e.g., protein supplements or fiber) and studying the effects of processing on bread structure.

Sample Preparation for Microscopy

Burhans and Clapp (1942) pointed out the difficulty involved in preparing flour, dough, and bread samples for TLM. The use of usual botanical techniques of sectioning and photographing was not satisfactory for bread films (Butterworth and Colbeck 1938; Verschaffelt and van Teutem, 1915). Wetting and drying of the samples called for in these methods caused structural distortions that may lead to misinterpretation (Sandstedt *et al.* 1954). Burhans and Clapp (1942) cut sections from samples frozen with carbon dioxide, fixed and stained with osmic acid vapor, dehydrated with calcium oxide, embedded with paraffin, and cut with a sliding microtome. Since bread crumb will absorb about five times its weight of water, samples as observed in water or stained by a water solution of stain present a distorted image of the crumb structure (Sandstedt *et al.* 1954). Dehydrating the samples in preparation for sectioning and staining with stains in the dehydrating solvents is not desirable because the shrinkage of the starch and protein causes breaks between starch granules and protein (Verschaffelt and van Teutem 1915). Sandstedt *et al.* (1954) developed a sample preparation procedure based on geological methods. The dough or bread is embedded in a liquid plastic containing a hardening catalyst. Infiltration of the plastic into the sample is facilitated by removal of gas by reduced pressure. The plastic embedded sample is sectioned by grinding. The exposed surface is stained with appropriate stains dissolved in 80% glycerin and viewed with reflected light.

The advantages of Sandstedt's *et al.* (1954) procedure are (1) the need for thin sections is avoided, and (2) the use of 80% glycerine as a stain solvent limits the shrinkage or swelling of the starch or protein present in the sample. Recent work using TEM and SEM has reaffirmed problems involved with the investigation of bread dough structure (Chabot 1979; Chabot *et al.* 1979; Varriano-Marston 1977, 1981; Barlow *et al.* 1973; Aranyi and Hawrylewicz 1969; Crozet 1977; Buttrose 1973).

There have been few micrographs of bread structure published (Chabot 1979). When an aqueous fixative is used to stabilize structures for dehydration and/or plastic embedding, certain physical changes immediately occur. Bread crumb samples swell rapidly when placed in aqueous fixatives (Chabot 1979). The starch granules imbibe water and become swollen (Chabot *et al.* 1979; Chabot 1979). The protein matrix shrinks away from the starch granules altering the functional relationship between these components (Chabot 1979; Chabot *et al.* 1979; Varriano-Marston 1977, 1981). Fixation with glutaraldehyde and post fixation with osmium tetroxide in buffers followed by several dehydration procedures all induce major physical changes in bread crumb structure (Chabot 1979). The gas cells are still identifiable, but the gas cell walls are distorted. The gluten film becomes fragmented, exposing the starch granule surfaces.

From the data presented in the literature, one can conclude that the fixation procedures commonly used for TEM induce major physical changes in bread crumb morphology. However, it should be stressed that as long as the artifacts produced by various sample preparation schemes are clearly defined and reproducible, these methods can still be used to gain useful information.

Results of Bread Microscopy Studies

Extensive studies into the process of mixing, gas cell formation during proofing, and crumb structure of the baked product were conducted by Burhans and Clapp (1942). These workers observed that the whiteness in baked bread is enhanced by the presence of a large number of small gas cells in the finished loaf.

Baker (1941) noted that a bread film was not stained by iodine unless the film had been cut or ruptured and concluded that the films of the bread were composed of impervious glutenous material. The starch granules were completely covered by a protein layer and were segregated to the interior of the film.

Sandstedt *et al.* (1954) studied cross sections of the gas cell wall of dough and bread. These authors found that there is no segregation of starch or protein to isolated positions in the films. In general, starch granules are separated from each other by protein, which is the continuous phase of the film (Sandstedt *et al.* 1954). The starch granules are oriented by the pull of the expanding gas bubble during fermentation. After gelatinization during baking, further expansion of gas bubbles caused a distortion of the flexible granules. Under normal baking conditions, the granules did not disintegrate but retained their structural integrity (Sandstedt *et al.* 1954). More recently, the structure of wheat flour doughs and the development of the doughs into bread have been studied with TEM by Simmonds (1975) and Khoo *et al.* (1975).

Simmonds (1975) described the changes that occurred during the conversion of flour to dough and suggested that two types of inclusions occurred in the protein phase of the dough: (1) type I inclusions were irregular, densely stained and assumed to be formed from the endoplasmic reticulum; (2) type II inclusions were spherical, had not been formed in dough from defatted flours, and were, therefore, assumed to be lipid-rich. Khoo *et al.* (1975) describe the stages of breadmaking as (1) freshly mixed dough, (2) fermented and proofed dough, and (3) fully baked bread. During baking, the protein fraction changed little (in microscopically visible structure), but the starch granules, particularly the large ones, became gelatinized. Data of these investigators indicate that during the dough formation (mixing), the wheat gluten is hydrated and its film forming (surface active) properties are "activated."

With adequate mixing, the gluten in dough is transformed into strands or sheets, which form a matrix in which starch and other components are embedded and dispersed (Bechtel *et al.* 1978) (Fig. 8.1). Fermentation of a dough produces gas cells (vacuoles). After the ovenspring, the protein strands are thin and contain small vacuoles (Bechtel *et al.* 1978).

The starch granules in the bottom center of the loaf vary widely in the degree of gelatinization after ovenspring. They start to gelatinize from the interior of the granule and appear fibrous (Bechtel *et al.* 1978). In a baked bread, most of the starch is gelatinized and appears as fibrous strands interwoven with the thin protein strands (Bechtel *et al.* 1978).

TLM Studies. Varriano-Marston *et al.* (1980) and Varriano-Marston (1981) reported that the starch granules observed with TLM in frozen thin sections of bread showed a distribution of birefringence in the loaf (Fig. 8.2). Those granules close to the junction of the crust and crumb retained the greatest degree of birefringence (few granules gelatinized)

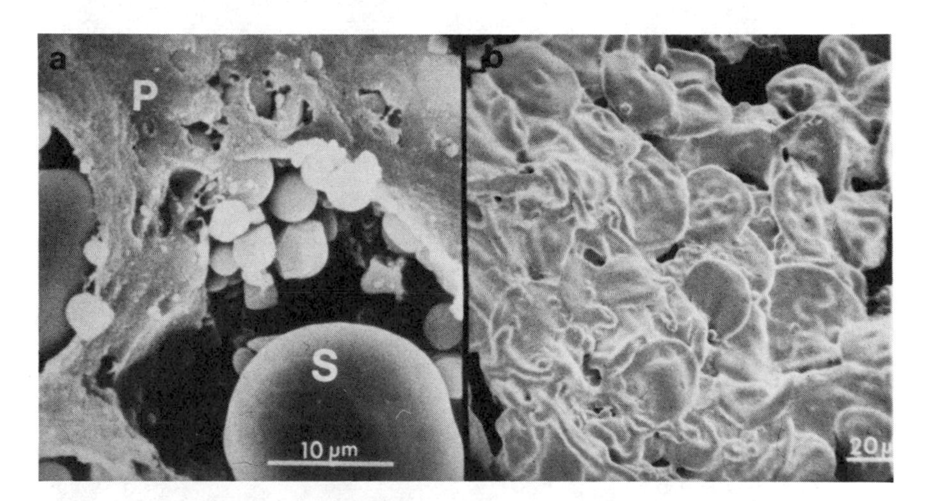

Fig. 8.1. Scanning electron micrographs. (a) Dough sample prepared for the scanning electron microscope using glutaraldehyde and OsO_4 fixatives. (b) Starch granules on gas "cell walls" in baked bread. P, protein matrix; S, starch granule.
Courtesy of Varriano-Marston (1981).

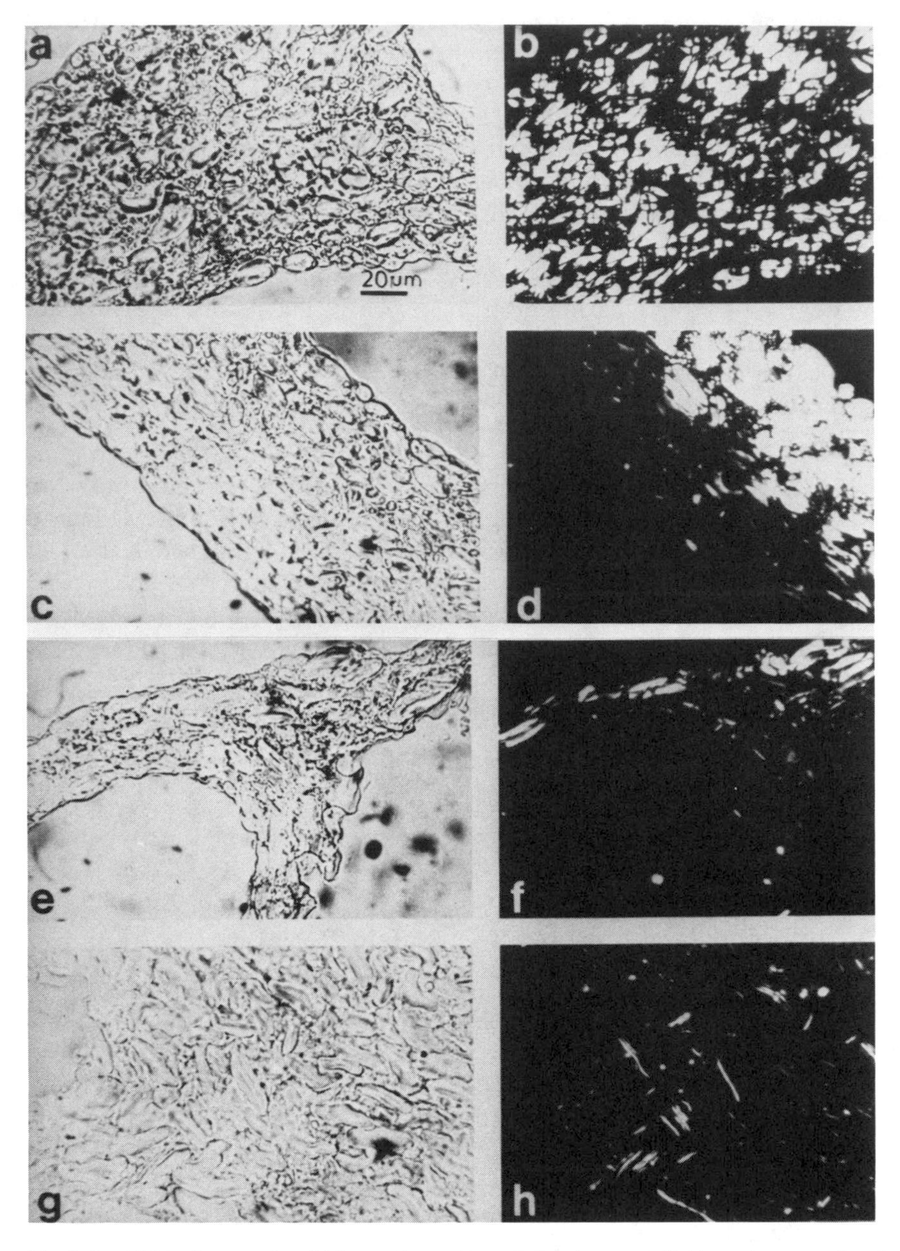

Fig. 8.2. Light micrographs of thin sections of bread showing same fields under normal and polarized light, same magnification. (a,b) Bread crust. (c,d) Junction of crust and crumb. (e,f) Gas cells in outer 2 cm section of crumb. (g,h) Bread crumb center.
Courtesy of Varriano-Marston (1980).

(Fig. 8.2c, d). The number of gelatinized granules increased as samples were taken closer to the interior of the loaf. Virtually all the starch granules were gelatinized in the center of the loaf (Fig. 8.2g, h). The data of Varriano-Marston *et al.* (1980), Varriano-Marston (1981), and Bechtel *et al.* (1978) indicate that during the baking process, the starch granules imbibed the water released from the protein network as it was heat denatured. This finding was also reported by other researchers (Sandstedt 1961; Medcalf and Gilles 1968; Dennett and Sterling 1979). One can explain the variation in the number of gelatinized starch granules from the crust to the crumb interior by taking into account the rapid dehydration that occurs in the outer layers of a loaf during the baking process. The rate of drying is so rapid in the outer layers that there is an insufficient amount of water to gelatinize the starch even though the product temperature is at or above the gelatinization onset temperature.

Moss (1974, 1975) used specific stains and TLM to study the effects of cysteine and oxidizing agents on bread structure and found that the gas cell walls increased in thickness as the result of the presence of cysteine. One of the artifacts noted by Moss (1975) was that the hydration necessary for embedding the sample in an aqueous medium ("Tissue-Tek") caused disruption of the protein network due to swelling of the gelatinized starch.

SEM Studies. Direct examination of bread with SEM provides information on gas cell size and distribution in the bread crumb and indicates the size and shape of starch granules in the gas cell walls (Varriano-Marston 1981; Angold 1979, Hoseney 1982) (Figs. 8.3 and 8.4). The gas cell walls exhibit a continuous protein phase with the starch granules embedded in the protein film (Chabot 1979). The starch granules are difficult to observe because of the protein film (Varriano-Marston 1981). The effect of overmixing the dough on the structure of the baked loaf was studied by Angold (1979) (Fig. 8.5). The breakdown of the gluten film made it easy to identify the starch granules since they were no longer enmeshed in the film.

These SEM studies corroborate the earlier findings of Sanstedt *et al.* (1954). The general consensus is that the starch granules retain their integrity in the bread crumb, but they are deformed and flattened so that their long axes lie parallel to the surface of the gas cells within the bread. Using a "synthetic" wheat gluten–starch bread system, Dennett and Sterling (1979) found that gelatinized wheat starch was uniformly oriented in fibrous strands of the crumb and in the protein films of the gas cell walls.

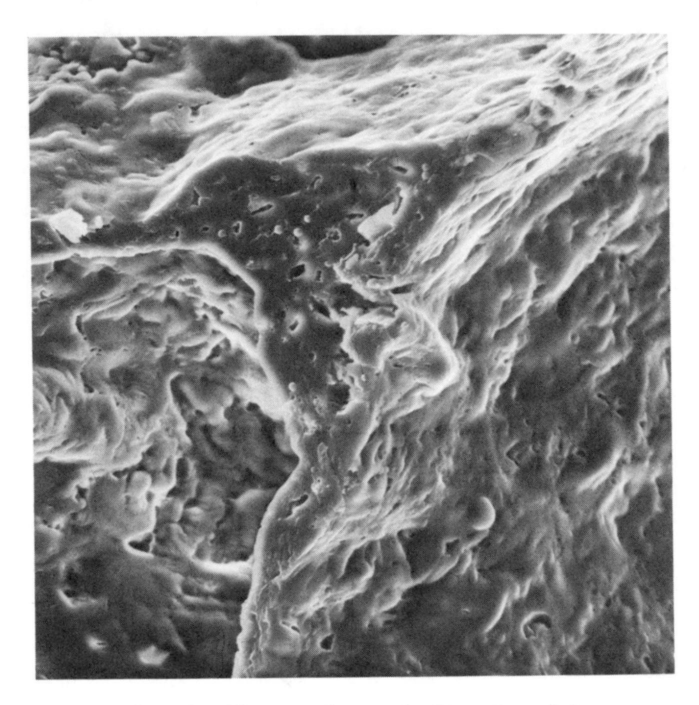

Fig. 8.3. Scanning Electron micrograph of bread crumb from a pup loaf prepared using the straight dough method. Low magnification view showing the crumb structure.
Courtesy of Hoseney (1982).

Most of the SEM work on the microstructure of bread crumb has been performed on samples that were air-dried or vacuum-dried. Angold (1979) reported that the shrinkage of starch granules during sample preparation can be avoided if the sample is freeze dried. The starch granules showed patterns of voids in the locations of the ice crystals formed during freezing. Wasserman and Dorfner (1974) also examined freeze-dried bread with SEM. These authors studied crumb samples made from wheat and rye flours and found that the protein film lining the gas cells in rye bread was spongy, with numerous pores occupying about half the surface area of the cells. Angold (1979) noted similar perforations in the gluten film of the gas cells in air-dried wheat bread (Fig. 8.6). He speculated that the small perforations allow a pressure equilibration to take place during and after baking.

Belderok (1975) used SEM to study the effects of bread improvers on crumb structure. He concluded that the microstructure of the crumb of baked wheat bread was altered by the use of Emulthin®. Flours with a low protein content could be used and they still gave the same fine

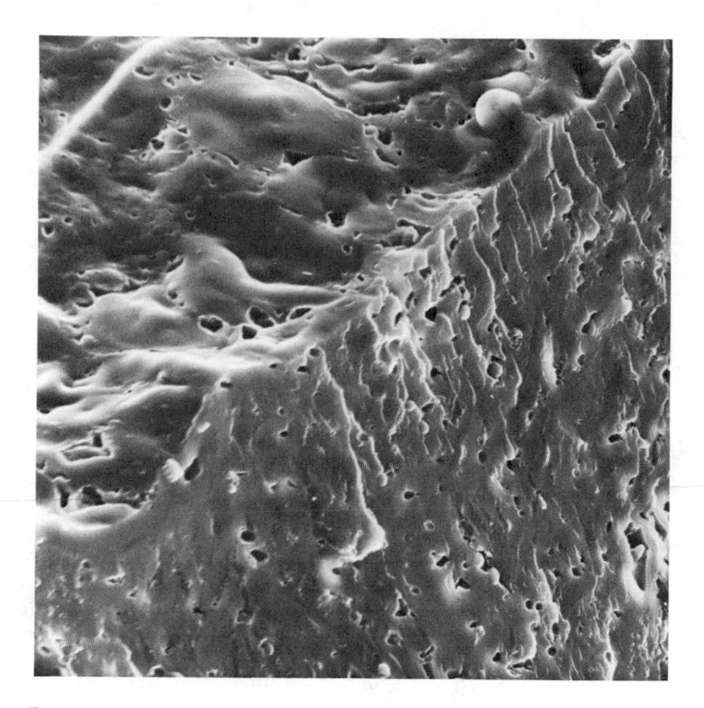

Fig. 8.4. Scanning electron micrograph of bread crumb from a pup loaf prepared using the straight dough method. Higher magnification view showing details of gas cell wall.
Courtesy of Hoseney (1982).

structure as bread made from a high protein ("strong") flour. The distinguishing characteristics noted in both breads were a fine, well-differentiated crumb structure with a distinctly discernible, even distribution of starch granules within the protein matrix.

When wheat flour is fortified with foreign proteins and subsequently baked into bread, the resultant products exhibit lower loaf volumes, more dense or coarse crumb grains, and firmer textures (Fleming and Sosulski 1978). Using TLM and SEM, Fleming and Sosulski (1978) showed that the supplemental proteins disrupted the well-defined protein–starch complex typical of wheat flour breads. Small pores were also observed in the thick cell walls of the supplemented breads, and these may have allowed the gas to escape during the baking process.

Pomeranz *et al.* (1977) used SEM in their study of the effects of different fiber sources on the breadmaking process and final crumb structure. The control bread prepared in this study had a fine crumb structure composed of thin sheets and filaments (Fig. 8.7). The gas cells varied in size from the top to the bottom of the loaf. The experimental

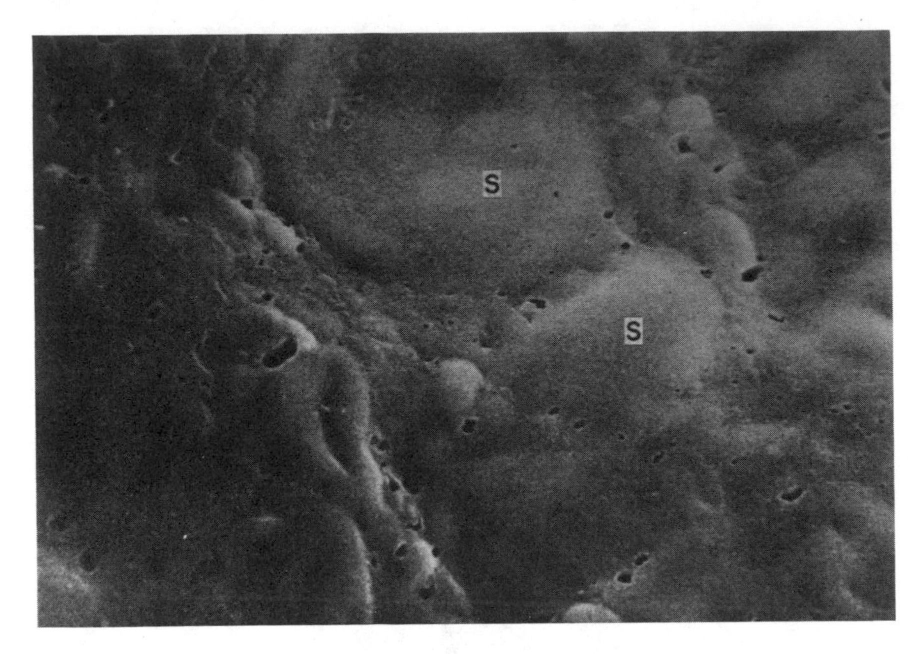

Fig. 8.5. Scanning electron micrograph of bread made from wheat flour, 990×, air dried. This bread was baked from a grossly overmixed dough (8 watt hr/lb or 63 kJ/kg at 800 rpm in a Tweedy mixer). Many of the starch granules (S) in the crumb lie upon the gluten film and are not enmeshed within it. This crumb lacked elasticity and the loaf was of poor volume.
Courtesty of Angold (1979).

fiber containing breads showed a wide variation in gas cell size and shape and in the number and sizes of holes (Figs. 8.8 and 8.9). The added fiber apparently caused a substantial modification of the crumb's fine structure whose filaments and sheets also became coarse and massive (Fig. 8.10). These authors were not able to identify the mode of the fine structure disruption, but they speculated that it could not be merely a dilution effect.

Bread Structure–Texture Interrelationships

The consistency and texture of bread have been qualitatively assessed using a wide variety of testing methods and related testing devices. The most common method to measure bread crumb firmness is compression with a testing device commonly referred to as a compressimeter.

The original version of the compressimeter (Platt 1930) consisted of a balance with one of its pans located directly over a plunger, in the form

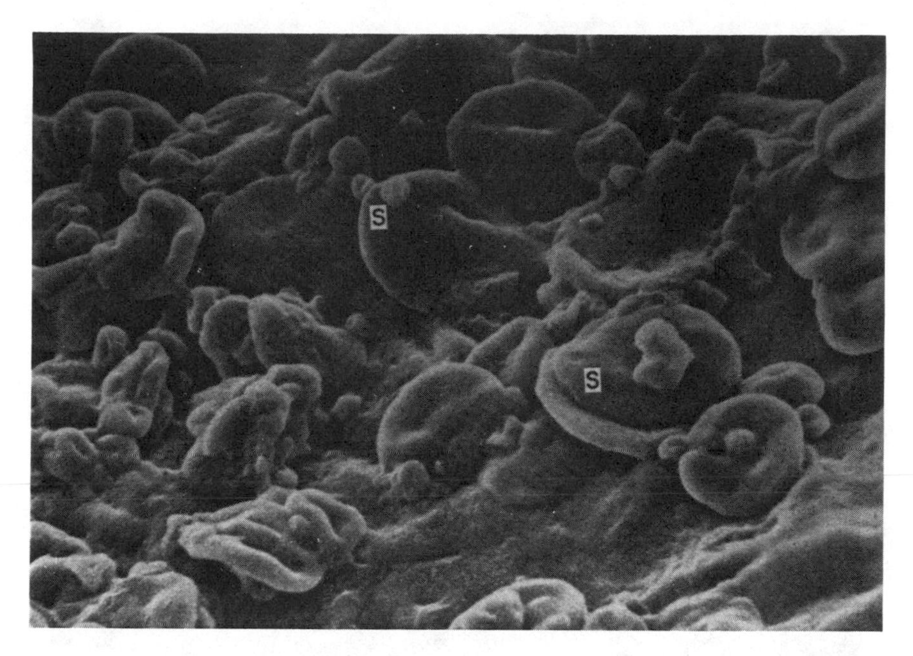

Fig. 8.6. Scanning electron micrograph of bread crumb from wheat flour, 990× air dried. The region shown is part of the wall separating the component gas cells. The starch granules (S), although gelatinized and expanded, are still recognizable. The continuous phase is the gluten and it appears to cover the starch granules as a thin film. There are numerous small perforations in the gluten film through which pressure equilibration takes place during and after baking.
Courtesy of Angold (1979).

of a rod with a larger diameter solid core at the other end, and an adjustable platform that supported the sample. A suitable weight was placed on the first pan and a chain of the same weight on the second. A bread sample was placed on the adjustable platform and then raised so that the surface of the sample just touched the base of the plunger. The chain was removed gently from its pan, so that the weight acted on the sample. The deformation (compression) after 1 min was recorded although it was noticed that most of the compression took place during the first 15 sec. A "degree of elasticity" was determined from the shape recovery when the load was removed.

A more sophisticated version of the compressimeter was designed by Cornford (1963). In this design, the sample was compressed between two circular brass plates. The lower plate was stationary and the upper plate, which was attached to a platform by a verticle rod could be loaded

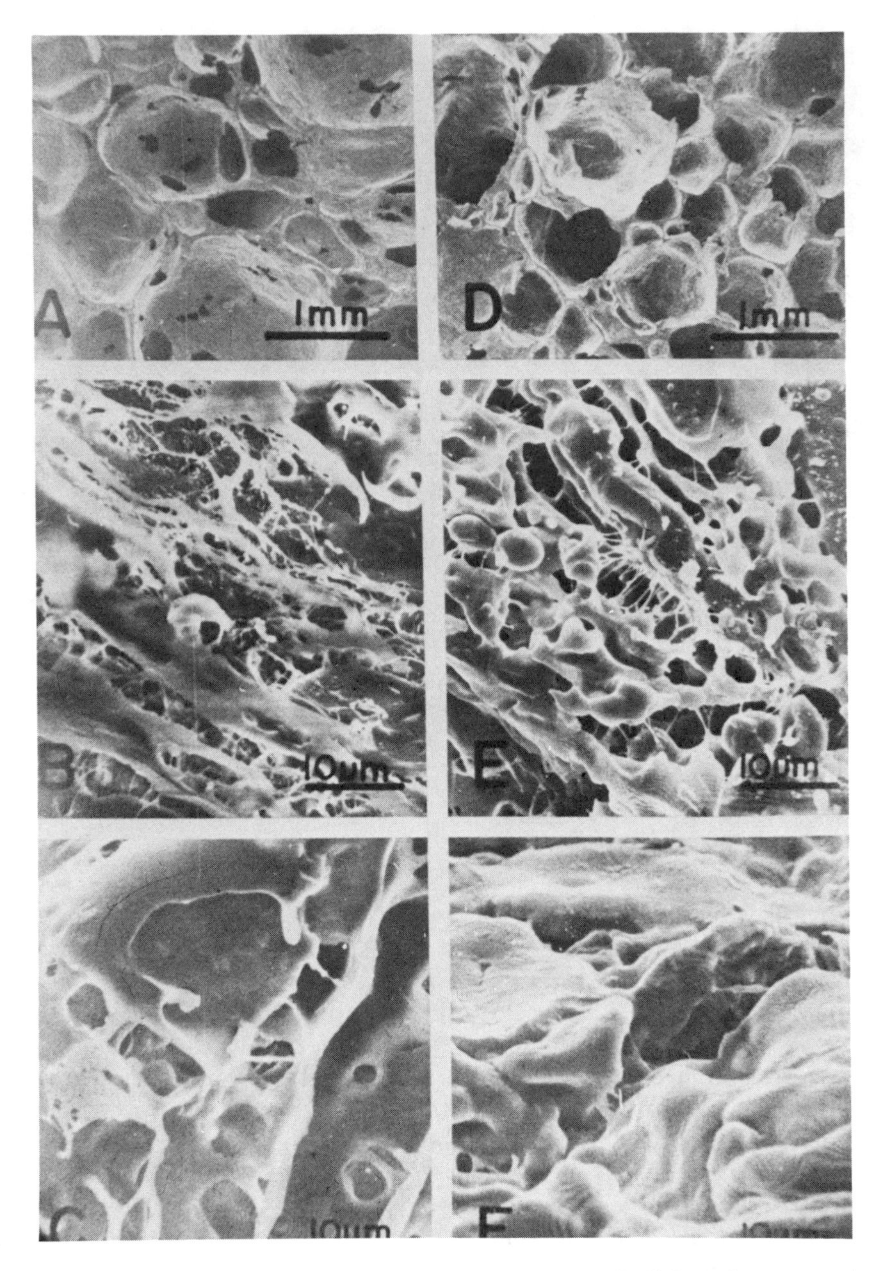

Fig. 8.7. Scanning electron micrographs of control bread. (A, B, C) From the top part of the loaf. (D, E, F) From the bottom part of the loaf B, C and E are cross-sections of crumb; F is the surface of a gas cell.

Courtesy of Pomeranz et al. (1977).

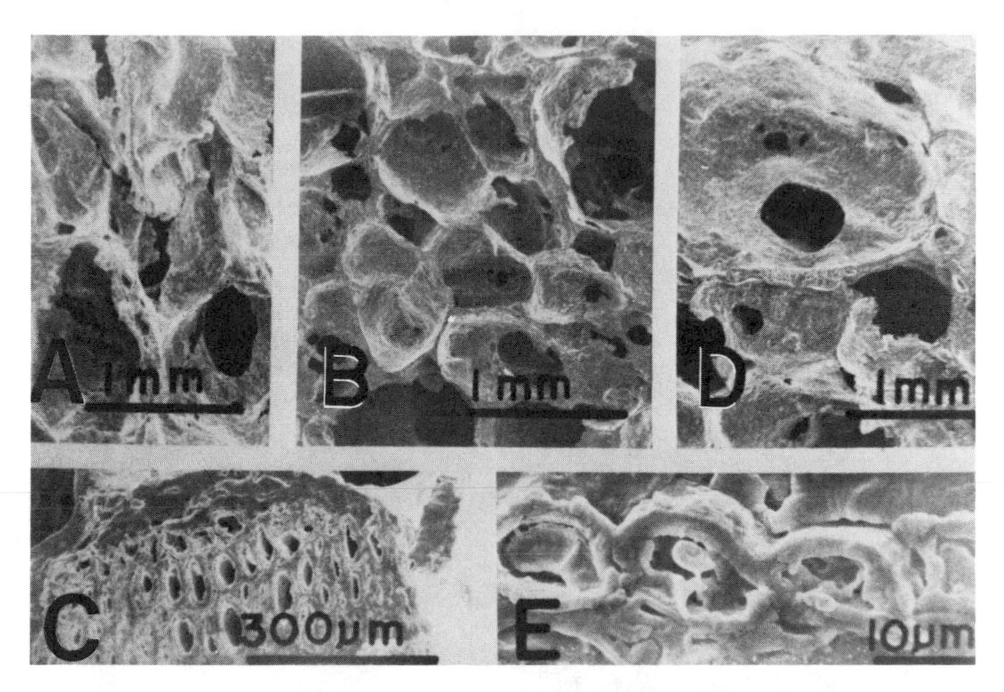

Fig. 8.8. Scanning electron micrographs of bread baked with 15% of the flour replaced by an experimental cellulose. (A) Finely ground oat hulls. (B) Finely ground fine wheat bran. (D) No cellulose particles were found in the cellulose-bread. However, oat hull particles were found imbedded in the bread crumb of the oat hull-bread (C) and wheat bran particles were found imbedded in the bread crumb of the wheat bran bread (E).
Courtesy of Pomeranz et al. (1977).

by weights. The platform was attached to a mechanical system that enabled recording the deformation through a moving pointer. Cornford's (1963) compressimeter was well-suited for bread or sponges, but not for testing cakes with crumbly texture or freshly baked bread (Sherman 1970). Today, the compressibility of bread is most commonly measured with more modern devices such as Universal testing machines.

The factors that affect the quality of bread crumb other than staling were studied by Platt and Powers (1940). These authors noted that flour strength (i.e., in reference to the gluten quality) is of primary importance. The rate of staling in terms of compressibility loss was found to be a nonlinear function of the flour protein content (Steller and Bailey 1938). Compressibility of freshly prepared bread increases with the shortening content, but is little affected by sugar addition (Sherman,

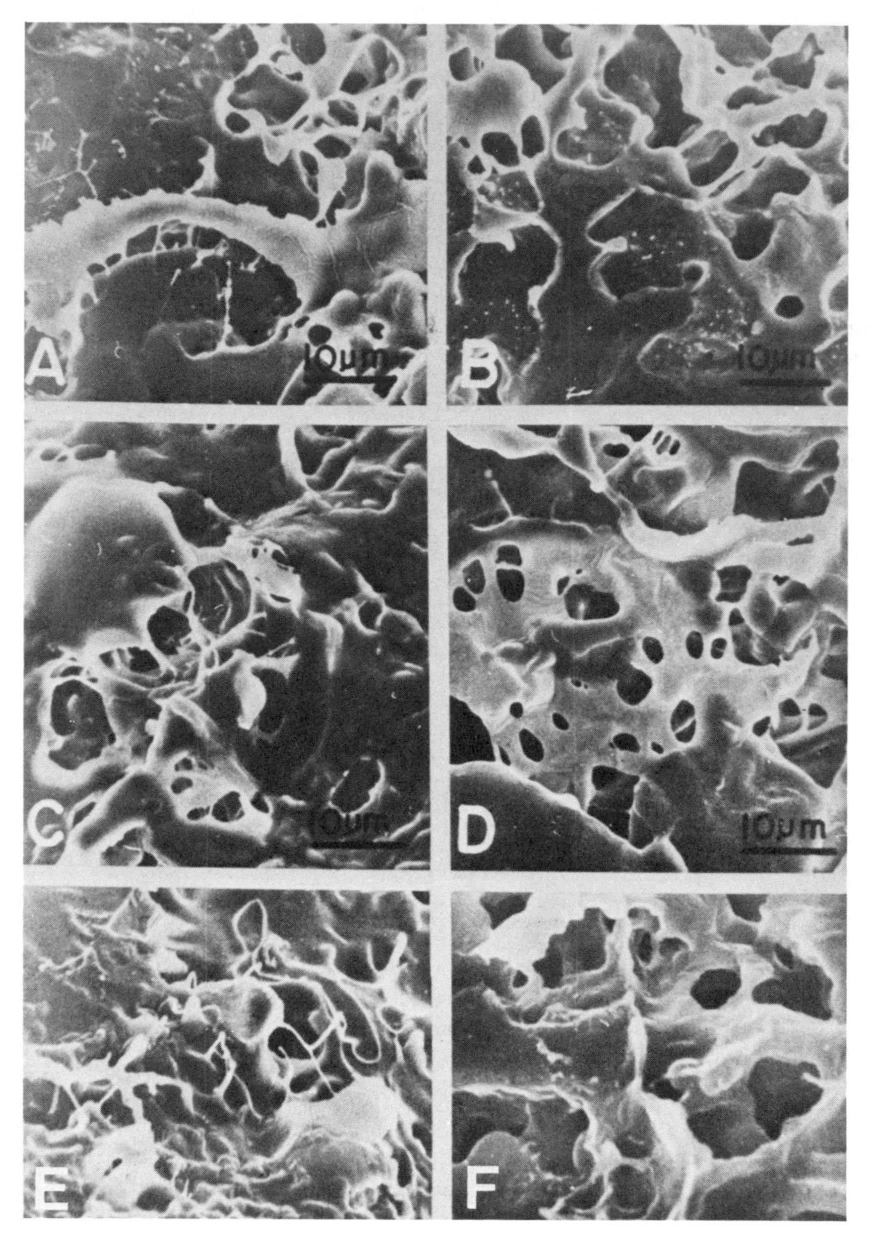

Fig. 8.9. Higher magnification scanning electron micrographs of the gas cell walls of bread baked from flour in which 15% was replaced (A, B) by an experimental cellulose; (C, D) by finely ground oat hulls; (E, F) by finely ground fine wheat bran. Note the substantial reduction in fine structure from adding the fibrous materials.
Courtesy of Pomeranz et al. (1977).

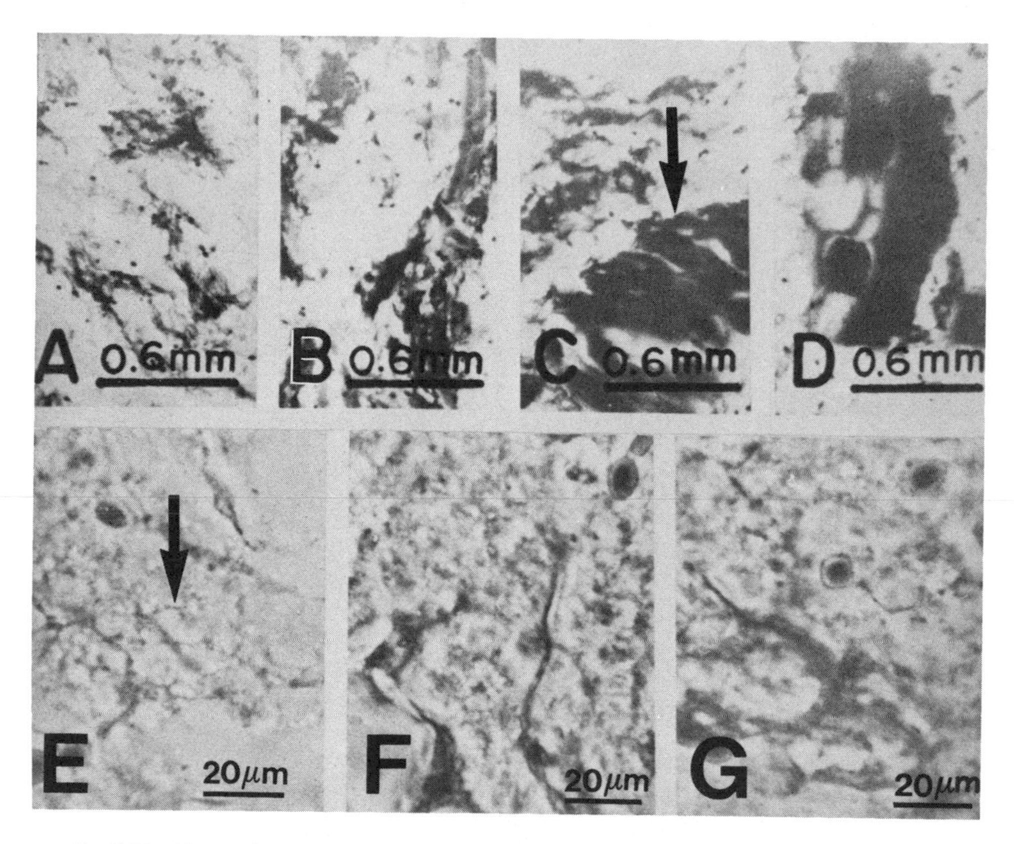

Fig. 8.10. Transmitted light micrographs of bread. (A) Control bread. (B) Cellulose bread. (C) Oat hulls-fine bread; the arrow points to a bran particle that shows aleurone cells. (E) Control bread viewed under oil immersion; the arrow points to fine strands. (F) Cellulose bread viewed under oil immersion. (G) Oat hull-fine bread viewed under oil immersion. Note the lack of fine (thin) filaments in the experimental breads.
Courtesy of Pomeranz et al. (1977).

1970). Baking time has a considerable effect on bread compressibility. Its maximum level, however, appears to be associated with conditions at the fermentation stage (Sherman 1970).

Ponte *et al.* (1962), in their study on bread firming, reported that the rate of firming was significantly correlated with proof time. They found no significant correlation with the usual flour quality indices such as protein and ash contents, maltose value or amylograph parameters. The most important finding of Ponte *et al.* (1962) was the systematic variation in intraloaf firming (Fig. 8.11): loaf centers were significantly firmer than the ends. Ponte *et al.* (1962) concluded that this firmness distribution was mainly attributed to differences in specific volume

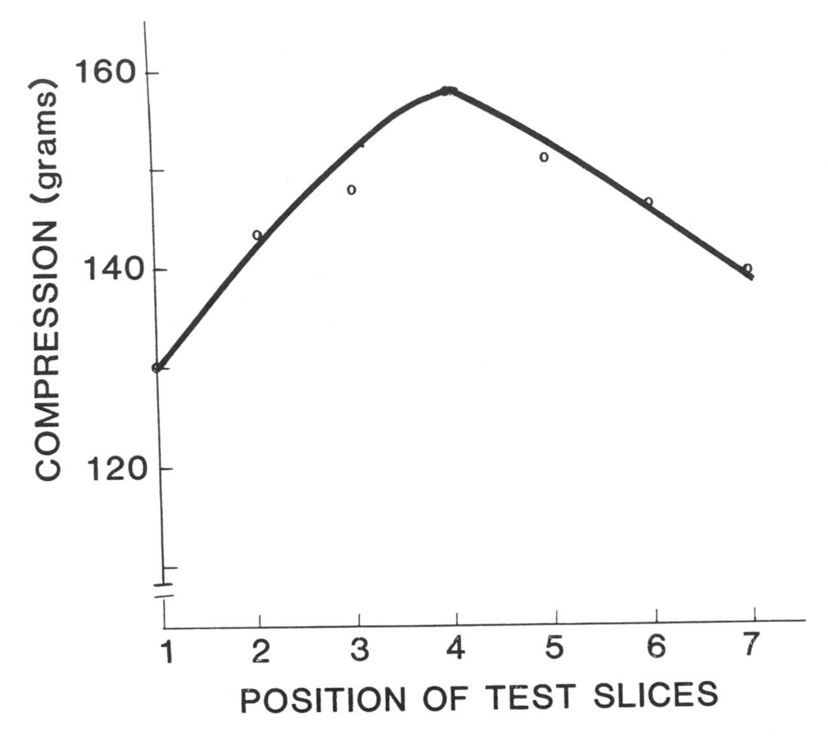

Fig. 8.11. Variation in crumb compressibility within loaves of bread. The first two slices from one end of each loaf were discarded. The firmness of the next seven two-slice sections was measured. The position of these sections, beginning with an arbitrarily selected end, forms the abscissa of this plot. Compression values are averages of 80 loaves.
Courtesy of Ponte et al. (1962).

within the loaves (Fig. 8.12). Their findings were confirmed by Short and Roberts (1971) who also reported that the firmness of bread crumb varies with position within a loaf with a maximum at the center regardless of the loaf size (between 0.3–1.5 kg), the baking procedure, and the presence of surfactants.

It has been shown by Waldt (1968) that the use of bacterial α-amylase in white bread retards product firming and staling, imparts smoother, silkier and moist textures that are retained, ensures easy slicing and less crumbling, lengthens overall keeping quality, and improves retention of fresh flavor and aroma. These effects have been verified in our laboratory (DeStefanis and Turner 1981). The exact mechanisms by which the enzyme induces these effects, however, are still not fully understood. Micrographs of starch extracted from a series of breads

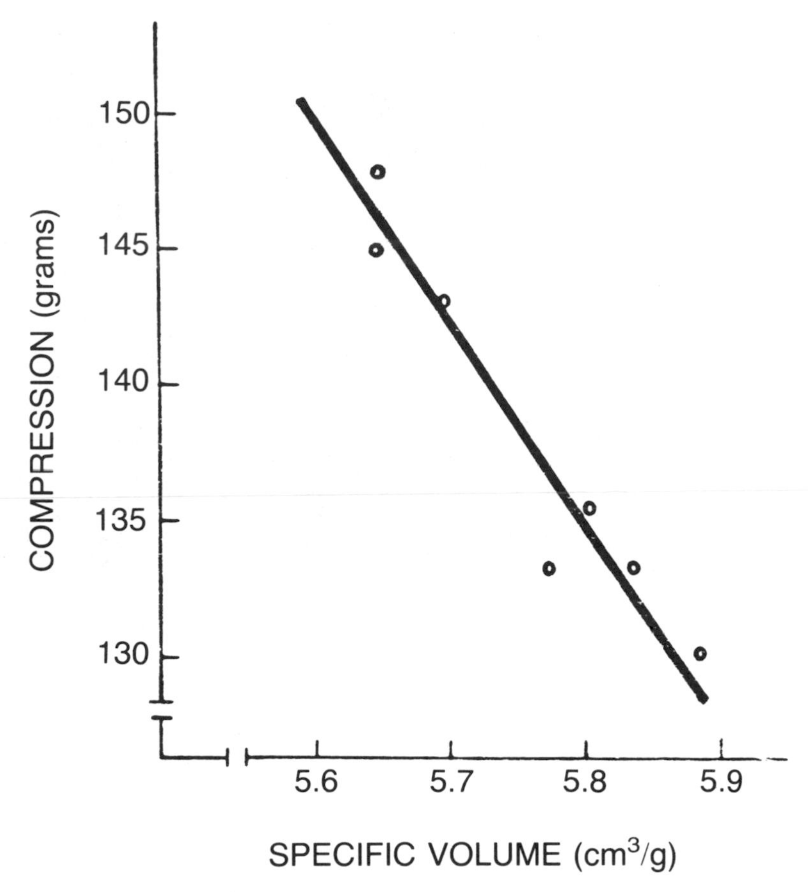

Fig. 8.12. Effect of specific volume on compression values of seven two-slice sections within a loaf of bread. (r = 0.93).
Courtesy of Ponte et al. (1962).

baked with and without α-amylase or calcium stearate are shown in Figs. 8.13–8.16. They do not indicate that these agents induced any major differences in starch granule morphology.

The moisture required for starch gelatinization during baking is drawn from the surrounding gluten matrix (Sandstedt 1961; Medcalf and Gilles 1968). This dehydration and the heat of baking denatures the gluten. Affinity tests indicated that denaturation of the gluten reduces the attraction between gluten and gelatinized starch by 30–50% (Dennett and Sterling 1979). A reduction in the number of hydrophilic bonding sites on the gluten molecule would limit bonding with the primarily hydrophilic gelatinized starch surface. The general, reduction in bonding during baking may be essential to the formation of a

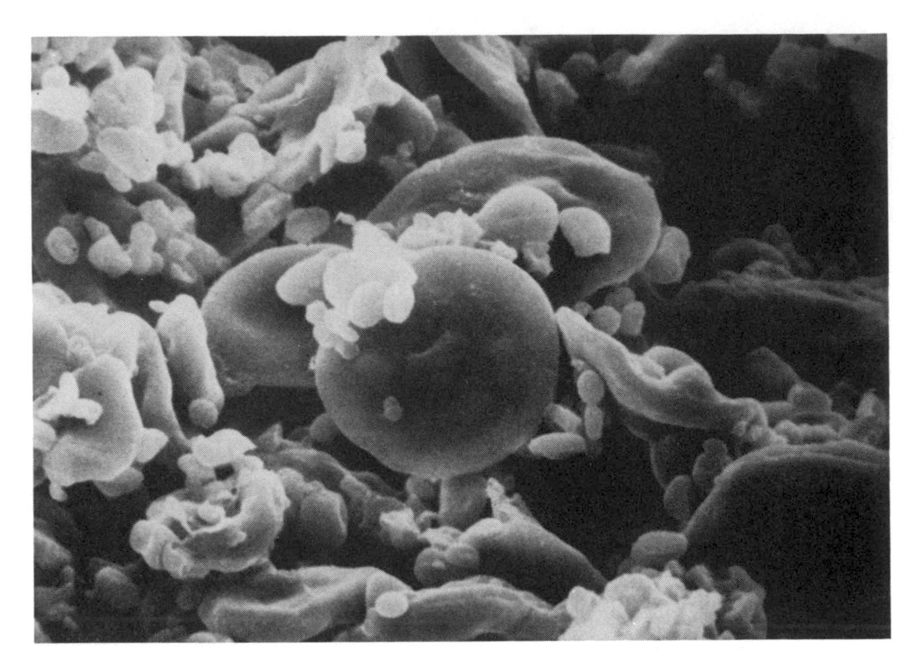

Fig. 8.13. Scanning electron micrograph of starch extracted from the control bread. No calcium stearate or α-amylase.

softer, more flexible crumb (Dennett and Sterling 1979). The reduced affinity may also induce a weakening of the gas cell walls making them more susceptible to rupture.

These authors also noted that wheat starch, compared to other starches, formed the firmest bread, because in its gelatinized state it was more responsive to the tensions developed in the baking dough by lining up in a more closely parallel array. The large wheat starch granules are discoid with a median plane of weakness (Melchior and Fuerber 1954; Evers 1971; Gallant and Guilbot 1973). When they are gelatinized, radial contraction and limited tangential expansion of the granules lead to splitting and to the formation of two flat, narrow, platelike disks (Melchoir and Feuerberg 1954).

Rye starch also contains large, discoid granules similar to those of wheat (Evers 1971). Bread loaves prepared from rye flour have a firm, spongy texture and structural characteristics similar to wheat breads (Wasserman and Dorfner 1974). This baking behavior of rye starch (flour) lends support to the interperation of the origin of wheat bread firmness.

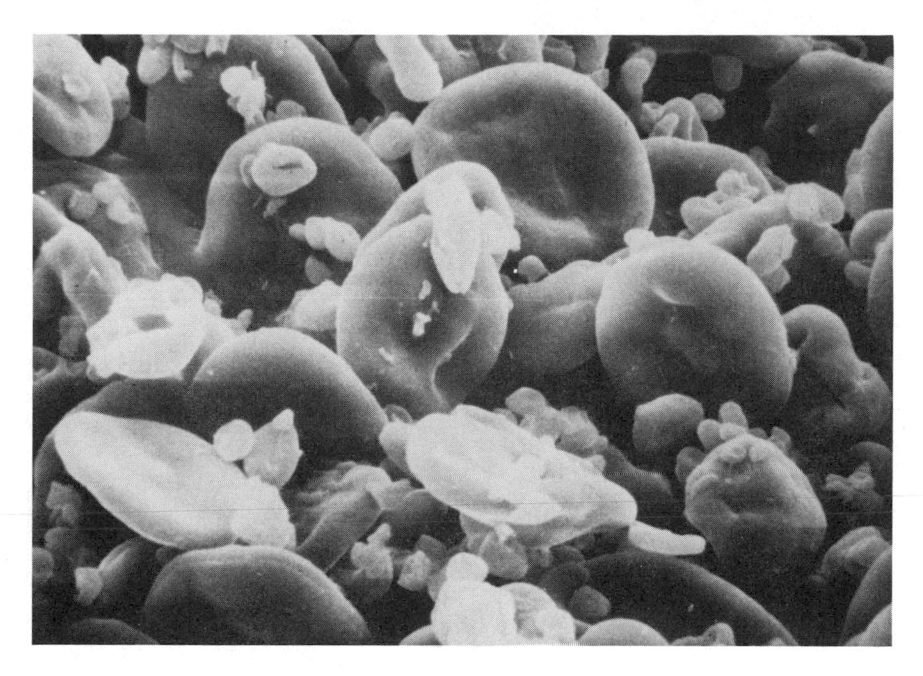

Fig. 8.14. Scanning electron micrograph of starch extracted from the bread containing calcium stearate but no α-amylase.

Fleming and Sosulsi (1978) used SEM to demonstrate that concentrated plant proteins caused small pores and a ruptured cell structure in bread. The cell walls of the supplemented breads were thick, complex structures compared to the thin, sheeted walls in the wheat breads, a factor that was largely responsible for these loaves depressed volume and excessive firmness.

CAKE BATTER AND CRUMB STRUCTURE

Sample Preparation for TLM and SEM

Cake consists of a complex mixture of water- and fat-soluble materials in addition to insoluble materials held in a relatively flimsy texture. The risk of distortion and other artifacts in microscopic evaluation is therefore very high.

The preparation of a cake crumb specimen for examination by TLM requires that the protein matrix be protected during sectioning. This can be done by sectioning in an ultracryostat at −20°C after the cake

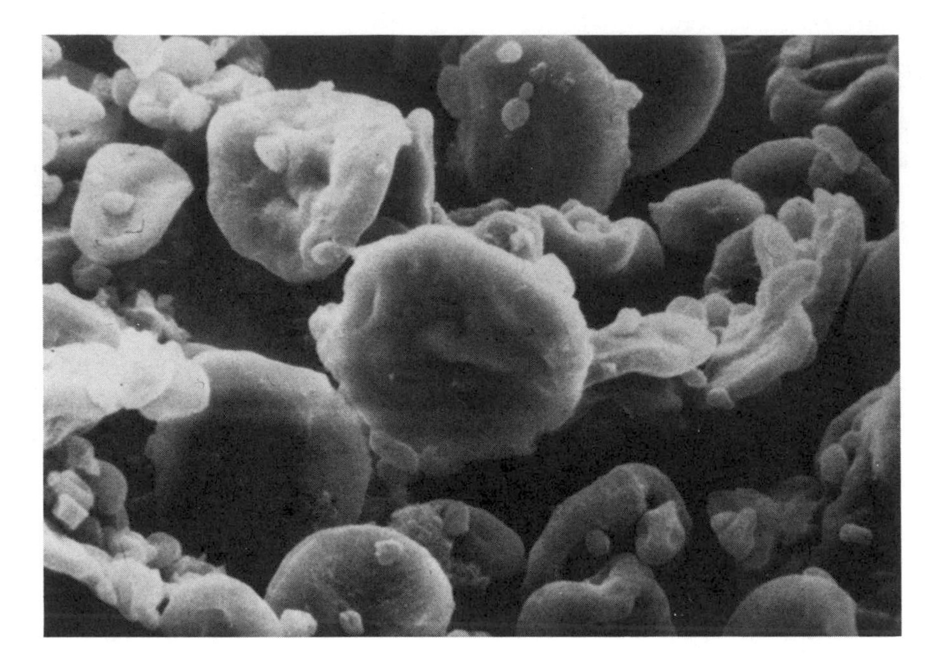

Fig. 8.15. Scanning electron micrograph of the starch extracted from bread containing α-amylase but no calcium stearate.

crumb sample has been flooded with water (Shephard and Yoell 1976) or sectioning with a conventional microtome of a cake crumb sample embedded in wax (Shephard and Yoell 1976) or gelatin (Francis and Groves 1962).

The ice crystals formed when the flooded cake sample is frozen provide the mechanical support for the cake matrix (Shephard and Yoell 1976). It should be noted, however, that the sample, in the presence of liquid water, will swell thus causing certain structural modifications (Francis and Groves 1962). In addition, sugars and other water-soluble components are dissolved and therefore, cannot be detected in the final sections (Francis and Groves 1962; Shephard and Yoell 1976).

Plant tissues are normally fixed prior to embedding. However, for baked goods, the baking process itself acts as a fixative (Francis and Groves 1962) because the heat sets the final structure and denatures the proteins. These researchers therefore suggested that for cake and other baked goods, it is not strictly necessary to employ a fixative. However, other workers have reported the need for sample fixation prior to examination (Burhans and Clapp 1942; Butterworth and Colbeck 1938; Shephard and Yoell 1976). In general, aqueous fixatives are used. Pro-

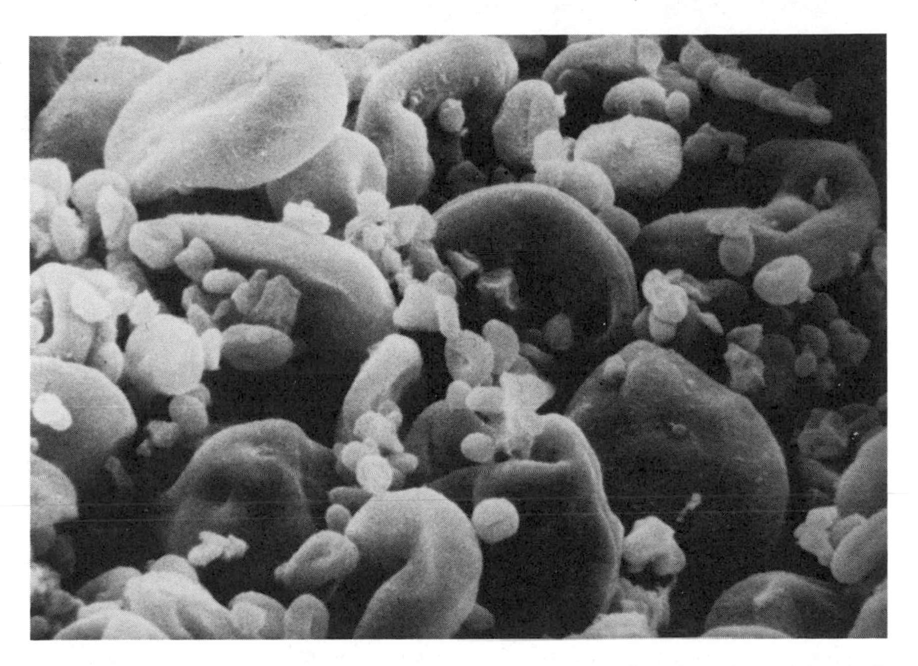

Fig. 8.16. Scanning electron micrograph of starch extracted from bread containing both α-amylase and calcium stearate.

teins are fixed with aldehydes (e.g., formaldehyde or glutaraldehyde). Fats, when necessary, are fixed either in an aqueous solution of osmium tetroxide or by a vapor phase fixation at 40°C (Shephard and Yoell 1976). Fixation is usually followed by dehydration and embedding. In some cases, the sample is embedded directly after fixation (as is the case with gelatin), sectioned and then dehydrated to fix the section on the glass slide. This means that the supporting gelatin part of the section shrinks; this can cause the disruption of the very fine gluten strands and may also result in separation of the gluten strands from the starch granules (Francis and Groves 1962). Samples so prepared can be used to identify specific structural components by selective staining. Oil Red O, Sudan III and Ponceau have been used to stain fat (Francis and Groves 1962; Shephard and Yoell 1976). Protein is usually stained with Fast Green FCF or acidic Ponceau 2R (Francis and Groves 1962; Shephard and Yoell 1976). Lugol's iodine solution has been used to identify starch (Shephard and Yoell 1976). These authors also studied the effects of osmium tetroxide fixation and reported that in samples treated with 2% aqueous osmium tetroxide solution the fat was fixed in the inner matrix in a form of irregular pieces, and also at the air-cake interface. Cake samples fixed in osmium tetroxide vapor showed no interior matrix fat. However, an almost continuous thin line of fixed fat at the air-cake interface was observed. They therefore speculated that the osmium vapor complexed with the surface fat and formed an impenetrable barrier to further fixation. The interior fat was therefore lost during the solvent treatment prior to wax embedding. The marked difference in the appearance of the fat between the two osmium fixation procedures was resolved by a SEM study of cake samples examined with and without fixation (Shephard and Yoell 1976). They found that it was easy to distinguish starch in varying degrees of gelatinization and distortion in the glutaraldehyde-fixed sample. The sample fixed with glutaraldehyde plus aqueous osmium tetroxide exhibited similar structures but also contained nearly spherical shapes adhering to the air–cake interface. These structures correlated with the structures stained with Ponceau in the TLM sections of the same sample. Based on this evidence, Shephard and Yoell (1976) concluded that the treatment with aqueous osmium tetroxide caused the surface fat, which forms a continuous layer of varying thickness at the air−cell interface, to roll up into balls.

This section was not intended to be an exhaustive summary of all the possible structural artifacts caused by sample preparation procedures. It is hoped that this brief summary points out the need to bear in mind the following points when interpreting the results of microscopic studies reported in the literature.

1. The use of aqueous media can result in swelling of the starchy components.
2. Shrinkage of the specimen can occur due to rupture of attenuated protein films.
3. The role of sugars and other soluble materials is indeterminable since they may (totally or partially) be removed in the process of the specimen preparation.
4. The localization and structural role of fat is confounded by certain fixation procedures.

Cake Batter and Baking Process

Cake batters are complex macroemulsion systems. Morris (1929) concluded that the air cells present in lean (no fat) cake batters are surrounded by a film of sugar syrup, egg protein, or both. The air cells are emulsified in the fat if it is present in the formulation. Grewe (1937) reported that creamed mixtures of sugar, fat, and eggs are a water-in-oil emulsion.

Dunn and White (1939) concluded that half of the increase in volume of a lean pound cake, containing no chemical leavening agents, is due to the thermal expansion of the air in the batter. When the air in a pound-cake batter is completely exhausted and the batter is baked, there is no volume increase. The incorporation of air in commercial shortenings improves the texture and volume of cake (Dunn and White 1939).

Based on a microscopic examination of cake batters, Sunderlin and Collins (1940) concluded that *thin* batters are oil-in-water emulsions, whereas *thick* batters are water-in-oil emulsions. They also concluded that gas bubbles are present in batter in grapelike clusters.

Lowe (1943) concluded that fats are dispersed in cake in several ways: oil-in-water emulsion; water-in-oil emulsion; films, pools or lakes throughout the cake ingredients; adsorbed onto the starch and protein of the crumb as a mono-molecular layer, at the cake–air interface, or a combination of all or some of the above. The application of heat to a slide during the TLM observation of the cake batter has been used in an effort to understand more about the baking process. Carlin (1944) used this technique along with fat-soluble dyes to follow the behavior of fat prior to and during baking. He found that layer- and pound-cake batters appear to be suspensions of air bubbles in fat distributed in a medium of flour and liquid. Very small amounts of liquid were emulsified in the fat and the soluble ingredients were dissolved in the water phase of the batter. The air spaces in layer- and pound-cake batters were surrounded by fat. He also noted that during baking the fat melts rapidly and releases its suspended air to the flour–water medium. He claimed that the gas produced by the baking powder finds its way into the air bubbles already existing in the batter and demonstrated that the completion of the baking process can be determined microscopically by the loss of birefringence in the starch granules when viewed by means of polarized light. Carlin also showed that at all times during the baking process there is movement of air bubbles, which follows a convection pattern until the end of the baking process. At this point, the movement becomes violent and without definite direction.

More recently, Bell *et al.* (1975) used the heated slide (stage) technique and fat staining coupled with cine and video recording to study the baking process in sponge cake, Madeira cake and high-ratio cake batters. Time lapse photography was used to follow the development of structure within these systems. Their photographic results indicated that the air cells in cake batters form in the lipid portion of the system and that the air cells in the finished cakes are lined by the lipid components in the batter.

Bell's *et al.* (1975) results confirmed earlier reports of Carlin (1944) and

Russo (1972) that air in fat-containing batters is initially located in the fat phase of the system. Bell *et al.* (1975) also pointed out that the restriction of their model cake system between two planes of glass is an artificial situation, and they conceded that the gas cells in their microslide bakes are only one-sixth the size of those found in a full size cake.

Most recently, Hsieh *et al.* (1981) examined the applicability of cryofixation freeze-etch methods as used with TEM to study cake batters and concluded that the technique is sensitive enough to detect the state of the batter components (including both lipids and starch) during heating. They also examined the appearance of individual batter components after dispersion in 42% sucrose (Figs. 8.17 and 8.21). The sucrose portion of the replica was granular in appearance (Fig. 8.17a) and therefore they concluded that it was made up of small vitrified crystals. The starch granules were found to be fractured either at the surface or through the granule (Fig. 8.17). Oil droplets were found to be spherical in shape (Fig. 8.18) and were characterized by irregular cross-fracture patterns. The crystalline baking powder was easily distinguishable from the other batter components as can be seen in (Fig. 8.19).

After heating to 102°C, starch granule swelling became evident (Fig. 8.20) and (8.21) and cross fractures were seen more frequently than surface fractures (Hsieh *et al.* 1981). They therefore concluded that this was due to weakening of the granule structure. The oil droplets were found to be collected into pools which appeared to separate the swollen starch granules as shown in Fig. 8.21.

Cake shortenings with improved aerating and creaming properties were first introduced in the United States in 1934 (Birnbaum 1978). These shortenings contained mono- and diglycerides and were based on a patent application of R.B. Harris in 1933 (Harris 1978). Carlin (1944), who studied the effect of emulsifiers on cake batters using his hot stage TLM technique, noted that the use of monoglycerides produced a finer dispersion of fat throughout the cake batter. Wootton *et al.* (1967) reported that surface-active lipids have been found to act as enhancers of air incorporation in cake batters when used as emulsifiers for shortening. The ability of the surface-active lipids to enhance air incorporation is related to their unique interfacial behavior at oil–water boundaries. They also found that these surface-active lipids crystallize in an alpha-tending form if the concentration at the interface exceeds the solubility limit. The interfacial film so formed possesses waxlike properties of the alpha-crystalline phase. Wootton *et al.* (1967) suggested that this film encapsulates the dispersed oil droplets within a protective coating. Therefore, the liquids normally detrimental to the foaming properties of soluble proteins are prevented from migrating into the aqueous phase and interfering with air incorporation.

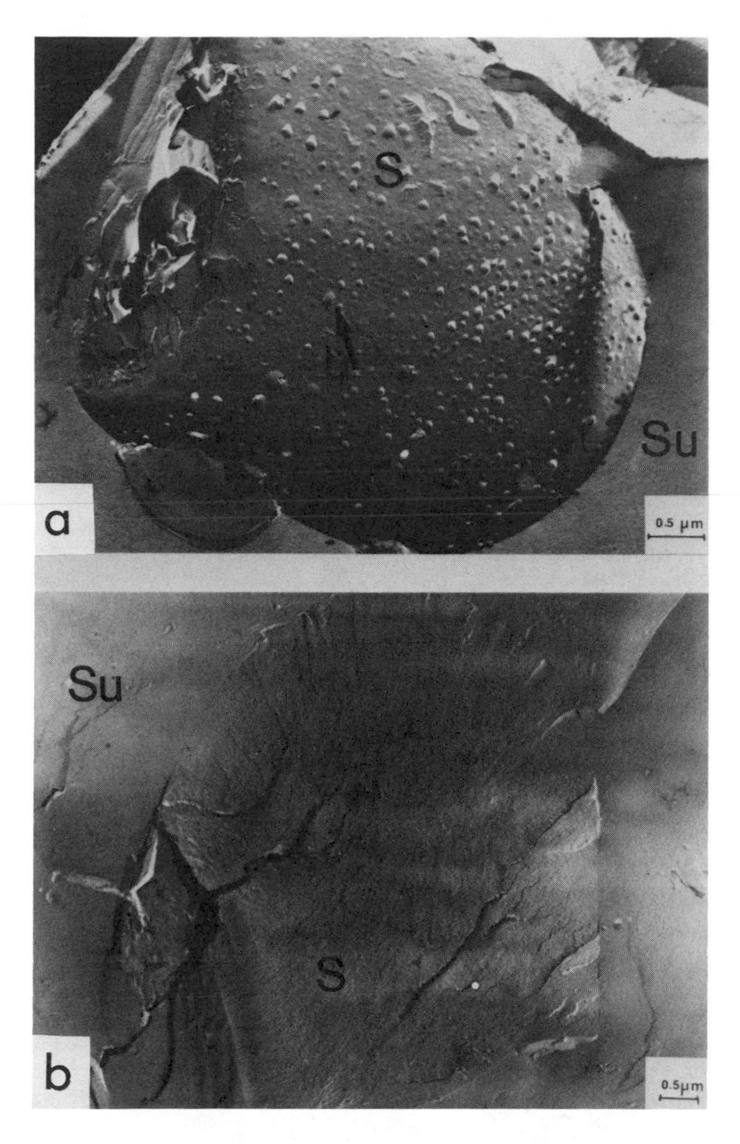

Fig. 8.17. Chlorinated flour in 42% sucrose. (a) Surface fracture of starch granule. (b) Cross fracture of starch granule. S = starch granule; SU = sucrose; B = bumpy structures.
Courtesy of Hsiesh et al. (1981).

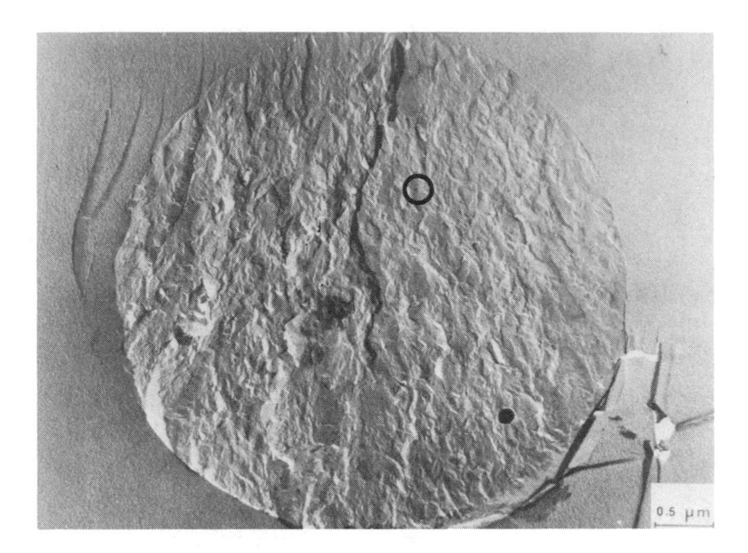

Fig. 8.18. Oil (O) in 42% sucrose.
Courtesy of Hsiesh et al. (1981).

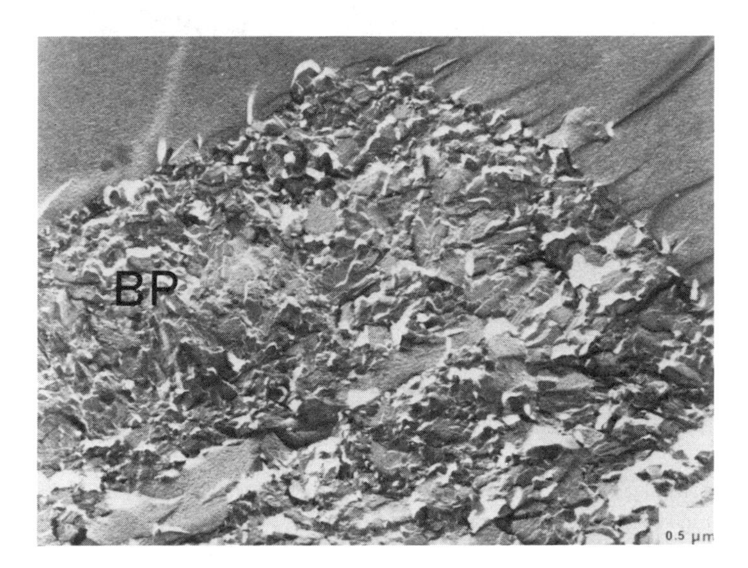

Fig. 8.19. Baking powder (BP) in 42% sucrose.
Courtesy of Hsiesh et al. (1981).

Fig. 8.20. Batter before heating. S = starch granule; O = oil droplet.
Courtesy of Hsiesh et al. (1981).

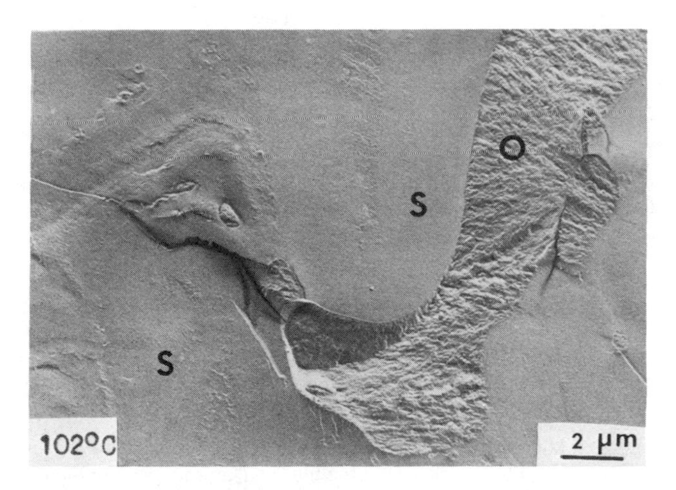

Fig. 8.21. Batter after heating to 102°C. S = starch granule; O = oil pool.
Courtesy of Hsiesh et al. (1981).

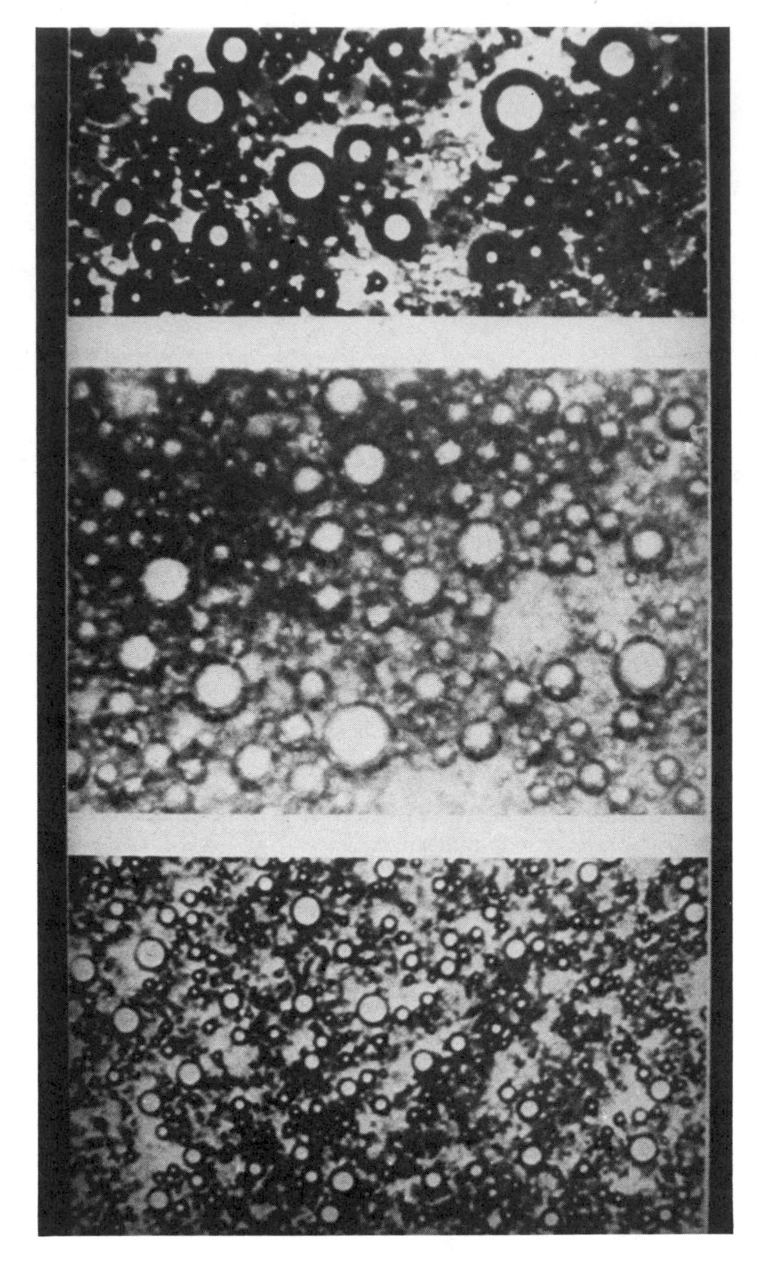

Fig. 8.22. Micrographs of cake batter showing effect of increasing levels of emulsifying agents. As air cells become smaller and more numerous, the resultant cake gains in volume, grain and tenderness. (*Top*) No emulsifier; (*middle*) below optimum level of emulsifier; (*bottom*) optimum level of emulsifier.

Without an emulsified shortening, cakes made with more sugar than flour are prone to collapse because of the batter's structural weakness (Birnbaum 1978). An emulsified shortening imparts strength to the batter and as a result the cake has a better air distribution. The fineness of the air distribution is a function of the emulsifier concentration (Birnbaum 1978). An increase in the emulsifier level reduces the size of individual air bubbles and increases their number (Fig. 8.22).

Although this effect was verified by Moncrieff (1970) using TLM, it was pointed out by Birnbaum (1978) that beyond a critical concentration of emulsifier in the shortening a reversal of quality sets in. The batter viscosity under such condition drops and the cake volume decreases because too fine a dispersion of air and shortening weakens the crumb structure.

Cake Crumb Structure

Morr (1939) made microscopic examinations of baked cakes and found that the starch and fat of cakes are imbedded in or at the surface of the protein matrix. He also showed that hydrogenated fats collected in clumps of fairly large size at the cake–air interface. Butter was found to be more finely divided than hydrogenated fats and was distributed throughout the entire crumb. Liquid fats collected in pools at the cake–air interface with a small portion distributed within the crumb. These objections were later confirmed by Lowe (1943) and Carlin (1944).

According to Hoseney *et al.* (1978), the major role of starch in cakes is to act as a water-sink and to set the structures in the baking process. Howard *et al.* (1968) demonstrated the essentiality of intact granular starch in the thermal-setting process of cake baking. They defined the thermal setting stage as that time at which the batter changes from a fluid, aerated emulsion to a solid, porous structure that will not shrink or collapse when the cake is removed from the oven. In order for the above physical transformation to occur, the free water in the system must be absorbed. Therefore, at this point the water-absorption properties of the starch granule actually control the final physical characteristics of the baked cake although other components (e.g., proteins) can compete with the starch for the water present in the system. This is evident from the fact that a characteristic cake structure cannot develop unless there is a sufficient amount of granular starch.

Gordon *et al.* (1979) used SEM to study the structure of "lean" experimental batters and cakes in which pure wheat starch replaced the cake flour at levels from 0–50% on a w/w basis. They found that the starch was suspended as lumps in the batter matrix and that the air incorporated in the first mixing stage was probably the source of the bubbles

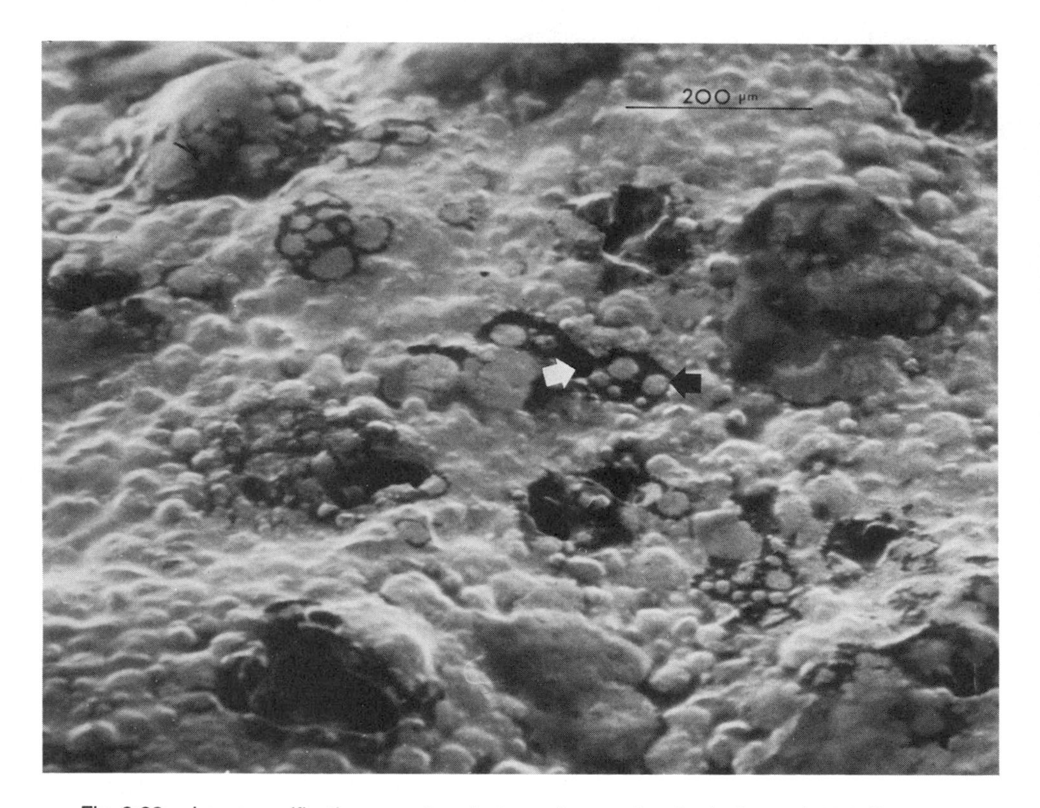

Fig. 8.23. Low magnification scanning electron micrographs of cake batter (100:0, flour-to-starch ratio). White arrow, air bubbles; black arrow, fat-starch pool.
Courtesy of Gordon et al. (1979).

that appeared near the fat–starch pools (Fig. 8.23). They noted no difference in the batter appearance over the tested range of flour-to-starch ratios. Although the crumb structure of the finished cakes was found to be similar at all tested levels of starch substitution, there were differences in the starch granules and the cake matrix (Fig. 8.24). Gordon *et al.* (1979) attributed these structural variations to differences in temperature and hydration gradients during baking. As can be seen in Fig. 8.24 the outer edge of the cakes exhibited starch granules in various stages of swelling with an underdeveloped matrix. From 0.5 to 3.5 cm from the edge, there was extensive starch swelling while the granules in the center of the cake were more swollen than those at the edge, but less than those 0.5–3.5 cm from the edge. It was also noted that the cake matrix was more fully developed at the center of the cake than at the edge. Gordon *et al.* (1979) also showed that the starch granules and matrix in the top of the cakes were compressed (Fig. 8.25). The area

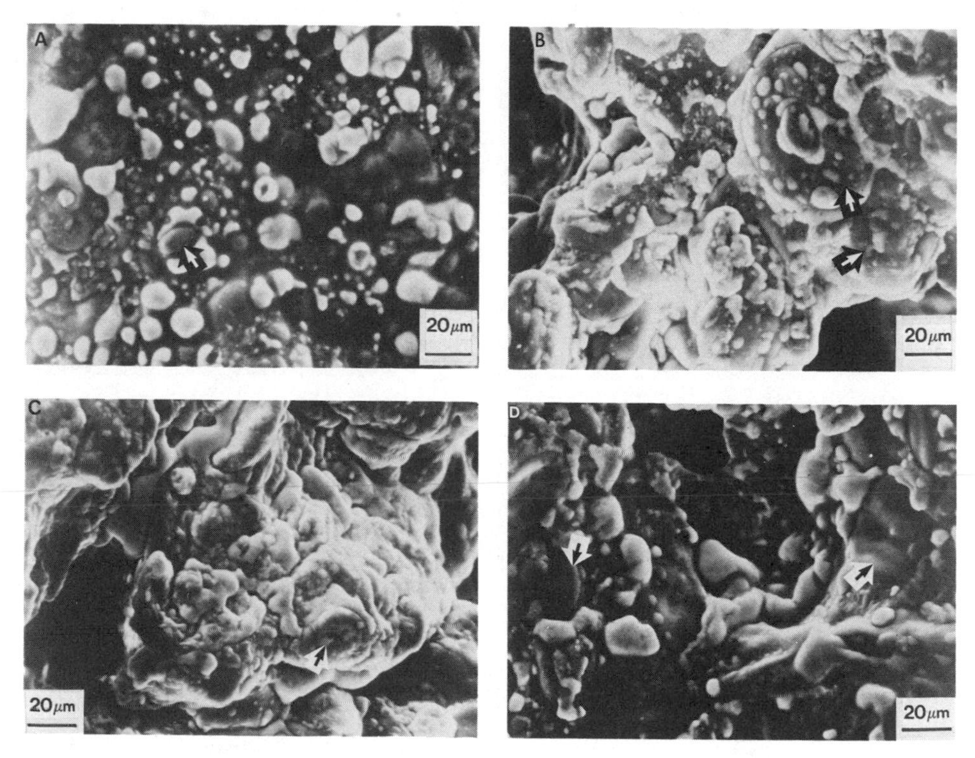

Fig. 8.24. High magnification scanning electron micrographs of cake crumb from cakes made with 100% cake flour. The cakes were sampled from the middle of the edge to the center of the cake. Arrows locate the starch granules. (A) outer edge; (B) 0.5 cm from the edge; (C) 3.5 cm from the edge; (D) center of cake.
Courtesy of Gordon et al. (1979).

immediately below the top exhibited less swollen and plasticized granules embedded in a less developed matrix whereas plasticized starch granules next to the bottom were embedded in a slightly developed matrix. Starch granules in the bottom of the cakes were compressed into a dense, compact layer with no evidence of a gluten matrix.

Martin and Tsen (1981) used SEM to compare the crumb structure of high-ratio white layer cakes baked in a conventional oven and with microwave energy. The cell structure in the center of the cake baked with microwave energy was coarser than that baked by conventional processes. The air cells of the microwave cake were irregular in size and had thicker cell walls. At 2.5 cm from the center of the cake, the microwave baked cake had more regular and finer cells than in the center, but these air cells were still coarser than those found in the conventional cake.

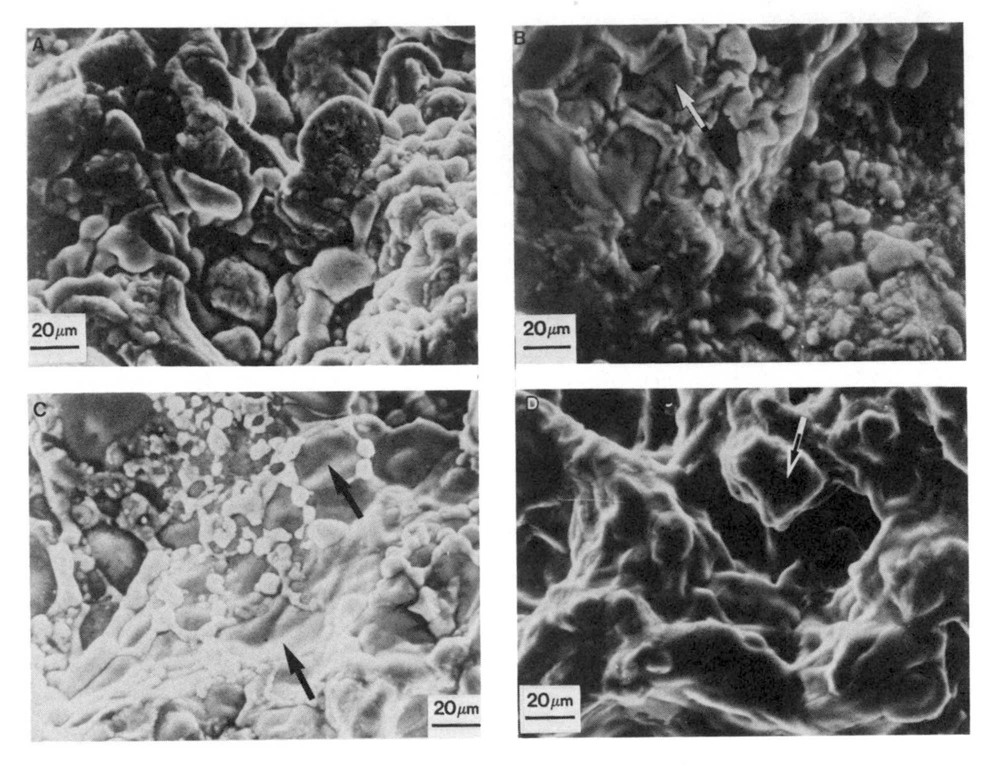

Fig. 8.25. High magnification scanning electron micrographs of cakes made from a 50:50 flour-to-starch ratio mixture. The cakes were sampled from the center of the cake, top to bottom. (A) top; (B) 0.5 cm from the top; (C) 0.5 cm from the bottom; (D) bottom of cake. *Courtesy of Gordon et al. (1979).*

Cake Texture

Francis and Groves (1962) reported that the starch in high-ratio cake appeared to be more formless and more gelatinized than in bread crumb. These authors, however, were not able to distinguish whether the starch or protein was the principal structure-forming material. Since a large amount of air was entrapped in the starch–protein matrix, they concluded that no structure, comparable to the protein networks in bread or cracker biscuits, could be found in a high-ratio cake. In their study on madeira and fat-free sponge cakes, Shephard and Yoell (1976) noted that the air cell walls of the madeira cake exhibited a much greater average thickness. In both types of cakes, however, the starch granules had distorted shapes due to the gelatinization, although in the fat-free sponge cake to a lesser degree. They also found varying degrees of starch gelatinization within the crumb of madeira cake, according to position,

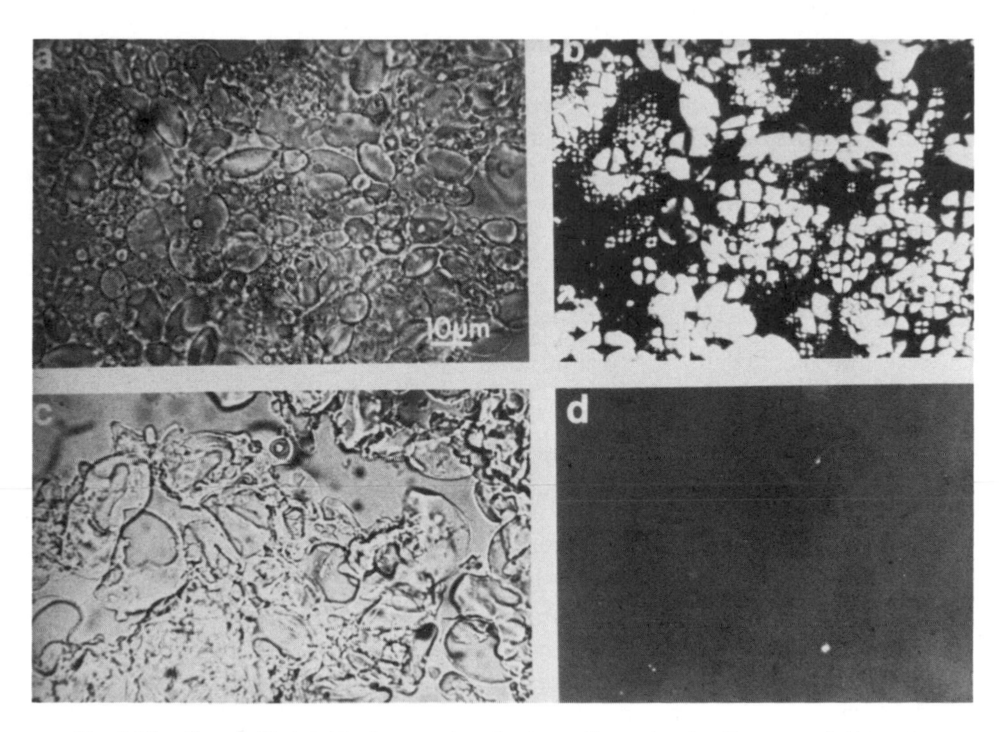

Fig. 8.26. Transmitted light micrographs of cake portions showing the same fields under normal and polarized light. All samples are at the same magnification. (a, b) Cake crust; (c, d) Center of cake crumb.
Courtesy of Varriano-Marston et al. (1980).

that were similar to those found in yellow layer cake, chocolate layer cake, and cake donuts (Varriano-Marston *et al.* 1980) (see Fig. 8.26).

Paton *et al.* (1981) studied the influence of various ingredients on the structure "cohesiveness" during baking. They found that the cohesive force results from the opposing action of toughening by flour and egg proteins and tenderizing by sugar and shortening. Their work suggests that this cohesive force has an optimal value and that its measurement could provide useful indication of the influence of individual ingredients on the cake structure.

Hodge (1977) in his study on cake staling (plain slab cake) reported that there is a rapid increase in crumb firmness (measured by a cone indenter) during the first few hours after baking. The crumb continued to firm during storage (70°F) even in the absence of any moisture loss, but at a reduced rate. He also reported that sponge cakes have a similar firming pattern, although the crumb was softer than that of the slab cake throughout the test. It is well known that frozen storage extends the shelf-life of baked cakes and bread. However, Hodge (1977) found that there is no increase in the rate of staling (evaluated by sensory panel) in cakes stored at refrigerator temperatures. He also found that the rate of staling decreases as the storage temperature decreases (even at refrigerator temperatures). His findings were confirmed in the ITT Continental Baking Co. laboratories and the results are shown in Table 8.1.

TABLE 8.1. EFFECTS OF STORAGE TEMPERATURE ON FIRMNESS (COMPRESSIBILITY) OF SPONGE CAKES[1]

Storage Temperature	Firmness (g force ± Standard Deviation)		
	Day 1	Day 7	Day 11
72°F (22°C)	90 ± 2.6	167 ± 3.0	199 ± 5.3
40°F (4.5°C)	75 ± 2.5	123 ± 3.2	151 ± 2.1

[1]Firmness was measured with the Instron Universal Testing Machine.

According to Hodge (1977) reheating of stale cake [7-day-old cake stored at 70°F (21°C)] at 122°F (50°C) for 2 hours did not reduce the firmness as measured by a cone indenter. One day after reheating, the reheated cakes were measureably firmer than the controls, and this difference was maintained throughout the next 7 days of storage at 70°F. He therefore concluded that cakes cannot be refreshed by a simple heat treatment. Using differential thermal analysis to study the degree of starch crystallization in slab and sponge cakes, Hodge found no evidence of starch crystallization in cakes that had been stored and allowed to stale for up to 6 weeks. In light of the above and his unsuccessful attempts to refreshen stale cake by heating, he suggested that starch crystallization is a minor factor in cake staling and speculated that the main factor responsible for cake firming is a migration of water from gluten to starch, thus hardening the gluten network. At this time, however, there is still insufficient evidence to draw any firm conclusions concerning the accurate mechanisms that regulate cake staling.

CONCLUSIONS

The scanning electron microscope will be used more and more to study the relationship between bread and cake texture and microstructure, but it is critical to have an understanding of the effects of sample preparation procedures on the material under study. It is also essential to assure that a sufficient number of specimens are examined to assure that the micrographs are indeed representative. The use of TLM and TEM with the appropriate histochemistry, and TEM with cryofixation and freeze-etching to identify components of the product is a necessary adjunct to any SEM study.

BIBLIOGRAPHY

ANGOLD, R.E. 1979. Cereals and bakery products. *In* Food Microscopy. J.G. Vaughan (Editor). Academic Press, New York.

ARANYI, C., and HAWRYLEWICZ, E.J. 1969. Application of scanning electron microscopy to cereal specimens. Cereal Sci. Today *14*(7), 230–233, 253.

BAKER, J.C. 1941. The structure of the gas cell in bread dough. Cereal Chem. *18*, 34–41.

BARLOW, K.K., BUTTROSE, M.S., and SIMMONDS, D.H. 1973. The nature of the starch-protein interface in wheat endosperm. Cereal Chem. *50*, 443–454.

BECHTEL, D.B., POMERANZ, Y., and DE FRANCISCO, A. 1978. Breadmaking studied by light and transmission electron microscopy. Cereal Chem. *55*, 392–401.

BELL, A.V., BERGER, K.G., RUSSO, J.V., WHITE, G.W., and WEATHERS, T.L. 1975. A study of the micro-baking of sponges and cakes using cine and television microscopy. J. Food Technol. *10*, 147–156.

BIRNBAUM, H. 1978. Surfactants and shortenings in cake making. Bakers Dig. *52*, 29–38.

BURHANS, M.E., and CLAPP, J. 1942. A microscopic study of bread and dough. Cereal Chem. *19*, 196–216.

BUTTERWORTH, S.W., and COLEBECK, W.J. 1938. Some photographic studies of dough and bread structure. Cereal Chem. *15*, 475–488.

BUTTROSE, M.S. 1973. Rapid water uptake and structural changes in inbibing seed tissues. Protoplasma *77*, 111–122.

CARLIN, G.T. 1944. A microscopic study of the behavior of fats in cake batters. Cereal Chem. *21*, 189–199.

CHABOT, J.F. 1979. Preparation of food science samples for SEM. Scanning Electron Microsc. *3*, 279–286.

CHABOT, J.F., HOOD, L.F., and LIBOFF, M. 1979. Effect of scanning electron microscopy preparation methods on the ultrastructure of white bread. Cereal Chem. *56*, 462.

CORNFORD, S.J. 1963. Cited by P. Sherman. 1980. *In* Industrial Rheology, p. 211. Academic Press, New York.

CROZET, N. 1977. Ultrastructural changes in wheat flour protein during fixation and embedding. Cereal Chem. *54*, 1108–1114.

DENNETT, K., and STERLING, C. 1979. Role of starch in bread formation. Staerke *31*(6), 209–213.

DE STEFANIS, V.A., and TURNER, E.W. 1981. Modified enzyme system to inhibit bread firming. Method for preparing same and use of same in bread and other bakery products. U.S. Patent 4,299,848. November 10.

DUNN, J.A., and WHITE, J.R. 1939. The leavening action of air included in cake batter. Cereal Chem. *16*, 93–100.

EVERS, A.D. 1971. Scanning electron microscopy of wheat starch. Part III. Staerke *23*, 157–162.

FLEMING, S.E., and SOSULSKI, F.W. 1978. Microscopic evaluation of bread fortified with concentrated plant proteins. Cereal Chem. *55*(3), 373–382.

FRANCIS, B., and GROVES, C.H. 1962. A note on the application of histological technique to the study of the structure of baked goods. J. R. Microsc. Soc. *81*, 53–59.

GALLANT, D., and GUILBOT, A. 1973. Development in knowledge on wheat starch ultrastructure. Staerke *25*, 335–342.

GORDON, J., DAVIS, E.A., and TIMMS, E.M. 1979. Water-loss rates and temperature profiles of cakes of different starch content baked in a controlled environment oven. Cereal Chem. *56*(2), 50–57.

GREWE, E. 1937. The emulsion-foam produced by agitating butter, sugar and egg: A method for testing the stability of the emulsion and the effect of the conditioning temperature of the fat. Cereal Chem. *14*, 802–818.

HARRIS, B.R. 1978. Cited by H. Birnbaum. 1978. Surfactants and shortening in cake making. Bakers Dig. *52*, 29–38.

HODGE, D. 1977. A fresh look at cake staling. Baking Ind. J. April 14–19.

HOSENEY, R.C. 1982. Personal communication. Kansas State Univ., Manhattan, Kansas.

HOSENEY, R.C., LINEBACK, D.R., and SEIB, P.A. 1978. Role of starch in baked foods. Bakers Dig. *52*, 11–18, 40.

HOWARD, N.B., HUGHES, D.H., and STROBEL, R.G.K. 1968. Function of the starch granule in the formation of layer cake structure. Cereal Chem. *45*, 329–338.

HSIEH, S.I., DAVIS, E.A., and GORDON, J. 1981. Cryofixation freeze-etch of cake batters. Cereal Foods World *26*(10), 562–564.

HYDER, M.A., HOSENEY, R.C., FINNEY, K.F., and SHOGREN, M.D. 1974. Interactions of soy flour fractions with wheat flour components in breadmaking. Cereal Chem. *51*, 666–675.

KHOO, U., CHRISTIANSON, D.D., and INGLETT, G.E. 1975. Scanning and transmission microscopy of dough and bread. Bakers Dig. *49*(4), 24–26.

LOWE, B. 1943. Experimental Cookery, 3rd Ed. John Wiley and Sons, New York.

MARTIN, D.J., and TSEN, C.C. 1981. Baking high-ratio white layer cakes with microwave energy. J. Food Sci. *46*, 1507–1513.

MEDCALF, D.G., and GILLES, K.A. 1968. The function of starch in dough. Cereal Sci. Today *13*(10), 382–385, 388, 392–393.

MELCHIOR, H., and FEUERBERG, H. 1954. Ber Dtsch. Bot. Ges. *67*, 394.

MONCRIEFF, J. 1970. Shortenings and emulsifiers for cakes and icings. Bakers Dig. *44*(5), 60–63.

MORR, M.L. 1939. Distribution of fat in plain cakes. M.S. Thesis. Iowa State College, Ames. Cited by G.T. Carlin. 1944. A microscopic study of the behavior of fats in cake batters. Cereal Chem. *21*, 189–199.

MORRIS, R. 1929. Shortening in cake making. Bakers Weekly *63*(4), 60.

MOSS, R. 1974. Dough microstructure as affected by the addition of cysteine, potassium bromate, and ascorbic acid. Cereal Sci. Today *19*(12), 557–561.

MOSS, R. 1975. Bread microstructure as affected by cysteine, potassium bromate and ascorbic acid. Cereal Foods World *20*(6), 289–292.

PATON, D., LAROCQUE, G.M., and HOLME, J. 1981. Development of cake structure: Influence of ingredients on the measurement of cohesive force during baking. Cereal Chem. *58*(6), 527–529.

PLATT, W. 1930. Staling of bread. Cereal Chem. *7*, 1–34.

PLATT, W., and POWERS, R. 1940. Compressibility of bread crumb. Cereal Chem. 17, 601–621.

POMERANZ, Y., SHOGREN, M.D., FINNEY, K.F., and BECHTEL, D.B. 1977. Fiber in breadmaking—Effects on functional properties. Cereal Chem. *54*(1), 25–41.

PONTE, J.G., TITCOMB, S.T., and COTTON, R.H. 1962. Flour as a factor in bread firming. Cereal Chem. *39*, 437–444.

RUSSO, J.C. 1972. BFMIRA Symp. Proc. *14*, 24. Citied by A.V. Bell, K.G. Berger, J.V. Russo, G.W. White, and T.L. Weathers. 1975. A study of the micro-baking of sponges and cakes using cine and television microscopy. J. Food Technol. *10*, 147–156.

SANDSTEDT, R.L. 1961. The function of starch in the baking of bread. Bakers Dig. *25*(3), 36–44.

SANDSTEDT, R.M., SCHAUMBURG, L., and FLEMING, J. 1954. The microstructure of bread and dough. Cereal Chem. *31*, 43–49.

SHEFFER, W. 1916. Z. Gesamte Getreidewes *8*, 6.

SHEPHARD, I.S., and YOELL, R.W.1976. Cake emulsions. *In* Food Emulsions. S. Friberg (Editor). Marcel Dekker, New York.

SHERMAN, P. 1970. Industrial Rheology. Academic Press, New York.

SHORT, A.L., and ROBERTS, E.A. 1971. Pattern of firmness within a bread loaf. J. Sci. Food Agric. *22*, 470–472.

SIMMONDS, D.H. 1972. The ultrastructure of the mature wheat endosperm. Cereal Chem. *49*, 212–222.

SIMMONDS, D.H. 1975. Wheat grain morphology and its relationship to dough structure. Cereal Chem. *49*, 324.

STELLER, W.R., and BAILEY, C.H. 1938. The relation of flour strength, soy flour and temperature of storage to the staling of bread. Cereal Chem. *15*, 291–401.

SUNDERLIN, G., and COLLINS, O.D. 1940. The viscosity of cake batters as related to batter structure. Unpublished Thesis, Purdue University, Lafayette, IN. Cited by G.T. Carlin. 1944. A microscopic study of the behavior of fats in cake batters. Cereal Chem. *21*, 189–199.

VARRIANO-MARSTON, E. 1977. A comparison of preparation procedures for scanning electron microscopy of doughs. Food Technol. *31*(10), 32–36.

VARRIANO-MARSTON, E. 1981. Integrating light and electron microscopy in cereal science. Cereal Foods World *26*(10), 558–561.

VARRIANO-MARSTON, E., HUANG, G., and PONTE, J. 1980. Comparison of methods to determine starch gelatinization in bakery foods. Cereal Chem. *57*(4), 242–248.

VERSCHAFFELT, E., and VAN TEUTEM, F.E. 1915. Hoppe-Seyler's Z. Physiol. Chem. *95*, 130–135.

VON BELDEROK, B. 1975. Investigations of bread crumb structure by means of electronic polarizing microscope. Muehle *19*, 251–254.

WALDT, L. 1968. The problem of staling: Its possible solution. Bakers Dig. *42*, 64–66, 73.

WASSERMAN, L., and DORFNER, H.H. 1974. Scanning electron microscopy of baked products. Getreide, Mehl Brot. *28*(12), 324–328.

WOOTTON, J.C., HOWARD, N.B., MARTIN, J.B., MCOSKER, D.E., and HOLME, J. 1967. The role of emulsifiers in the incorporation of air into layer cake batter systems. Cereal Chem. *44*, 333–343.

9

Physical Properties of Synthetic Food Materials

J.M. de Man[1]

INTRODUCTION

For the purpose of this discussion, synthetic foods will be defined as those products obtained by extensive processing, which substantially alters their physical properties. It will further be necessary to limit the area to the relatively new products that are mostly derived from plant proteins. Strictly speaking, synthetic foods would be derived from chemical synthesis of simpler compounds. This would probably apply only to synthetic chemical compounds such as flavors. When more broadly defined as those products manufactured by certain processing techniques from various raw materials, the field would cover a large variety of manufactured foods, including such well-known items as bread, cheese, chocolate, and margarine. The basic principle involved in both groups of foods is the same. Processing brings about changes in chemical and physical properties that make the end product more attractive and useful as a food. Plant proteins available as bland powders have little appeal as foods. Much effort has gone into research to change these protein foods into products with more interesting physical characteristics. This chapter will explore some of the achievements of the past few decades in this area and highlight the advances made in the study of physical properties and their measurement.

---

[1] Department of Food Science, University of Guelph, Guelph, Ontario, Canada NIG 2WI.

MICROSTRUCTURE

The basic objective of the production of synthetic foods is to achieve texture formation through chemical and/or physical means. The various processes used for this purpose may include application of one or more of the following: heat, pressure, freezing, and chemical modification. Many of these processes have been developed in an empirical way, but much basic information has been generated through studies of microstructure formation. Given the interrelationship of chemical composition, physical structure, and physical properties (Fig. 9.1), it is evident that for a particular chemical composition physical properties may be altered by appropriate changes in structure. To fully understand and control the physical characteristics of a food requires a good knowledge of what happens to the microstructure during processing. Much of the research in this area has been done with soy products, either defatted soy meal or protein concentrates and isolates. Initial studies on the microstructure of unprocessed soybeans by electron microscopy provided the basic information on the nature of the protein bodies in soybeans (Saio and Watanabe 1966; Wolf 1970; Wolf and Baker 1972). The protein bodies range in size from 5 to 8 μm with some in the wider range of 2–20 μm and can be visualized by electron microscopy (Fig. 9.2).

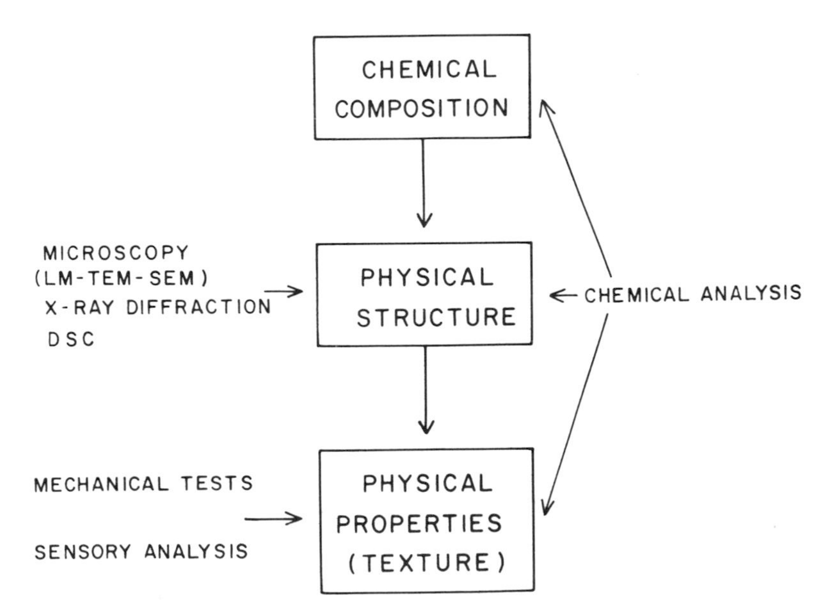

Fig. 9.1. Interrelationship of chemical and physical properties of materials.

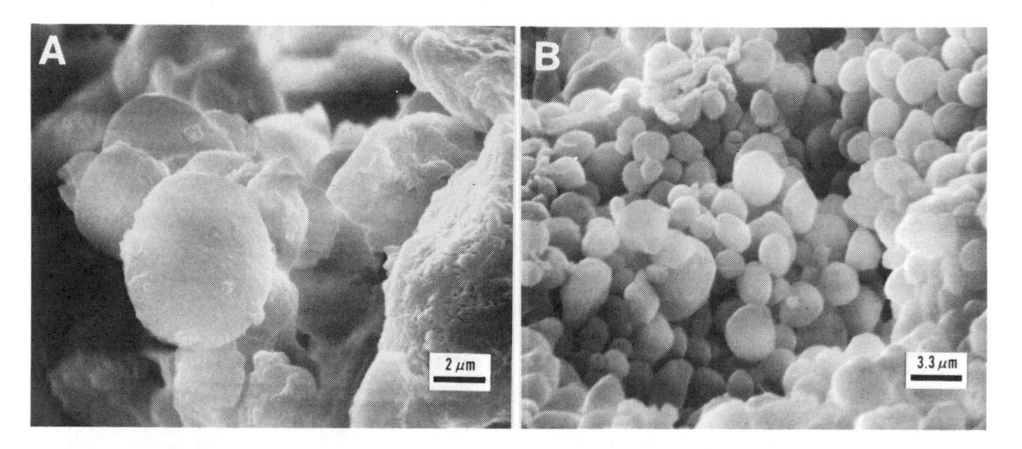

Fig. 9.2. Scanning electron micrographs of protein bodies: (A) in undefatted soybean flour; (B) isolated by sucrose density gradient centrifugation.
From Wolf (1970).

It was obvious from early electron microscopic studies that extensive changes in structure occur during texturization. Stanley *et al.* (1972) showed by scanning electron microscopy (SEM) that spun soy protein fibers are thread-like with longitudinal fissures and transverse scales (Fig. 9.3). Extruded defatted soy meal was shown by Cumming *et al.* (1972) to take on a distinctly fiber-like structure (Fig. 9.4). The extrusion temperature was found to be an important process parameter and in a subsequent report (Maurice *et al.* 1976), evidence was presented that as the temperature increased, the product became less compact and more spongy in texture (Fig. 9.5). Aguilera *et al.* (1976) investigated the process of fiber formation as the material moved down the extruder barrel. Fiber formation occurred only in the last part of the barrel. Most of the fiber formation occurred at the die, and it was suggested that high shear rates and compression forces at this location combined to texturize the plasticized material. The expansion upon leaving the die would be responsible for the expansion of the product and the sponge-like appearance shown in Fig. 9.5.

Maurice and Stanley (1978) investigated the effect of various process variables (extruder pressure, extruder torque, extruder throughput, product moisture level) on the microstructure of extruded soymeal. They found that at a low protein level (41%), little texture formation took place and that increasing protein content to 51 and 61% resulted in more fibrous structures.

Although soy protein products have been studied most extensively, there has been increasing interest in other protein sources and their

Fig. 9.3. Scanning electron micrographs of spun soy fibers.
From Stanley et al. (1972).

mixtures with soy protein. Stanley and deMan (1978) described extrusion products of mixtures of soy and corn, and mixtures of soy and minced cod fish muscle. These mixtures produced spongy, fibrous extrudates, and the soy–fish mixture also contained fish muscle fibers as an integral part of the structure (Fig. 9.6). Taranto *et al.* (1975) and Cegla *et al.* (1978) have reported on studies of extrusion texturization of cottonseed flour and cottonseed flour–rice flour blends. In these studies, transmission electron microscopy (TEM) was used to demonstrate that both defatted native glandless and deglanded cottonseed flours produced extrudates with highly puffed, expanded structures. Extrudates from steam heated (denatured) cottonseed flours yielded products with higher density, more closely resembling soy extrudates. Addition of reducing salts prior to extrusion resulted in the formation of a more uniform, less disrupted protein matrix and the insoluble carbohydrates were more evenly dispersed. In another study, Taranto *et al.* (1978) used SEM to study the microstructure of soy and cottonseed flours texturized by extrusion and nonextrusion processing. The latter was done with a hand-operated heated press. It was found that similar

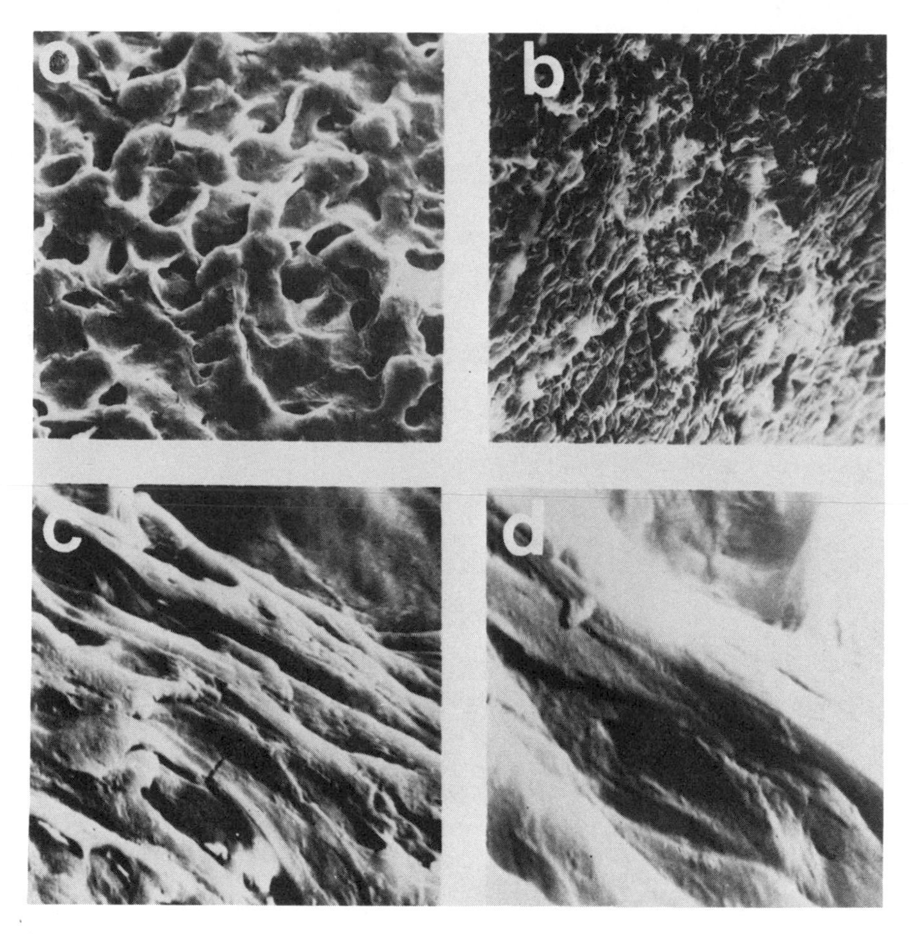

Fig. 9.4. Scanning electron micrographs of soy meal at various stages of processing. (a) Raw soybean. (b) Unprocessed meal. (c) Texturized meal, 1100×. (d) Texturized meal, 5500×.
From Cumming et al. (1972).

surface morphology was obtained with both processes (Fig. 9.7). On the basis of SEM and TEM studies, it was concluded that neither of the two processes was completely efficient in disruption of the cells and fusion of the cellular contents. However, in each case, the level of fusion and cell disruption was sufficient to create texture. These authors also suggested that texture formation is not necessarily obtained by extrusion only and, in fact, the usual extrusion process may even partially destroy the texture created by the multiple effect of pressure, heat, and

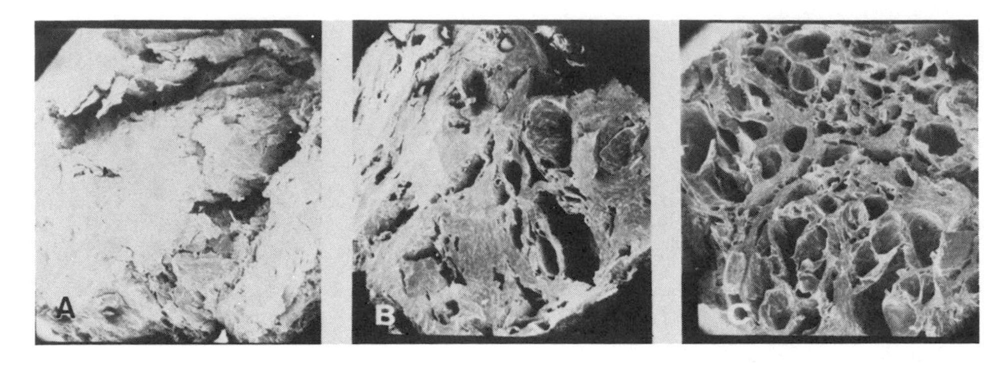

Fig. 9.5. Scanning electron micrographs of extruded soy meal. Extrusion temperature: (A) 135°C. (B) 165°C. (C) 180°C.
From Maurice et al. (1976).

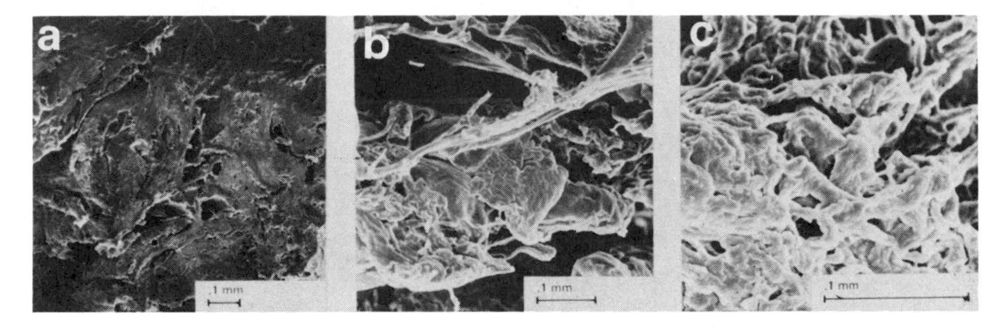

Fig. 9.6. Scanning electron micrographs of co-extrudates of soy and fish. (a) Cross-section. (b) and (c) Longitudinal sections.
From Stanley and de Man (1978).

time. They also state that the extrusion process does not create fibers, but a sheeted protein matrix with insoluble carbohydrates dispersed throughout it. The uniformity of the protein matrix and the evenness of the insoluble carbohydrate dispersion appeared to be correlated with the rheological properties of the product.

It appeared at this point that a difference of opinion had developed between researchers ascribing a "fiber structure" to extruded proteinacous products and those who maintained that their structure was not fibrous in the true sense of the word. This matter was dealt with in a paper by Shen and Morr (1979). It was suggested that the molecular arrangement in a fiber may consist of a random arrangement of molecules resulting in amorphous filaments, the molecules may be either

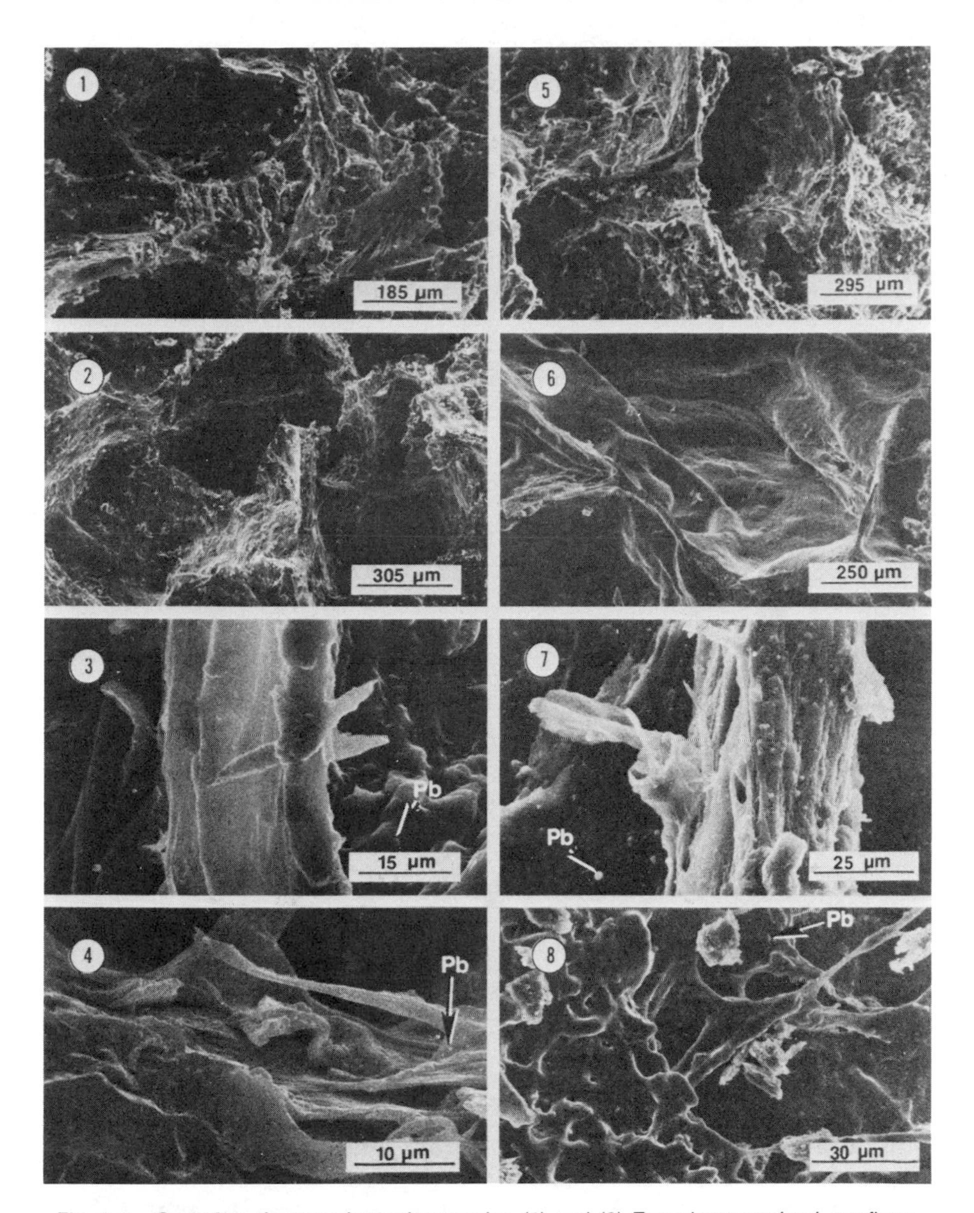

Fig. 9.7. Scanning electron photomicrographs. (1) and (3) Extrusion texturized soy flour. (2) and (4) Nonextrusion texturized soy flour. (5) and (7) Extrusion texturized cottonseed flour. (6) and (8) Nonextrusion texturized cottonseed flour. (1, 3, 5 and 7 are surfaces oriented perpendicular to the direction of extrusion; 2, 4, 6 and 8 are cross-sections.) The spherical particles (Pb) have tentatively been identified as protein bodies.
From Taranto et al. (1978).

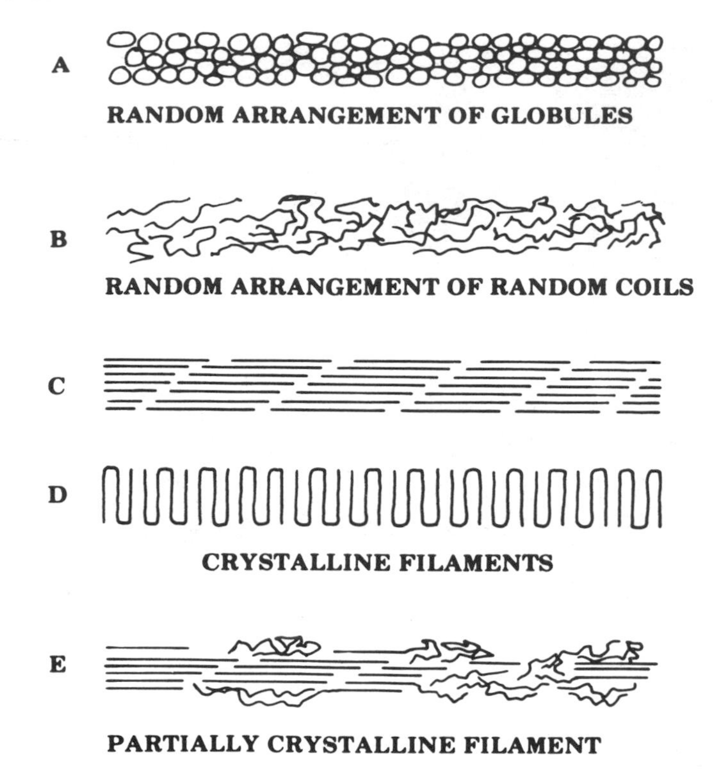

Fig. 9.8. Proposed molecular arrangements for drawn filaments.
From Shen and Morr (1979).

globular or coiled as shown in (Fig. 9.8). In filaments with ordered molecular arrangement, crystalline fibers are obtained. These may be either partially or completely crystalline as indicated in the figure. Shen and Morr proposed to limit the use of the term fiber to only those filaments that have an appreciable degree of crystallinity along their filament axis. The requirement of considering crystallinity in the textural assessment of protein products indicates the need for additional methods of investigation. Since SEM currently has a resolution of 15 nm, it can only be used to look at the gross structure of large fibers and fiber bundles. Individual microfibrils require the resolution of TEM which is about 2 nm. In addition, crystal spacings can be analyzed by X-ray diffraction methods, which enable the measurement of spacings of molecules in the crystal lattices. Shen and Morr (1979) presented a schematic representation of the events taking place in the transformation of globular proteins into fibers (Fig. 9.9). The process consists of five major steps. In the first step, the protein must be dissolved and dena-

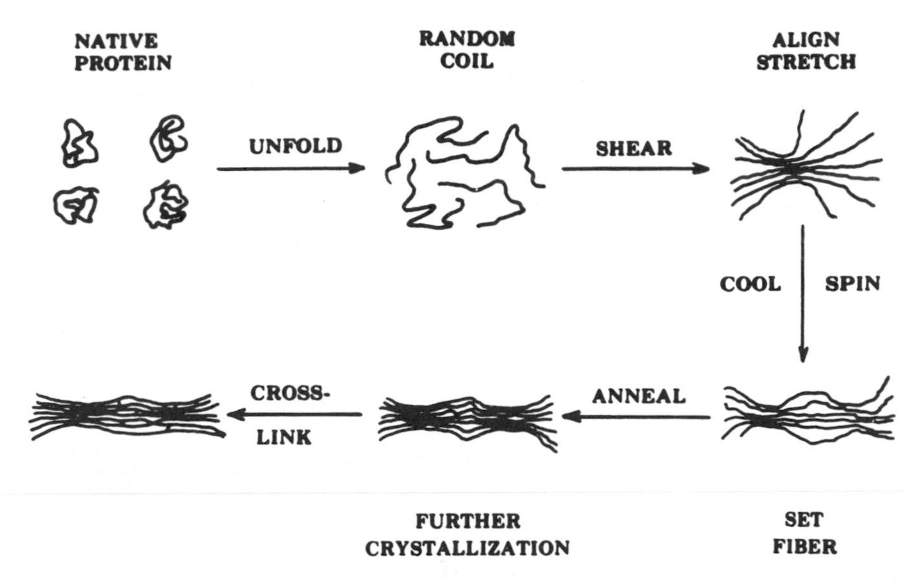

Fig. 9.9. Schematic representation of the postulated fiber forming process.
From Shen and Morr (1979).

tured to produce a solution of random coils. In the second step, the coils, as a result of the shear forces are stretched and aligned. In the third step, the structure must be fixed to prevent the molecules from returning to a random orientation. In the fourth step, the fiber is annealed to increase the degree of crystallinity. In the final step, the so-formed fiber is cured by crosslinking. This last stage is important, stabilizing the texture, so that such products can withstand a canning process, for example.

Burgess and Stanley (1976) proposed a possible mechanism for the thermal texturization of soybean protein. It involves rupturing previously denatured protein molecules into subunits under the conditions of extrusion. These subunits subsequently reaggregate into the characteristic textural elements, mainly by intermolecular amide bonds and to a lesser extent by hydrophobic interactions and hydrogen and disulfide bonds.

Additional information on the mechanism of texturization has been obtained by modification of the protein molecules themselves. Choi *et al.* (1981) found that succinylation of cottonseed flour resulted in increased solubility of its protein. Simonsky and Stanley (1982) used acetic or succinic anhydride to acylate soy protein. This treatment resulted in a significant loss of extrudate texture, but it increased its solubility. SEM study of the soy extrudate and acylated protein extrudate showed (Fig.

9.10) that the acylated products had a denser texture with fewer air cells than the nonacylated product. Simonsky and Stanley concluded that good texture is associated with a cellular structure. They also showed that ninhydrin, which irreversibly binds free amino groups, inhibits texture formation. Maximum texturization took place at pH 8 where the amino groups are unprotonated and therefore more reactive. All of these findings support the hypothesis that amino group interactions play a role in texture formation.

As already mentioned, texturization processes are not necessarily restricted to extrusion, although this is the most widely used commercial method. Shen and Morr (1979) list a number of alternative proce-

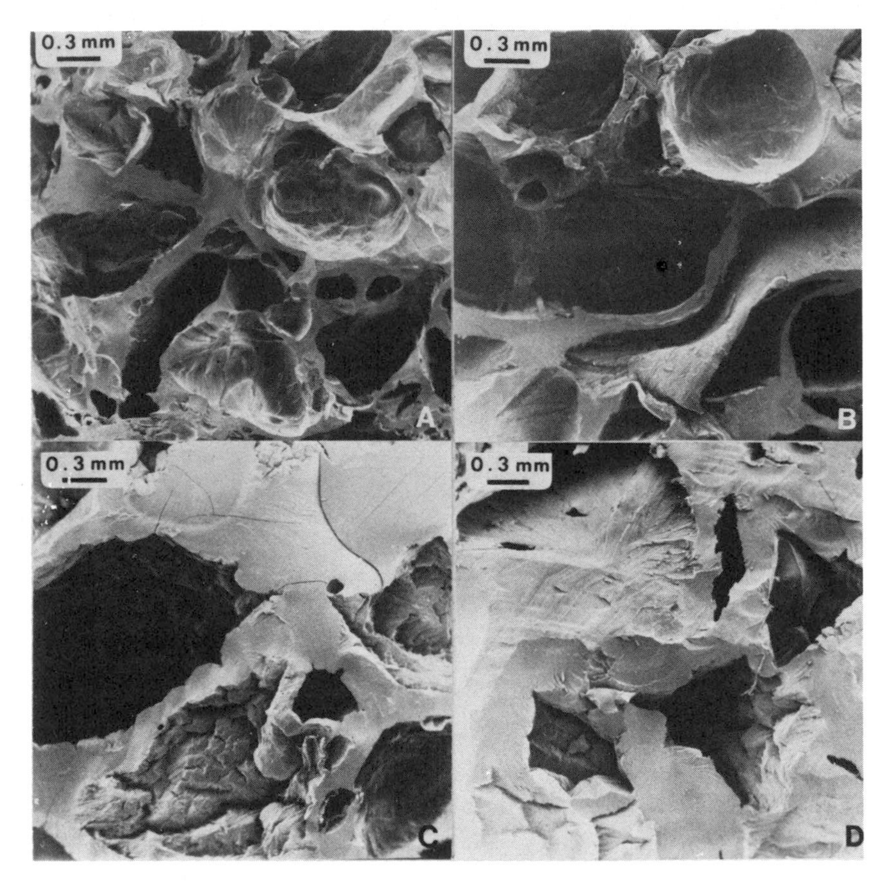

Fig. 9.10. Scanning electron micrographs. (A) and (B) Texturized soy protein isolate. (C) and (D) Acylated soy protein
From Simonsky and Stanley (1982).

dures, including spinning of protein solutions. Texture can also be obtained by addition of certain salts to protein solutions to produce gels; gels can also be formed by heating. In addition, texture can be produced by freezing.

Furukawa *et al.* (1979) studied the texture–structure relationships in heat-induced soy protein gels. The gels were prepared by heating 20% soy protein pastes for 30 minutes at temperatures ranging from 25° to 130°C. It was suggested that the gel network was formed through crosslinking by disulfide, hydrogen, and hydrophobic bonds. This was confirmed by testing the solubility of the solid in phosphate buffer containing 2-mercaptoethanol and/or urea. The degree of network formation was controlled by the heating temperature. The gels produced at 40°, 80°, and 120°C were examined by SEM, and the micrographs are presented in Fig. 9.11. The gels made at 40°C were soft with only limited formation of porous structure. The gels produced at 80°C were hard by comparison and had a well-developed porous structure with membranous walls of a thin compact film type. The gels produced at 120°C were fragile and their structure was partially collapsed.

Lee and Abdollahi (1981) have described the structure of fish protein gels obtained by comminution of fish muscle with added salt. Freezing and thawing of cooked fish gels caused a sponge-like texture development. The texture of the fish protein gels could be modified by incorporation of plastic fat. Freeze texturization of soy protein is a commonly employed process in Japan and the product, *kori tofu*, can be produced in a variety of textural forms depending on processing conditions used (Stanley and deMan 1978). Scanning electron microscopy indicates that freeze texturization produces a porous sponge-like structure (Fig. 9.12). Freeze texturization can also be applied directly to soymilk (Lugay and Kim 1981). The frozen soymilk is freeze-dried, and the dry material is stabilized by moist heat treatment.

Products in which two or more raw materials are co-extruded are receiving increasing attention. The following are a few examples of this trend. Murray *et al.* (1980) produced co-extrudates of mixtures of soy and fish proteins. The microstructure as examined by SEM was similar to well-textured vegetable protein extrudates and exhibited a definite structural network and a fibrous or sheet-like appearance (Fig. 9.13). Co-extrusion of soy protein and rice flour was investigated by Noguchi *et al.* (1981). The microstructure was examined by SEM and revealed that the soy protein isolate formed fine strings in the starch matrix of the extrudate. The extrudates showed the usual sponge-like structure. The protein bodies of rice, which are reported to be in the 1–4 μm size range, could not be identified in the extruded material. The extruded rice flour

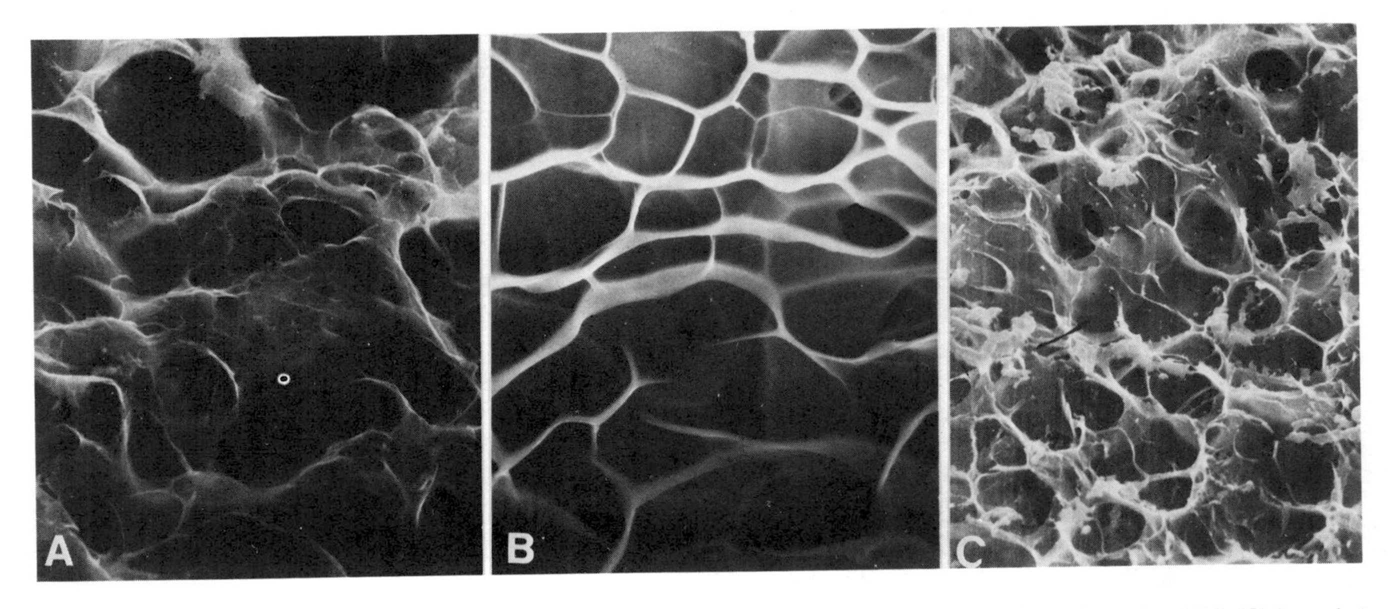

Fig. 9.11. Scanning electron micrographs of heat-induced soy protein (HISP) gels: (A) formed at 40°C; (B) formed at 80°C; (C) formed at 120°C.

From Furukawa et al. (1979).

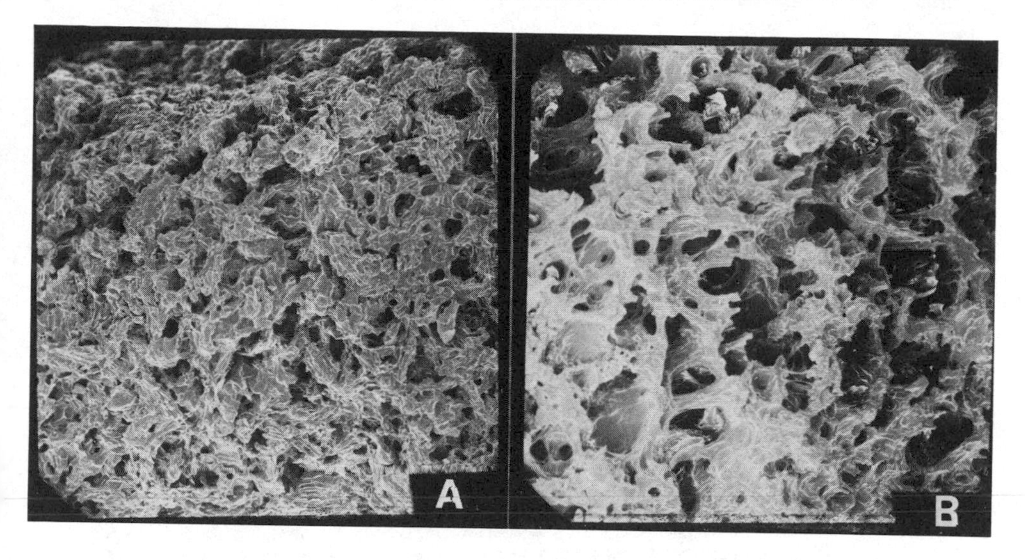

Fig. 9.12. Scanning electron micrograph. Microstructure of (A) freeze-texturized soy product. (B) Japanese *kori-tofu*.

had large air cells, whereas the product containing soy protein isolate was more rigid and denser. The difference in microstructure was examined after pulverization and rehydration (Fig. 9.14). The protein fortified product had a finer structure, presumably because of a fine protein network.

TEXTURE

The major objective of the production of many synthetic food materials is the creation of desirable textural attributes. The measurement and evaluation of texture, therefore, is of great importance. Texture can be measured by instrumental or sensory methods, and both have found widespread application in the area of synthetic foods. Because meat-like texture has been aimed for in many of these products, methods identical or similar to those used in the study of meat texture have been widely employed. Stanley *et al.* (1972) investigated the texture of spun soy fibers by extension testing under cycling in the Instron universal testing machine. The load-extension curves obtained for spun soy fibers were different than those normally observed with meat. The breaking strength of spun soy fiber was found to be higher than that of meat.

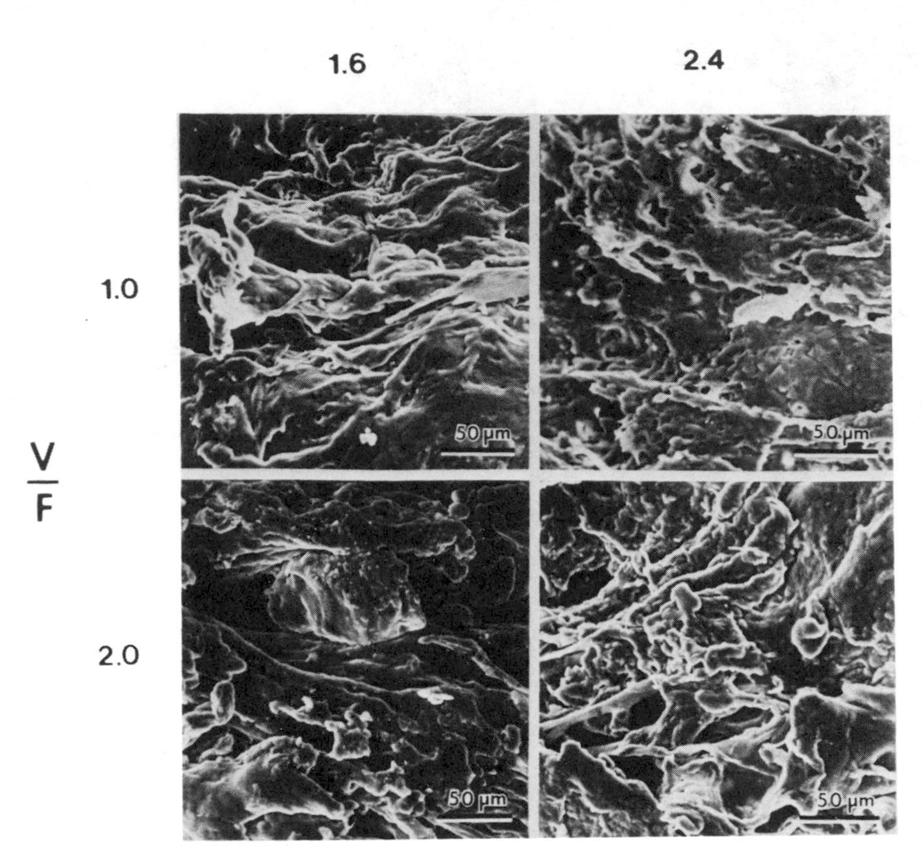

Fig. 9.13. Scanning electron micrographs. Microstructure of soy and fish co-extrudates at different vegetable protein-to-fish (V/F) ratios and protein-to-water (P/W) ratios.
From Murray et al. (1980).

Other parameters measured were break elongation and loss of strength during extension cycling. Results of the latter technique are indicated in Fig. 9.15, which gives the force required to produce constant elongation as a function of cycling. These results were interpreted to indicate that during mastication, spun soy would experience a much greater loss of structural integrity than meat. In a subsequent paper, Cumming *et al.* (1972) examined the texture of an extruded soy product by Warner–

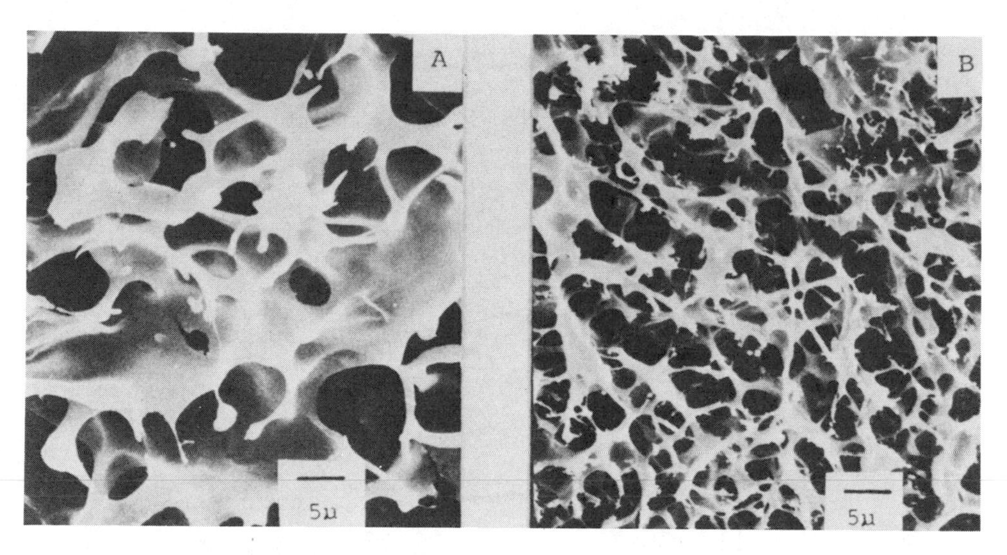

Fig. 9.14. Scanning electron micrographs. Microstructure of pulverized and rehydrated extrudates. (A) Rice. (B) Rice and soy protein isolate.
From Noguchi et al. (1981).

Bratzler shear, the Kramer Shear Press, and by extension testing with the Instron universal testing machine. It was found that the density adjusted shear force in the Warner–Bratzler shear test gave a sigmoid shaped curve when extrusion temperature was increased from 110° to 195°C (Fig. 9.16). This indicates a rapid change in product characteristics between 130° and 170°C. Examination of the tensile strength in the Instron machine yielded the nonlinear results presented in Fig. 9.17. When the same products were examined by the Kramer Shear Press, a linear relationship between extrusion temperature and shear force and shear work was obtained (Fig. 9.18). Such results indicate the usefulness of employing several different methods for texture evaluation, since any single method may not provide sufficient information about the mechanical properties of the products. Spun soy fibers were also examined by Aguilera *et al.* (1975). They used tensile strength measurements with the Instron universal testing machine, Warner–Bratzler shear, the Kramer Shear Press, and a modified version of the Volodkevich tenderometer. It was observed that fracture in the tensile test occurred primarily at a macrocavity or weak point in the fiber. Compression and flow at the zone of rupture were induced by the Volodkevich test and to a lesser extent by the Warner–Bratzler shear. Fracture by the Kramer Shear Press was the closest to pure

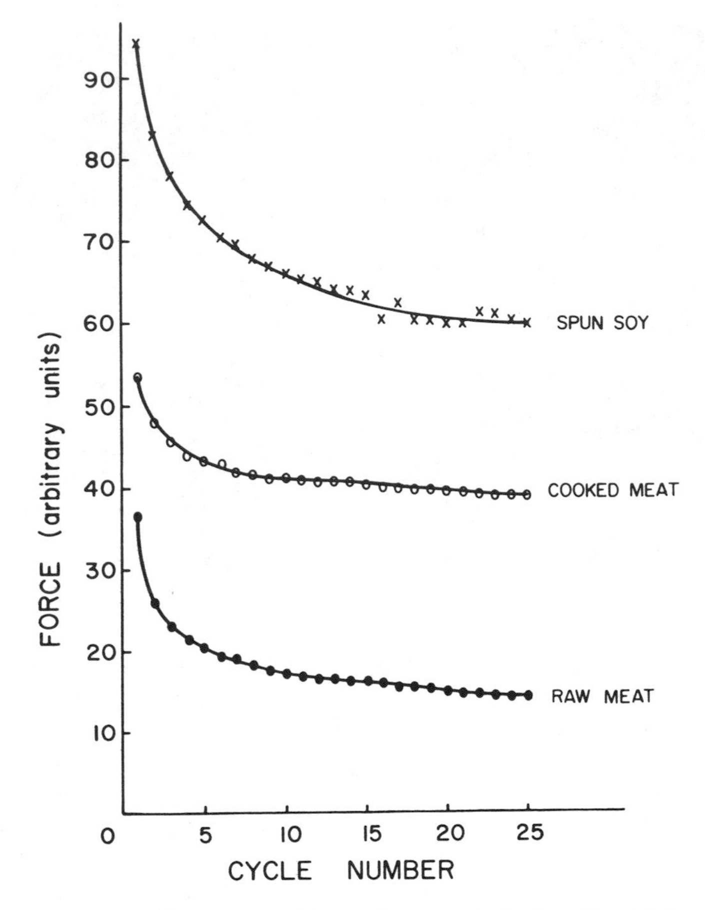

Fig. 9.15. Force required to produce constant elongation as a function of cycling.
From Stanley et al. (1972).

shear. This again indicates that the use of multiple methods may be desirable.

Taranto *et al.* (1978) and Cegla *et al.* (1978) examined the textural properties of extruded and nonextruded texturized soy and cottonseed products. The products were placed in an Ottawa Texture Measurement System test cell with a wire grid attachment, and the cell was operated in conjunction with an Instron universal testing machine. The samples were hydrated before the measurement. The force–deformation curves

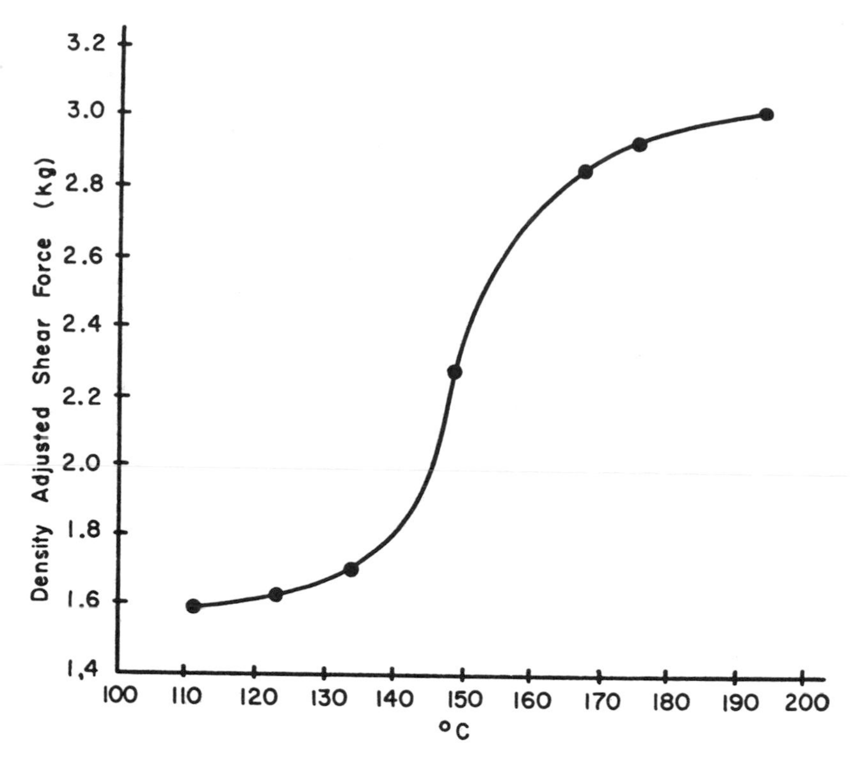

Fig. 9.16. Effect of extrusion temperature on Warner–Bratzler shear values of an extruded soy product.
From Cumming et al. (1972).

were evaluated in terms of point of inflection and "resilience." The resilience was defined as the area under the force–deformation curve up to the rupture point as described earlier by Lee and Toledo (1976).

This type of testing was described earlier by Breene and Barker (1975) who also used a 10 cm^2 Ottawa Texture Measuring System cell with an eight wire grid operated in the Instron universal testing machine. A typical force–distance curve is presented in Fig. 9.19. The interpretation of this type of curve was suggested by Voisey (1971) and Voisey *et al.* (1972) and involves the following parameters:

Hardness, the force required to attain a given deformation. This can be obtained from the average slope of the initial, approximately linear, portion of the curve, and expressed in kg/cm.

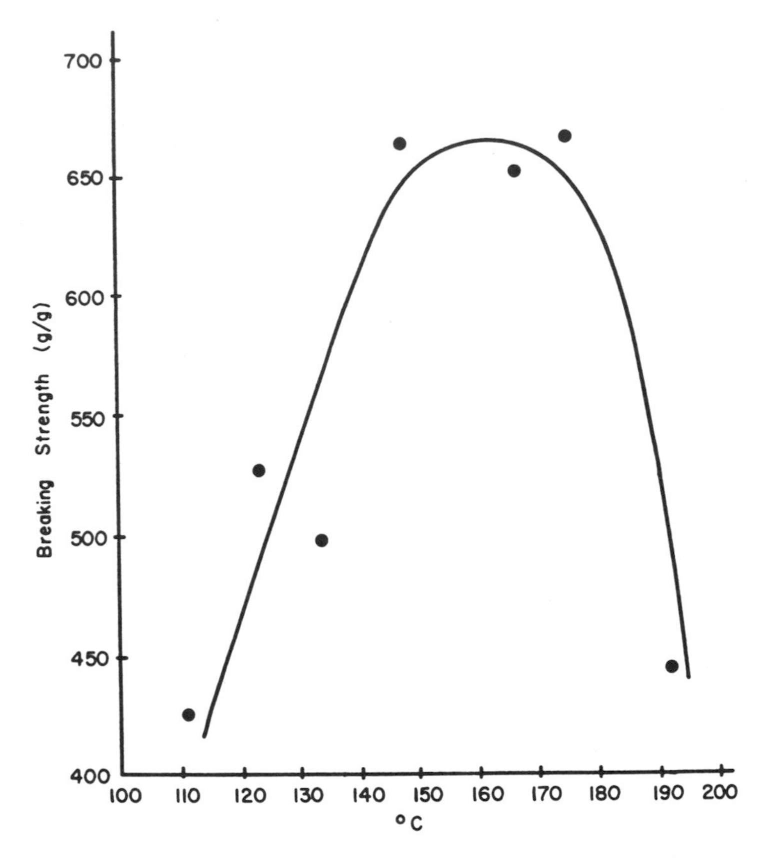

Fig. 9.17. Effect of process temperature on breaking strength of a soy product.
From Cumming et al. (1972).

Extrudability, a function of hardness and cohesiveness. It is expressed as the average slope of the curve in kg/cm after the onset of shear and extrusion.

Chewiness, the energy required to masticate a solid food to a state ready for swallowing. It is measured as the area under the curve in cm^2 and represents the energy used to compress, shear and extrude the sample.

Maximum force, the maximum force registered during extrusion.

Average maximum force, the mean force in kg during extrusion.

Cohesiveness, the strength of the internal bonds making up the body

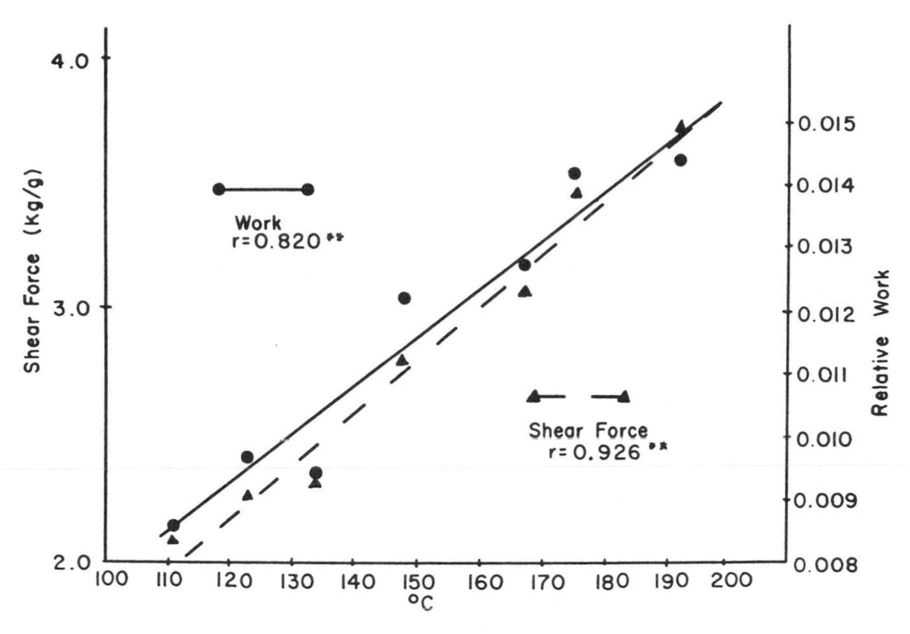

Fig. 9.18. Kramer shear press results as a function of extrusion temperature of a soy product.
From Cumming et al. (1972).

of the product. It is expressed as the force in kg required to initiate shear and extrusion.

Packability, the distance in cm traveled by the plunger before an average linear slope is reached.

From this system, a standard testing procedure was developed (Minnesota Texture Method) which involved measurement of maximum force, average maximum force, chewiness, cohesiveness, and hardness. These parameters were found to indicate differences between textural classes of a variety of textured vegetable protein products and their mixtures with ground beef. In a subsequent paper, Loh and Breene (1977) reported on the relation between instrumental and sensory evaluation methods. For the sensory evaluation, a trained panel and an eight-point unstructured scale were used. It was found that the sensory parameters of springiness, tenderness, and chewiness were well correlated with instrumental results in mixtures of 1:3 texturized soy protein–beef, but the sensory parameters of cohesiveness and juiciness

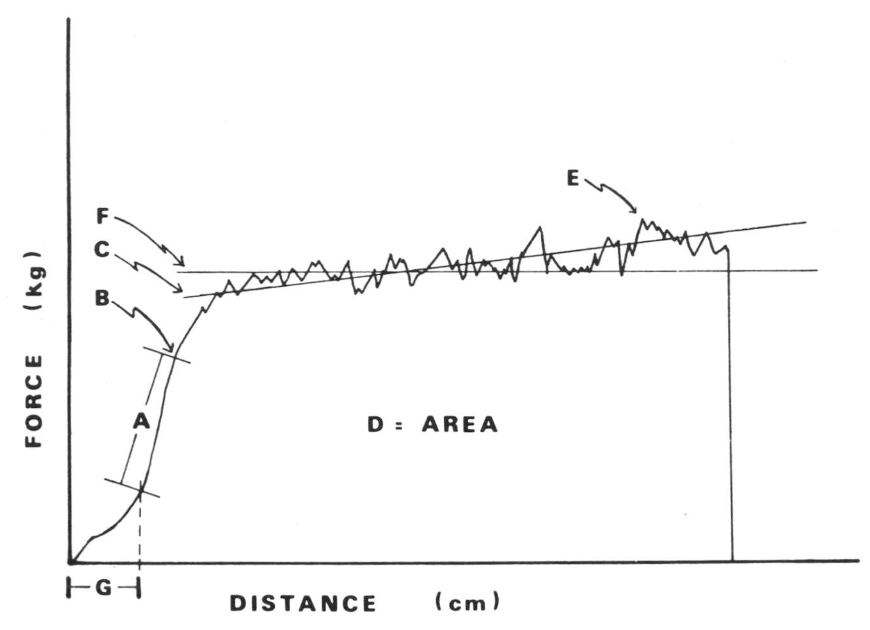

Fig. 9.19. Typical force-distance curve for compression of a soy protein product in an Ottawa texture measuring system wire extrusion cell. (A) Hardness. (B) Cohesiveness. (C) Extrudability. (D) Chewiness. (E) Maximum force. (F) Average maximum force. (G) Packability.
From Breene and Barker (1975).

showed poorer correlation. In mixtures of equal quantities of texturized soy protein–beef, springiness correlated best with instrumental results. Sensory cohesiveness and juiciness correlated only with instrumental chewiness, average force, and cohesiveness. Sensory mealiness was independent of any instrumental parameter. For texturized soy protein without added beef, all of the relationships between instrumental and sensory parameters were highly significant. On the basis of regression analysis performed on these data it was concluded that it is possible to predict sensory responses under certain conditions from instrumental analysis. It was also suggested that it might be possible to predict sensory response to cooked texturized vegetable protein–ground beef mixtures from instrumental analysis of the raw rehydrated texturized vegetable protein.

The interest in mixtures of texturized vegetable proteins and meat as well as other foods is evidenced by the many studies dedicated to texture measurements of these products. Only a few of these will be mentioned as examples.

Sofos *et al.* (1977) studied the effect of incorporation of texturized soy protein and soy protein isolate on the properties of wieners. Texture was evaluated by instrumental and sensory methods. Products were compressed in the Instron universal testing machine and force–distance curves were used to determine hardness, brittleness, elasticity, cohesiveness, gumminess, and chewiness. It was found that texturized soy protein, when used at levels greater than 25–30% on a hydrated basis, resulted in texture softening and decreased emulsion stability.

Berry *et al.* (1979) prepared dry fermented salami with incorporation of structured soy protein fiber. Texture measurements in this study were done with two methods. One involved use of the Precision penetrometer. This instrument was used for compression tests as well as for puncture resistance with a blunt-tipped needle. The other method made use of the Instron universal testing machine provided with the Slice Tenderness Evaluator shear and the single blade shear. The results of these tests were expressed in kilograms of force as well as in energy or work (cm g). The results indicated that 15% incorporation of structured soy protein fiber was acceptable, but 30% incorporation resulted in a product of lower acceptability.

Terrell *et al.* (1979) investigated the effect of replacement of beef by oilseed protein products. The effect on texture was measured by using a Kramer shear cell in an Instron universal testing machine as well as by sensory panel. All-meat frankfurters had significantly better texture than that of the products containing replacements, with the exception of those containing textured soy flour. Frankfurter-like products containing soy concentrate, soy isolate, and textured soy flour had greater bioyield values than did those products containing cottonseed proteins.

The effect of the addition of non-meat proteins on the binding strength and cook yield of a beef roll was reported by Moore *et al.* (1976). Binding strength was measured by placing the beef rolls on an adjustable breaking apparatus attached to an Instron machine and breaking the rolls with a breaking bar attached to the crosshead. The binding strength was expressed as maximum force per cross-sectional area.

The effect of addition of textured soy protein in ground beef formulations was described by Seideman *et al.* (1979). Textural properties were measured by sensory panel and by using a pneumatically operated tension tester (Suter *et al.* 1976). Force–deformation curves were obtained from which parameters such as bioyield point and area under the curve could be measured. A typical stress deformation curve obtained with this technique is shown in Fig. 9.20. The bioyield value was defined as the maximum stress value recorded in the test. The area under the curve was taken as a measure of cohesiveness of the beef patties. Addition of 10 or 20% of textured soy protein increased stress values over those obtained with control all-beef patties.

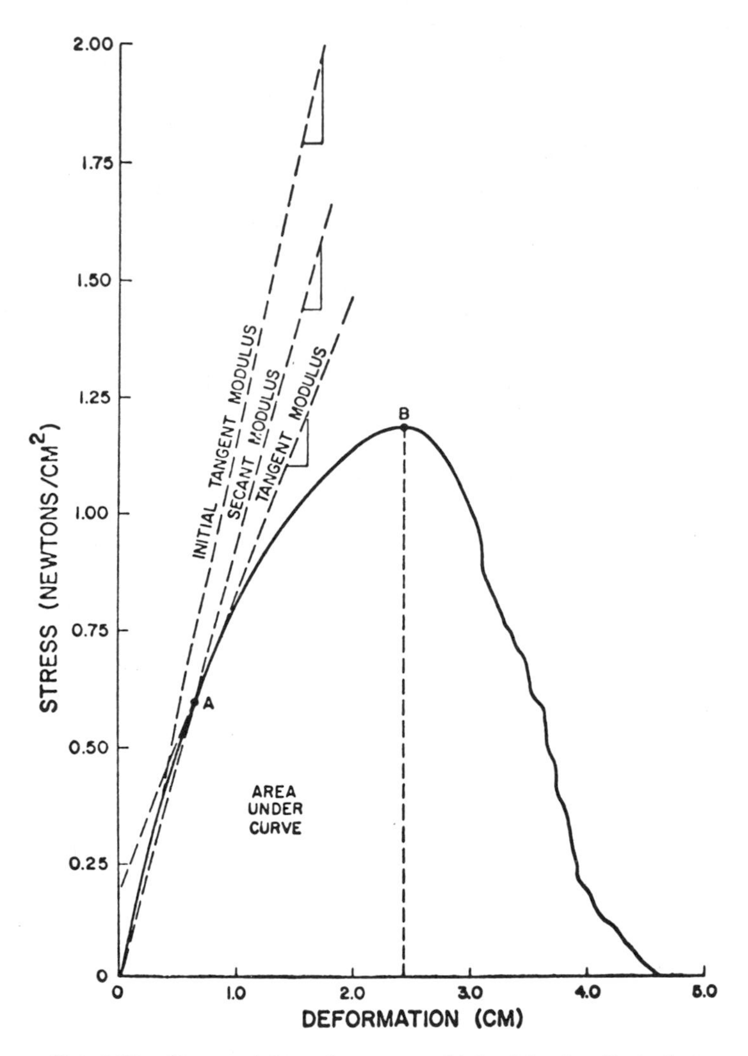

Fig. 9.20. Stress–deformation curve obtained by tensile test conducted on cooked ground beef patties. A, ½ force at bioyield point; B, bioyield point.
From Suter et al. (1976).

Although much of the literature concerning synthetic foods deals with meat-like products, there have been many reports dealing with other products. One of these concerns the effect of plant protein addition on the textural properties of pasta (Hanna *et al.*, 1978). One of the instrumental texture tests involved cutting macaroni in an Instron machine with a metal bar indenter and with a blunted 60° wedge-type metal indenter. Force–deformation curves were obtained, and the area under the curve was designated as toughness. Another test procedure consisted of loading the macaroni with a dead load for 4 min and then unloading. The deformation after loading and 1 min after unloading was recorded as a function of time. The results of this creep and recovery test were fairly well correlated with the sensory panel evaluations.

In the foregoing examples the area under a force–deformation curve has alternately been associated with chewiness, cohesiveness, and toughness. In the multitude of methods that has been developed to measure textural properties of synthetic foods a need for uniformity and standardization is obvious. As time goes on our knowledge about the fundamental principles involved in these tests will increase and a measure of standardization should evolve which will make interlaboratory comparisons more simple.

ACKNOWLEDGMENT

Financial support was provided by the Ontario Ministry of Agriculture and Food and the Natural Sciences and Engineering Research Council of Canada.

BIBLIOGRAPHY

AGUILERA, J.M., KOSIKOWSKI, F.V., and HOOD, L.F. 1975. Ultrastructure of soy protein fibers fractured by various texture measuring devices. J. Texture Stud. *6*, 549–554.

AGUILERA, J.M., KOSIKOWSKI, F.V., and HOOD, L.F. 1976. Ultrastructural changes occurring during thermoplastic extrusion of soybean grits. J. Food Sci. *41*, 1209–1213.

BERRY, B.W., CROSS, H.R., JOSEPH, A.L., WAGNER, S.B., and MAGA, J.A. 1979. Sensory and physical measurements of dry fermented salami prepared with mechanically processed beef product and structured soy protein fiber. J. Food Sci. *44*, 465–468, 474.

BREENE, W.M., and BARKER, T.G. 1975. Development and application of a texture measurement procedure for textured vegetable protein. J. Texture Stud. *6*, 459–472.

BURGESS, L.D., and STANLEY, D.W. 1976. A possible mechanism for thermal texturization of soybean protein. Can. Inst. Food Sci. Technol. J. *9*, 228–231.

CEGLA, G.F., TARANTO, M.V., BELL, K.R., and RHEE, K.C. 1978. Microscopic structure of textured cottonseed flour blends. J. Food Sci. *43*, 775–779.

CHOI, Y.R., LUSAS, E.W., and RHEE, K.C. 1981. Succinylation of cottonseed flour: Effect on the functional properties of protein isolates prepared from modified flour. J. Food Sci. *46*, 954–955.

CUMMING, D.B., STANLEY, D.W., and deMAN, J.M. 1972. Texture-structure relationships in texturized soy protein. II. Textural properties and ultrastructure of an extruded soybean product. Can. Inst. Food Sci. Technol. J. *5*, 124–128.

FURUKAWA, T., OHTA, S., and YAMAMOTO, A. 1979. Texture-structure relationships in heat-induced soy protein gels. J. Texture Stud. *10*, 333–346.

HANNA, M.A., SATTERLEE, L.D., and THAYER, K.W. 1978. Sensory and selected textural properties of pasta fortified with plant proteins and whey. J. Food Sci. *43*, 231–235.

LEE, C.M., and ABDOLLAHI, A. 1981. Effect of hardness of plastic fat on structure and material properties of fish protein gels. J. Food Sci. *46*, 1755–1759.

LEE, C.M., and TOLEDO, R.T. 1976. Factors affecting textural characteristics of cooked comminuted fish muscle. J. Food Sci. *41*, 391–397.

LOH, J., and BREENE, W.M. 1977. Texture analysis of textured soy protein products: Relations between instrumental and sensory evaluation. J. Texture Stud. *7*, 405–419.

LUGAY, J.C., and KIM, M.K. 1981. Freeze alignment: A novel method for protein texturization. *In* Utilization of Protein Resources. D.W. Stanley, E.D. Murray, and D.H. Lees (Editors). Food and Nutrition Press, Inc., Westport, CT.

MAURICE, T.J., and STANLEY, D.W. 1978. Texture-structure relationships in texturized soy protein. IV. Influence of process variables on extrusion texturization. Can. Inst. Food Sci. Technol. J. *11*, 1–6.

MAURICE, T.J., BURGESS, L.D., and STANLEY, D.W. 1976. Texture-structure relationships in texturized soy protein. III. Textural evaluation of extruded products. Can. Inst. Food Sci. Technol. J. *9*, 173–176.

MOORE, S.L., THENO, D.M., ANDERSON, C.R., and SCHMIDT, G.R. 1976. Effect of salt, phosphate and some nonmeat proteins on binding strength and cook yield of a beef roll. J. Food Sci. *41*, 424–426.

MURRAY, B.P., STANLEY, D.W., and GILL, T.A. 1980. Improved utilization of fish protein—Co-extrusion of mechanically deboned salted minced fish. Can. Inst. Food Sci. Technol. J. *13*, 125–130.

NOGUCHI, A., KUGIMIYA, W., HAGUE, Z., and SAIO, K. 1981. Physical and chemical characteristics of extruded rice flour and rice flour fortified with soybean protein isolate. J. Food Sci. *47*, 240–245.

SAIO, K., and WATANABE, T. 1966. Preliminary investigation on protein bodies of soybean seeds. Agric. Biol. Chem. *30*, 1133–1138.

SEIDEMAN, S.C., SMITH, G.C., CARPENTER, Z.L., and DILL, C.W. 1979. Plasma protein isolate and textured soy protein in ground beef formulations. J. Food Sci. *44*, 1032–1035.

SHEN, J.L., and MORR, C.V. 1979. Physico-chemical aspects of texturization: Fiber formation from globular proteins. J. Am. Oil Chem. Soc. *56*, 63A–70A.

SIMONSKY, R.W., and STANLEY, D.W. 1982. Texture-structure relationships in textured soy protein. V. Influence of pH and protein acylation on extrusion texturization. Can. Inst. Food Sci. Technol. J. *15*, 294–301.

SOFOS, J.N., NODA, I., and ALLEN, C.E. 1977. Effects of soy proteins and their levels of incorporation on the properties of wiener-type products. J. Food Sci. *42*, 879–884.

STANLEY, D.W., and deMAN, J.M. 1978. Structural and mechanical properties of textured proteins. J. Texture Stud. *9*, 59–76.

STANLEY, D.W., CUMMING, D.B., and deMAN, J.M. 1972. Texture-structure relationships in texturized soy protein. I. Textural properties and ultrastructure of rehydrated spun soy fibers. Can. Inst. Food Sci. Technol. J. *5*, 118–123.

SUTER, D.A., SUSTEK, E., DILL, C.W., MARSHALL, W.H., and CARPENTER, Z.L. 1976. A method for measurement of the effect of blood protein concentrates on the binding forces in cooked ground beef patties. J. Food Sci. *41*, 1428–1432.

TARANTO, M.V., MEINKE, W.W., CATER, C.M., and MATTIL, K.F. 1975. Parameters affecting production and character of extrusion texturized defatted glandless cottonseed meal. J. Food Sci. *40*, 1264–1269.

TARANTO, M.V., CEGLA, G.F., and RHEE, K.C. 1978. Morphological, ultrastructural and rheological evaluation of soy and cottonseed flours texturized by extrusion and nonextrusion processing. J. Food Sci. *43*, 973–979.

TERRELL, R.N., BROWN, J.A., CARPENTER, Z.L., MATTIL, K.F., and MONAGLE, C.W. 1979. Effects of oilseed proteins, at two replacement levels, on chemical, sensory and physical properties of frankfurters. J. Food Sci. *44*, 865–868.

VOISEY, P.W. 1971. Use of the Ottawa Texture Measuring System for Testing Fish Products. Report No. 7022. Engineering Research Service, Agriculture Canada, Ottawa, Ontario, Canada.

VOISEY, P.W., MACDONALD, D.C., KLOEK, M., and FOSTER, W. 1972. The Ottawa Texture Measuring System. An operational manual. Report No. 7024. Engineering Research Service, Agriculture Canada, Ottawa, Ontario, Canada.

WOLF, W.J. 1970. Scanning electron microscopy of soybean protein bodies. J. Am. Oil Chem. Soc. *47*, 107–108.

WOLF, W.J., and BAKER, F.L. 1972. Scanning electron microscopy of soybeans. Cereal Sci. Today *17*, 124–126, 128–130, 147.

10

Physical Characteristics of Food Powders

Micha Peleg[1]

INTRODUCTION

Food powders are a large group of different kinds of powders that have little in common, except for being used as (or in) foods. Classification criteria of food powders may, therefore, vary for the purpose of convenience or according to any particular practical application. Following are a few possibilities for food powders classification (with incomplete listings):

Classification by Usage

1. Flours
2. Instant beverages, babyfood and soups
3. Spices
4. Sweetners
5. Pre-mixed convenience foods (e.g., cake and ice-cream mixes)
6. Minor ingredients or additives (e.g., salts, pigments and vitamin C)

Classification by a Major Chemical Component

1. Starchy (e.g., wheat flour)
2. Proteinaceous (e.g., soy isolate)
3. Crystalline sugar (e.g., sucrose, glucose)
4. Amorphous sugars (e.g., dehydrated fruit juices)
5. Fatty (e.g., soup mix)

[1] Department of Food Engineering, University of Massachusetts, Amherst, MA 01003.

Classification by Process

1. Ground powders (e.g., spices, powdered sugar)
2. Spray dried powders (e.g., powdered egg)
3. Drum dried powder (e.g., mashed potato)
4. Agglomerated powders (e.g., instant coffee and milk)
5. Precipitated powders (e.g., protein isolate)
6. Crystalline powders (e.g., salt, sugars)
7. Mixtures of powders (e.g., dry fruit drinks)

Classification by Size

1. "Fine" powders (e.g., flours)
2. Coarse powders (e.g., granulated sugar)

Classification by Moisture Sorption Pattern (and Effects)

1. Extremely hygroscopic (e.g., dehydrated fruit juices)
2. Hygroscopic (e.g., spray dried coffee)
3. Moderately hygroscopic (e.g., flours)

Classification by Flowability

1. Free flowing (e.g., granular sugar)
2. Moderately cohesive (e.g., flour)
3. Very cohesive (almost all food powders when let to absorb moisture)

The above classifications demonstrate the difficulties in treating food powders as a group at a generalization level that will not make the analysis too vague and consequently impractical. Furthermore, some of the more interesting and potentially useful criteria, e.g., hygroscopicity or flowability are not easy to quantify because they represent the combined effect of different sorts of physical and physicochemical phenomena. It should also be remembered that the composition and properties of many food powders may be variable to different degrees and may also change with time. Therefore, it is not uncommon that the same free-flowing powder, for example, becomes sticky during storage, or that a relatively nonhygroscopic powder (e.g., salt) becomes highly hygroscopic in the presence of impurities. Realizing these problems, and with the understanding that exceptions to the discussion are not only possible but sometimes unavoidable, this chapter is an attempt to evaluate the factors that determine or influence the physical properties of food powders with special emphasis on their specific or unique characteristics.

PHYSICAL CHARACTERIZATION OF POWDERS

Powders are usually characterized at two levels, that of the individual particles and that of the powder in bulk. Although it is self-evident that the bulk properties are primarily influenced by the particles properties, the relationship between the two is by no means simple and involves external factors such as the system geometry and the mechanical and thermal histories of the powder. It is therefore practically impossible, in most cases, to predict bulk properties directly from those of the particles even if the latter could completely and accurately be defined. This, however, is not necessarily true with respect to influence trends. In many systems, the kind of changes that a powder will undergo, as a result of modification in its particles characteristics, can well be anticipated although their exact magnitude cannot.

PARTICLES CHARACTERISTICS

The physical characteristics of the individual particles are mainly determined by the material from which they are made and the process by which they are formed (see examples in Fig. 10.1 and 10.2). Obviously, changes in the particles features may occur during storage or handling as a result of moisture absorption, chemical reactions (e.g., browning), or mechanical attrition.

As previously mentioned, the mean particle size of the food powders is in a range of several orders of magnitude, that is, between single microns (e.g., individual starch granules) to several hundreds or even thousands of microns (e.g., instant coffee). It ought to be mentioned here, however, that conventional sieve analyses, especially of cohesive food powders, do not always provide accurate accounts of the real size distribution because the fine particles may adhere to the surface of the larger particles as shown in Fig. 10.1a. (Other implications of this phenomenon will be discussed later.)

Similarly, shape variations in food powders are enormous and they range from extreme degrees of irregularity (in ground materials) to an approximate sphericity (starch) or well-defined crystalline shapes (granulated sugar, salt).

Density

Most food particles have a similar *solid density* (i.e., the density of the solid material from which they are made disregarding any internal pores) of about 1.4–1.5 g/cm^3 depending on the moisture content. This is mainly due to the similar density of the main indgredients as shown in the following table:

Glucose	1.56 g/cm^3
Sucrose	1.59 g/cm^3
Starch	1.50 g/cm^3

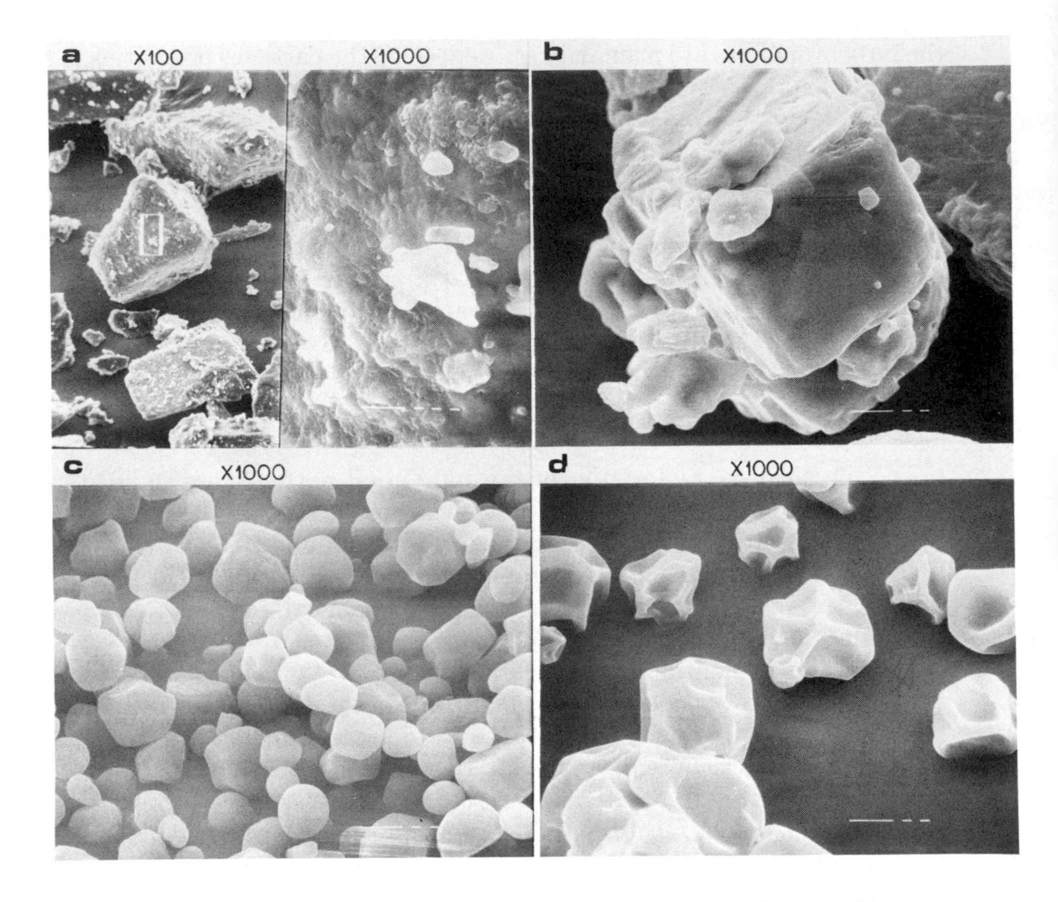

Fig. 10.1. Particles shape in food powders. (a) Ground sugar (note the adherence of the fines to the surface of the larger particles), (b) powdered salt, (c) corn starch, (d) soy protein.

Cellulose	1.27–1.61 g/cm^3
Protein (globular)	~1.4 g/cm^3
Fat	0.9–0.95 g/cm^3
Salt	2.16 g/cm^3
Citric acid	1.54 g/cm^3
Water	1.00 g/cm^3

Notable exceptions are salt-based or fat-rich powders whose density may vary considerably according to composition.

Another density characteristic is the *particle density* which is defined by

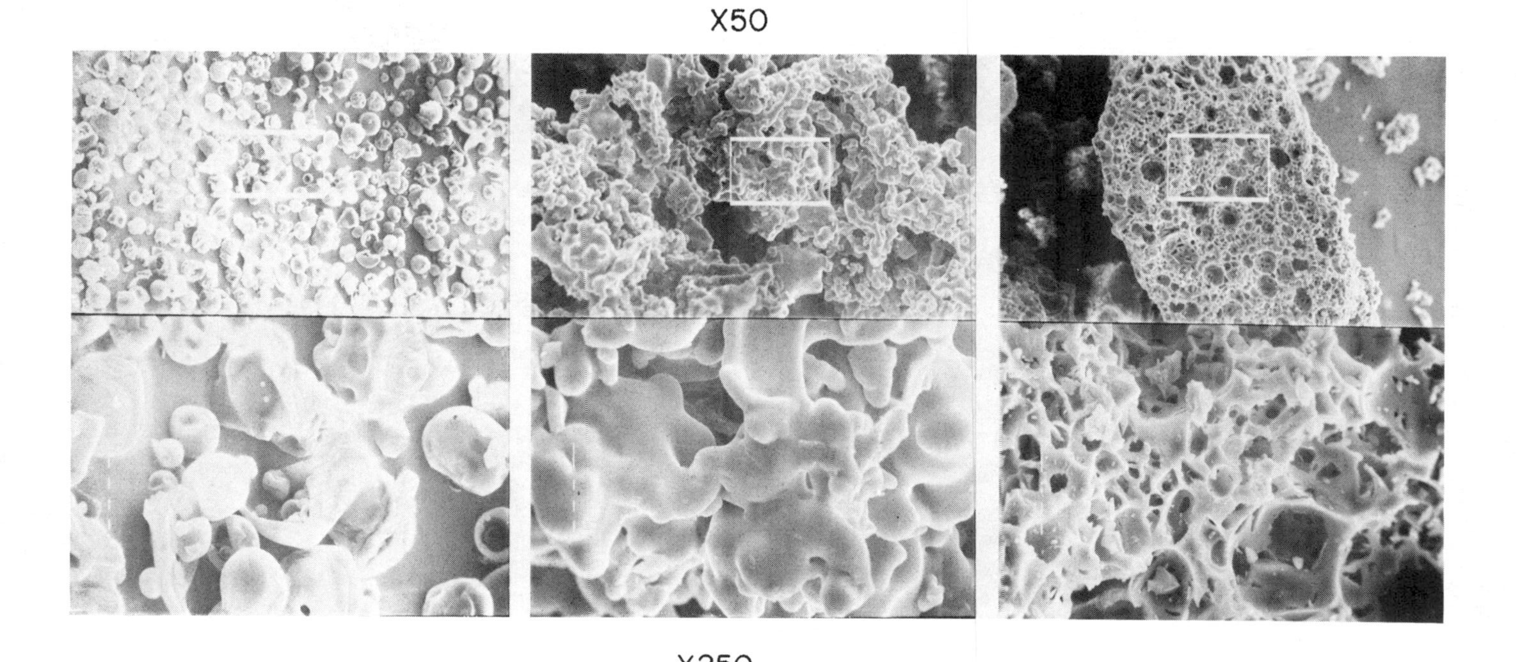

Fig. 10.2. Effect of process on the particle characteristics illustrated in instant coffee. (*Left*) spray-dried (experimental), (*middle*) agglomerated spray-dried (commercial), (*right*) freeze-dried (commercial).

$$\text{Particle density} = \frac{\text{Particle actual mass}}{\text{Particle actual volume}}$$

This parameter does account for the existence of internal pores and therefore can be considered as a measure of the true density of the particles. (As we shall see later, this parameter is more relevant to situations where the relationship between particle weight and interparticle forces is of concern.) However, this parameter does not provide any information regarding the shape of the internal pores and their position in the particle structure. The latter can have distinctly different forms whose character is established by the type of process and the conditions under which the particle was formed. (Schematic demonstration of various types of porous structures in particles are shown in Fig. 10.3.)

Surface Activity

Since the water vapor sorption phenomenon in food has been extensively studied and discussed in the literature, it needs not be rediscussed here. Less information is available on the capacity of many kinds of food surfaces to absorb fine solid particles or to interact with other particles and equipment surfaces. These interactions are not limited to particles of same or similar chemical species, although there is evidence to suggest that surface affinity can considerably differ among materials (e.g., as in the case of certain anticaking agent–powder systems). The mechanisms by which particles surfaces interact are also different, and they include liquid bridging by surface moisture or melted fat, electrostatic charges (as in dust), molecular forces, or crystalline surface energy.

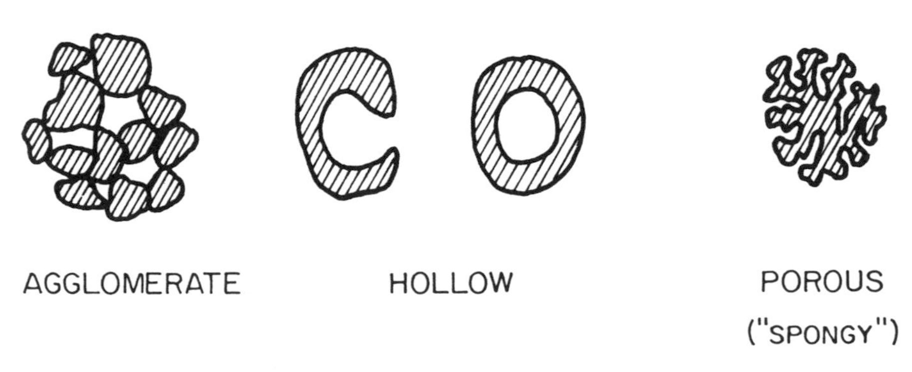

Fig. 10.3. Schematic presentation of the different kinds of particles porosity.

Detailed theoretical discussions and mathematical analyses of such interactions and their implications in powder technology have been published by Rumpf (1961), Pietsch (1969), and Zimon (1969).

Mechanical Strength and Dust Formation

Many solid food materials, especially when dry, are brittle and fragile. [Their hardness on the Moh scale is in the order of 1–2 (Carr 1976.)] Since, particularly in small objects, surface or shape irregularities are normally associated with mechanical weakness (due to stress concentration, for example), dry food particles have the tendency to wear down or disintegrate. Mechanical attrition of food powders usually occurs during handling or processing, when the particles are subjected to impact and frictional forces. The result is frequently a dust problem that may also develop into a dust explosion hazard. (The incidence of dust explosion mainly depends on the dust particle size, the dust/air ratio and the avilability of a triggering spark.) Carr (1976) lists potentially explosive agricultural dusts and ranks them according to their explosibility in the following descending order; starches (50), sugar (13.2), grains (9.2), wheat flour (3.8), wheat (2.5), skim milk (1.4), cocoa (1.4), and coffee (<0.1). (The numbers in parentheses are the "explosibility indices" where severe hazard is ranked by an index of 10 and above, strong by 1–10, moderate by 0.1–1, and weak by less then 0.1.)

Other implications of dust in the food industry include human health and safety, plant and equipment maintenance, and material loss.

When the size of the particles produced through mechanical attrition is larger than that of dust particles, the process can still influence the product's bulk density or cause a segregation problem that may affect the product's appearance, as in the case of instant coffee. Other aspects of these phenomena are discussed subsequently.

BULK PROPERTIES

The bulk properties of fine powders, always interdependent, are determined by both the physical-chemical properties of the material (e.g., composition, moisture content), the geometry, size and surface characteristics of the individual particles, and the history of the system as a whole. The shape of the container can affect flowability, and the powder density usually increases as a result of vibration for example. Numerical values assigned to such properties therefore ought to be regarded as useful only under the conditions under which they have been determined or as indicators of an order of magnitude only.

Bulk Density and Porosity

Bulk (or apparent) density is the mass of particles that occupies a unit volume of the bed. It is usually determined by weighing a container of a known volume and dividing the net weight of the powder by the container's volume. Porosity is the fraction of volume not occupied by a particle or solid material and therefore it can be expressed as either

$$\text{Total porosity} = 1 - \frac{\text{Bulk density}}{\text{Solid density}}$$

or

$$\text{Interparticle porosity} = 1 - \frac{\text{Bulk density}}{\text{Particle density}}$$

Because powders are compressible (see later) their bulk density is usually given with an additional specifier: loose bulk density (as poured), tapped bulk density (after vibration), or compact density (after compression).

Another way to express bulk density is in the form of a fraction of its particles' solid density, that is sometimes referred to as the "theoretical density." This expression, as well as the use of porosity instead of density, enables and facilitates the unified treatment and meaningful comparisons of powders that may have considerably different solid or particle densities.

Loose Bulk Density

Approximal values of loose bulk density of a variety of food powders are listed in Table 10.1. The table shows that with very few exceptions food powders have apparent densities in the range 0.3–0.8 g/cm^3. As previously mentioned, the solid density of most food powders is about 1.4, and therefore these values are an indication that food powders have high porosity (i.e., 40–80%) that can be internal, external, or both. There are many published theoretical and experimental studies of porosity as a function of the particle size, distribution, and shape. Mostly they pertain to free-flowing powders or models (e.g., steel shots, metal powders), where porosity can be treated as primarily due to geometrical and statistical factors only (Gray 1968; McGeary 1967). Even though in such cases porosity can vary considerably, depending on such factors as the concentration of fines, it is still evident that the exceedingly low density of food powders cannot be explained by geometrical considerations only.

TABLE 10.1 APPROXIMATE BULK DENSITY AND MOISTURE OF VARIOUS FOOD POWDERS

Powder	Bulk Density (g cm^3)	Moisture Content (%)
Baby formula	0.40	2.5
Cocoa	0.48[1]	3–5[2]
Coffee (ground and roasted)	0.33[3]	7[3]
Coffee (instant)	0.33[3]	2.5[3]
Coffee Creamer	0.47	3
Cornmeal	0.66[1]	12[2]
Cornstarch	0.56[1]	12[2]
Egg (whole)	0.34[1]	2–4[2]
Gelatin (ground)	0.68	12
Microcrystalline cellulose	0.68	6
Milk	0.61[1]	2–4
Oatmeal	0.43[1]	8[2]
Onion (powdered)	0.51	1–4
Salt (granulated)	0.96[1]	0.2[2]
Salt (powdered)	0.95	0.2[2]
Soy protein (precipitated)	0.28	2–3
Sugar (granulated)	0.80[1]	0.5[2]
Sugar (powdered)	0.48	0.5[2]
Wheat flour	0.48[1]	12[2]
Wheat (whole)	0.80[1]	12[2]
Whey	0.56[1]	4.5[2]
Yeast (active dry Baker's)	0.52	8
Yeast (active dry wine)	0.82	8

[1] Data from Carr 1976.
[2] Data from Watt and Merrill 1975.
[3] Data from Schwartzberg 1982.

Most food powders are known to be cohesive (Carr 1976), which means that their attractive interparticle forces are significantly high relative to the particle's own weight. Therefore (see Fig. 10.4), open bed structure, supported by interparticle forces is not only possible but also very likely (Peleg and Mannheim 1973; Moreyra and Peleg 1981; Scoville and Peleg 1980; Gerritsen and Stemerding 1980; Dobbs *et al.* 1982).

Since the bulk density of food powders depends on the combined effects of interrelated factors—the intensity of attractive interparticle forces, the particle size, and the number of contact points (Rumpf 1960) —it is clear that a change in any one of the powder characteristics may result in a significant change in the powder bulk density, of a magnitude that cannot always be anticipated. The reason for this situation becomes clear when one realizes the intricate relationships between the parameters shown schematically in Fig. 10.5 and the number of major factors that determine the particles surface activity and cohesion as shown in Fig. 10.6.

Effect of Moisture Content and Anticaking Agents

In general, moisture sorption is associated with increased cohesiveness, mainly due to interparticle liquid bridges. Therefore, especially in

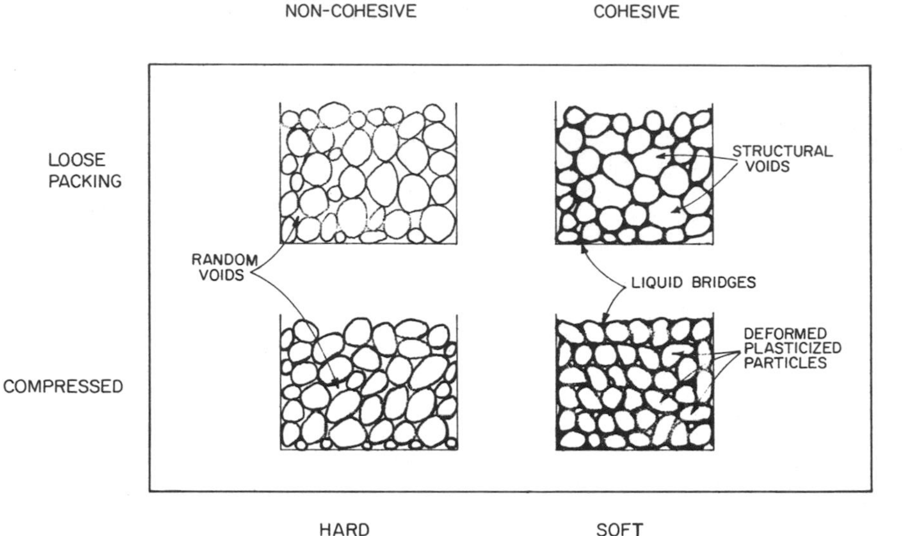

Fig. 10.4. Internal structure of cohesive and noncohesive powders in loose packing and after compression.
From Moreyra and Peleg (1981).

hygroscopic food powders, higher moisture ought to result in lowering the loose bulk density (see Fig. 10.4), as indeed is the case in powdered sugar and salt, for example. It should be mentioned, however, that this decrease will only be detected in freshly sieved or in flowing powders, where these same interparticle forces are not allowed to cause caking of the mass (see below).

Another notable exception to this trend is the case of fine powders that are very cohesive even in their dry form (baby formula, coffee creamer). In such cases it appears that the bed array has reached maximum "openness" at low moisture contents, and therefore further lowering of the density becomes impossible. It is also worthwhile to remember that excessive moisture levels, especially in powders containing soluble crystalline compounds (such as sugars or salt), may result in liquefaction of the powder, and consequently in an increase in its density. Need less to say, at this stage the powder most probably has already lost its utility, and therefore the phenomenon has little practical importance. [A more detailed discussion of the effect of moisture on the density and other bulk properties of food powders has recently been published by Scoville and Peleg (1980) and Moreyra and Peleg (1981).]

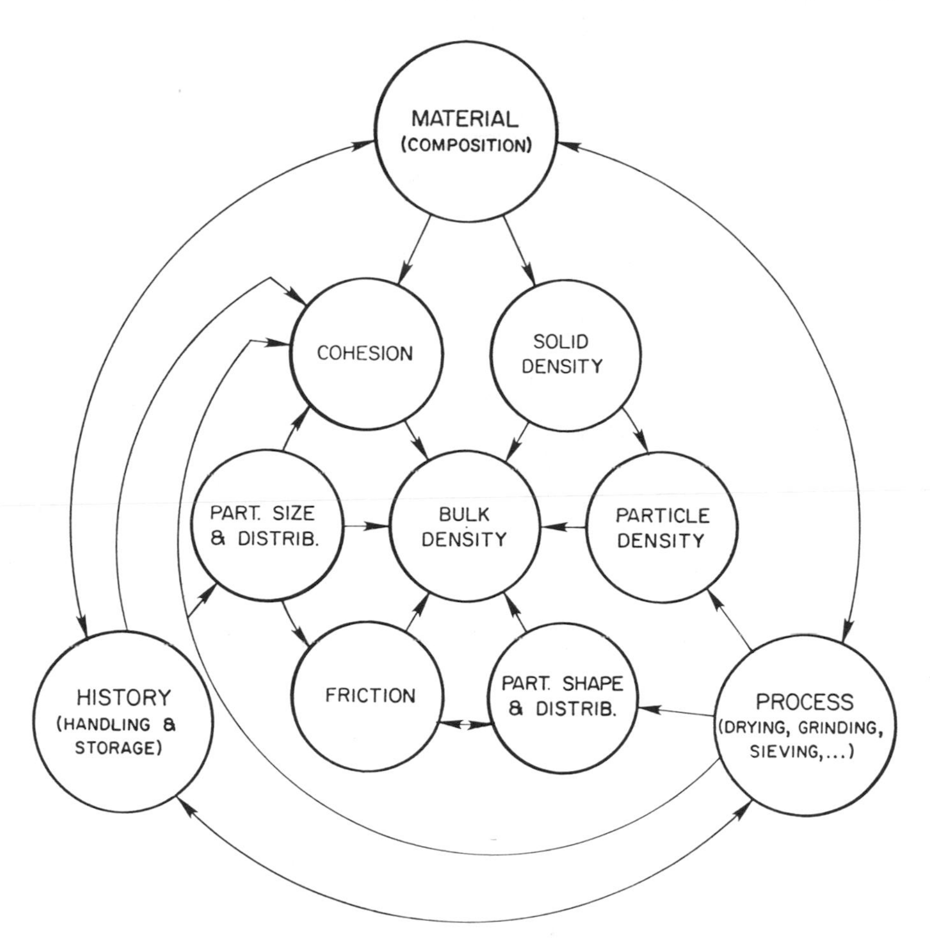

Fig. 10.5. Factors that affect powders bulk density.
From Peleg et al. (1982).

Anticaking agents (or flow conditioners) are supposed to reduce interparticle forces and as such they are expected to increase the bulk density of powders (Peleg and Mannheim 1973). It has been observed though that there may be an optimal concentration beyond which the effect will diminish (Nash *et al.* 1965; Hollenbach *et al.* 1982) or will be practically unaffected by the conditioner concentration (Hollenbach *et al.* 1982). It can also be observed that for a noticeable effect on the bulk density (i.e., an increase in the order of 10% or more) the agent and host particles must have surface affinity (Hollenbach *et al.* 1983). Other-

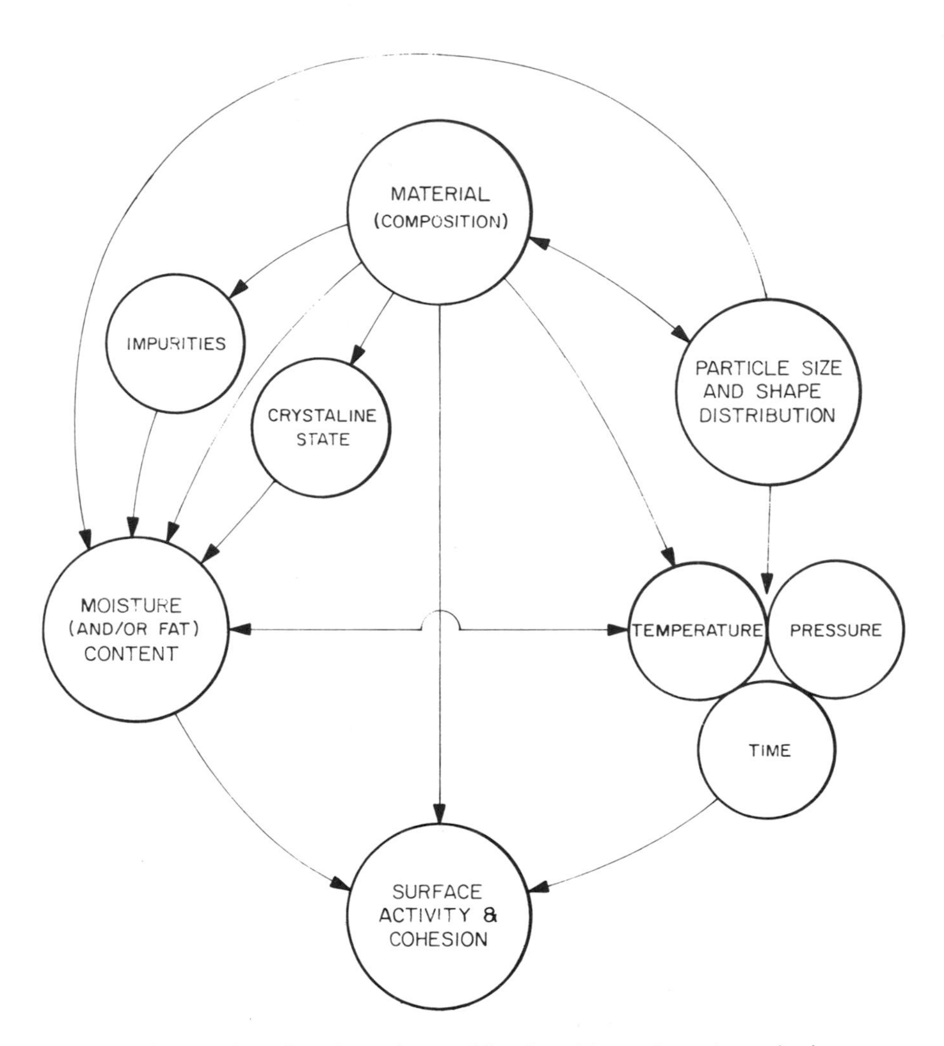

Fig. 10.6. Factors that affect the surface activity of particles and powders cohesiveness.

wise the agents particles may segregate and, instead of reducing interparticle forces, will only fill interparticle space. It seems, however, that there is very little information on the exact nature of these surface interactions and the mechanism by which they affect the bed structure. Examples of the effects of moisture and anticaking agents on the bulk properties of selected food powders are given in Tables 10.2 and 10.3.

TABLE 10.2 EFFECT OF MOISTURE CONTENT ON THE MECHANICAL CHARACTERISTICS OF SELECTED FOOD POWDERS

Powder	Moisture (%)	Loose Bulk Density (g/cm^3)	Compress-ibility[1]	Cohesion[2] (g cm^{-2})	Angle of Internal Friction[2] (deg)	Reference
Glass beads	dry	1.72	~0	~0	23	Scoville and Peleg (1979)
(175 μm)	1.0	1.20	0.23	15	21	
Powdered salt	dry	1.26	0.02	~0	40	Moreyra and Peleg (1981)
(100/200 mesh)	0.6	0.78	0.12	50	36	
Powdered sucrose	dry	0.62	0.152	~10	39	Peleg and Mannheim (1973)
(60/80 mesh)	0.1	0.50	0.185	~14	37	
Starch	dry	0.81	0.12	~6	33	Peleg (1971)
	18.5	0.69	0.15	~13	30	
Powdered onion	dry	0.51	0.03	5	—	Peleg (1971)
(80/120 mesh)	5.2	0.51	0.05	15	—	
Baby formula	dry	0.52	0.08	37	31	Moreyra and Peleg (1981)
(commercial)	2.7	0.41	0.08	TC[3]	TC	
Coffee creamer	dry	0.46	0.08	49	37	Moreyra and Peleg (1981)
(commercial)	7	0.45	0.19	32	38	
Active dry	5.2	0.52	0.05	~0	41	Dobbs *et al.* (1982)
Baker's yeast	8.4	0.52	0.08	14	42	
	13	0.49	0.26	TC	TC	

[1] Defined as the constant b in the equation $\rho_B = a + b \log \sigma_n$ (see text).
[2] Determined by the Jenike Flow Factor Tester at consolidation levels of 0.2–0.5 kg/cm^2
[3] TC is an indication that the powder was too cohesive for measurement by the Flow Factor Tester.

TABLE 10.3 EFFECT OF ANTICAKING AGENTS ON THE BULK DENSITY AND COMPRESSIBILITY OF SELECTED FOOD POWDERS

Powder	Agent	Concentration (%)	Loose Bulk Density (g/cm^3)	Compressibility[1]	Reference
Sucrose (powdered)	None	—	0.70	0.066	Hollenbach *et al.* (1982)
	Calcium stearate	0.5	0.87*	0.039*	
	Silicon oxide	0.5	0.75*	0.052*	
	Tricalcium phosphate	0.5	0.76*	0.044*	
Salt (powdered)	None	—	1.01	0.080	Hollenbach *et al.* (1983)
	Calcium stearate	0.1	1.14*	0.032*	
	Silicon oxide	0.1	1.10*	0.045*	
	Tricalcium phosphate	0.1	1.16*	0.025*	
Soup Mix	None	—	0.70	0.27	Peleg (1971)
	Aluminum silicate	2	0.75*	0.15*	
	Calcium stearate	2	0.63	0.27	
Gelatin (powdered)	None	—	0.68	~0	Peleg (1971)
	Aluminum silicate	1	0.70	0.016	
Microcrystalline cellulose	None	—	0.35	0.017	Peleg (1971)
	Aluminum silicate	1	0.36	0.030	
Corn starch	None	—	0.62	0.109	Hollenbach *et al.* (1983)
	Calcium stearate	1	0.59	0.099	
	Silicon oxide	1	0.67	0.077*	
	Tricalcium phosphate	1	0.61	0.062*	
Soy protein	None	—	0.27	0.040	Hollenbach *et al.* (1983)
	Calcium stearate	1	0.27	0.041	
	Silicon oxide	1	0.27	0.036	
	Tricalcium phosphate	1	0.31*	0.024*	

[1] Defined as the constant b in the equation $\rho_B = a + b \log \sigma_N$ (see text). * Indicates significant change relative to the untreated powder.

Compressibility

Powders can be compacted in two ways: by tapping, i.e., by the application of vibrations, or by mechanical compression. Both processes can occur unintentionally (vibrations during transport or handling or pressure developed as a result of storage in high bins) or intentionally (vibrating the powder in order to increase its weight in a given container or tableting as in the case of certain soups and candies). It ought to be mentioned, however, that the stresses that develop in storage are usually a few orders of magnitude smaller than those applied in tableting or similar forming operations. (The latter involve factors that are outside the scope of this chapter and therefore this topic will not be discussed here.)

Tapping

The theoretical and empirical considerations in vibratory compaction have mainly been studied in nonfood powders (Hausner *et al.* 1976). Sone (1972) reported that in a variety of food powders the relationship between the volume reduction fraction (γ_n) and the number of taps (n) is given by

$$\gamma_n = \frac{V_0 - V_n}{V_0} = \frac{abn}{1 + bn}$$

where V_0 is the initial volume, V_n the volume after n taps, and a and b are constants (see Fig. 10.7).

The applicability of this equation was tested through the fit of its linear form

$$\frac{n}{\gamma_n} = \frac{1}{ab} + \frac{n}{a}$$

which was indeed successful.

It ought to be mentioned that the constant a in this equation represents the asymptotic level of the volume change or, in other words, the level that will be obtained after a large number of tappings or a long time in vibration. The constant b is a representative of the rate at which this compaction is achieved, i.e., $1/b$ is the number of vibrations necessary to reach half of the asymptotic change. In general, this form of data presentation is very convenient for systems comparisons since it only involves two experimental constants. The magnitude of these constants,

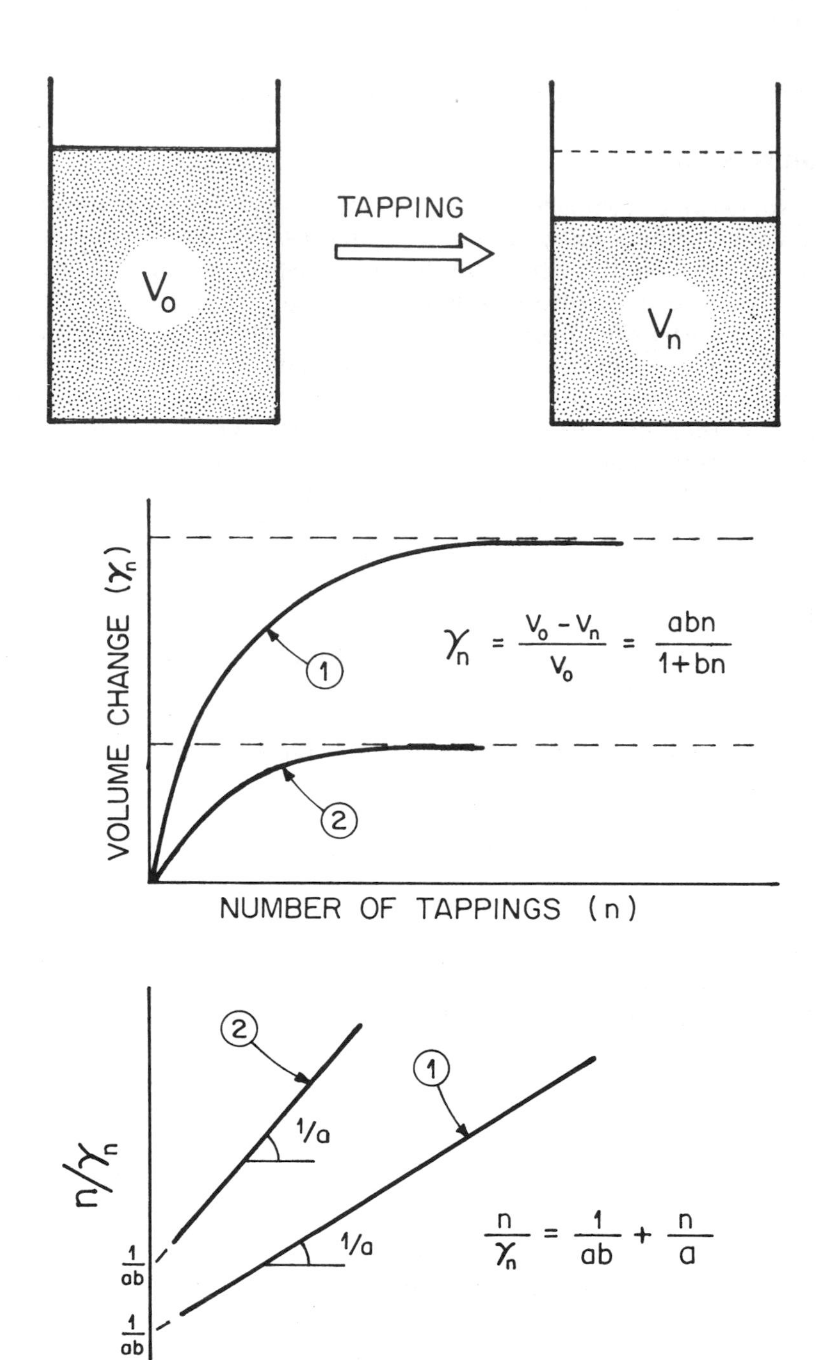

Fig. 10.7. Effect of tapping on powders volume. The mathematical description of the relationships between the number of tappings and the volume reduction as based on Sone (1972).

especially the constant representing the asymptotic compaction must depend on both the conditions by which the vibrations are administered (e.g., amplitude and frequency) and the powder internal friction and cohesion. The latter has led to the development of the commonly referred to "Hausner ratio," given as the ratio between tapped and loose bulk density (Hausner 1967). It was shown that this ratio could be used as a flowability index (Gray and Bedow 1968/9) in powders where friction is the major obstacle to flow. There is no evidence that such an index must also be useful for cohesive food powders.

Compaction Under Compressive Load

The relationship between applied stresses and powders density (or porosity) has been a topic of extensive studies. Kawakita and Lüdde (1970/1) list 15 published mathematical expressions to describe such relationships. The existence of such a large number of models is due to the distinctly different mechanisms by which a powder bed deforms. The main modes are particles spatial rearrangement (without particle deformation), the filling of interparticle voids through particle (mainly plastic) deformation, and filling voids through particles fragmentation (Gray 1968). It is obvious though, that the relative contribution of each mechanism mainly depends on the particles properties (e.g., size, shape, and hardness), the magnitude of the applied pressure, and the stress distribution within the compacted specimen. Therefore, the compressibility pattern of the same powder may be totally different at different load ranges. At the low end of the pressure range, i.e., normal stress in the order of up to about 1–5 kgf/cm^2, the first mechanism seems to be dominant and with few exceptions like brittle coffee agglomerates or fatty powders, most of the change in density is due to the bed structural rearrangement. As previously mentioned, the original structure (Fig. 10.4) is largely influenced by the presence of interparticle forces. Therefore, it is expected that an open structure supported by such forces will readily collapse under small stresses, resulting in high apparent compressibility at such stresses range. This indeed has been observed in a large variety of food powders (Peleg and Mannheim 1973; Peleg *et al.* 1973; Moreyra and Peleg 1981). It was also shown that compressibility under small loads is a sensitive index of the powder's cohesiveness (Carr 1976) and can be used to detect potential flow problems.

It was also demonstrated that in this low stresses range the compressive deformability of food powders can be described by

$$\rho_B = a + b \log \sigma_N$$

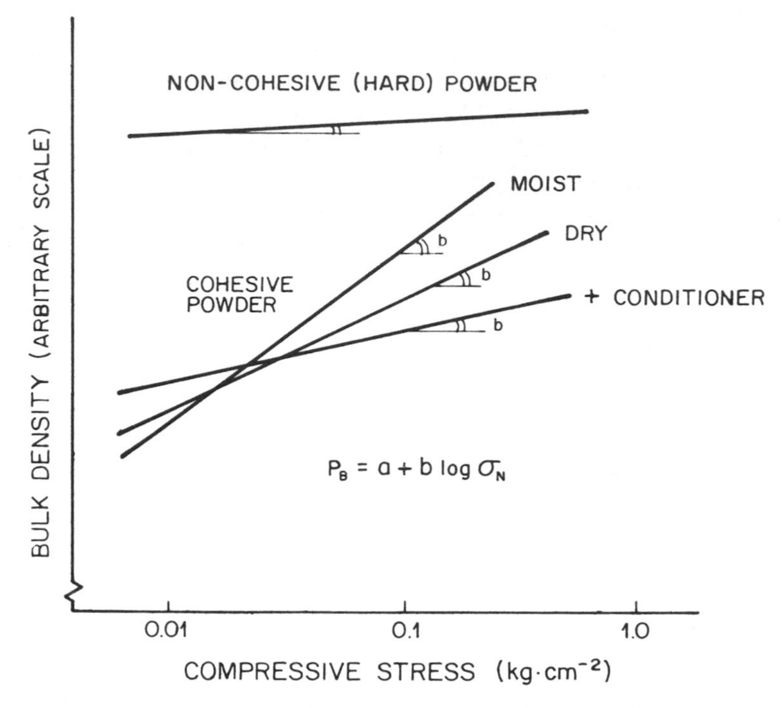

Fig. 10.8. Schematic presentation of the relationship between food powders bulk density and the compressive stress. The slopes of the lines are referred to as "compressibility" in Tables 10.2 and 10.3.

where ρ_B is the bulk density, σ_N the applied stress, a and b are constants. The constant b has been defined as "compressibility." Its values for selected food powders are listed in Tables 10.2 and 10.3. Schematic representation of the ρ_B vs σ_N relationship of different kinds of powders is shown in Fig. 10.8.

Angle of Repose

The angle of repose (Fig. 10.9) is an indispensible parameter in the design of the powder's processing, storage, and conveying systems. Its actual magnitude, however, depends on the way in which the powder heap is formed (e.g., the impact velocity), and therefore published values of the angle are not always comparable (Brown and Richards 1970). In cohesive powders the measurement of the angle itself is sometimes difficult because of the irregular shapes that the heaps can assume. The angle of repose is sometimes confused with the angle of internal friction.

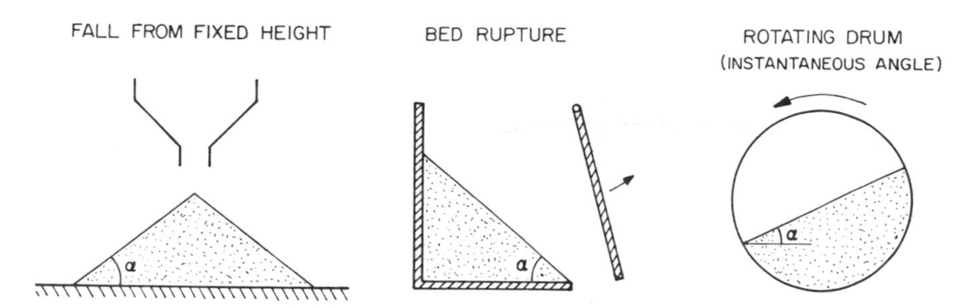

Fig. 10.9. Examples of different kinds of angle of repose. These and other types are discussed by Brown and Richards (1970) and Carr (1976).

Although its magnitude is certainly influenced by frictional forces (especially in free-flowing powders) it is also affected by interparticle attractive forces—a factor that becomes dominant in wet and cohesive powders. Mainly for this reason, the angle of repose (regardless of how it is measured) can be used as a rough flowability indicator. According to Carr (1976) angles of up to about 35 deg indicate free flowability, 35–45 some cohesiveness, 45–55 cohesiveness (loss of free flowability), and 55 and above very high cohesiveness and very limited flowability, if at all.

A special food-related implication of the angle of repose is the volume of powder (e.g., dry beverage) carried in a spoon (Figs. 10.10 and 10.11). In such a situation, the concentration or strength of the reconstituted

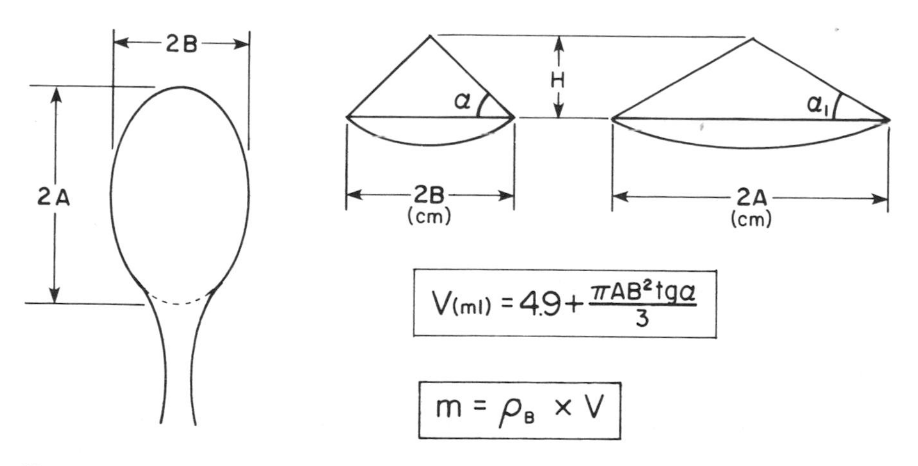

Fig. 10.10. Approximate geometry of a teaspoon filled with powder. The standard volume of a teaspoon is 4.9 ml. α is the angle of rest. The mass contained in the spoon (m) is also determined by the bulk density (ρ_B).

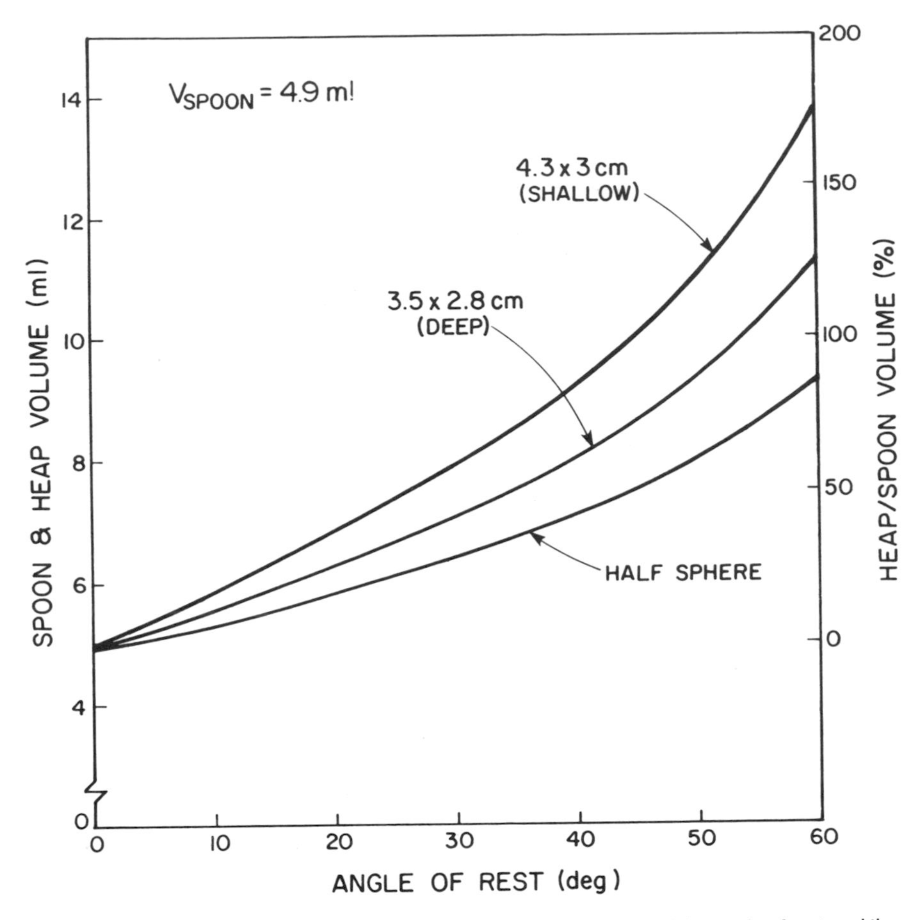

Fig. 10.11. Approximate volume of a filled teaspoon as a function of the angle of rest and the spoon dimensions.

beverage will depend on the combined effect of the bulk density and the angle of repose. (As can be seen from Fig. 10.5, the two may be interrelated and therefore require a delicate process control.)

Flowability

Despite few superficial similarities, the flowability of liquids and powders are different physical phenomena. The main distinctions are as follows.

1. The flow rate of powders (Fig. 10.12) is practically independent of the head (height) above the aperture if the powder head is more

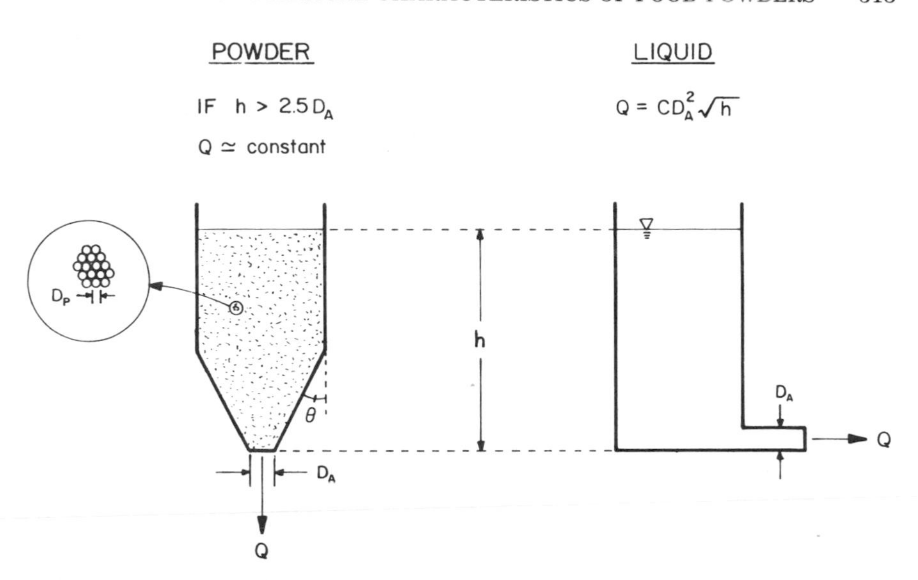

Fig. 10.12. Difference between particulates and liquids gravitational flow. (See also Fig. 10.13.)

than about 2.5 times that of the aperture diameter (Brown and Richards 1970).

2. Powders can resist appreciable shear stresses. Once compacted, under their own weight or by external pressure, they can also form mechanically stable structures (e.g., arches) that will halt flow all together despite the existence of head.

For these reasons, treatment of a powder's gravitational flowability must be based on solid mechanics theories (Jenike 1964) and not on hydrodynamics. The general principle on which such analyses are based is that, from a physical point of view, powder flow is equivalent to solid failure in shear. In ideal free-flowing powders or granular materials, the resistance to flow is primarily through friction, and therefore flowability evaluation is fairly straightforward. In cohesive powders (as are most food powders) interparticle forces are enhanced by compaction (through the increase in the number of contact points, for example) with the result that the "compact" can develop appreciable mechanical strength. Therefore, even under small pressure, many food powders may cause serious flow problems. It ought to be mentioned that, especially for cohesive powders, the system geometry (e.g., the bin angle and aperture diameter) plays a decisive role in establishing the flow regime, i.e., mass or funnel flow (Fig. 10.13) and its rate and stability. A detailed description and analyses of the methodology involved can be found in

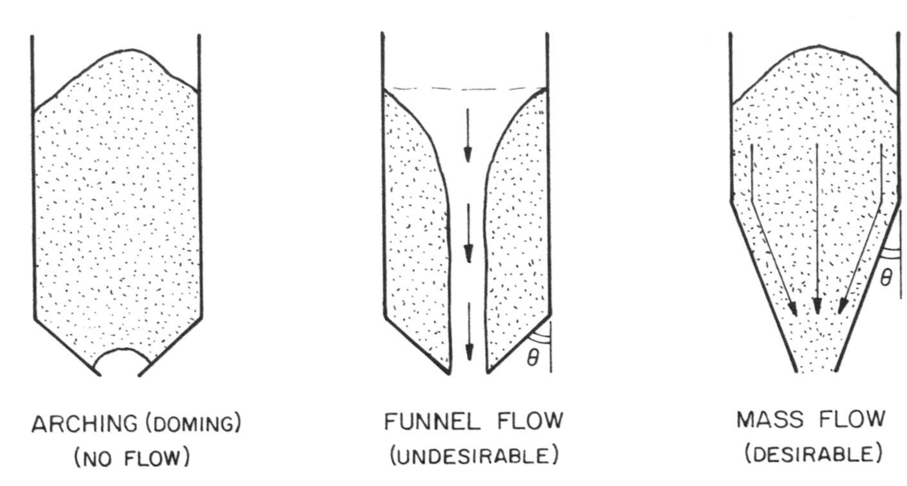

Fig. 10.13. Demonstration of the different flow (and no-flow) patterns in gravitational flow of powders. The actual pattern is established by the powder properties, the bin angle (θ), the aperture diameter, and the friction along the walls. The mechanical aspects of these situations and other considerations for proper bin design are discussed in detail by Jenike (1964).

the work of Jenike (1964) as well as in numerous more recent publications (Brown and Richards 1970; Stanforth and Berry 1973, 1975; Morelus 1975; Schwedes 1975). Results of shear analyses in several food powders have been reported by Peleg *et al.* (1973). It was also shown, however, that this type of analysis may not be appropriate for extremely cohesive food powders (Moreyra and Peleg 1981; Dobbs *et al.* 1982) because of theoretical and instrumental considerations (Peleg 1977), or that their sensitivity may be insufficient to detect changes in cohesiveness that can be detected by other mechanical means (Hollenbach *et al.* 1982; Peleg *et al.* 1982). It seems justified to argue, however, that under such circumstances (i.e., where shear analysis is inapplicable), the possibilitiy of stable gravitational flow is very doubtful and that special design considerations must be taken if such powders need to be conveyed or processed.

It is also worth mentioning that in cohesive food powders, internal friction has very little influence on flowability. Most food powders have angle of internal friction of the order of 30–45 deg (Peleg *et al.* 1973, 1982), and it usually decreases slightly as moisture is absorbed. This is mainly because of reduction in the particle's surface roughness through dissolution and lubrication. It is obvious that the reduction in internal friction does not result in improved flowability. Fine dry sand, for example, may have an angle of friction of about 40 deg, but it is clear that its flowability is by far greater than that of baby food formula (30 deg) or corn starch (31–33 deg), despite their considerably smaller angle of friction.

Caking

Many food powders, especially those containing soluble components or fats tend to agglomerate spontaneously when exposed to moist atmosphere or elevated storage temperature. The phenomenon can result in anything between small soft aggregates that break easily, to rock hard lumps of variable size, or solidification of the whole powder mass. In most cases, the process is initiated by the formation of liquid bridges between particles that can later solidify by drying or cooling. This mechanism, known as "humidity caking" (Burak 1966), is schematically described in Fig. 10.14. Incidentally, the attraction stage shown in the figure is not a hypothetical stage. It can actually be observed that

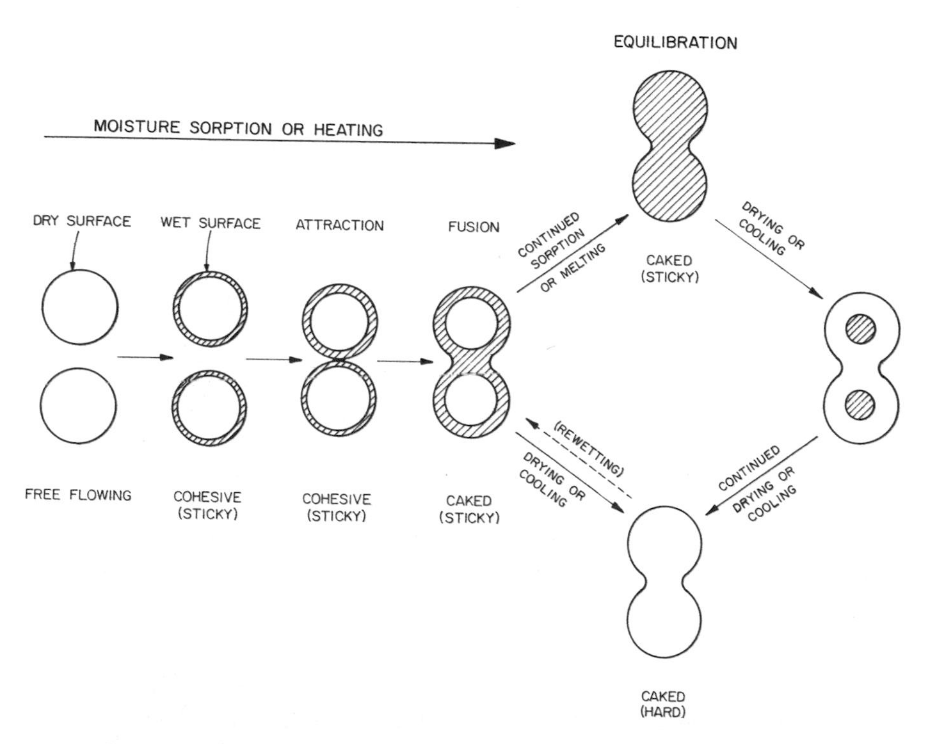

Fig. 10.14. Schematic presentation of the most common caking mechanism in food powders. It is very similar in kind to that of many (intentional) agglomeration processes.

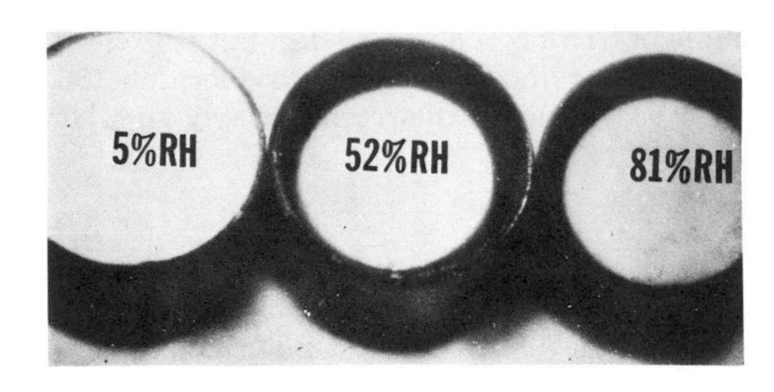

Fig. 10.15. Humidity shrinkage in onion powder. *From Peleg and Mannheim (1977).*

moisture absorption is accompanied by shrinkage of the powder, as demonstrated in Fig. 10.15.

In food powders the source of the liquid bridges is a sticky or liquefied particles surface mainly produced by one of the following.

(a) Moisture sorption, accidental wetting, or moisture condensation that causes dissolution of the surface, and/or the presence of a liquid film around the particles.

(b) Liquefaction of the surface itself as a result of exceeding the temperature at which amorphous sugars become thermoplastic, without the addition of external moisture. (A detailed study of the physicochemical aspects of this phenomenon is reported by To and Flink (1978 A, B, C). This temperature, also known as the "sticky point" has strong dependency on the powder's moisture content (Notter *et al.* 1959; Lazar and Morgan 1966). The sticky point is especially low in fruit juice powders, and this is the main reason for their physical instability. This temperature can easily be exceeded in other sugar-rich powders (e.g., onion, certain spices) during grinding (and therefore in such cases, refrigerated mills are recommended) or during storage when the temperature or moisture are not tightly controlled.

(c) Liberation of absorbed moisture when amorphous sugars crystallize. (A notable example is the crystallization of lactose in milk powders (Berlin *et al.* 1968)—one of the main reasons for a variety of instantization processes).

(d) Melting of fats.

It should be mentioned that both the onset and rate of the caking phenomenon does not necessitate the liquefaction of the whole particle or the powder bed. In fact, it is enough that only part of the surface becomes wet to initiate the agglomeration. Furthermore, the intensity and the spread of the phenomenon within a given bed depend on the moisture absorption rate, the moisture diffusion rates into the particles interior, and its penetration rate into the bed. (Fig. 10.16). The hardness of the aggregates will depend on the material—crystalline, glassy or fatty— and on the temperature history of the particles—range and fluctuations frequency. It can also happen that because of insufficient drying, or mixing of ingredients with different moisture contents, caking will only occur after moisture equilibration in the package (Fig. 10.17).

It is clear that under such circumstances, the meaning of the terms "caking tendency" and "caking intensity" is fairly vague. Despite this limitation, however, it can be shown that most powders that are known to be cohesive (in terms of bulk properties and flowability) also tend to cake readily, especially if under static pressure (Peleg and Mannheim 1973).

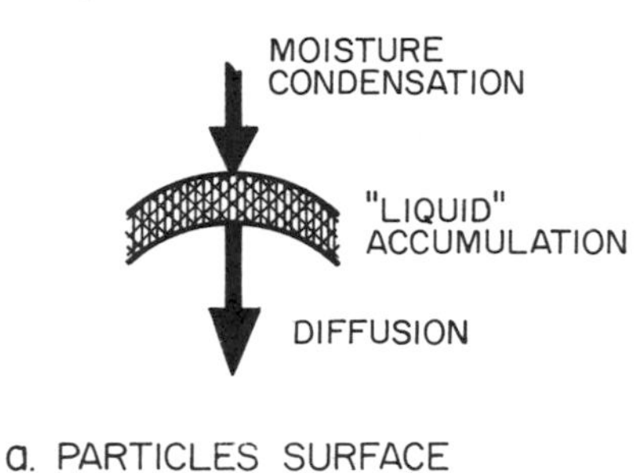

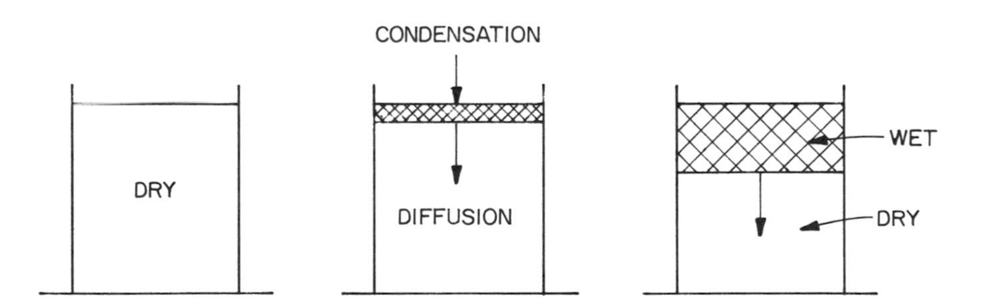

Fig. 10.16. Schematic representation of moisture diffusion in an individual particle and in a powder bed.

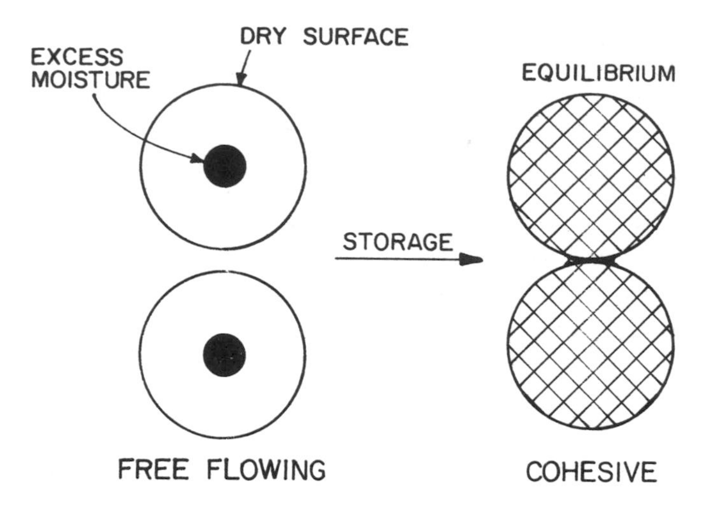

Fig. 10.17. Schematic view of a hypothetical situation where free-flowing powder can cake during storage without added moisture.

Theoretical aspects of the different kinds of bridging mechanisms and the factors that affect the agglomerate's strength can be found in the works of Rumpf (1961), Zimon (1969), Schubert (1975), and Klinzing (1981).

Effect of Anticaking Agents

Anticaking agents (also known as flow conditioners, glidants free-flowing agents) are very fine powders (particle size 1–4 μm) of an inert or fairly inert chemical substance that are added to powders with much larger particle size in order to inhibit caking and improve flowability. Studies in sucrose and onion powders (Peleg and Mannheim 1973, 1977) show that such agents are effective (in both roles) only at a limited relative humidity range. The explanation is that the agent particles coverage of the host particles surface is sufficient to reduce interparticle attraction and perhaps to interfere with the continuity of liquid bridges. Their presence is not sufficient, however, to cover moisture sorption sites. Therefore if moisture sorption is undisturbed, liquid bridges will eventually be formed and caking will occur as if the agent did not exist.

Additional aspects of the use of anticaking agents in food and other powders are discussed by Nash *et al.* (1965), Irani and Callis (1960), Irani *et al.* (1961), Burak (1966), Peleg and Mannheim (1969), Carr (1976) and Hollenbach *et al.* (1982).

Segregation

According to Gray (1968) "Segregation of particles occurs when particles of different properties are distributed preferentially in different parts of the bed." The main differences responsible for segregation are differences in particle size, density, shape, and resilience (Williams 1976). It has been shown, however, that in practice the difference in particle size is by far the most important factor (Williams 1976; Johanson 1978). The segregation process usually occurs when free-flowing powders having significant range of particle size are exposed to vibration or other types of mechanical motion. Under such conditions, the smaller particles will migrate to the bottom of the bed so that their concentration decreases as the function of the bed height (Fig. 10.18). The phenomenon is not limited to mixtures of particles of different

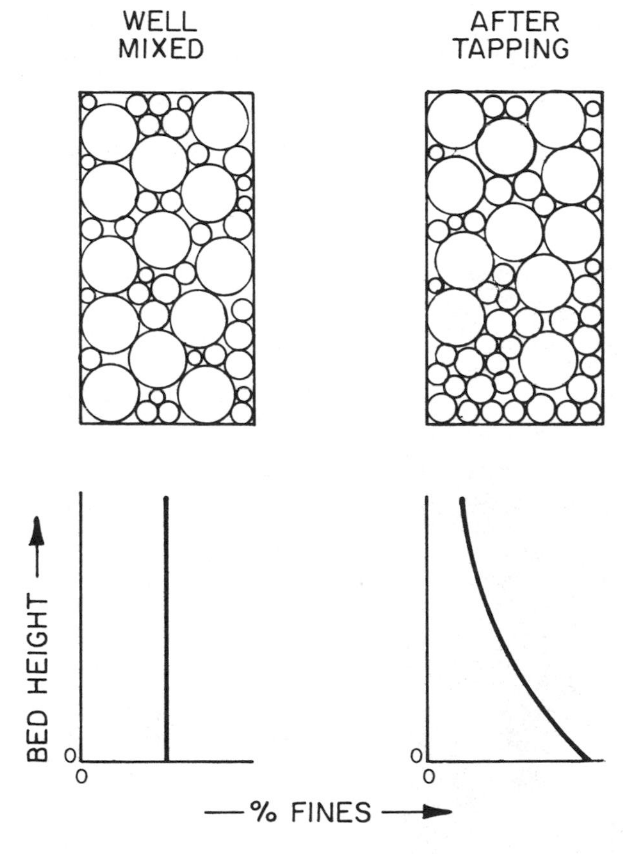

Fig. 10.18. Schematic of powders segregation by particle size. Note that in tapping the fines will always migrate to the bottom of the container.

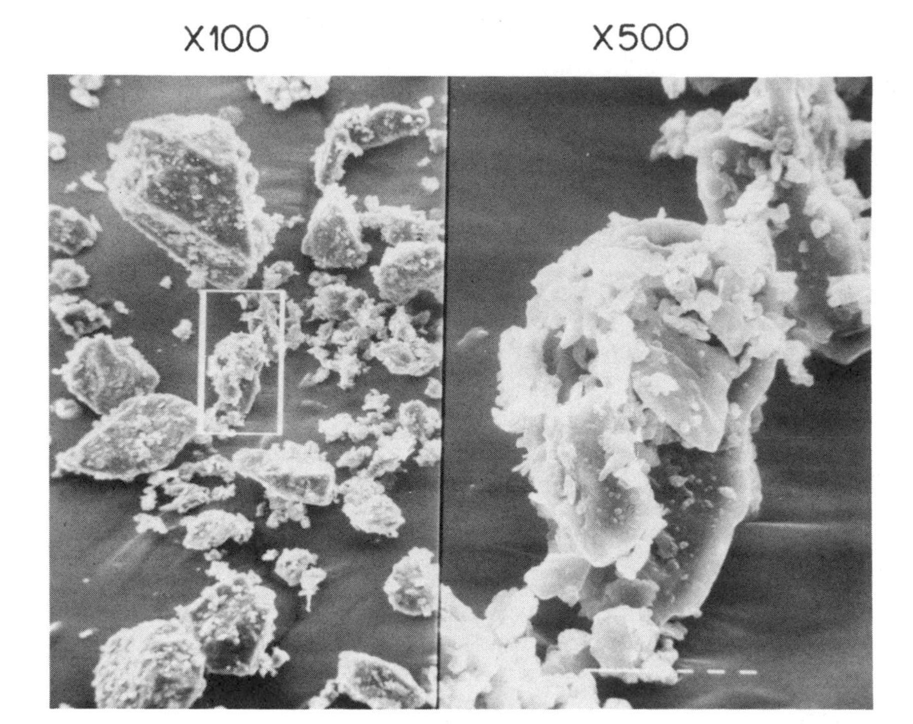
X100
X500

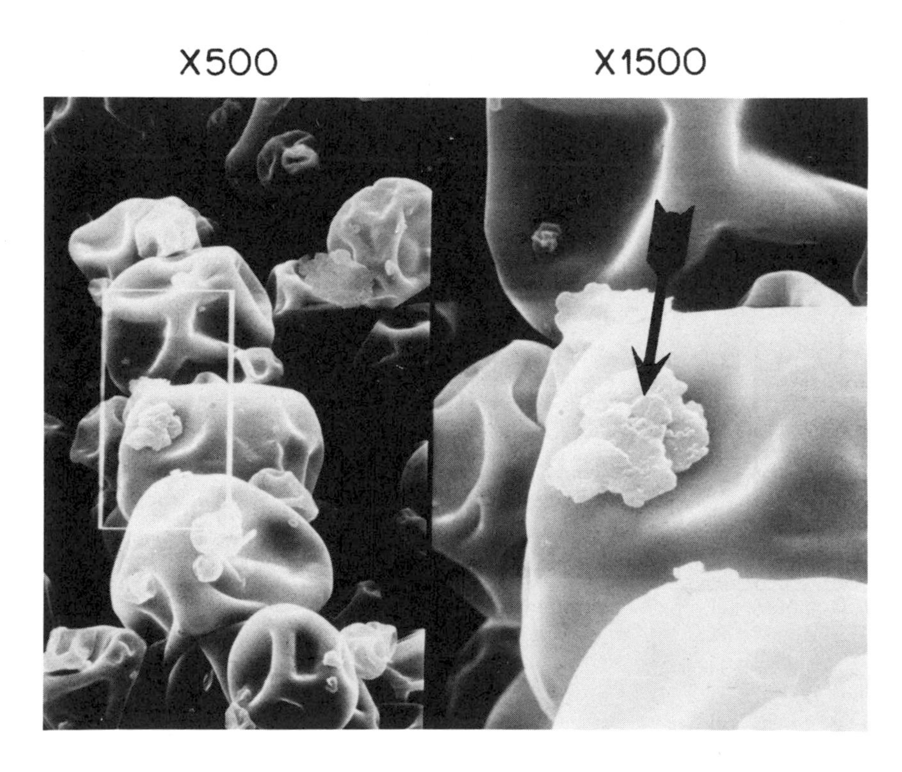
X500
X1500

species. It can and does occur in chemically uniform powders whenever significant size differences exist. The segregation phenomenon is particularly noticeable when the powder contains a considerably amount of fines. The source of these fines is unimportant. They can be an ingredient (e.g., colorants in drink powders) or the product of mechanical attrition (e.g., fines at the bottom of an instant coffee jar). Since the occurrence of this type of segregation depends on the mobility of the fines and their physical ability to sift through the coarse particles array, it is limited to free-flowing powders where the particles can easily move. The severity of the problem (in any given powder) depends on the intensity of the mechanical motion and on the geometrical features of the conveying, mixing, or storage systems (Johanson 1978).

In cohesive powders, segregation of fines is less likely to occur. The reason is that in such powders the fines usually adhere to the surface of the larger particles to form what is known as "ordered mixtures" (Yeung and Hersey 1979; Egermann 1980). Examples of such mixtures are shown in Figs. 10.1a and 10.19. Here again the mixture can be of particles belonging to different chemical species (Fig. 10.19) or of one species of particles having large difference in size (Fig. 10.1a). In both cases, the mechanical mobility of such fines is considerably decreased, and consequently their tendency to segregate is greatly reduced.

ACKNOWLEDGMENT

The author expresses his thanks to Dr. Robyn Rufner for the scanning electron micrographs and Mr. Richard J. Grant for his graphical aid.

BIBLIOGRAPHY

BERLIN, E., ANDERSON, B.A., and PALLANSCH, M.J. 1968. Comparison of water vapor sorption by milk powder components. J. Dairy Sci. *51*, 1912–1915.

BROWN, R.L., and RICHARDS, J.C. 1970. Principles of Powder Mechanics. Pergamon Press, Oxford, England.

BURAK, N. 1966. Chemicals for improving the flow properties of powders. Chem. Ind. 844–850.

CARR, R.L. 1976. Powder and granule properties and mechanics. *In* Gas-Solids Handling in the Processing Industries. J.M. Marchello and A. Gomezplata (Editors). Marcel Dekker, Inc., New York.

Fig. 10.19. Ordered and random mixtures. (*Top*) Sucrose and 2% calcium stearate is an ordered mixture, i.e., the fine stearate particles adhere to the larger sucrose particles. (*Bottom*) No such adherence is evident in the mixture of soy protein and 2% calcium stearate and the protein particles surface is visibly uncoated. (The arrow points to an aggregate of the stearate particles.)

DOBBS, A.J., PELEG, M., MUDGETT, R.E., and RUFNER, R. 1982. Some physical characteristics of active dry yeast. Powder Technol. *32*, 75–81.
EGERMANN, H. 1980. Effects of adhesion on mixing homogeneity. Part 1. Ordered adhesion-random adhesion. Powder Technol. *27*, 203–206.
GERRITSEN, A.H., and STEMERDING, S. 1980. Crackling of powdered materials during moderate compression. Powder Technol. *27*, 183–188.
GRAY, W.A. 1968. The packing of solid particles. Chapman and Hall, Ltd., London.
GRAY, R.D., and BEDDOW, J.K. 1968/9. On the Housner ratio and its relationships to some properties of metal powders. Powder Technol. *2*, 323–326.
HAUSNER, H.H. 1967. Friction conditions in a mass of metal powder. Int. J. Powder Metall. *3*, 7–13.
HAUSNER, H.H., ROLL, K.H., and JOHNSON, P.K. 1976. Vibratory Compaction—Principles and Methods. Plenum Press, New York.
HOLLENBACH, A.M., PELEG, M., and RUFNER, R. 1982. Effect of four anticaking agents on the bulk characteristics of ground sugar. J. Food Sci *47*, 538–544.
HOLLENBACH, A.M., PELEG, M., and RUFNER, R. 1983. Interparticle surface affinity and the bulk properties of conditioned powders. Powder Technol. *33*, (In press).
IRANI, R.R., and CALLIS, C.F. 1960. The use of conditioning agents to improve the handling of cereal products. Cereal Sci. Today 5(7), 1968–1972.
IRANI, R.R., VANDERSALL, H.L., and MORGENTHALER, W.W. 1961. Water vapor sorption on flow conditioning and cake inhibition. Ind. Eng. Chem. *53*, 141–142.
JENIKE, A.W. 1964. Storage and flow of solids. Bull. No. 123, Utah Engineering Exp. Stn., Univ. of Utah, Salt Lake City.
JOHANSON, J.R. 1978. Particle segregation and what to do about it. Chem. Eng. *85*, 182–188.
KLINZING, G.E. 1981. Gas-solid Transport. McGraw Hill, Inc., New York.
KAWAKITA, K., and LÜDDE, K-H. 1970/1. Some considerations of powder compression equations. Powder Technol. *4*, 61–68.
LAZAR, M.E., and MORGAN, A.J. 1966. Instant apple sauce. Food Technol. *20*, 179–181.
McGEARY, R.K. 1967. Mechanical packing of spherical particles. *In* Vibratory Compacting. H.H. Hausner, K.H. Roll, and P.K. Johnson (Editors). Plenum Press, New York.
MOREYRA, R., and PELEG, M. 1981. Effect of equilibrium water activity on the bulk properties of selected food powders. J. Food Sci. *46*, 1918–1922.
MORELUS, O. 1975. Theory of yield of cohesive powders. Powder Technol. *12*, 259–275.
NASH, J.H., LEITER, R., and JOHNSON, A.P. 1965. Effect of antiagglomerant agents on physical properties of finely divided solids. Ind. Eng. Chem. (Prod. Res. Dev.) *4*(2), 140–145.
NOTTER, G.K., TAYLOR, D.H., and DOWNES, N.J. 1959. Orange juice powders—Factors affecting stability. Food Technol. *13*, 113–118.
ORR, C. 1966. Particulate Technology. MacMillan Co., New York.
PELEG, M. 1971. Measurements of cohesiveness and flow properties of food powders. D.Sc. Thesis. Technion, Haifa, Israel.
PELEG, M. 1977. Flowability of food powders and methods for its evaluation—A review. J. Food Proc. Eng. *1*, 303–328.
PELEG, M., and MANNHEIM, C.H. 1973. Effect of conditions on the flow properties of powdered sucrose. Powder Technol. *7*, 45–50.
PELEG, M., and MANNHEIM, C.H. 1977. The mechanism of caking of powdered onion. J. Food Proc. Preserve. *1*, 3–11.
PELEG, M., MANNHEIM, C.H., and PASSY, N. 1973. Flow properties of some food powders. J. Food Sci. *38*, 959–964.

PELEG, M., MOREYRA, R., and SCOVILLE, E. 1982. Rheological characteristics of food powders. AIChE Symp. Ser., in press.
PELEG, Y., and MANNHEIM, C.H. 1969. Caking of onion powder. J. Food Technol. *4*, 157–160.
PIETSCH, W.B. 1969. Adhesion and agglomeration solids during storage flow and handling. Trans. ASME Ser. B5, 435–449.
RUMPF, H. 1961. The strength of granules and agglomerates. *In* Agglomeration. W.A. Knepper (Editor). Industrial Pub., New York.
SCHUBERT, H. 1975. Tensile strength of agglomerates. Powder Technol. *11*, 107–119.
SCHWARTZBERG, H.G. 1982. Private Communication. Univ. of Massachusetts, Amherst.
SCHWEDES, J. 1975. Shearing behavior of slightly compressed cohesive granular materials. Powder Technol. *1*, 59–67.
SCOVILLE, E., and PELEG, M. 1980. Evaluation of the effects of liquid bridges on the bulk properties of model powders. J. Food Sci. *46*, 174–177.
SONE, T. 1972. Consistency of foodstuffs. D. Reidel Pub.Co., Dordrecht, Holland.
STANFORTH, P.T., and BERRY, P.E.R. 1973. A general flowability index for powders. Powder Technol. *8*, 243–251.
STANFORTH, P.T., and BERRY, P.E.R. 1975. Flow properties analysis of irregular powders. Powder Technol. *12*, 29–36.
TO, E.C., and FLINK, J.M. 1978A. 'Collapse' a structural transition in freeze-dried carbohydrates: Evaluation of analytical methods. J. Food Technol. *13*, 555–565.
TO, E.C., and FLINK. J.M. 1978B. 'Collapse' a structural transition in freeze-dried carbohydrates: Effect of solute composition. J. Food Technol. *13*, 567–581.
TO, E.C., and FLINK, J.M. 1978C. 'Collapse' a structural transition in freeze-dried carbohydrates: Prerequisite of recrystallization. J. Food Technol. *13*, 583–594.
WATT, B.K., and MERRILL, A.L. 1975. Composition of foods. Agriculture Handbook No. 8. U.S. Dept. of Agriculture, Washington, DC.
WILLIAMS, J.C. 1976. The segregation of particulate materials—A review. Powder Technol. *15*, 245–251.
YEUNG, C.C., and HERSEY, J.A. 1979. Ordered powder mixing of coarse and fine particulate systems. Powder Technol. *22*, 127–131.
ZIMON, A.D. 1969. Adhesion of Dust and Powders. Plenum Press, New York.

11

Large Deformations in Testing and Processing of Food Materials

E.B. Bagley[1]

To the food scientist and technologist the importance of physical tests to describe the properties and behavior of foodstuffs has long been evident (Scott Blair 1953). Over the years the sophistication of the approaches has increased (deMan *et al.* 1976). The polymer industry, which, like the food industry, needs to characterize and understand the physical behavior of complex systems, has developed approaches and concepts (Bagley and Schreiber 1969) whose applicability to food systems is being increasingly recognized (Rha 1975).

Figure 11.1 shows a relationship between process and end use applications where both empirical tests and basic measurements play a role. Industrial processes result ultimately in a consumer product. Empirical tests play a vital role in controlling the process and in predicting properties in end use application. Such empirical tests are the result of experience and insight but suffer from a weakness in providing little predictive capability beyond relatively narrow ranges.

Basic measurements, which can relate to fundamental process or product variables, provide opportunities for new concepts and new approaches for process and product innovation. A thorough understanding of the basic measurements, which often of necessity are slow and tedious, nevertheless allows the design of simple test procedures that often turn out to be more effective and more general than the laboriously developed empirical tests, allowing more effective control and characterization of both a process and a product. The situation is well

[1]Northern Regional Research Center, Agricultural Research Service, U.S. Department of Agriculture, Peoria, IL 61604.

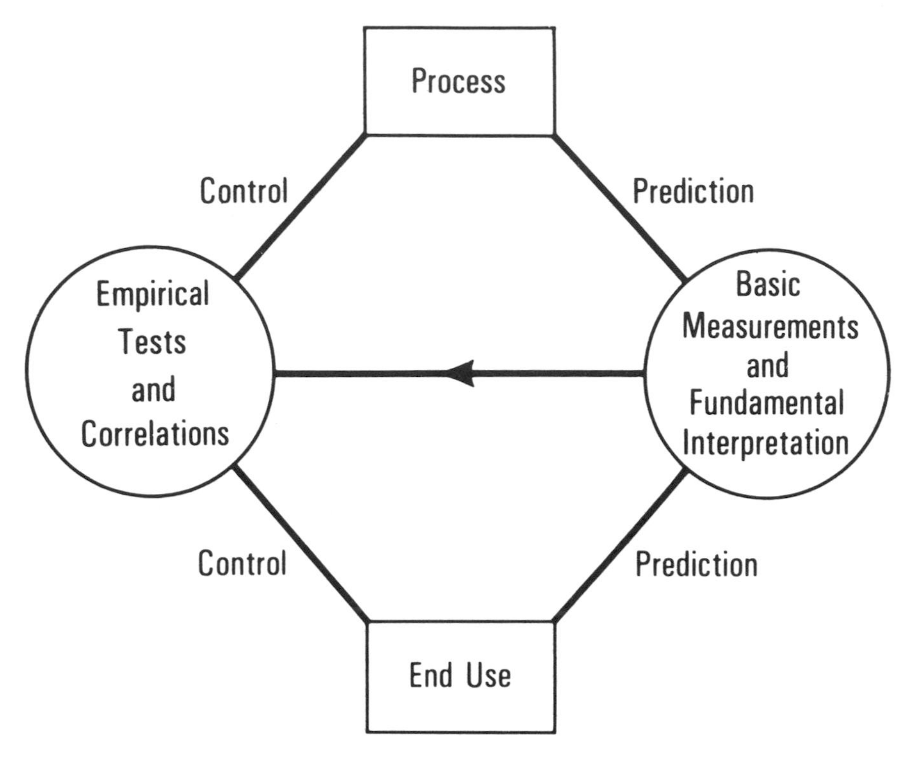

Fig. 11.1. Roles of empirical and fundamental measurements in relating process conditions to product properties. Empirical methods should be based, as far as possible, on a fundamental understanding of material behavior.

summarized by Szczesniak (1977) as follows: "Today, a large number of empirical instruments such as penetrometers, compressimeters, shearing devices, cutting devices and different types of viscometers are used to quantify the various parameters of foods related to maturity, quality, and functional properties . . . these instruments are now being subjected to critical scrutiny and theoretical analysis."

In following and understanding this critical scrutiny and theoretical analysis, the food scientist and technologist face some real difficulties. Even the treatment of thermally, chemically, and biologically stable synthetic polymers results in a vast and mathematically complex literature. Further, foods may be liquids, semi-solids, or solids. The liquid or semi-liquid food systems may show yield points rarely encountered in synthetic polymers but of considerable practical importance in foods. Food systems, particularly solids, are probably anisotropic and may well be "composites." (The study of synthetic polymeric composites, e.g., epoxy-fiber glass, is a specialized area in itself.)

The purpose of this chapter is to present some ways of looking at the response of food systems to applied forces or imposed strains; to indicate the physical considerations that can guide work in this area; and to reference some published work that can be helpful in considering basic concepts in the physical testing of foods.

VISCOELASTICITY

It is helpful in considering the response of a food material to use a simple model. The use of such models is well established for polymers and is summarized by Alfrey (1948). A model I find useful to remind me of what occurs in the physical testing of a complex material is the so-called Model A of Alfrey shown in Fig. 11.2. This model is more commonly termed the Maxwell–Voigt model. The behavior of a material in a creep experiment is shown in Fig. 11.3. If a constant stress is applied to a viscoelastic material, the response in general is threefold:

1. An instantaneous elastic deformation corresponding to the extension of the spring labeled G_1 in Fig. 11.2.
2. A retarded elastic response associated with the deformation of the spring G_2 at a rate determined by the viscous component, η_2, the rate characterized by a relaxation time $\theta = \eta_2/G_2$.
3. Viscous flow in the system at a rate governed by the third element, η_3.

The response curve, the sum of the three effects, is shown in Fig. 11.3. At times long compared to the relaxation time θ, both the instantaneous elastic (ε_i) and retarded elastic (ε_r) response will be complete, and the resultant elastic strain for these deformations will be

$$\varepsilon_r + \varepsilon_i = \frac{\sigma_0}{G_1} + \frac{\sigma_0}{G_2} \tag{1}$$

The response curve at these long times will then have a constant slope, $d\varepsilon/dt$, given by

$$\eta_3 = \frac{\sigma_0}{(d\varepsilon/dt)} \tag{2}$$

If the stress is removed at time t, there will be a recovery that will show two components—an instantaneous recovery due to the response of G_1

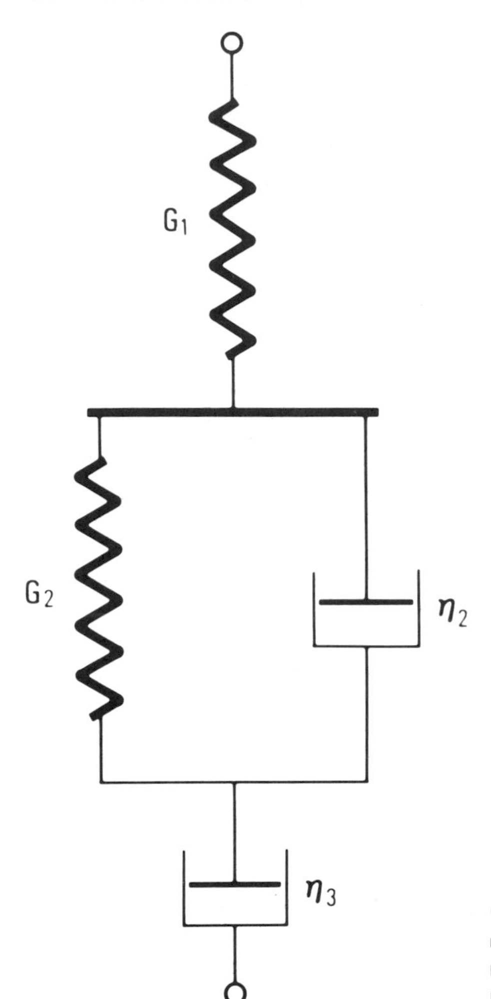

Fig. 11.2. The Maxwell–Voigt element, useful in emphasizing the three basic elements in the response of a viscoelastic material to applied forces.

and a time dependent recovery due to G_2. There will, however, be permanent deformation of the system as a result of the flow associated with η_3.

Precisely this creep experiment has been carried out for many systems. For a simple example of application to a starch gel, the work of Collison and Elton (1961) is worth examining. They found for a 17% gel cooked at 68°C a value of $(G_1 + G_2)$ of 5×10^5 dynes/cm^2 and η_3 of 2×10^8 poise. Creep and relaxation studies continue to be an area of interest to food scientists, as evidenced by recent work by Gross *et al.* (1980).

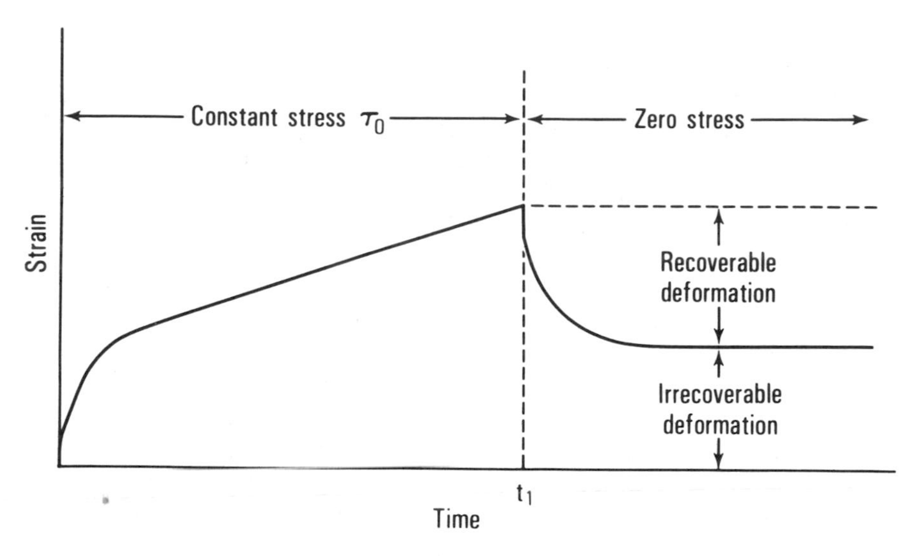

Fig. 11.3. The response of a Maxwell-Voigt element to an applied stress—a creep experiment and recovery response when the applied stress is reduced to zero.

From Fig. 11.3 it is evident that in examining large deformation behavior a critical concern is how much permanent deformation has resulted due to flow. A first consideration, then, is a knowledge of how much recoverable and how much irrecoverable deformation has been imposed on the sample.

Secondly, it must be recognized that the response of polymer and food systems will seldom be governed by a single relaxation time (η_2/G_2). Rather, a spectrum or distribution of relaxation times, represented by a series of elements (η_i/G_i) in series in Fig. 11.2, will be needed to describe the system.

Finally, depending on stress, strain level, time, and particular system studied, there may be permanent, irrecoverable deformation resulting from changes in the material itself. Such material property changes can be investigated by resubjecting the tested material to subsequent creep experiments to determine whether the parameters of the model change with testing. Changes in the material induced by the stresses and deformations can be most informative in understanding the system behavior, particularly on the macroscopic and microscopic level. This change in properties will in general be expected for foods, as exemplified by the Texture Profile Analysis force-time curves as discussed by Bourne in deMan *et al.* (1976).

These material changes can occur under large deformations, and foodstuffs can exhibit yield effects along with work softening and stiffening. A method for dealing with yield stresses in models is discussed by Chen and Rosenberg (1977), in Instron studies of American cheese. The yield stress was introduced into the Voigt component of the Maxwell–Voigt model of Fig. 11.2. The result of this is a discontinuity in the stress rate, which made possible the model simulation of the yield behavior of the cheese.

Extensive and valuable discussions of the applications of rheological models to foods have been given by Peleg (1980A). He emphasizes that "... mechanical analogues do not represent any specific structural element. They are purely a visual representation of strain and strain rate related components of the stress. It should also be remembered that tensors are the appropriate representation of real stresses and strains, and therefore any unidimensional mathematical model cannot fully account for the actual changes that occur during deformation."

In this same reference, Peleg emphasizes that models constructed from only elastic and viscous elements cannot always account for the observed rheological response. Additional types of elements are needed to account for "such phenomena as plasticity, internal fracture and increase in strength (as a result of compressibility, strain hardening or biochemically induced contraction)" (Peleg 1980A). The need for models that show such effects has also been alluded to by A. Szczesniak in Chapter 1 of this volume. Further discussion of the special model elements pertinent to the mechanical response of foods are provided by Peleg (1977A,B; 1979). Application of these concepts to mastication is provided by Peleg (1978) and Peleg and Normand (1982) and to the relation of hardness to sensory evaluation by Peleg (1980B). Other useful references to food properties are given by Calzada and Peleg (1978) and Finkowski and Peleg (1981).

DEFINITIONS OF STRESS AND STRAIN

The use of the terms stress and strain can be confusing unless these quantities are very clearly defined. As noted by Tschoegl *et al.* (1970), the nominal or engineering stress σ_0 is given by the force per initial (undeformed) cross-sectional area, A_0, so

$$\sigma_0 = f/A_0 \tag{3}$$

with units for σ_0 of dynes/cm^2 or Pascals (Newtons/m^2). When the deformation is large the true stress σ differs significantly from the nominal or engineering stress. The true stress is the force divided by the area of the deformed specimen. In simple extension, if l_o is the original length, l the final length, the extension ratio, λ, is l/l_o and

$$\lambda = l/l_o = 1 + \varepsilon \tag{4}$$

$$\varepsilon = \Delta l/l_o \tag{5}$$

The actual cross-sectional area A is related to the original area A_0 by

$$A(1 + \varepsilon) = A_0 \tag{6}$$

so the true tensile stress is related to the engineering tensile stress by

$$\sigma = \sigma_0(1 + \varepsilon) \tag{7}$$

The true stress can thus be considerably different from the apparent engineering stress and, of course, in any fundamental theoretical interpretation of data the true stress will be required.

The definition of strain causes much confusion because of the many different approaches to the problem. The problem is complicated also because there are a variety of testing modes—simple elongation, simple shear, pure shear, torsion, biaxial tension, etc.—and the definition of strain depends on which testing mode is being investigated. The usual way in which such problems are handled is through the use of tensor analysis and, for those who become deeply involved in the analysis of stress–strain response, a knowledge of the methods of tensors is essential (see, for example, Middleman 1968; Han 1976). A useful discussion of deformation that is physically significant and clearly describes the factors invovled in stress–strain analysis applicable to large deformations is provided by Treloar (1958). He considers a unit cube of material deformed to dimensions λ_1, λ_2, λ_3 as shown in Fig. 11.4. The deformed sample can be described by a stored energy function W, which is the energy stored in the deformed sample and with $W = 0$ in the undeformed sample. The λ_1, λ_2, λ_3 values are used in W through the strain invariants, I_1, I_2, I_3 defined as

$$I_1 = \lambda_1^2 + \lambda_2^2 + \lambda_3^2 \tag{8}$$

$$I_2 = \lambda_1^2\lambda_2^2 + \lambda_2^2\lambda_3^2 + \lambda_3^2\lambda_1^2 \tag{9}$$

$$I_3 = \lambda_1^2\lambda_2^2\lambda_3^2 \tag{10}$$

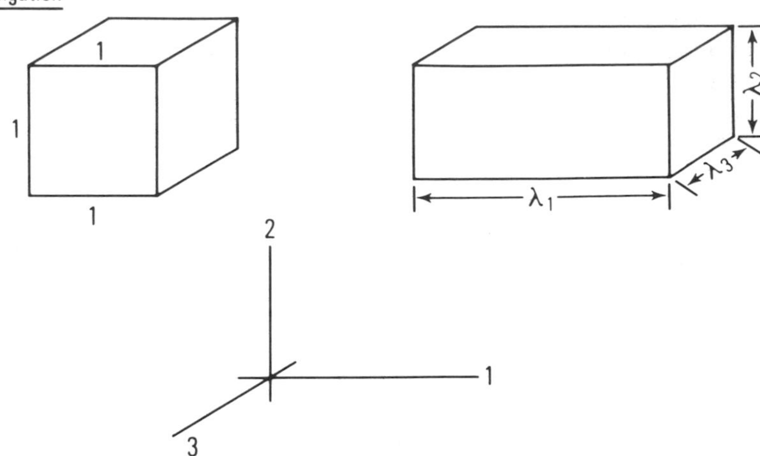

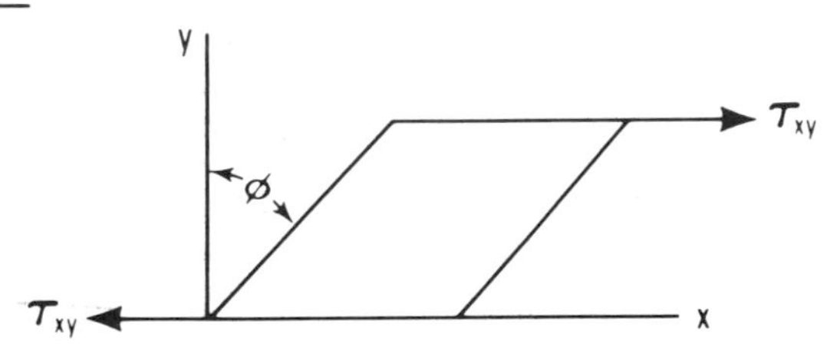

Fig. 11.4. Change in dimensions of a unit cube to a deformed structure of dimensions λ_1, λ_2, and λ_3.

Note that $\sqrt{I_3}$ is the volume of the deformed sample, and if no volume change occurs on deformation $I_3 = 1$. Although in the discussion Treloar gives a general expression for W, it turns out in practice that the following expression is of particular interest:

$$W = C_1(I_1 - 3) + C_2(I_2 - 3) \tag{11}$$

where C_1 and C_2 are material constants and the number 3 is included in each term of Eq. (11) so that $W = 0$ in the undeformed state, where $I_1 = 3, I_2 = 3$ as in Eqs. (8) and (9). Equation (11) can now be applied to specific deformations.

SIMPLE ELONGATION

In this experiment the unit cube is stretched in the l direction to a length $\lambda_1 = \lambda$. We have, if no volume change occurs, $\lambda_2 = \lambda_3 = \lambda^{-1/2}$. The resultant force f_1 per unit area of unstrained cross-section will be

$$f_1 = 2\left(\lambda - \frac{1}{\lambda^2}\right)\left(C_1 + \frac{C_2}{\lambda}\right) \tag{12}$$

In a simple elongation experiment at small deformations one would obtain results where

$$f_1 = E\,\frac{\Delta l}{l_o} = E\,(\lambda - 1) \tag{13}$$

which is Hooke's law with E a Young's modulus. This small deformation law of Eq. (13) is quite different from that relationship, Eq. (12),which is applicable to large deformations as well as small. A crosslinked network will often obey Eq. (12), and the modulus is given by C_1. In crosslinked systems, C_1 will relate to the crosslink density. The physical interpretation of C_2 is less clear, and often C_2 is much less than C_1 and can be neglected. In any event a plot of $[f_1/2(\lambda - 1/\lambda^2)]$ against λ^{-1} will be linear and C_1 and C_2 can be evaluated experimentally to characterize a given system.

Now Szczesniak (1977) has pointed out that "except for gelatin, hydrocolloid food gels do not obey the theory of rubber elasticity, a phenomenon that has slowed down progress in this area." More recently, Mitchell (1980) noted "the stiff and extended nature of polysaccharide chains make it unlikely that polysaccharide gels obey rubber elasticity, though it is possible that this theory holds for gelatin gels and also for ovalbumin gels in 6 M urea."

In spite of these "caveats," my view is that the applicability of a result as embodied in Eq. (12) has not been adequately tested in food systems. The general approach described above is, after all, the expected behavior in what could be termed an ideal material. A real material will deviate from this expected response for a variety of reasons. Just as the ideal gas laws provide a reference behavior against which real, nonideal gases are measured, so a stress–strain response as described by Eq. (12) provides a reference against which the behavior of the real material can be compared. The magnitude and direction of the deviations from what is expected can provide considerable insight into specific processes or material properties. Even in rubbers, there can be chemical changes—crosslinking reactions or material degradation via oxidation, for instance—which can be systematically examined as described by Treloar (1958). In food systems, it is evident that numerous mechanisms can be involved in giving apparent departures from expected behavior. For example, Rasper (1976) discusses the network of linear macromolecules formed in the process of dough development. Hydrogen bonds contribute to the properties of the dough as do the —SS— crosslinks in gluten. Such interactions can be regarded as labile crosslinks, the effectiveness of which can depend on such factors as the stresses or strains applied to the system. The fact that strain level affects the ratio of stress to strain means that the behavior is nonlinear. By definition, in linear viscoelasticity the ratio of stress to strain is a function of time only. Nonlinear viscoelasticity is a difficult subject but progress is being made, as evidenced by a recent Symposium on Nonlinear Viscoelasticity, Boulder, Colorado, 1982. The information being obtained in polymer and composite materials in particular will in the long run be of value in analyzing complex food materials.

SIMPLE SHEAR

The large deformation treatment of a simple elongational test gives a relation between stress and deformation (Eq. 12) more complicated than Hooke's law as expressed in Eq. (13). Oddly enough, the same approach to simple shear (Fig. 11.4) shows that the response of a viscoelastic material is Hookean, so

$$\tau_{xy} = G\,\sigma \qquad (14)$$

where τ_{xy} is a shearing stress, G is a shear modulus, and σ is the shear strain which, expressed in terms of λ is defined by

$$\lambda_1 = \lambda; \qquad \lambda_2 = 1; \qquad \lambda_3 = 1/\lambda \qquad (15)$$

and

$$\sigma = \tan\phi = \lambda - 1/\lambda \qquad (16)$$

This simple result has consequences in considering the extrusion of food materials. It has been known for some time that in the extrusion of viscoelastic polymers through pipes of length to radius ratio L/R that the pressure P required to give a constant output rate was related to L/R by

$$P = 2\tau\,(L/R) + 2\tau\, e \tag{17}$$

where τ is the shear stress in the pipe and e is a measure of the pressure drops occurring as the material flows into the pipe. Thus plots of pressure versus L/R are linear, the slope of the line giving τ and the intercept at $P = 0$ giving e. It further was found that the entry pressure drop is due to two factors. First, it is a measure of the viscous, irreversible deformation occurring as material flows towards and into the pipe; second, it is a measure of the elastic energy stored in the fluid. Fortunately, these two effects can be separated, and it was found that even in a flowing viscoelastic fluid Hooke's law in shear is obeyed as described by Eq. (14).

Typical P-(L/R) plots are shown schematically in Fig. 11.5 for a viscoelastic fluid. The intercept e increases with shear rate and is often found to be linear with the true shear stress, as shown in Fig. 11.7. The linearity of this plot means that Hooke's law in shear is obeyed (Eq. 14), and the steeper the slope of such a plot, the lower the shear modulus.

This behavior can serve as a reference for examining the behavior of food materials in extrusion, where the food is subjected to large deformations (Jasberg *et al.* 1982). The result of such a comparison is shown in Fig. 11.6, where pressure is plotted against L/R at several different output rates for a soy dough at 120°C. These plots are linear as in Fig. 11.5. However, it is immediately evident that the intercept e, instead of increasing with increasing output rate as in Fig. 11.5, decreases with increasing output rate.

The difference between the expected behavior of a viscoelastic fluid and the actual behavior of a viscoelastic food material is not unexpected. At the temperature of extrusion, material changes are only to be expected. Material properties, including both the viscosity and the elasticity, will change with time. Thus, in food systems, "the residence time" can be a critical factor, as emphasized, for example, by Bruin *et al.* (1978). We carried out experiments of the type indicated in Fig. 11.6, but residence time was varied by changing both the radius of the pipe and the length of the path from the end of an extruder to the pipe. For three different experimental arrangements the plots of e versus shear stress were as shown by curves G1, G2, G3 in Fig. 11.7. This looks like a far more complex result than expected, but if values of e are compared at the same residence time, the result is the straight line plot indicated by the dotted line of Fig. 11.7. This result is tentative and requires further

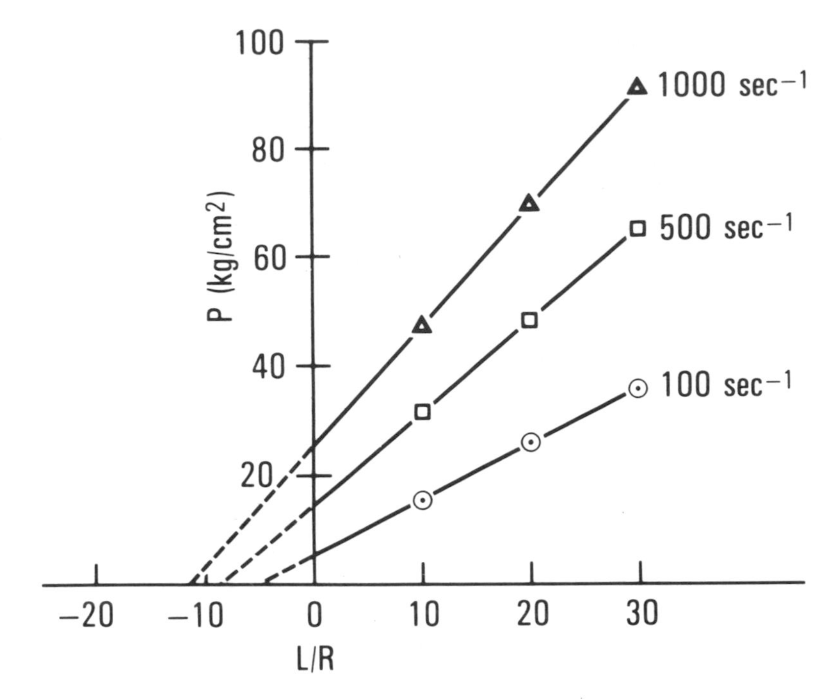

Fig. 11.5. Flow of material from a reservoir through a pipe of length *L* and radius *R*. The pressure required to obtain a given output rate (shear rate) is plotted against *L*/*R* for a typical thermoplastic synthetic polymer system.

substantiation. Nevertheless, the results do serve to emphasize that while food materials may be complex, consideration of experimental results in terms of theoretical expection can well serve to remove some of the complexity and clarify factors (in the extrusion case, the residence time) that must be considered in particularly situations.

OTHER TESTING MODES

The simple elongation and simple shear deformations have been discussed to illustrate the approach to large deformation studies. Other modes will not be considered here. There is a considerable literature on torsion tests, biaxial tests, etc., which have been extensively studied in the polymer and polymer composite literature. The extent to which this can be carried in polymeric testing is described by Clinard (1982). The approach described in the article is not suggested here as a recommendation for application to food systems. However, the article does empha-

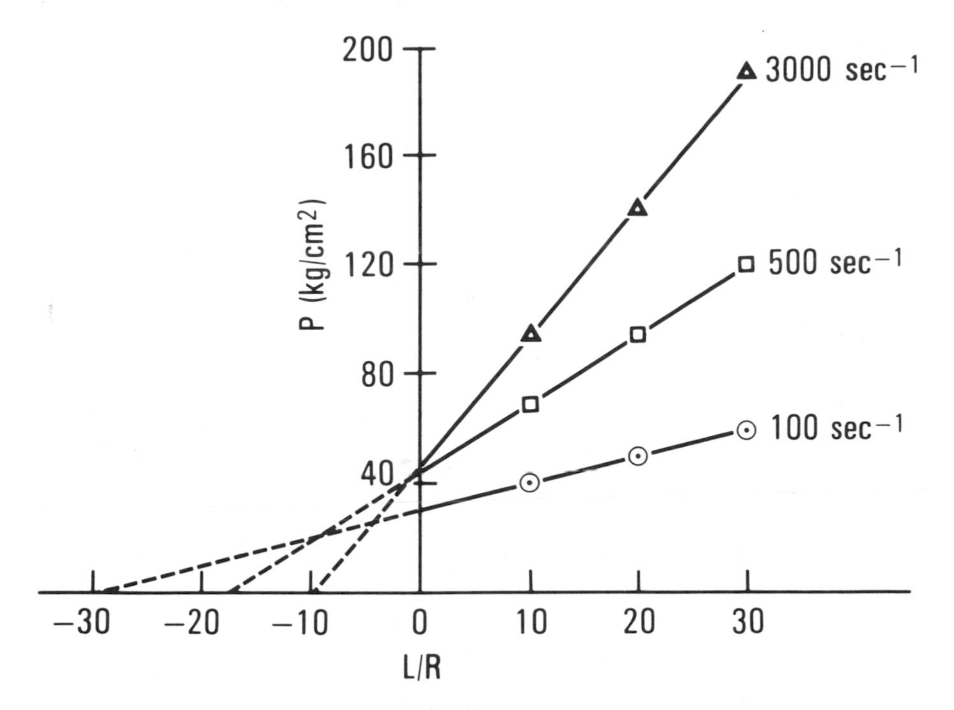

Fig. 11.6. Pressure versus *L/R* for a soy dough extruded at 120°C.

size the need in testing complex materials to "simulate the various forces and conditions a part will encounter in actual use" and the usefulness of computer systems in monitoring the tests and collecting performance data, guided by the operators knowledge of the types of forces and deformations being applied.

NONLINEAR BEHAVIOR

Allusion has already been made to problems of nonlinearity in large deformations. This topic is of considerable current interest, as shown by the recent work of Hibberd and Parker (1979). These authors provide an informative discussion (along with references to earlier pertinent work) that is worth examining. The essence of the approach is as follows.

For linear viscoelastic materials, the compliance (the reciprocal of a modulus) is independent of the stress or strain applied (although it will be a function of time in general) and completely defines the mechanical behavior. For nonlinear materials the compliance or modulus are functions of both stress (or strain) and time. For their nonyeasted wheat

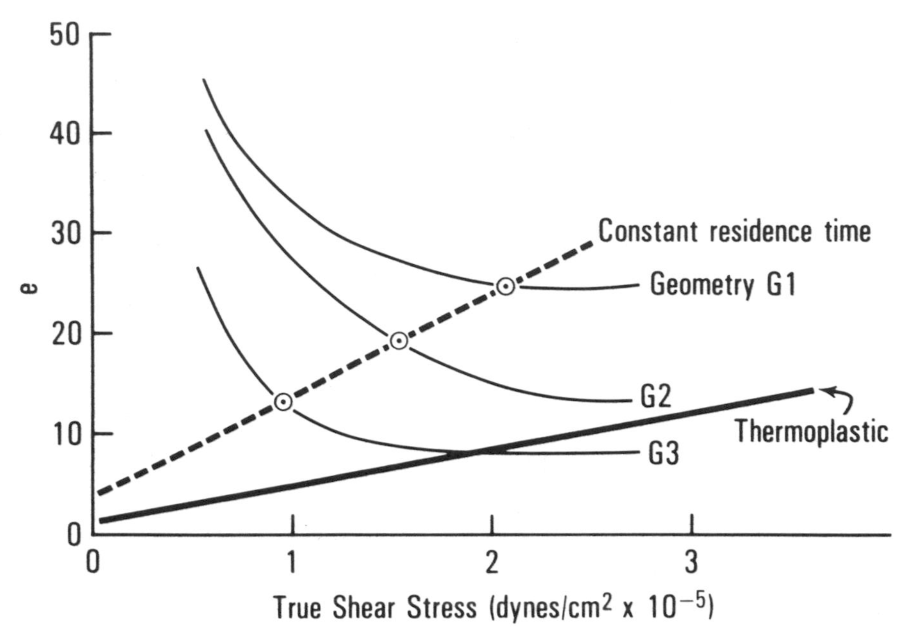

Fig. 11.7. End corrections versus shear stress for a typical thermoplastic material and also for soy doughs extruded in different geometries giving a range of residence times.

flour doughs it was shown that the nonlinear creep behavior could be represented by independent functions of stress and time. Thus, in Smith's terms (1962)

$$\sigma(\lambda,t) = \Gamma(\lambda)\, E(t) \tag{18}$$

where $\sigma(\lambda,t)$ is a stress that depends on deformation λ, and time t, $\Gamma(\lambda)$ is a function of the deformation only and $E(t)$ is a time dependent modulus.

Hibberd and Parker (1979) note that their results "clearly show that it is possible to represent the mechanical properties of a dough to a step in the deforming force over a limited range of stress and time by accepted rheological functions even though it is a nonlinear material." Although the statement is true, it is also limiting. The conclusion could be broadened because the simplification of the data, such as they obtained on their wheat flour doughs, by separating time and strain effects, offers hope that the factors governing behavior can be resolved. That is, from such analysis of large deformation, nonlinear behavior, the effects of composition, structural rearrangement, and molecular interactions can be elucidated and the results applied in both improving the quality of accepted food materials and introducing new concepts in fabrication of foods.

DYNAMIC MEASUREMENTS

Dynamic measurements have long been used in studying polymers (Ferry 1961). Applications to food systems and particularly to doughs have been reported, for example, by Hibberd and Wallace (1966), Smith *et al.* (1970), and Baird (1981). Since Professor Baird's chapter follows, the only point to be emphasized here is, again, that large deformations can lead to strain effects that must be separated from time effects to provide fundamental insight into the factors determining the material properties.

In dynamic measurements, a sinusoidal strain (or stress) is imposed and the resulting stress (or strain) is measured. When the material behaves in a linear fashion, the response is sinusoidal and is in two parts. For sinusoidal strain, the stress will be sinusoidal, part in phase with the strain (giving a storage or elastic modulus) and part out of phase (giving a loss modulus). In principle, the small strain response to any strain–time history can be calculated. When the deformations are large, the response is nonlinear, so that although the imposed strain may be sinusoidal the resultant stress is not. This can occur at strains as low as 0.2% and can also depend on the frequency of the sinusoidal strain imposed.

The existence of such "large deformation effects," far from being merely a complication, can lead to more fundamental understanding of a system. Thus Smith *et al.* (1970) note that the modulus dependence on amplitude which they observed in wheat doughs is also a characteristic of elastomers filled with carbon black. In the elastomer case, the carbon black particles form aggregates and, in large enough concentration, can form a "structure" that breaks down progressively with increases in strain. Thus nonlinear effects can serve to provide insight into interactions that result in a "structure."

A number of valuable insights into doughs or starch dispersion structure and behavior are given in works such as Smith *et al.* (1970). In comparing dough behavior with the carbon black elastomer, they emphasize that there are added complexities in the foods because of the distribution of water between the gluten and starch phase. As water content of their doughs decreased from 58 to 43%, the modulus rose by a factor of 40. It is not clear how much of this change is due to modulus changes in the gluten phase and how much is due to the reinforcing capability of the starch particles. Answers to such questions are critically needed to understand more fully the properties of food systems, and such knowledge will provide better use of food materials, particularly vegetable protein, in designing acceptable foods of the future.

FINAL COMMENTS

To illustrate the recent activity both in experiment and theory, mention will be made of two recent papers. Elongational viscosity is one active area of research. The concepts here date back to about 1900 but have been discussed by Middleman (1968) and by Münstedt and Middleman (1981). Testing in the elongational, large deformation mode is particularly pertinent to dough systems or in spin extrusion processes. This paper is mentioned to bring to your attention the concept of elongational flow and the instrumentation available for studies in this area. The results being obtained appear sensitive to the particular system being examined, but awareness of this area of research is important to anyone involved in material response to large deformation studies.

An interesting theoretical paper is by Doi and Kuzuu (1980). This work dealt with the finite deformation elasticity of a gel consisting of stiff, rod-like polymer chains. Xanthan gum molecules in a food gel could be an example of a system to which this Doi–Kuzuu analysis might be applied. Unlike flexible molecule systems, where finite deformations result in entropic effects that determine stress–strain response, the rigid rods undergo energetic changes associated with the bending of the rods under macroscopic strain. The stress–strain curves for the rigid rod systems are nonlinear even if the rod bending is small, giving an S-shaped curve unlike that of a rubber.

SUMMARY

Foods consist of complex mixtures of interacting components. These interactions can be on a microscopic scale (e.g., hydrogen bonding) or on a macroscopic scale where a material structure is set up as a result of physical interactions. To characterize the organizational structure, physical tests are carried out in which the response of the food to imposed deformations is examined. The material response, in terms of stress–strain–time relations, can be complicated. Often information on the processes occurring within the sample can be analyzed, relative to a theoretical model, to provide insights into the factors affecting material properties. Such information is essential to characterize, improve, control, and modify the properties of food materials.

BIBLIOGRAPHY

ALFREY, T., JR. 1948. Mechanical Behavior of High Polymers. Wiley (Interscience) Publishers, Inc., New York.

BAGLEY, E.B., and SCHREIBER, H.P. 1969. Elasticity effects in polymer extrusion. *In* Rheology, Theory and Applications. Vol. V. F.R. Eirich (Editor). Academic Press, New York.

BAIRD, D.G. 1981. Dynamic viscoelastic properties of soy isolate doughs. J. Texture Stud. *12*, 1–16.

BRUIN, S., ZUILICHEM, V., and STOLP, W. 1978. A review of fundamental and engineering aspects of extrusion of biopolymers in a single screw extruder. J. Food Process Eng. *2*, 83–95.

CALZADA, J. F., and PELEG, M. 1978. Mechanical interpretation of compressive stress–strain relationships of solid foods. J. Food Sci. *43*, 1087–1092.

CHEN, Y., and ROSENBERG, J. 1977. Nonlinear viscoselastic model containing a yield element for modeling a food material. J. Texture Stud. *8*, 477–485.

CLINARD, R.L. 1982. Rubber-metal parts testing simulates service conditions. Ind. Res. Dev., March, 130–133.

COLLISON, R., and ELTON, G.A.H. 1961. Some factors which influence the rheological properties of starch gels. Staerke *13*, 164–173.

deMAN, J.M., VOISEY, P.W., RASPER, V.F., and STANLEY, D.W. 1976. Rheology and Texture in Food Quality. AVI Publishing Co., Inc., Westport, CT.

DOI, M., and KUZUU, N.Y. 1980. Nonlinear elasticity of rodlike molecules in condensed state. J. Polym. Sci., Polym. Phys. Ed. *18*, 409–419.

FERRY, J.D. 1961. Viscoelastic Properties of Polymers (New Ed. 1980). John Wiley and Sons, Inc., New York.

FINKOWSKI, J.W., and PELEG, M. 1981. Some rheological characteristics of soy extrudates in tension. J. Food Sci. *46*, 207–211.

GROSS, M.O., RAO, V.N.M., and SMIT, C.J.B. 1980. Rheological characterization of low-methoxy pectin gel by normal creep and relaxation. J. Texture Stud. *11*, 271–290.

HAN, C.D. 1976. Rheology in Polymer Processing. Academic Press, New York.

HIBBERD, G.E., and PARKER, N.S. 1979. Nonlinear creep and creep recovery of wheat doughs. Cereal Chem. *56*, 232–236.

HIBBERD, G.E., and WALLACE, W.J. 1966. Rheol. Acta *5*, 193.

JASBERG, B.K., MUSTAKAS, G.C., and BAGLEY, E.B. 1982. Comparison of extrusion and capillary flow of thermoplastics with soy doughs. Food Process Eng. *5* (1), 43–56.

MIDDLEMAN, S. 1968. The Flow of High Polymers. Wiley (Interscience) Publishers, New York.

MÜNSTEDT, H., and MIDDLEMAN, S. 1981. A comparison of elongational rheology as measured in the universal extensional rheometer and by the bubble-collapse method. J. Rheol. *25*, 29–40.

MITCHELL, J.R. 1980. The rheology of gels. J. Texture Stud. *11*, 315–337.

PELEG, M. 1977A. Contact and fracture elements as components of the rheological memory of solid foods. J. Texture Stud. *8*, 67–76.

PELEG, M. 1977B. Operational conditions and the stress–strain relationship of solid foods—theoretical evaluation. J. Texture Stud. *8*, 283–295.

PELEG, M. 1978. Some mathematical aspects of mastication and its simulation by machines. J. Food Sci. *43*, 1093–1095.

PELEG, M. 1979. A model for creep and early failure. Mater. Sci. Eng. *40*, 197–205.

PELEG, M. 1980A. Theoretical analysis of the relationship between mechanical hardness and its sensory evaluation. J. Food Sci. *45*, 1156–1160.
PELEG, M. 1980B. Rheological models for solid foods. *In* Food Process Engineering. Vol. 1, Chapter 30, p. 250–256. P. Linko, Y. Malki, and J. Olkku (Editors). Applied Science Publishers, Ltd., London.
PELEG, M., and NORMAND, M.D. 1982. A computer assisted analysis of some theoretical rate effects in mastication and in deformation testing of foods. J. Food Sci. *47*, 1572–1578.
RASPER, V.F. 1976. Texture of dough, pasta and baked products. *In* Rheology and Texture in Food Quality. J.M. deMan, P.W. Voisey, V.F. Rasper, and D.W. Stanley (Editors). AVI Publishing Co., Inc., Westport, CT.
RHA, C.K. (Editor). 1975. Theory, Determination and Control of Physical Properties of Food Materials. D. Reidel Publ. Co., Dordrecht, Holland.
SCOTT BLAIR, G.W. 1953. Food Stuffs, Their Plasticity, Fluidity and Consistency. Wiley (Interscience) Publishers, New York.
SMITH, J.R., SMITH, T.L., and TSCHOEGL, N.W. 1970. Rheological properties of wheat flour doughs. III. Dynamic shear modulus and; its dependence on amplitude, frequency and dough composition. Rheol. Acta *9*(2), 239–252.
SMITH, T.L. 1962. Nonlinear viscoelastic response of amorphous elastomers to constant strain rates. Trans. Soc. Rheol. *6*, 61–80.
SZCZESNIAK, A.S. 1977. Rheological problems in the food industry. J. Texture Stud. *8*, 119–133.
TRELOAR, L.R.G. 1958. The Physics of Rubber Elasticity. Oxford University Press, London and New York.
TSCHOEGL, N.W., RINDE, J.A., and SMITH, T.L. 1970. Rheological properties of wheat flour doughs. I. Method for determining the large deformation and rupture properties in simple tension. J. Agric. Food. Sci. *21*, 65–70.

12

Food Dough Rheology

Donald G. Baird[1]

INTRODUCTION

This chapter is concerned with rheological properties (i.e., the flow and deformation characteristics) and techniques for measuring rheological properties of food doughs. Food doughs are defined here to be low moisture mixtures of water and wheat, corn, oat, semolina or soy flours and mixtures of these flours. Other ingredients can also be added such as flavoring and oil. These doughs may be processed by methods as common as baking or by extrusion to produce snacks, breakfast cereal, and textured vegetable protein. In this chapter we emphasize the measurement of rheological properties of food doughs in equipment which leads to quantities with engineering significance. It is these types of measurements that are needed for both quality control and for process modeling and scale-up.

Certainly rheological measurements of food doughs have been carried out for many years in the food industry (Shuey 1975). The farinograph, amylograph, and the Braebender torque rheometer are several of the instruments that have been used to provide some indication of the deformation characteristics of doughs. However, although these instruments have proved to be quite useful in obtaining qualitative information about dough properties, they do not give dough properties in quantities defined in engineering or scientific units. This is because the flow field created in these devices is usually so complicated that basic material properties cannot be obtained. With the need to model food processes for the purpose of process optimization and scale-up it

[1]Department of Chemical Engineering, Virginia Polytechnic Institute and State University, Blacksburg, VA.

becomes imperative that we carry out measurements in devices that provide material properties in engineering/scientific units. This chapter is concerned with methods for measuring dough rheology, the types of rheological experiments that should be carried out, and the application of these measurements to the processing of food doughs.

Types of Rheological Experiments

There are basically two types of flow used in rheological measurements: unidirectional simple shear and simple extensional flow. These flows are shown in schematic form in Fig. 12.1. For steady simple shear flow it can be shown that there are at most three independent quantities of stress: the shear stress, τ_{yx}, the primary normal stress difference, $N_1 = p_{xx} - p_{yy}$, and the secondary normal stress difference, $N_2 = p_{yy} - p_{zz}$. From these three stress quantities we define three material functions: viscosity, the primary normal stress difference coefficient (ψ_1), and the secondary normal stress difference (ψ_2).

$$\tau_{yx} = -\eta \dot{\gamma}_{yx} \tag{1}$$

$$N_1 = -\psi_1 \dot{\gamma}^2_{yx} \tag{2}$$

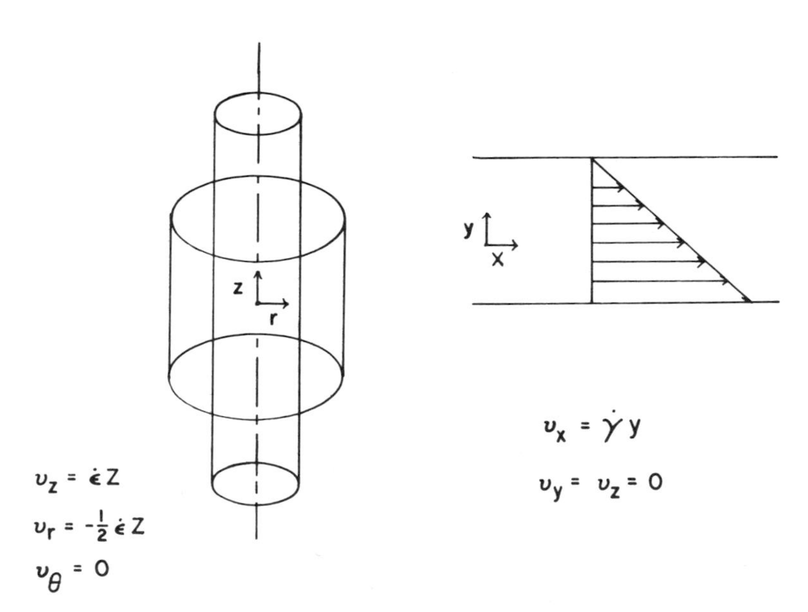

Fig. 12.1. Schematic drawing of elongational (*left*) and simple shear (*right*) flows.

$$N_2 = -\psi_2 \dot{\gamma}_{yx}^2 \tag{3}$$

The shear experiment can also be carried out in transient conditions leading to material functions that depend on both $\dot{\gamma}_{yx}$ and time. We have listed these in Fig. 12.2 along with the corresponding material functions.

Simple extensional flow is shown in Fig. 12.1 also. In this case there is only one independent quantity of stress measured, $p_{zz} - p_{rr}$, which leads to the definition of the extensional viscosity

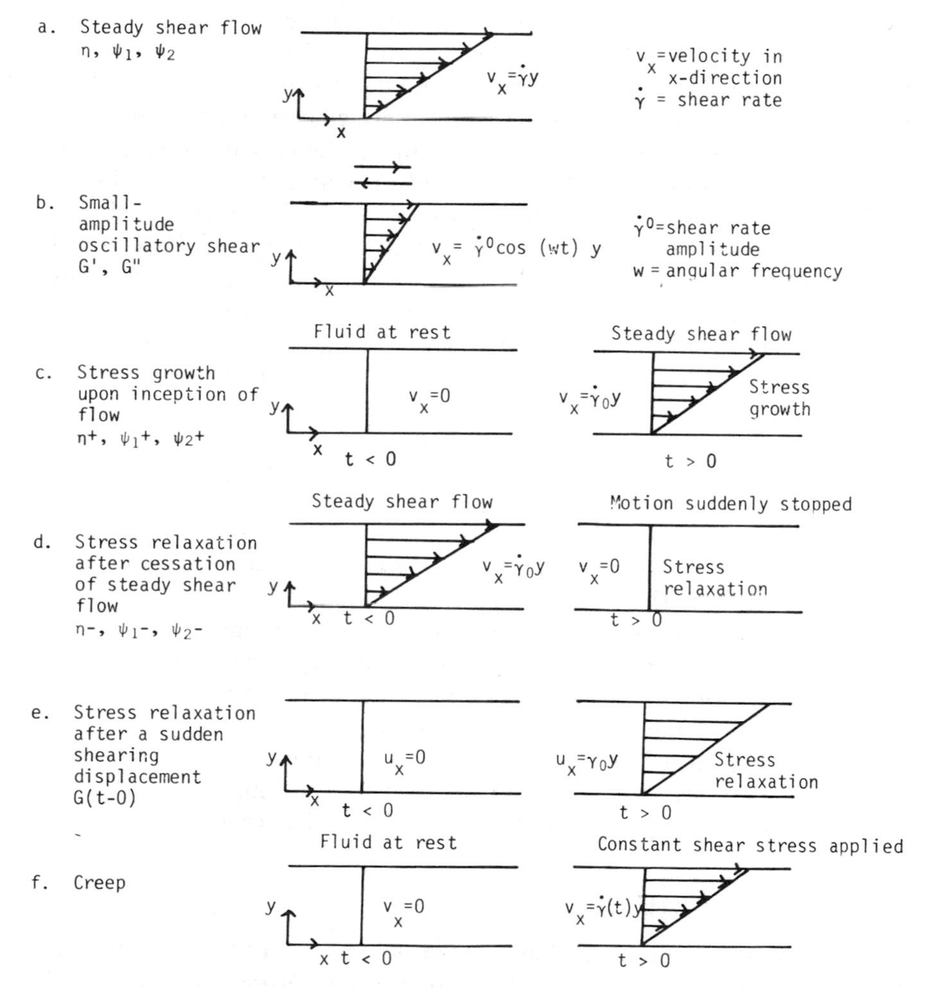

Fig. 12.2. Types of unidirectional shear flow experiments commonly used in rheology and the corresponding material functions.

$$\tau_{zz} - \tau_{xx} = - \bar{\eta}(\dot{\epsilon})\,\dot{\epsilon} \tag{4}$$

where $\dot{\epsilon}$ is the extension rate. Similar to the case of shear flow we can describe the same set of transient flow experiments for elongational flow.

METHODS FOR MEASURING DOUGH RHEOLOGICAL PROPERTIES

In order to obtain the well-defined flows described above that lead to material functions, one must employ certain test geometries. The material functions defined for shear flow are most directly obtained in a cone-and-plate rheometer but can also be determined in a plate-plate rheometer (see Fig. 12.3). The cone-and-plate geometry leads to a uniform shear rate ($\dot{\gamma}$) throughout the sample. Hence from torque and normal thrust measurements one can obtain η and N_1 as a function of $\dot{\gamma}$. In the plate-plate rheometer $\dot{\gamma}$ varies with the radial position, which requires extra calculations to obtain η and N_1 as a function of $\dot{\gamma}$. These test geometries can also be used to carry out the transient shear experiments although the cone-and-plate is the preferred geometry. η can be obtained at higher shear rates using a capillary rheometer rather than a rotary rheometer. However, the shear rate also varies with radial position and is a function of the viscosity. The wall shear rate can be obtained using a procedure to correct for the nonparabolic velocity profile. Unfortunately there is no established way to obtain N_1 from a capillary rheometer at present, but methods are under investigation (Boger and Denn 1980). Two companies, Rheometrics, Inc. and Sangamo Ltd. presently manufacture and sell rotary rheometers suited to carry out the various shear flow experiments.

Elongational viscosity is obtained most often by extending the end of a cylindrical specimen exponentially with time that leads to values of $\dot{\epsilon}$ independent of position in the sample or at a constant rate that requires a knowledge of the diameter profile to calculate $\dot{\epsilon}$. Methods for generating both biaxial extension and planar extension are also available. An instrument for carrying out the various unidirectional elongational flow experiments is manufactured by Rheometrics.

FOOD DOUGH RHEOLOGY

A complete review of studies of food doughs using test geometries leading to well-defined material functions is given elsewhere (Baird and Labropoulos 1982). We present here several examples of rheological measurements on food doughs to illustrate their value. The dynamic

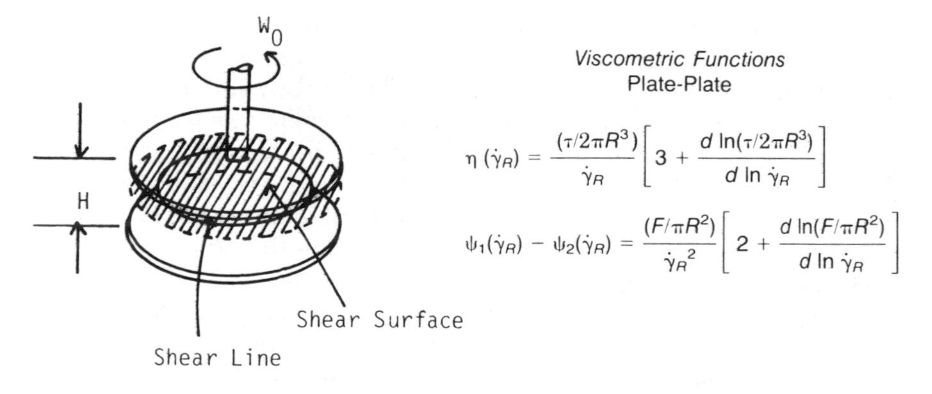

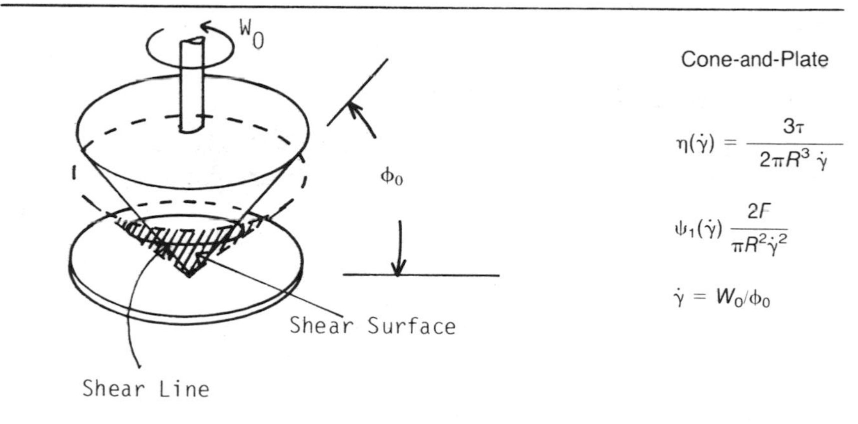

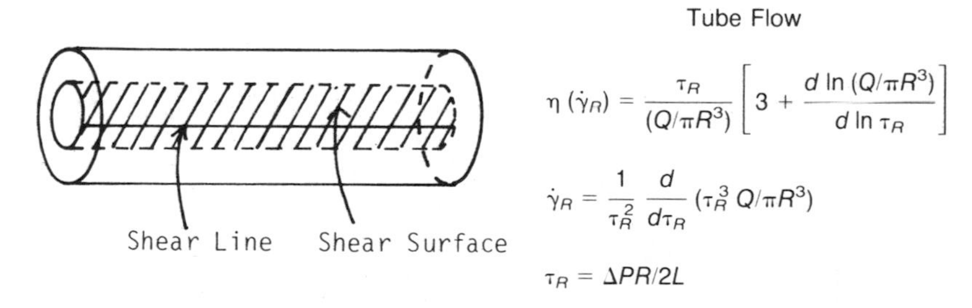

Fig. 12.3. Various types of rheometer geometries used to measure rheological properties of food doughs.

oscillatory experiment has been used most frequently to obtain information about the linear viscoelastic response of food doughs. The performance of wheat flour dough in bread making processes has been linked directly to their linear viscoelastic properties (Bloksma 1972).

Baird (1982) has measured the dynamic moduli (i.e., the loss (G'') and storage (G') moduli) during the cooking and heating of defatted soy doughs in order to follow the kinetics of cooking. Representative results from this study are presented in Fig. 12.4. The steady shear response of defatted soy doughs has also been measured to provide physical properties for the modeling of extrusion cooking processes. Representative data for a defatted soy flour dough are presented in Fig. 12.5. Here we see that η is highly shear rate dependent but that N_1 is relatively independent of $\dot{\gamma}$. Transient flow properties of some food doughs have been measured and are reviewed elsewhere (Baird and Labropoulos 1982). There is only one report of extensional flow measurements for food doughs, and this was concerned with extensional creep behavior of wheat flour doughs.

EXPERIMENTAL DIFFICULTIES

Obtaining accurate and repeatable results for food doughs from rheological experiments is a difficult task. There are many difficulties to

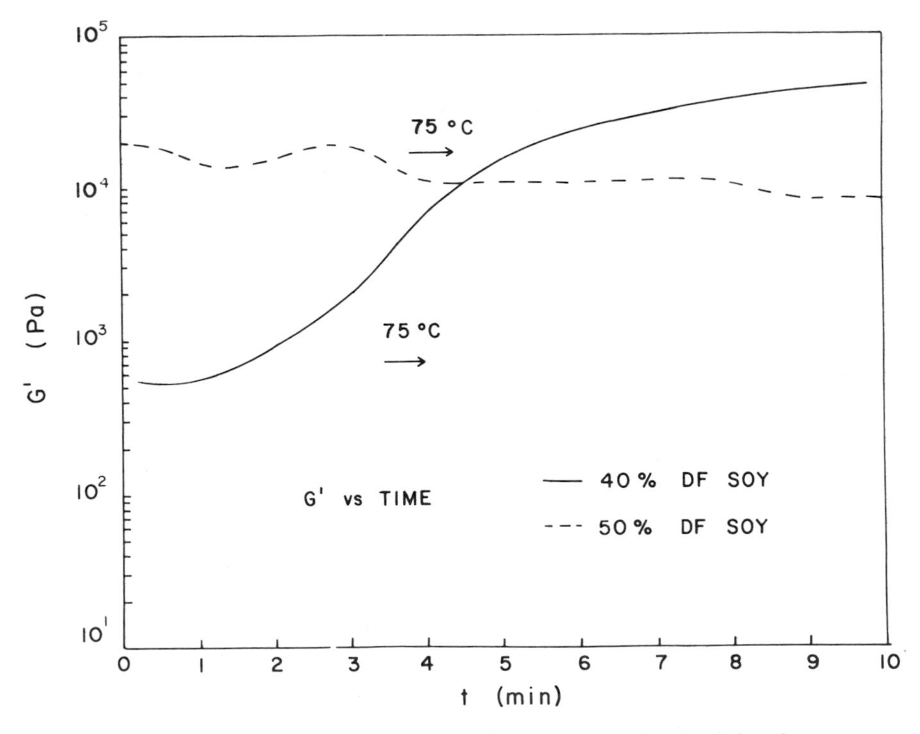

Fig. 12.4. Following of the cooking reaction of defatted soy doughs using the storage modulus.

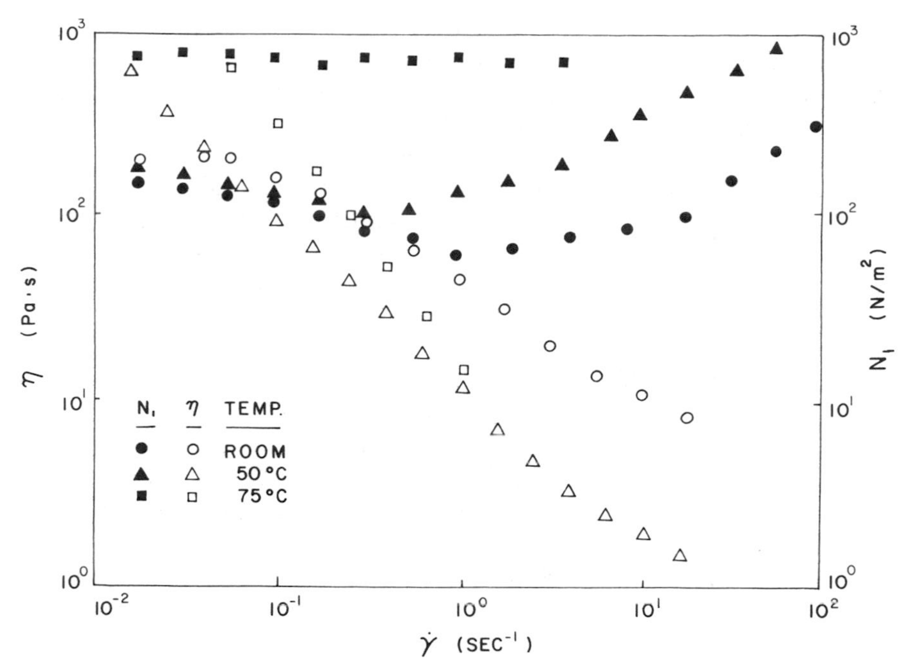

Fig. 12.5. Viscosity and normal stess difference of a defatted soy flour dough.

overcome when handling food doughs. First, the rheology of a dough can change during the process of loading the dough in a rheometer. In the cone-and-plate or parallel plate devices residual stresses are left in the sample after loading, which affect the measurements. In rotary rheometers it is possible to measure rheological properties only at low shear rates because the highly viscous doughs come out of the gap. At temperatures above 25°C the free edge of the sample tends to dry leaving a hard crust that also affects measurements. At temperatures above 100°C, moisture flashes off from the sample making it impossible to measure the rheological properties above 100°C in rotary rheometers. Other problems exist in making measurements in highly viscous materials besides those directly related to food doughs and these are discussed elsewhere (Walters 1975). Methods for overcoming some of the difficulties associated with food doughs are also reported elsewhere (Baird and Labropoulos 1982).

CONCLUDING REMARKS

This presentation is concluded by trying to answer the question: which material functions should we measure and under what circumstances do we measure them? Obviously, the answer to this question is long and complex. Here we only select several experiments to illustrate the importance of several different material functions. Viscosity (η) is one of the key parameters required to model extrusion processes and predicting die pressure drops. The measurement of N_1 is an important quantity since it is sensitive to changes in molecular structure of food doughs and may be useful in predicting extrudate swell in some extrusion processes. The linear viscoelastic properties can be used to provide information about food dough structure and to monitor any reactions in doughs during processing. The stress growth functions can be useful in situations where the processing time is very short. The viscosity of a dough during short processing times could be several times higher than the steady state value. Likewise, the stress growth functions are sensitive to dough structure and could provide qualitative information about these changes that correlate well with the quality of the dough. The true value of each of these material functions cannot be realized until further studies are carried out and empirical correlations are developed relating dough rheology to processing performance of food doughs.

BIBLIOGRAPHY

BAIRD, D.G. 1982. The effect of heat and shear on a cooking defatted soy flour dough. J. Food Process. Engr., to be published.

BAIRD, D.G., and LABROPOULOS, A.E. 1982. A review of food dough rheology. Chem. Engr. Commun., to be published.

BLOKSMA, A.H. 1972. Rheology of wheat flour doughs. J. Texture Stud. *3*, 3.

BOGER, D.V., and DENN, M.M. 1980. Capillary and slit methods of normal stress measurement. J. Non-Newtonian Fluid Mech. *6*, 163–185.

SHUEY, W.C. 1975. Practical instruments for rheological measurements on wheat products. Cereal Chem. *52*(3), 42r.

WALTERS, K. 1975. Rheometry. Chapman and Hall, London.

13

Structural Failure in Solid Foods

D.D. Hamann[1]

INTRODUCTION

The casual but attentive reader in the field of food science, seeing a title involving structural failure in solid foods, may wonder what is meant and why is it important. In this chapter, structural failure means breaking the structural bond between some portion of food and a neighboring portion on the macroscopic level. Knowing stress and strain conditions causing such breaking are important because they relate to sensory texture. A relevant discussion of structural failure in foods and food texture is given by Jowitt (1977).

If we are to truly engineer foods we should be able to discuss the deformation, fracture, and flow of food materials in terms of the fundamental units of physics. This is food rheology in its classic sense and will provide basic information on which to build a food concept. Specifications for formed foods are an example of an application. Using basic units we are not tied to a specific test geometry thus allowing comparison of data obtained using different machines and test specimen shapes. Although it is likely that empirical machines currently in use will remain the best choice for quality control, they do not give the fundamental parameters of physics necessary to understand the basics of food texture.

Texture of processed foods is usually much different than that for raw or unprocessed forms. Sometimes it is desired to retain as much of the raw texture as possible, while in other cases texture is superior in the processed form. Once optimum structural failure parameters are known,

[1] Food Science Department, North Carolina State University, Raleigh, NC 27650.

process variables can be varied to obtain the best products within imposed constraints.

This chapter is not an exhaustive review but presents some of the basics plus some current information on structural failure of protein gel systems, raw fruits and vegetables, butter and margarine, and a few comments on cheese. The foods discussed are fairly homogeneous and isotropic and thus are a good starting point in quantifying and understanding conditions at failure.

SOME TEST METHODS PRODUCING RESULTS IN UNITS OF BASIC PHYSICS[2]

Uniaxial (axial) compression or tension of specimens of known size and shape are the most common tests used that yield stresses and strains at failure. If the specimen is in the form of a cylinder, a rectangular parallelopiped, or other shape of uniform stressed cross section the computations are simple with the maximum axial (normal) compressive stresses and strains being

$$\sigma_{max} = F/A \tag{1}$$

$$\varepsilon_{max} = (\Delta L)/L \tag{2}$$

At levels for failure to occur, however, the change in length can be large as can the change in cross-sectional area. Continuously modifying the L in the denominator of Eq. (2) for each change in length, ΔL, is used to calculate a strain called "true strain." Peleg (1977B) has a good discussion of this. Appropriate equations for true stress and true strain are given in Table 13.1 for axial compression or tension of a cylinder.

Stresses and strains are often separated into those called "dilatational" and those called "deviatoric." Those called dilatational cause volume changes and those called deviatoric change shapes but not volumes. This becomes important in studying material failure because some foods, particularly those that are incompressible, may not be sensitive to volume-changing stresses but be very sensitive to shape-changing stresses. To illustrate, if we subject nearly incompressible materials like butter or gelled egg white to hydrostatic pressure there will be no change in shape, and they tend to be unaffected. If, however, they are deformed in shape they eventually break. Shear stresses are deviatoric and if we have a condition called "pure shear" the overall effect is a change in shape with negligible change in volume. Torsion of a

---

[2] See list of symbols at end of chapter.

specimen having circular cross sections produces pure shear (Table 13.1), and the sandwich shear shown at the bottom of Table 13.1 approximates pure shear if t is small compared to the dimensions of the surfaces where F is applied.

The above concepts can often be best understood by graphical representations called "Mohr's" circles. Consider a cylinder in uniaxial compression. If compressive stress is considered negative and tensile positive a plot of τ versus σ will appear as shown in Fig. 13.1. The distance b represents the axial stress. The diametral stress is zero. The maximum (absolute value) positive and negative shear stresses equal one-half the axial stress. If the strain shortening the cylinder is considered negative and the increase in diameter positive we obtain the $\gamma/2$ versus ε plot shown in Fig. 13.1. The distance c is the axial strain and d the strain increasing the diameter. Note that $\nu=d/c$ (Poisson's ratio). The directions on the graph and on the specimen are also important. Starting at σ_z or ε_z on the circles and going around the circles the radian measure represents one-half of the radian measure on the specimen. Thus, the maximum (absolute value) shear stresses and strains are $\pi/4$ radians (45°) from the z axis on the specimens.

The pure shear state is shown in Fig. 13.2 along with one test (torsion) for producing it. An enlarged surface element on the twisted shaft is also shown. Note that one dashed line represents a plane on which compressive stress acts and the other line represents a plane on which tensile stress acts. We note symmetry of the Mohr circles around the origin so that $\pm\ \tau_{max} = \pm\ \sigma_{max}$ and $\pm\ (\gamma_{max}/2) = \pm\ \varepsilon_{max}$. It is important to remember that maximum shear stresses and maximum normal stresses are equal in pure shear.

In order to generalize the stress and strain conditions in food specimens, hydrostatic pressure is sometimes superimposed on uniaxial compression (e.g., Miles and Rehkugler 1973). This shifts the circles in Fig. 13.1 to the left but does not change the τ_{max} or $\gamma_{max}/2$ values. Triaxial stress states have been used that do vary the magnitudes of τ_{max} as well as σ_{max} (Segerlind and Dal Fabbro 1978).

Other test geometrics (Mohsenin 1977) can be used that yield results in stress and strain units. The requirement is that it must be possible to mathematically model the stress and strain in the specimen.

FAILURE PROPERTIES OF SELECTED FOODS

Protein Gels

The rheology of various gel systems has been studied by a number of researchers. The excellent review by Mitchell (1980) discusses several

TABLE 13.1. EQUATIONS FOR CALCULATION OF STRESSES AND STRAINS AT STRUCTURAL FAILURE

Test	τ_{max}	γ_{max}	σ_{max}	$\sigma_{true\ max}$	ϵ_{max}	$\epsilon_{true\ max}$
Uniaxial compression of a cylinder	$\pm \frac{\sigma^a_{max}}{2}$	$\pm \epsilon^a_{max}(1 + \nu)$	$\frac{F}{\pi R^2}$	$\frac{F}{\pi R^2 (1 + \nu\epsilon)^2}$	$\frac{\Delta L}{L}$	$-\ln(1 - \epsilon_{max})$
Torsion						
M vs ψ linear to failure	$\frac{2 K^b M}{\pi r^3_{min}}$	$\frac{2K\psi}{\pi r^3_{min} Q^c}$	τ_{max}	τ_{max}	$\frac{\gamma_{max}}{2}$	$\frac{1}{2}\ln\left[1 + \frac{\gamma^2_{max}}{2} + \gamma_{max}\left(1 + \frac{\gamma^2_{max}}{4}\right)^{1/2}\right]$
M vs ψ in the form $M = b_1\psi + b_2\psi^2$	$\frac{K(4b_1\psi + 5b_2\psi^2)}{2\pi r^3_{min}}$	$\frac{2K\psi}{\pi r^3_{min} Q}$	τ_{max}	τ_{max}	$\frac{\gamma_{max}}{2}$	
Slope of M vs ψ is zero at failure	$\frac{3KM}{2\pi r^3_{min}}$	$\frac{2K\psi}{\pi R^3_{min} Q}$	τ_{max}	τ_{max}	$\frac{\gamma_{max}}{2}$	
Sandwich shear	$\frac{F}{A}$	$\frac{\delta}{t}$	τ_{max}	τ_{max}	$\frac{\gamma_{max}}{2}$	

[a]For true τ_{max} or γ_{max} use $\sigma_{true\ max}$ or $\epsilon_{true\ max}$.
[b]$K = 1.08$ for specimen shape described by Diehl *et al.* (1979).
[c]$Q = 8.45 \times 10^6\ m^{-3}$ for specimen shape described by Diehl *et al.* (1979).

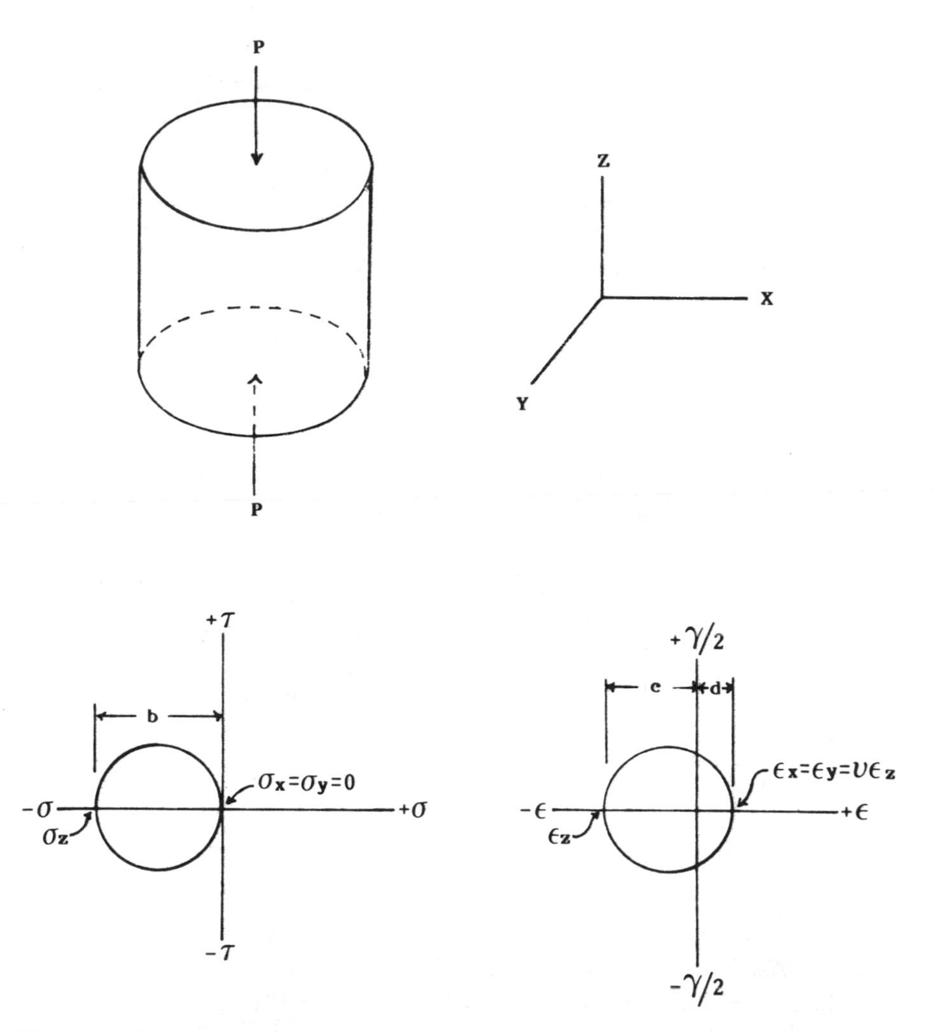

Fig. 13.1. Stress and strain Mohr's circles for axial compression.

important aspects, one of which is the rupture strength. He emphasizes that "rupture strength of a gel is not necessarily related to its apparent elastic modulus." Two factors affecting gel strength are suggested from the data. The first is the number of crosslinks in the network. This also influences the apparent elastic modulus so some correlation between gel strength and apparent modulus would be expected. The second factor

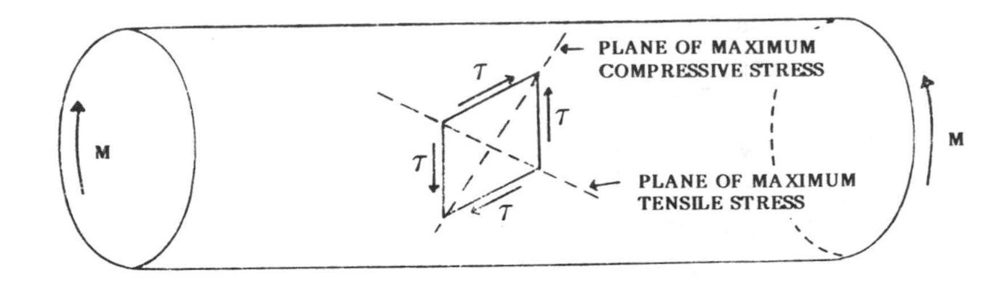

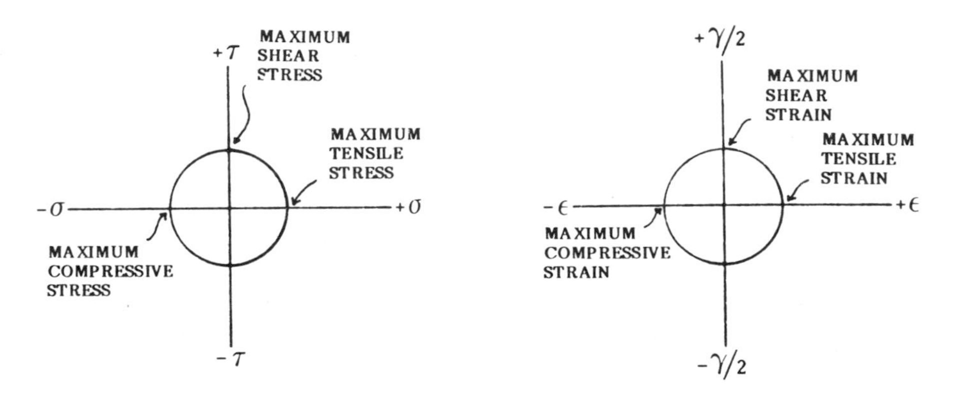

Fig. 13.2. Stress and strain Mohr's circles for torsion.

affecting gel strength seems to be the primary molecular weight (Mitchell 1976). Above a specific molecular weight the apparent elastic modulus becomes relatively constant, whereas rupture strength continues to increase with molecular weight. Mitchell (1980) cites work of Ward and Cobbett (1968) on gelatin gels, McDowell (1975) on calcium alginate gels, Kim *et al.* (1978) on low ester pectin gels, and Mitchell (1979) on calcium alginate gels as showing the molecular weight dependency. Reduction of molecular weight in fish protein gel systems by proteolytic activity greatly reduces gel strength, as shown by Cheng *et al.* (1979), Su *et al.* (1981), and Lanier *et al.* (1981A).

Heat-induced protein gels are important foods and lend themselves to structural failure analysis because they are quite homogeneous and isotropic. Two such gels will be discussed in detail; egg white gels and fish tissue gels.

Egg white proteins produce firm gels on heating and are effective texture modifiers in foods. Research is underway to further improve their functional character and stability (Ball and Winn 1982). Toda *et al.* (1978) evaluated the textural characteristics of various gel foods, including egg white gels. They found sensory "hardness" was highly correlated with breaking strength. Sensory "brittleness" and "springiness" were correlated with deformation at failure.

Fish gel products have long been of interest in Japan where *kamaboko* and similar products are food staples (Tanikawa 1971). These products are heat-induced gels composed primarily of fish muscle with some fat and most water-soluble proteins removed. Such tissue has excellent binding ability and is more bland in flavor than unwashed tissue. Sensory texture profile evaluation of fish gels tends to correlate much better with destructive mechanical tests than small deformation tests yielding rigidity ratings (Hamann and Webb 1979).

Textural properties of gels made from washed fish tissue are very dependent on species (Armstrong 1981), postcatch handling and eviscerating procedures (Su *et al.* 1981), and thermal processing temperatures (Lanier *et al.* 1981B). Improper procedures increase alkaline protease activity which is very detrimental (Lanier *et al.* 1981A). An important property of fish gels is low temperature setting (termed "*suwari*" in Japan) at about 20°C to 50°C. During *suwari* a strong binding takes place, some of which is retained during processing at higher temperatures resulting in a product that has higher strength compared to that processed only at the higher temperature. Figure 13.3 shows compression force values (hardness) from compressing 12 mm diameter cylinders diametrically to 75% deformation, which usually produces failure. Note that unwashed tissue gels are weakest and that a washed tissue gel held at 40°C for 1 hr followed by 80°C for the specified times was strongest.

Tests Yielding Fundamental Failure Parameters. Axial compression and torsion tests (torsion specimen shape shown in Fig. 13.4) have been used by Montejano *et al.* (1983A, B) to evaluate both egg white and fish gels. The equations of Table 13.1 for uniaxial compression and for torsion are applicable. Before axial compression and torsion results could be compared values for Poisson's ratio (ν) had to be known, so they could be used in the equations for axial compression. Cylindrical specimens with diameters from 9.8 to 18.6 mm and lengths of 16.7 and 19.7 mm were stressed axially and diameter changes monitored with a measuring microscope. At least 20 specimens of each size were tested. It was found that all specimen sizes gave the same values for ν. Results

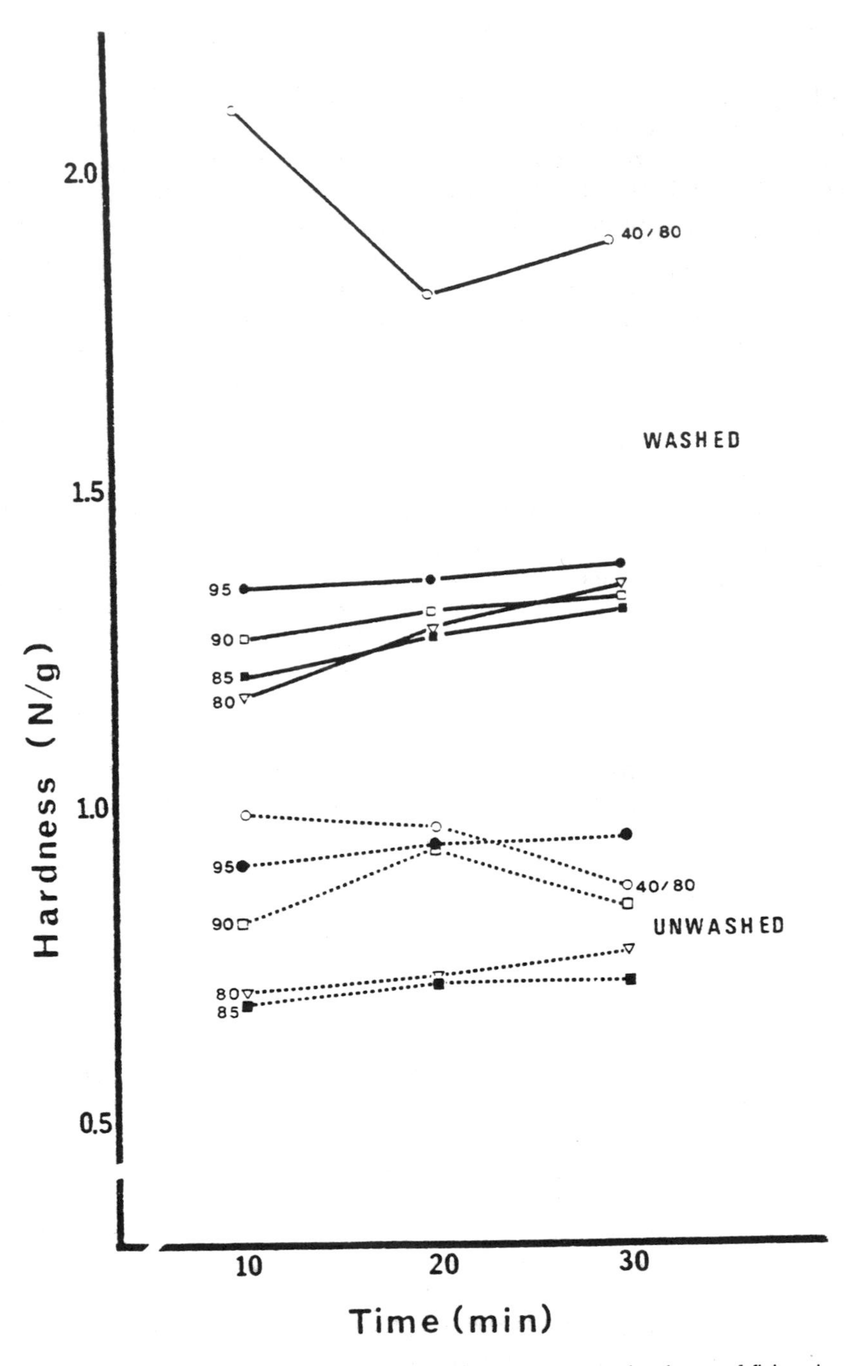

Fig. 13.3. Influence of processing time and temperature on hardness of fish gels prepared from washed and unwashed minced croaker. 40/80 was preprocessed at 40°C for 1 hr prior to processing at 80°C for the specified time.
From Lanier et al. (1981B).

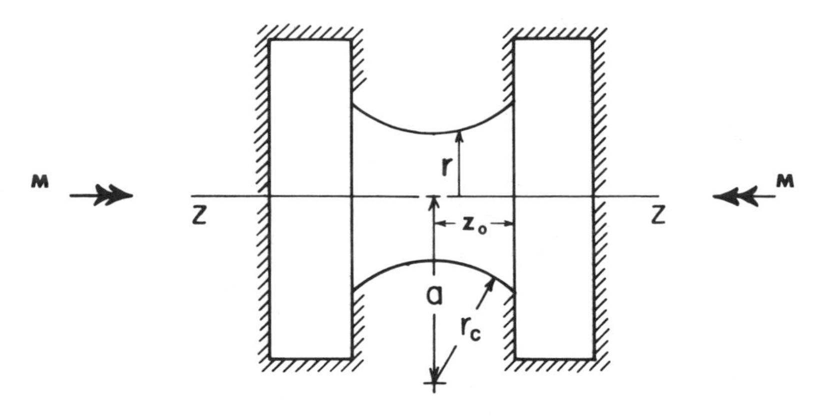

Fig. 13.4. Critical dimensions of the torsion specimen.
From Diehl et al. (1979).

showed fish gels to be nearly incompressible, having a Poisson's ratio of 0.48. Similarly, Poisson's ratio for egg white gels was found to be 0.49.

Identical fish gels and identical egg white gels were tested in both axial compression and torsion to determine if results were comparable. Failure of both egg white and fish gels in torsion involved both shear and tension based on observations of failure surface orientations. Often failure would start along a maximum shear or tension surface and then shift to the other. In axial compression failure was at approximately an angle of $\pi/4$ radians from the cylinder axis in the compressed form indicating a shear failure. Table 13.2 shows results for the fish gels and Table 13.3 shows results for a native egg white gel and an egg white acetylated by adding 25 moles acetic anhydride per 50,000 g of egg white protein by the method of Ball and Winn (1982). There were no significant differences between the torsion and axial compression in the parameters listed, except in the case of fish gel C. Values of shear stress at failure and shear modulus from axial compression were higher in this case for two likely reasons. (1) These specimens were used full diameter after cooking in tubes and a skin like formation next to the tube was present. (2) The diameter was larger than the length, which tends to elevate the compression stress required for failure (Peleg and Brito 1977).

Inspection of Table 13.2 and 13.3 reveals that failure shear stresses and strains for all the gels, both fish and egg, are all approximately the same. It is important to note, however, that if strain results were given in terms of the true maximum normal strain the results from torsion and axial compression would not agree. Considering the native egg

TABLE 13.2. MEAN VALUES OF MECHANICAL FAILURE PARAMETERS FROM TORSION AND COMPRESSION TESTS[1]

	True shear stress		True shear strain		Shear modulus	
Material	Torsion (kPa)	Comp. (kPa)	Torsion	Comp.	Torsion (kPa)	Comp. (kPa)
Fish gel A (L = 1.67 cm, D = 0.98 cm)[2]	12.74[a] (2.30)	14.21[a] (1.70)	1.088[a] (0.09)	1.095[a] (0.07)	12.96[a] (2.30)	12.97[a] (1.14)
Fish gel B (L = 1.97 cm, D = 0.98 cm)	13.47[a] (2.03)	14.16[a] (1.35)	1.096[a] (0.09)	1.024[a] (0.05)	13.09[a] (2.10)	13.81[a] (1.03)
Fish gel C (L = 1.67 cm, D = 1.86 cm)	12.22[a] (2.43)	20.00[b] (2.38)	1.073[a] (0.07)	1.024[a] (0.09)	13.41[a] (2.03)	14.51[b] (3.18)

Source: From Montejano *et al.* (1983B)

[1] Values given in parenthesis represent standard deviations. The same letter superscripts indicate that the associated numbers in the two columns are not significantly different ($P > .05$).

[2] Values given in parenthesis represent the dimension of the cylindrical specimens used in compression tests. All torsion specimens had a dumbbell shape with circular cross sections with minimum diameter = 1.0 cm.

TABLE 13.3. MEAN VALUES OF MECHANICAL FAILURE PARAMETERS FROM TORSION AND AXIAL COMPRESSION TESTS[1]

Material	True shear stress		True shear strain		Shear modulus	
	Torsion (kPa)	Comp. (kPa)	Torsion	Comp.	Torsion (kPa)	Comp. (kPa)
Native egg white	12.96 (2.90)	13.91 (2.83)	1.090 (0.132)	1.159 (0.134)	10.57 (1.76)	11.53 (2.02)
Acetylated egg white	11.54 (2.41)	12.27 (2.01)	1.048 (0.112)	1.113 (0.045)	12.53 (2.32)	12.61 (1.17)

Source: From Montejano *et al.* (1983A)
[1] Values given in parenthesis represent standard deviations. There were no significant differences in the parameters listed between torsion and axial compression results ($P > .05$).

white, for axial compression $\varepsilon_{true\ max} = 1.159/1.5 = 0.77$ and from torsion a close approximate value is $\varepsilon_{true\ max} = 1.1090/2 = 0.55$. This suggests that these gels fail at the same shear strain (γ). This observation concerning strains is especially important because it will be shown later that for fruits and vegetables having turgor pressure (fluid pressure within individual cells) the observation seems reversed (i.e., true normal strains agree but shear strains do not).

One of the problems when testing gels in axial compression is that required strains can be so large as to make material failure impossible. Testing in tension or torsion then become the logical choices. Table 13.4 gives results for native egg white gel and several modified egg white gels. Modification of the second and fifth gels was by addition of 15 moles and 25 moles of succinic anhydride per 50,000 g of egg white protein, respectively. The other two modifications were by adding appropriate reagents at the 25 mole level. The succinylated and oleic acid-treated egg whites were very rubbery and required true shear strains about 2.5 times native egg white. The level of succinic anhydride added affected the shear stress but not the shear strain. Quantitative information such as illustrated by these gel systems should prove valuable for similar foods.

TABLE 13.4. COMPARISON OF MEAN VALUES OF TORSIONAL FAILURE PARAMETERS FOR GEL MATERIALS[1]

Material	Shear stress (kPa)	True shear strain	Shear modulus (kPa)
Native egg white	12.96[a]	1.090[a]	10.07[b]
Succinylated egg white (15m)	14.88[a]	2.488[b]	5.72[a]
Acetylated egg white	11.54[a]	1.048[a]	12.53[c]
Oleic Acid treated egg white	19.75[b]	2.784[b]	5.62[a]
Succinylated egg white (25m)	19.36[b]	2.548[b]	6.01[a]

Source: From Montejano *et al.* (1983A)
[1] Means within a column with different superscripts are significantly different ($P > .05$).

Some Tentative Conclusions on Selected Protein Gels. A specific fish tissue or egg white gel will tend to fail at a specific true shear strain independent of test mode (torsion or axial compression). These levels are approximately 100% true shear strain for heat-coagulated fish tissue (*surimi*) gels or native egg white gels. Modifying egg white with succinic anhydride or oleic acid increases strain to failure by a factor of about 2.5. Although strain rate effects were not considered in the protein gel work cited, it has been shown by Kamel and de Man (1977) in penetration studies of gelatin gel that depth of penetration at rupture did not change as crosshead speeds were varied from 0.04 to 0.12 cm/sec. Such results suggest that a decade or two of change in strain rate may not significantly influence failure strain in gel systems.

It remains to be seen if the shear strain level failure criteria have general application to gel systems. From the limited work cited it looks rather promising.

Raw Fruits and Vegetables

Fruits and vegetables as thought of in the food industry are those edible fleshy foods of plant origin (Bourne 1976). They are composed primarily of parenchyma cells, which are relatively simple in structure when compared to other plant cells. An isodiametric polyhedral form is a good approximation (Sterling 1963). Intercellular space varies considerably, being approximately 25% for some apples and only about 1% for the Irish potato (Sterling 1963). Volume compressibility, which in turn affects some aspects of texture, depends to a large part on the intercellular space. Parenchyma cells generally are less than 500 μm in diameter and tend to increase in size as the distance from the skin increases. Tissue consisting of small cells with little intercellular space tends to have a firm, dense texture while tissue with large cells and larger intercellular spaces have a more spongy feel (Reeve 1970).

Cell walls are important structures affecting texture. The outside wall is made up of cellulose microfibrils woven together and embedded in a pectic and hemicellulose matrix. Inside this wall is a secondary wall in which the microfibrils are embedded in a matrix of hemicellulose and lignin (Bourne 1976). Living cell membranes allow small molecules such as water to pass in and out of the cell but restrict large molecules. Resulting turgor pressure changes influence the texture of raw fruits and vegetables.

Between the cells is the middle lamella composed of pectic materials that glue the cells together. The strength of bonding determines whether cells separate when mechanically stressed or fracture passes through cells.

The models of Nilsson *et al.* (1958) and Holt and Schoorl (1976) qualitatively picture cells as approximately spherical liquid-filled membranes. Applied forces distort cells in critical regions resulting in cell wall distension. Visual observations show that failure is usually in the form of regional cell bursting (Holt and Schoorl 1976) or failure planes passing through cells (Huff 1967; Diehl *et al* 1979). Separation between planes of cells has not been evident in the cited work but may occur if pectic cementing materials degrade. Many thermally processed fruits and vegetables normally do break down by cell separation. If cooked potato cells rupture releasing starch, a sticky texture results (Reeve 1977).

Irish Potato Flesh. De Baerdemaeker *et al.* (1978) tested Irish potato tissue (Bintje cultivar stored 6 months) and apple tissue (Chieftain and Granny Smith cultivars) equilibrated to specific osmotic potentials by soaking in manitol solutions. Uniaxial compression of short cylindrical specimens and uniaxial tension of ring shaped specimens was performed. A displacement rate of 1 mm/sec was used for both tests.

Figure 13.5 and Table 13.5 show axial compression strength decreases as water potential increases (increased turgidity) while the

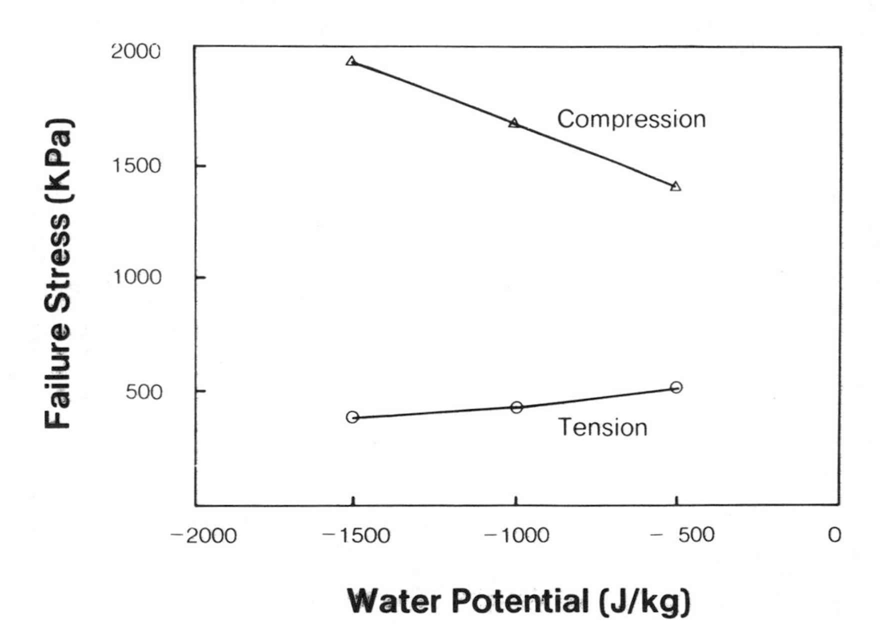

Fig. 13.5. Compressive and tensile failure stress of potato tissue as affected by water potential.
From De Baerdemaeker et al. (1978).

TABLE 13.5. SELECTED LITERATURE RESULTS FROM SUBJECTING SPECIMENS OF RAW IRISH POTATO FLESH TO CONDITIONS CAUSING STRUCTURAL FAILURE

Potato description and location of specimen if known	Test mode	Mode of failure	True shear strain	True normal strain	Approx. shear rate (sec^{-1})	True shear stress (kPa)	Reference
Kennebec							
Outside	Tension[1]	Tension	0.26	0.17	variable	204	Huff (1967)
Mid	Tension	Tension	0.53	0.35	variable	251	
Center	Tension	Tension	0.21	0.14	variable	490	
Red skinned							
Center	Torsion[2]	Tension	0.68	0.34	0.26	530	Diehl *et al.* (1979)
	Axial comp.[3]	Shear	0.54	0.36	0.29	454	
Eastern white							
Center	Torsion	Tension	0.68	0.34	0.26	560	
	Axial comp.	Shear	0.54	0.36	0.29	468	
Russet							
Center	Torsion	Tension	0.72	0.36	0.26	620	
	Axial comp.	Shear	0.54	0.36	0.29	476	
Russet (flaccid)							
Center	Torsion	Tension	0.92	0.46	0.26	870	
	Axial comp.	Shear	0.51	0.34	0.29	507	
Idaho baking	Axial comp.		0.84	0.56	150[5]	522	Dal Fabbro *et al.* (1980)
	Axial comp.		0.84	0.56	550	450	
	Axial comp.		0.84	0.56	1000	430	
		Water Potential (J/kg)					
Bintje stored 6 months	Axial comp.	WP = –500			0.15	526	DeBaerdemaeker *et al.* (1978)
	Tension[4]	WP = –500			0.15	293	
	Axial comp.	WP = –1000			0.15	634	
	Tension	WP = –1000			0.15	160	
	Axial comp.	WP = –1500			0.15	735	
	Tension	WP = –1500			0.15	146	

[1] These tension specimens were dumbbell shaped with rectangular cross sections.
[2] These torsion specimens were dumbbell shaped with circular cross sections.
[3] Axial compression specimens were cylindrical in shape.
[4] These specimens were in the form of hollow cylinders loaded radially.
[5] Stress rate in kPa/min.

reverse is true for uniaxial tensile strength. This is also true for apples (Table 13.6). Note in Tables 13.5 and 13.6 that stresses given are shear stresses which, in the case of the uniaxial tests, can be multiplied by two to obtain the maximum compression or tensile stress magnitude. In torsion the maximum compression stress magnitude is the same as the shear stress.

Diehl *et al* (1979) tested Irish potatoes in torsion and uniaxial compression (Table 13.5). The two groups of Russet potatoes are identical, except that one was stored at room temperature and low humidity becoming noticeably flaccid. This desiccated flesh required a higher stress for failure, confirming the results of De Baerdemaeker *et al.* (1978). It seems then that higher turgor pressure produces lower uniaxial compression failure stress levels.

Let us now consider strains. In Table 13.5 the true normal strains are quite uniform within specific groups of potatoes tested in axial compression and torsion. Differences when tension specimens are taken from different locations in the potato are given by Huff (1967) and between axial compression and torsion results for flaccid potatoes by Diehl *et al* 1979). In this later work axial compression strain at failure was not significantly different comparing the flaccid desiccated potatoes with those of normal flesh. The question is why? Results of the torsion testing suggest an answer.

The torsion test produces the stress condition of pure shear in which for small strains the net effect of the strains is deviatoric (change in shape) but not dilatational (change in volume). The 30–40% strains given in Table 13.5 are certainly not small, but in torsion for twists producing shear strains up to 80% (normal strains up to 40%) calculated strains are within 3% of true strain (Nadai 1937). This constant volume characteristic of the torsion test seems a likely reason that low turgor potato flesh (flaccid Russets of Table 13.5) was strained in torsion about 1 1/3 times as far as the higher turgor potato flesh before failure. In uniaxial compression volume reduction would increase turgor and make the low turgor flesh behave as if it had higher initial turgor.

Comparison of true failure shear stress values in torsion with those for uniaxial compression in the results of Diehl *et al.* (1979) (Table 13.5) adds further insight. Information in the reference showed that the red skinned and eastern white potatoes had high turgor with little compressibility as indicated by *E*/*G* ratios approaching the theoretical value of three for incompressible homogeneous isotropic materials. In this case, stress at failure in axial compression was 85% of that for torsion. The typical Russets had a lower *E*/*G* ratio, indicating more compressibility and probably less turgor. Here failure stress in axial compression was

TABLE 13.6. SELECTED LITERATURE RESULTS FROM SUBJECTING SPECIMENS OF RAW APPLE FLESH TO CONDITIONS CAUSING STRUCTURAL FAILURE

Apple description	Test information		True normal strain	True shear stress (kPa)	Approx. shear rate (sec^{-1})	Reference
Northern Spy held at 0°C 130 days	Bx comp.[1], very		0.07	120	0.006	Miles and Rehkugler (1973)
	small effect		0.07	135	0.07	
	from hydrostatic		0.08	150	0.7	
	pressure, failure		0.09	325	7.0	
	along shear plane		0.10	430	110	
Chieftain, water potential (WP) adjusted	Ax comp., WP = −1000 J/kg			226	0.15	De Baerdemaeker *et al.* (1978)
	Tension[2], WP = −1000 J/kg			113	0.15	
	Ax comp., WP = −1500 J/kg			283	0.15	
	Tension, WP = −1500 J/kg			108	0.15	
	Ax comp., WP = −2000 J/kg			269	0.15	
	Tension, WP = −2000 J/kg			85	0.15	
Granny Smith, water potential (WP) adjusted	Ax comp., WP = −1000 J/kg			123	0.15	De Baerdemaeker *et al.* (1978)
	Ax comp., WP = −1500 J/kg			156	0.15	
	Ax comp., WP = −2000 J/kg			167	0.15	
	Tension, WP = −1300 J/kg			56	0.15	
	Tension, WP = −2000 J/kg			37	0.15	
Red Delicious	Tx. comp.					Segerlind and Dal Fabbro (1978)
	Rad. pressure (kPa)	Ax. press. (kPa)				
	0	320	0.11	160	variable	
	70	300	0.14	120	variable	
	140	260	0.13	60	variable	
	210	260	0.13	25	variable	
	280	220	0.13	30	variable	
	340	200	0.11	70	variable	
McIntosh						
	0	240	0.13	120	variable	
	70	230	0.14	80	variable	
	140	220	0.16	40	variable	
	210	220	0.11	5	variable	
	280	180	0.10	50	variable	

Jonathan	0	270	0.15	135	variable	
	70	260	0.14	95	variable	
	140	250	0.15	55	variable	
	210	220	0.13	5	variable	
	280	190	0.11	45	variable	
Red Delicious, ripe held at 5°C 7 weeks	Torsion[3], shear failure		0.10	127	0.26	Diehl *et al.* (1979)
	Ax. comp., regional failure		0.10	91	0.29	
Red Delicious, held at room temp. 7 weeks	Torsion, shear failure		0.11	99	0.26	
	Ax. comp., regional failure		0.11	83	0.29	
Red Delicious	Ax. comp.		0.13	182	35[4]	Dal Fabbro *et al.* (1980)
	Ax. comp.		0.13	165	100	
	Ax. comp.		0.13	150	150	
	Ax. comp.		0.13	134	200	
	Ax. comp.		0.13	107	300	
	Ax. comp.		0.13	85	500	
	Ax. comp.		0.13	65	800	
Jonathan, picked 9-29-78 stored at 2°C	Ax. comp. (10-1-78)		0.15	205	0.17	Shahabasi and Segerlind (1981)
	Ax. comp. (11-15-78)		0.12	118	0.17	
	Ax. comp. (12-31-78)		0.13	110	0.17	

[1] Biaxial (Bx), axial (Ax), and triaxial (Tx) specimens were all cylinders.
[2] Tension specimens were in the form of a hollow cylinder.
[3] Torsion specimens were dumbbell shaped with circular cross sections.
[4] Stress rates in kPa/min.

77% of that in torsion. The flaccid Russets had such low turgor that stress/strain ratios were low in magnitude and were not linear. The E/G values were not calculated. Stress required for failure in axial compression was only 58% of that in torsion.

Since both stress and strain of the flaccid potatoes were nearly one-third greater at torsion failure than the other groups, it can be concluded that low turgor potato flesh is more ductile and tougher than the higher turgor flesh if dilation is small. This could not be observed in uniaxial compression where volume reducing strains are large. Sensory evaluation of the same potato groups revealed a significant ($P < .01$) negative correlation between sensory moisture release from the flesh and shear stress at failure in torsion (Table 13.8), which is consistent with the interpretation of instrumental results.

It can be reasoned that increased toughness of low turgor potato flesh in torsion is because, instead of cell bursting, a tensile tear not due to interior fluid pressure occurs. Torsion failure of potato is along a plane of maximum tensile stress (Fig. 13.6) for both high and low turgor flesh. Cell walls in high turgor tissue could be expected to be prestressed by the internal pressure, so less additional stress and resulting strain would cause failure compared to low turgor flesh. De Baerdemaeker *et al.* (1978) found in uniaxial tension tests of potato flesh that tensile

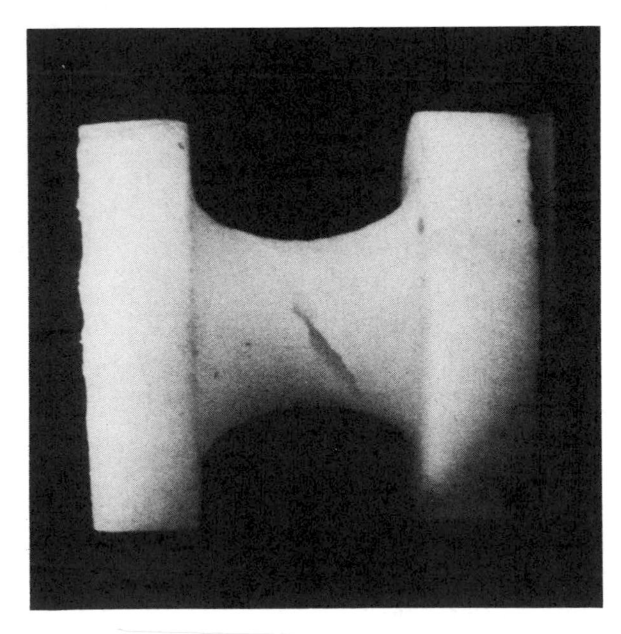

Fig. 13.6. Failure crack in potato tested in torsion. *From Diehl et al. (1979).*

stress at failure increased with higher turgor (Fig. 13.5). This seems contrary to the torsion test results of Diehl *et al.* (1979), but axial tension produces a volume increasing stress state in contrast to the constant volume state in torsion. Axial tension produces failure in a plane of maximum tensile stress passing through cells (Huff 1967). The tensile strength is much smaller than the strength in uniaxial compression. This is typical of brittle structural materials such as cast iron, and standard textbook reasoning seems applicable (e.g., Polakowski and Ripling 1966; Jowitt, 1977). Weakness in uniaxial tension occurs because stress concentrations form at voids, and these propogate forming a crack at relatively low stress levels. This is reduced where strains are not dilatory or dilatation is a volume reduction. The increase in uniaxial tensile strength with turgor shown by De Baerdemaeker *et al.* (1978) is probably because higher turgor decreases void volumes.

Potato flesh will fail in tension when subjected to uniaxial tension or torsion but fails in shear when stressed in uniaxial compression. Qualitatively, potato flesh is weakest in tension, strongest in compression with shear strength intermediate. Quantitatively, the torsion test produces a stress state in which the maximum tension, shear, and compression stresses are all equal although they act along or on different planes. The fact that potato flesh nearly always fails along a plane of maximum tensile stress tells us that it is weakest in tension. Of note, however, is the fact that values for shear stresses produced in torsion were larger than true shear stresses at failure due to uniaxial compression. This strongly suggests that shear strength without dilatation is greater than shear strength if there is a volume reduction.

Conclusions About Raw Potato Flesh. (1) It is strongest in resisting compression stresses, weakest in resisting tension stresses and intermediate in resisting shear stresses. (2) Increased turgor pressure reduces compressive strength and shear strength but increases strength in uniaxial tension. (3) Uniaxial compression of potato tissue produces lower true failure stress levels than the torsion (pure shear) test. (4) Potato flesh with typical turgor pressure subjected to a constant volume or volume reducing stress state can be expected to fail at true normal strains of about 30–60%. Within a specific group these strains are quite consistent.

Speculation. The low strength of potato flesh in uniaxial tension is because stress concentrations form at voids and cracks propogate. This is intensified by the volume increasing nature of the stress state. Higher turgor pressure decreases void volume and this is the cause of increased uniaxial tension strength.

Apple Flesh. Apple flesh differs from Irish potato flesh in that gas volume is much higher, carbohydrates of lower molecular weight are dominant, and a ripening process occurs. Changes in pectic material in the middle lamella are particularly important. In green fruit the pectic material is composed mainly of partially esterified polygalacturonic acid called protopectin, which imparts great strength to the tissue and is insoluble in water (Bourne 1976). With ripening, the chain length of the pectin polymer decreases, forming water-soluble pectin. The texture becomes softer and eventually mushy. Matz (1962) lists three factors thought to influence texture strongly. These are intercellular forces, mechanical strength of the cell wall, and cell turgidity. The ripening process and turgor pressure both affect the tissue with time, but, in contrast with potato flesh, the changes due to ripening are large.

Turgor pressure influences reported by De Baerdemaeker *et al.* (1978) are similar to those reported for potatoes with increased turgor increasing axial tensile strength and decreasing axial compression strength. Green apples with high turgor tend to fail along shear planes when subjected to uniaxial compression as do firm apples such as the Northern Spy cultivar (Miles 1971). Very ripe Red Delicious apples, on the other hand, tend to fail by cell rupture at one end of the specimen when tested in axial compression, but in torsion they fail in shear (Diehl *et al.* 1979). In the case of the ripe Red Delicious apples, calculated shear stresses at failure tended to be higher for torsion (as were potatoes) by up to 30%; and normal strains at failure were not significantly different, comparing axial compression and torsion tests performed at similar shear rates (also true for potatoes of average turgor). Observing the true normal strain values in Table 13.6 it seems that values are quite consistent within a group even when test variables are changed by large magnitudes. Note in the results of Segerlind and Dal Fabbro (1978) that as the radial pressure (stress) in a triaxial compression test was changed from 0 to 340 kPa in incremental steps, the shear failure stress changed greatly, whereas the normal strain did not. Overall, normal strains at failure seem to range from about 7–16%, which is less than half the strain at failure for Irish potatoes. Poisson's ratio has been shown to be about 0.3 for apples (Chappell and Hamann 1968), and they can be classed as a rigid solid (Van Wazer *et al.* 1963), whereas Irish potato flesh and the gels previously discussed are rubbery solids.

Other Fruit. Approximate results are given for a number of other fruit in Table 13.7. We note for the pears tested by Dal Fabbro *et al.* (1980) that true normal strain at failure is independent of stress rate but shear stress at failure tends to increase with stress rate. This again

TABLE 13.7. SELECTED LITERATURE RESULTS FROM SUBJECTING RAW FLESH SPECIMENS OF SEVERAL FRUITS TO CONDITIONS CAUSING STRUCTURAL FAILURE

Fruit description	TSS (°Brix)	Test information	True normal strain	Approx.[1] true shear stress (kPa)	Approx. shear rate (sec^{-1})	Reference
Mango	8.2	Axial comp.[2]	0.35[4]	200	0.11	Peleg *et al.* (1976)
	9.6	Note: Unripe mangoes	0.35	215	0.11	
	19.0	melon and papaya failed in shear.	0.20	10	0.11	
Melon (unspecified)	7.4	Ripe fruits	0.80	230	0.11	
	9.7	exhibited regional	0.50	50	0.11	
	10.2	failure followed by flow.	0.20	30	0.11	
Papaya	5.2		0.60	830	0.11	
	11.0		0.10	15	0.11	
Pineapple	13.0	Note: Pineapple failed	0.15	15	0.11	
	14.5	along fibers.	0.20	30	0.11	
	16.8		0.20	30	0.11	
Honey dew melons, ripe		Torsion[3], shear failure.	0.21	72	0.26	Diehl *et al* (1979)
		Axial comp. regional failure, same shear	0.23	58	0.29	
Pears		Axial comp.	0.17	52	35[5]	Dal Fabbro *et al.* (1980)
			0.17	63	100	
			0.17	71	150	
			0.17	77	200	
			0.17	83	300	
			0.17	100	500	
			0.17	86	800	
Plaintain, given in order of increasing ripeness		Axial comp.,	0.34	335	0.08	Peleg and Brito (1977)
		shear failure	0.39	230	0.08	
		Progressive	0.51	170	0.08	
		formation of	0.54	170	0.08	
		irregular failure	0.49	100	0.08	
		planes	0.43	90	0.08	
			0.49	55	0.08	

[1] Approx. true shear stress was calculated from data in the references assuming $\nu = 0.3$.
[2] Axial compression specimens were all cylinders. Data from highest *L/D* ratio are used if this is variable.
[3] Torsion specimens were dumbbell shaped with circular cross sections.
[4] These values were estimated from force-displacement graphs given in the reference by Peleg *et al.* 1976.
[5] Stress rates in kPa/min.

supports the normal strain failure criteria for specific groups of fruit. This is also supported by the results of Diehl *et al.* (1979) for torsion and axial compression testing of honeydew melon flesh. Note how some shear occurs in failure of honeydew melon flesh (Fig. 13.7).

The results given by Peleg *et al.* (1976) and Peleg and Brito (1977) show the effect of ripening for a number of tropical fruits. The juicy fruits (mango, melon, and papaya) decrease in strain to failure as they ripen. Pineapple increased slightly. Plaintain *(Musa paradisiaca)* which seems more dry as it ripens required more strain for failure when ripe. This agrees with the concept that high turgor correlates with lower strains to failure for specific fruits.

Most unripe fruits fail along shear planes, whereas ripe fruits fracture in a more complex pattern or show regional failure without distinct failure cracks.

Correlation of Fundamental Failure Parameters for Raw Fruits with Sensory Evaluations. When discussing potato flesh it was noted that Diehl and Hamann (1979) found a significant negative correlation between failure shear stress in torsion and sensory product moisture release during the first bite (Table 13.8). This was also true for honeydew melons, confirming the fact that decreased turgor tended to increase strength. Red Delicious apples, however, showed a positive correlation, suggesting that other factors are more dominant. Generally, apples that are just barely ripe are most juicy and require the most biting force. As they are stored or become overripe they decrease in structural strength as well as juiciness. The breakdown of the protopectin in the cell is a likely reason for this.

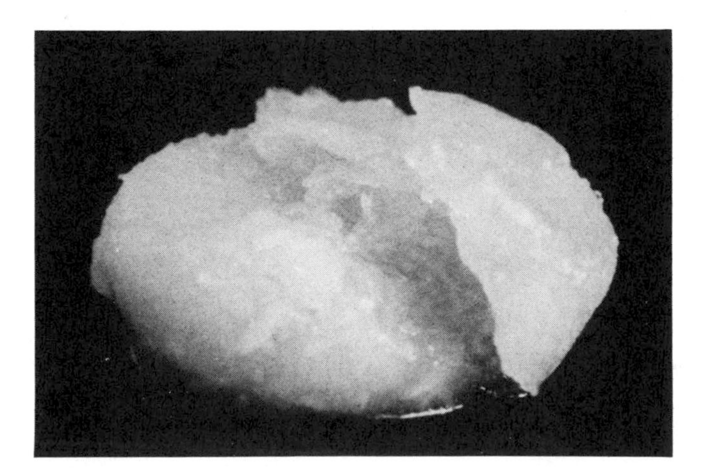

Fig. 13.7. Failed specimen of honeydew melon tested in axial compression.

It is particularly interesting in Table 13.8 that for apples the failure normal strain in torsion is negatively correlated with crispness. The apples actually failed in shear, and since shear strain is twice the normal strain during the torsion test, shear strain would show the same negative correlation with crispness as normal strain did. The above result confirms results reported by Sherman and Dehaidy (1978) who found high crispness of raw fruits and vegetables associated with small deformations to failure.

Firmness (hardness) correlated with failure stress for apples and melons but not potatoes. Sensory firmness is defined as the force to bring the teeth together (bite through). A possible reason for the lack of correlation for potatoes is that their strain to failure is larger than apples or melons, and this strain to failure becomes larger as turgor pressure decreases and failure stress increases; i.e., they deform more as they become stronger. In contrast apples fail at low strains possibly making sensory hardness more apparent to the panelist.

Butter and Margarine

Butter texture and rheology has been under study for many years. Prentice (1972) gives an excellent review of past work and interpretations. In discussing this he envisions a network of randomly oriented, needle-like crystals. Three distinct alterations of butter under stress are described. The first involves small strains and storage of potential energy. The second is rearrangement and fracture caused by a higher stress level. Finally, dissipated mechanical energy causes melting of the crystals. Effects of churning methods have been studied by Sone *et al.* (1966) and Vasic and de Man (1968). Effect of test temperature has been

TABLE 13.8. CORRELATION COEFFICIENTS RELATING FAILURE PARAMETERS AND SENSORY EVALUATIONS

	Firmness	Moisture release	Crispness
Failure shear stress in torsion			
Potatoes	NS[1]	-0.73	NS
Honey dew melons	0.73	-0.79	NS
Red Delicious Apples	0.94	0.74	NS
Failure shear stress in axial compression			
Potatoes	NS	NS	NS
Honey dew melons	0.60	-0.60	NS
Red Delicious apples	0.78	NS	NS
Failure normal strain in torsion			
Red Delicious apples	NS	NS	-0.82

Source: From Diehl and Hamann (1979).
[1] Indicates not significant ($P > .05$).

studied by a number of investigators, including Vasic and de Man (1968).

Some fundamental properties of four representative commercial butters stored at -30°C and 5°C were studied by Kawanari *et al.* (1981). These butters originated at four different plants and were made on four different churns (three continuous and one batch). Compositions were nearly identical with ranges as follows: moisture 17.7–18%, fat 80.3–81.0% and salt 1.52–1.64%. Shear stress at failure was evaluated by the sandwich shear method at a shear rate of 0.26 per second, and the equations of Table 13.1 are applicable. The actual apparatus is shown in Fig. 13.8. The butters were also evaluated by axial compression and by cone penetrometer. Shear stress was computed from the cone penetrometer data using an equation from the literature (e.g., Sherman 1970). The butters were tested at 5°, 10°, and 15°C.

Figures 13.9 and 13.10 show the mean shear strengths of the above four butters as affected by the variables. Three of the butters had very similar strengths and the other butter churned in a Cherry–Burrell continuous churn was over 2.5 times as strong at any one of the test temperatures. Because they are composed primarily of fat, butters are extremely temperature sensitive, and we note that strengths dropped by over 50% as test temperature went from 5° to 15°C. Comparing the shear, axial compression, and penetrometer results, the authors found that calculated shear stresses agreed very closely. Failure planes in both sandwich shear and axial compression were shear planes.

Shear failure stresses were compared with sensory profile panel evaluations at 10°C. The coefficient of determination value (R^2) relating to spreadability was 0.86 (negative correlation) and that relating to firmness was 0.79 (positive correlation). No other butter property (including flow properties) obtained in the above study explained sensory evaluations as well. It has been shown by Chari and Awasthy (1971) that yield stress of butter is also a better indicator of its overall texture quality than viscosity at tropical temperatures.

Spreadability of butter and margarine (conventional and soft) was studied by de Man *et al.* (1977) to establish the spreadability most desired by consumers. A sensory panel and constant speed cylindrical penetrometer were used. The maximum penetrometer force was called "hardness." Hardness correlated ($R^2 = 0.58$) with sensory evaluation of spreadability. The soft margarines and some of the conventional margarines were less hard and more spreadable than the butter, but other conventional margarines were close to the butter in these ratings. At the higher temperature (20° and 25°C) the softer margarines were judged too soft. Margarines, then, may have widely varying properties.

Fig. 13.8. Shear test fixture for butter and margarine.
From Kawanari et al. (1981).

Conventional margarine seems to fail in shear similar to butter. Evaluation of a widely marketed commercial soybean oil margarine purchased locally was done on the sandwich shear instrument described above for butter. Results are shown in Fig. 13.11. Shear failure stresses

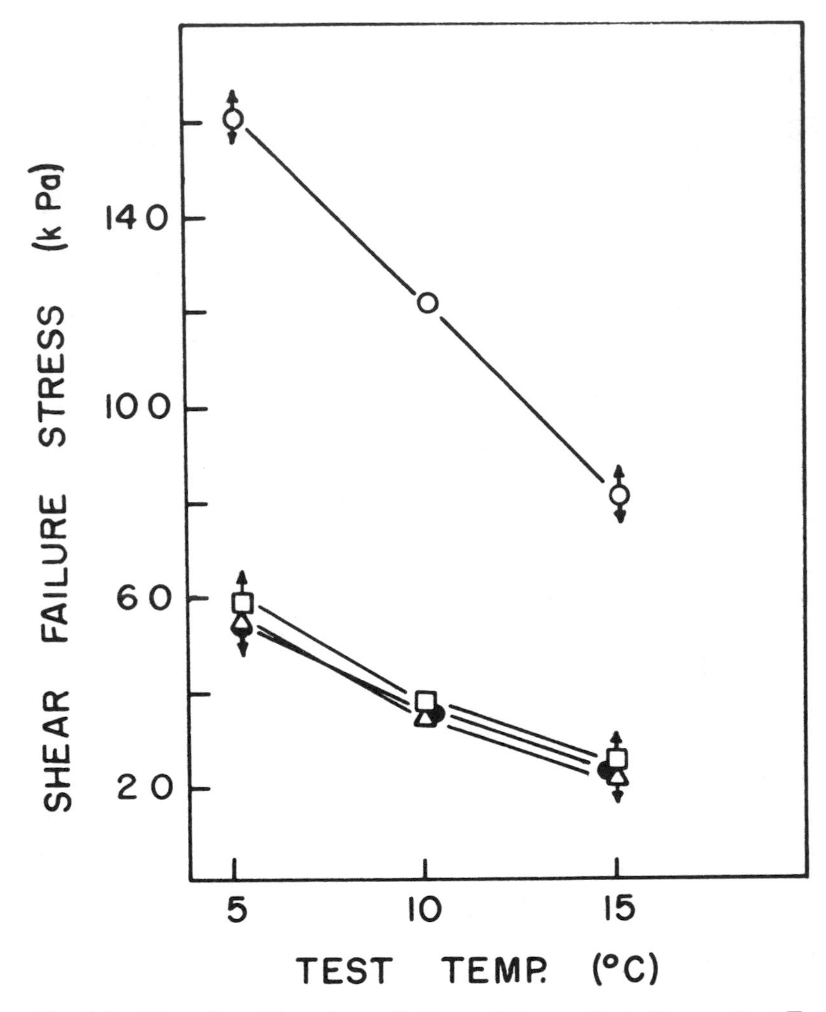

Fig. 13.9. Mean shear failure stress of butters as influenced by test temperature. □ Conventional butter; ● Westfalia; △ Contimab; ○ Cherry–Burrell.
From Kawanari et al. (1981).

were about 20, 30, and 40% less at 5°, 10°, and 15°C, respectively, than the weakest butters.

Temperature Sensitivity. Because high fat products are so temperature sensitive it may be of value to describe τ_{max} using the Arrhenius equation (e.g., Heldman and Singh 1981):

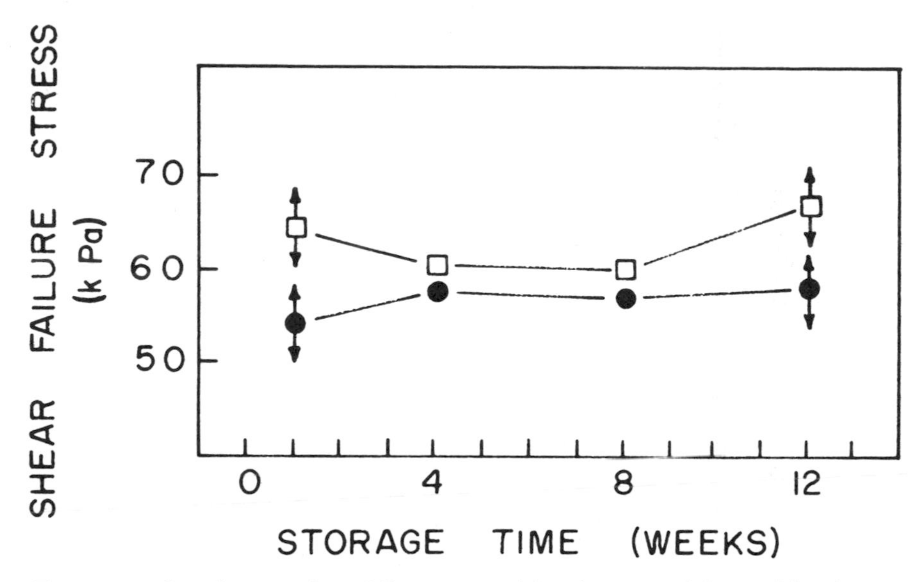

Fig. 13.10. Overall mean shear failure stress of four butters as influenced by storage temperature (●−30°C; □ +5°C) and time.
From Kawanari et al. (1981).

$$\ln \tau_{\max} = \ln B - (E_n / R_c)(1/T) \tag{3}$$

This equation, normally used in reaction kinetics, does seem to represent the relationship between shear stress and temperature quite well for both butter and margarine. For butters (Fig. 13.9) and margarine (Fig. 13.11), the apparent activation energies were −45.3 kJ/mole for the Cherry–Burrell butter, an average of −59.0 kJ/mole for the other three butters, and −96.4 kJ/mole for the margarine.

Cheese

There are many types of cheeses, so a study of failure properties of cheese could be very lengthy. A few observations from the literature will be discussed here.

Gouda cheese failure was studied by Culioli and Sherman (1976) using axial compression and noting the distorted shapes and failure planes. Influence of cheese maturity, temperature, specimen dimensions, strain rate, and lubrication of contact surfaces was noted and discussed. Gouda cheese tended to barrel out in the center or spread at the ends depending on end friction. Axial tension fractures opening at

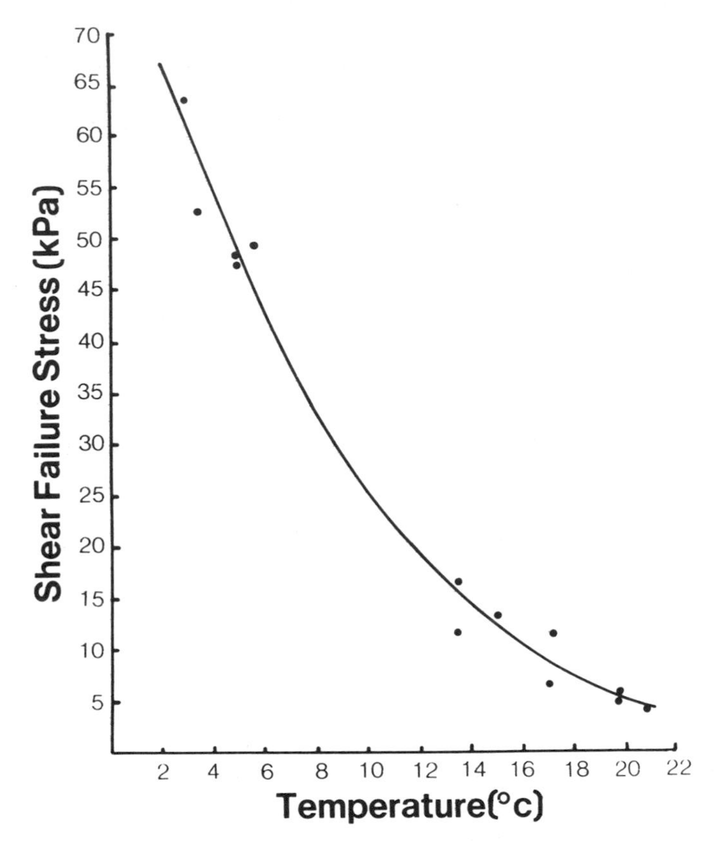

Fig. 13.11. Shear failure stress of margarine as influenced by test temperature.

the periphery were evident at 30–40% compression. This would be at true axial strains of about 36–51%. At ambient temperature (20°–21°C) a typical true compressive stress was about 75 kPa.

Rosenau *et al.* (1978) studied products produced by direct acidification and compared them with a commercial cheddar cheese and a commercial processed American cheese. In axial compression at ambient temperatures the commercial cheddar (averaging two strain rates) failed at about 50 kPa true axial stress and 0.90 true axial strain. The processed American cheese failed at about 25 kPa true axial stress and about 0.80 true axial strain.

From the limited observations cited it seems that cheeses similar to the above fracture at relatively large strains and modest stresses as we may have expected. Culioli and Sherman (1976) point out that Gouda cheese taken near the edge of the block can require twice the stress for failure compared to the interior. Variation in water content is cited as a major factor.

CONCLUDING REMARKS

The foods discussed here can be assumed to be homogeneous and isotropic. Relatively simple stress and strain theory can be applied to determine failure conditions with reasonable accuracy. More than small strain theory is required, however. The torsion test and analysis cited allowed for a curvilinear moment–displacement relationship. In the uniaxial tests, true stress and true strain were used to compensate for dimensional changes during deformation.

A specific failure criterion may only have application to a specific class of foods. The maximum shear strain criterion seems to apply to the protein gels tested. Raw fruits and vegetables respond differently in that the maximum normal strain criterion appears most appropriate. In this case turgor pressure in tissue cells is an important factor affecting tissue strength. High turgor pressure or a test mode that reduces volume (increasing fluid pressure in the cells) reduces the failure stress but has little effect on normal strain at failure. A deviatoric test such as torsion seems more sensitive to turgor pressure changes and reduced pressure results in an increase in failure strain level. Increased turgor tends to increase the tensile strength of parenchyma tissue, opposite to the effect on axial compression strength.

At low temperatures (<15°C) hard fat products represented by butter and margarine tend to fail in shear and shear failure stress correlates negatively with sensory spreadability. Failure stress levels are very temperature dependent and the Arrhenius equation represents the data well. Hard cheeses represented by Gouda and cheddar fail at moderate stress levels (room temperature) similar to the cold butters. Failure strains for these cheeses tend to be high (about 0.40–0.80 true axial strain).

Many foods like uncomminuted red meats are very anisotropic and heterogeneous so what has been discussed is inadequate. Foods that are homogeneous and isotropic but with less solid character require visco-elastic and/or elastic-plastic analysis. This chapter is, thus, only a beginning work dealing with relatively simple analyses. The fracture element models of Peleg (1977A) and Drake (1979) should be valuable in further work.

NOMENCLATURE

Symbol		*Units*
A	Area	m^2
a	Distance to center of axial curvature (torsion specimen)	m
B	A constant	Pa
b,c,d	Dimensions on Mohr circle diagrams	
b_1, b_2	Polynomial constants	
D	Specimen diameter	m
E	Axial stress/axial strain	Pa
E_n	Apparent activation energy	J
F	Force	N
G	Shear stress/shear strain	Pa
K,Q	Specimen shape factors	
L	Specimen length	m
ΔL	Change in specimen length	m
M	Twisting moment	N m
P	Pressure on a surface	Pa
R	Radius of circular cylinder	m
R_c	Gas constant	J/(mole°K
r	Radius of a specimen circular cross-section	m
r_{min}	Minimum value of r for a specimen	m
T	Temperature	°K
t	Dimension on a specimen	m
x,y,z	Directions in a coordinate system	
z_0	Distance from specimen center to end flange	m
γ	Shear strain	
γ_{max}	Shear strain of largest magnitude	
$\gamma_{true\ max}$	γ_{max} calculated considering dimensional changes during the test	
δ	Displacement	m
ε	Axial or normal strain	
ε_{max}	Normal strain of largest magnitude	
$\varepsilon_{true\ max}$	ε_{max} calculated considering dimensional changes during the test	
$\varepsilon_x,\varepsilon_y,\varepsilon_z$	Normal strains in coordinate directions	
σ	Axial or normal stress	Pa
σ_{max}	Normal stress of largest magnitude	Pa
$\sigma_{true\ max}$	σ_{max} calculated after correcting for dimensional changes	Pa
$\sigma_x,\sigma_y,\sigma_z$	Normal stresses in coordinate directions	Pa
τ	Shear stress	Pa
τ_{max}	Shear stress of largest magnitude	Pa
$\tau_{true\ max}$	τ_{max} calculated after correcting for dimensional changes	Pa
ν	Poisson's ratio (lateral strain/axial strain)	
ψ	Angle of specimen twist	Radians

BIBLIOGRAPHY

ARMSTRONG, J.E. 1981. The species-dependent factors affecting the texture of gels made from mechanically deboned fish meat. M.S. Thesis. Department of Food Science, North Carolina State University, Raleigh, NC.

BALL, H.R., JR., and WINN, S.E. 1982. Acylation of egg white proteins with acetic anhydride and succinic anhydride. Poultry Sci. *61*, 710–715.

BOURNE, M.C. 1976. Texture of fruits and vegetables. *In* Rheology and Texture in Food Quality. J.M., de Man, P.W. Voisey, V.F. Rasper, and D.W. Stanley (Editors). AVI Publishing Co., Westport, CT.

CHAPPELL, T.W., and HAMANN, D.D. 1968. Poisson's ratio and Young's modulus for apple flesh under compressive loading. Trans. ASAE *11*, 608–610.

CHARI, S.S., and AWASTHY, B.R. 1971. Instrument for determining rheological properties of butter at tropical temperatures. Indian J. Anim. Sci. *41*, 260–263.

CHENG, C.S., HAMANN, D.D., and WEBB, N.B. 1979. Effect of thermal processing on minced fish gel texture. J. Food Sci. *44*, 1080–1086.

CULIOLI, J., and SHERMAN, P. 1976. Evaluation of Gouda cheese firmness by compression tests. J. Texture Stud. *7*, 353–372.

DAL FABBRO, I.M., MURASE, H., and SEGERLIND, L.J. 1980. Strain failure of apple, pear and potato tissue. ASAE Paper No. 80-3048. Am. Soc. Agric. Eng., St. Joseph, MI.

De BAERDEMAEKER, J., SEGERLIND, L.J., MURASE, H., and MERVA, G. E. 1978. Water potential effect on tensile and compressive failure stresses of apple and potato tissue. ASAE Paper No. 78-3057. Am. Soc. Agric. Eng. St. Joseph, MI.

de MAN, J.M., DOBBS, J.E., and SHERMAN, P. 1977. Spreadability of butter and margarine. *In* Food Texture and Rheology. P. Sherman (Editor). Academic Press, New York.

DIEHL, K.C., and HAMANN, D.D. 1979. Relationships between sensory profile parameters and fundamental mechanical parameters for raw potatoes, melons and apples. J. Texture Stud. *10*, 401–420.

DIEHL, K.C., HAMANN, D.D., and WHITFIELD, J.K. 1979. Structural failure in selected raw fruits and vegetables. J. Texture Stud. *10*, 371–340.

DRAKE, B. 1979. A FORTRAN program "Framod" for simulation of large fractoviscoelastic models. J. Texture Stud. *10*, 165–182.

HAMANN, D.D., and WEBB, N.B. 1979. Sensory and instrumental evaluation of material properties of fish gels. J. Texture Stud. 10, 117–130.

HELDMAN, D.R., and SINGH, R.P. 1981. Food Process Engineering, 2nd Ed. AVI Publishing Co., Westport, CT.

HOLT, J.E., and SCHOORL, D. 1976. Bruising and energy dissipation in apples. J. Texture Stud. *7*, 421–432.

HUFF, E.R. 1967. Tensile properties of Kennebec potatoes. Trans. ASAE *10*, 414–419.

JOWITT, R. 1977. An engineering approach to some aspects of food texture. *In* Food Texture and Rheology. P. Sherman (Editor). Academic Press, New York.

KAMEL, B.S., and DE MAN, J.M. 1977. Some factors affecting gelatin gel texture evaluation by penetration testing. J. Texture Stud. *8*, 327–338.

KAWANARI, M., HAMANN, D.D., SWARTZEL, K.R., and HANSEN, A.P. 1981. Rheological and texture studies of butter. J. Texture Stud. *12*, 483–505.

KIM, W.J., RAO, V.N.M., and SMIT, C.J.B. 1978. Effect of chemical composition on compressive mechanical properties of low ester pectin gels. J. Food Sci. *43*, 572–575.

LANIER, T.C., LIN, T.S., HAMANN, D.D., and THOMAS, F.B. 1981A. Effects of alkaline protease in minced fish on texture of heat processed gels. J. Food Sci. *46*, 1643–1645.

LANIER, T.C., LIN, T.S., HAMANN, D.D., and THOMAS, F.B. 1981B. Gel formation in comminuted fish systems. Proc. 3rd National Technical Seminar on Mechanical Recovery and Utilization of Fish Flesh. R.E. Martin (Editor). National Fisheries Institute, Washington, D.C.

MATZ, S.A. 1962. Food Texture. AVI Publishing Co., Westport, CT.

MCDOWELL, R.H. 1975. New developments in the chemistry of alginates and their use in food. Chem. Ind. *3*, 391–397.

MILES, J.A. 1971. The development of a failure criterion for apple flesh. Ph.D. Thesis. Department of Agricultural Engineering. Cornell University, Ithaca, NY.

MILES, J.A., and REHKUGLER, G.E. 1973. A failure criterion for apple flesh. Trans. ASAE *16*, 1148–1153.

MITCHELL, J.R. 1976. Rheology of gels. J. Texture Stud. *7*, 313–339.

MITCHELL, J.R. 1979. Rheology of polysaccharide solutions and gels. *In* Polysaccharides in Food. J.M.V. Blanshard and J.R. Mitchell (Editors). Butterworths, London.

MITCHELL, J.R. 1980. The rheology of gels. J. Texture Stud. *11*, 315–337.

MOHSENIN, N.N. 1977. Characterization and failure in solid foods with particular reference to fruits and vegetables. J. Texture Stud. *8*, 169–193.

MONTEJANO, J.G., HAMANN, D.D., and BALL, H.R., JR. 1983A. Mechanical failure characteristics of native and modified egg white gels. Poultry Sci. (Submitted for publication).

MONTEJANO, J.G., HAMANN, D.D., and LANIER, T.C. 1983B. Rheological changes during processing of thermally induced fish muscle gels. J. Rheology (In press).

NADAI, A. 1937. Plastic behavior of metals in the strain-hardening range. J. Appl. Physics *8*, 205–213.

NILSSON, S.B., HERTZ, C.H., and FAULK, S. 1958. On the relation between turgor pressure and tissue rigidity. II. Physiologia Plantarum *11*, 818–837.

PELEG, M. 1977A. Contact and fracture elements as components of the rheological memory of solid foods. J. Texture Stud. *8*, 39–48.

PELEG, M. 1977B. Operational conditions and the stress-strain relationship of solid foods—theoretical evaluation. J. Texture Stud. *8*, 283–295.

PELEG, M., and BRITO, L.G. 1976. Textural changes in ripening plantains. J. Texture Stud. *7*, 457–463.

PELEG, M., BRITO, L.G., and MALEVSKI, Y. 1976. Compressive failure patterns in some juicy fruits. J. Food Sci. *41*, 1320–1324.

POLAKOWSKI, N.H., and RIPLING, E.J. 1966. Strength and Structure of Engineering Materials. Prentice-Hall, Inc., Englewood Cliffs, NJ.

PRENTICE, J.A. 1972. Rheology and texture of dairy products. J. Texture Stud. *3*, 415–458.

REEVE, R.M. 1970. Relationships of histological structure to texture of fresh and processed fruits and vegetables. J. Texture Stud. *1*, 247–284.

REEVE, R.M. 1977. Pectin, starch and texture of potatoes: some practical and theoretical implications. J. Texture Stud. *8*, 1–18.

ROSENAU, J.R., CALZADA, J.F., and PELEG, M. 1978. Some rheological properties of a cheese-like product prepared by direct acidification. J. Food Sci. *43*, 948–950, 953.

SEGERLIND, L.J., and DAL FABBRO, I.D. 1978. A failure criterion for apple flesh. ASAE Paper No. 78-3556. Am. Soc. Agric. Eng., St. Joseph, MI.

SHAHABASI, Y., and SEGERLIND, L.J. 1981. Determination of allowable delpths for apples stored in bulk. ASAE Paper No. 81-1064. Am. Soc. Agric. Eng., Joseph, MI.

SHERMAN, P. 1970. Industrial Rheology. Academic Press, New York.

SHERMAN, P., and DEGHAIDY, F.S. 1978. Force-deformation conditions associated with the evaluation of brittleness and crispness in selected foods. J. Texture Stud. *9*, 437–459.

SONE, T., OKADA, M., and FUKUSHIMA, M. 1966. Physical properties of butter made by different methods. Proc. 17th Int. Dairy Congr. pp. 225–235.

STERLING, C. 1963. Texture and cell-wall polysaccharides in foods. *In* Recent Advances in Food Science, J.M. Leitch and D.N. Rhodes (Editors). Butterworths, London.

SU, H.K., LIN, T.S., and LANIER, T.C. 1981. The contribution of retained organ tissues to the alkaline protease content of mechanically separated Atlantic croaker. J. Food Sci. *46*, 1650–1653, 1658.

TANIKAWA, E. 1971. Marine Products in Japan. Koseisha-Koseikaku Company. Tokyo, Japan.

TODA, J., WADA, T., and KONNO, A. 1978. Sensory evaluation of textural properties of gelatin: agar-agar and egg white gels. J. Agric. Chem. Soc. Japan *52*, 539–544.

VAN WAZER, J.K., LYONS, J.W., KIM, K.Y., and COLWELL, R.E. 1963. Viscosity and flow measurements—A Laboratory Handbook of Rheology. Wiley (Interscience) Publ, New York.

VASIC, I., and DE MAN, J.M. 1968. Effect of mechanical treatment on some rheological properties of butter. *In* Rheology and Texture of Foodstuffs. Monograph 27, Soc. Chem. Ind., London.

WARD, A.G., and COBBETT, W.G. 1968. Mechanical behavior of dilute gelatin gels under large strains. *In* Rheology and Texture of Foodstuffs. Monograph 27, Soc. Chem. Ind., London.

14

Rheology of Emulsions and Dispersions

Irvin M. Krieger[1]

INTRODUCTION

Definitions

Rheology is the study of deformation and flow of matter. *Colloids* are mixtures in which at least one of the components is present as particles which are large compared to ordinary molecules, yet so small that they cannot be resolved using a microscope of moderate power. *Emulsions* are colloidal dispersions of liquids in liquids. The term *dispersion*, which is often applied to all colloids in which the continuous phase is liquid, is here reserved for those in which the disperse phase is solid.

Other colloidal systems include foams, aerosols, and gels. *Foams* consist of particles of gas dispersed in a liquid, while *aerosols* are liquids or solids dispersed in a gas. *Gels* are solid-in-liquid colloids in which the solid phase forms a long-range network structure, immobilizing the liquid and conferring solid-like properties on the system. The sizes of polymers and other macromolecules qualify them to be considered as colloidal particles. Because of their many internal degrees of freedom, however, the rheological behavior of macromolecules is frequently more complex than that of other colloids.

---

[1]Departments of Chemistry and Macromolecular Science, Case Western Reserve University, Cleveland, OH 44106.

Relation to Food Rheology

Many foods are emulsions or dispersions in which the continuous phase is an aqueous solution. Starch pastes, bean pastes, and batters are dispersions, while milk, gravies, and salad dressings are emulsions. Whipped cream and beaten egg-white are foams. Natural macromolecular colloids include starch and protein solutions, while many synthetic macromolecules (e.g., hydroxyethylcellulose) have recently found their way into commercial recipes to modify their rheology.

Rheological Variables

The physical variables encountered in a rheological study are illustrated in Fig. 14.1. This is an idealized parallel-plane viscometer, consisting of two planes parallel to the $x-y$ plane, each of area A and separated by a distance h along the z axis. The sample fills the space between the planes, which are assumed to be so large that leakage and other edge effects are negligible. The lower plane is fixed, while a force F applied to the upper plane causes it to be displaced in the x direction. If the sample is solid, each particle will be displaced by a distance $x = zX/h$, where X is the displacement of the upper plane. The *shear stress* is defined as $\sigma = F/A$, while the *shear strain* is $\gamma = dx/dz = X/h$. The *shear modulus* is the ratio of stress to strain: $G = \sigma/\gamma$. The integral $\int \sigma d\gamma$ represents the strain energy stored in unit volume of the solid.

If the sample is a liquid, the force will produce laminar flow of the sample. Each particle takes on a velocity $v = Vz/h$, where V is the velocity of the upper plane. The *shear rate* is defined as $\dot{\gamma} = dv/dz = V/h$. The viscosity η is the ratio $\eta = \sigma/\dot{\gamma}$. The product $\sigma\dot{\gamma}$ represents the energy dissipated (as heat) in unit volume per unit time.

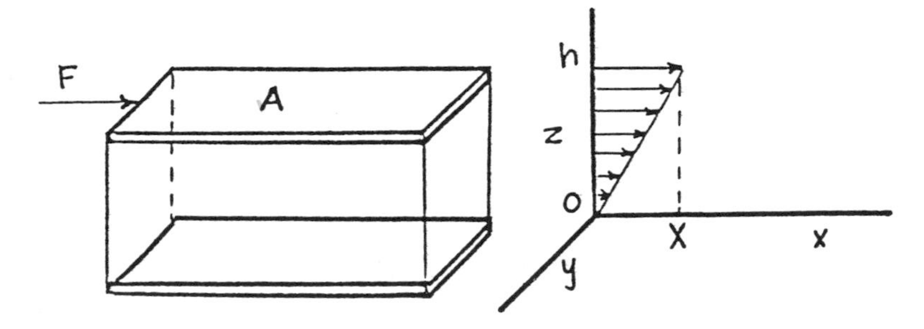

Fig. 14.1. Idealized parallel-plane viscometer, showing displacement (or velocity) as a function of distance from the stationary plane.

RHEOLOGICAL BEHAVIOR OF COLLOIDS

Colloidal Variables

For particulate systems such as emulsions and dispersions, the most convenient concentration variable is the volume fraction ϕ of the disperse phase. This will range from 0 to a maximum p called the *packing fraction*. It represents the highest concentration at which flow can occur without deformation of the particles themselves.

If we let η be the viscosity of the colloidal system and η_0 be that of its continuous phase, the relative viscosity η_r is the ratio of the two: $\eta_r = \eta/\eta_0$. The initial slope of the graph of η_r vs. ϕ, called the intrinsic viscosity $[\eta]$, is a function of the shape and flexibility of the particle.

Viscosity vs. Concentration

Figure 14.2 shows schematically the variation of relative viscosity with volume fraction for a typical colloidal dispersion. Both intrinsic

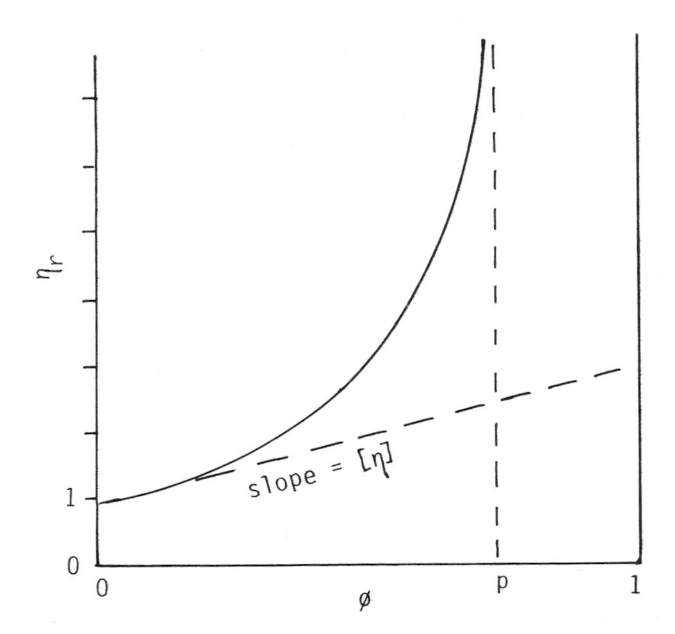

Fig. 14.2. Typical curve of relative viscosity as a function of volume fraction.

viscosity and packing fraction are indicated in the figure. Many equations have been proposed to describe the functional dependence of η_r on ϕ; an early review (Rutgers 1962) cited more than 100. Although there is no theory that is applicable over the entire concentration range, there are excellent theories of the intrinsic viscosity for the most important types of particles.

The rigid sphere is the simplest of colloidal particles. As a part of his doctoral dissertation on Brownian motion, Albert Einstein (Einstein 1905, 1906, 1911) showed that the intrinsic viscosity is precisely 2.5, independent of particle diameter. For emulsions, where the particles are liquid of viscosity η', the intrinsic viscosity is given by (Taylor 1932, 1934):

$$[\eta] = \frac{5}{2}\left[\frac{1 + (2\eta'/5\eta_0)}{1 + \eta'/\eta_0}\right]$$

When the viscosity of the inner phase becomes infinite, Taylor's formula reduces to that of Einstein for rigid spheres, as it should. The other limiting case is an inner-phase viscosity of zero, corresponding to a foam; here the intrinsic viscosity is precisely 1. What is remarkable is that the intrinsic viscosity is always positive, so that even the addition of particles of an inviscid gas will raise the viscosity of a colloidal system. This is not necessarily the case when the added substance goes into solution; adding water to molasses lowers the viscosity.

Intrinsic viscosities have been calculated (Simha 1940, 1945; Kuhn and Kuhn 1945) for dispersions of ellipsoidal particles. Letting $r = b/a$ be the ratio of major to minor semiaxes, the results of Kuhn and Kuhn can be expressed for oblate ellipsoids ($0<r<1$):

$$[\eta] = \frac{5}{2} + \frac{32(1 - r)}{15\pi r} - \frac{0.682(1 - r)}{1 - 0.75r}$$

and for prolate ellipsoids ($1<r<15$):

$$[\eta] = \frac{5}{2} + 0.4075(r - 1)^{1.508}$$

An asymptotic form for large axial ratio ($r>15$) was derived by Simha and is applicable to both prolate ellipsoids ($\lambda = 1.5$) and cylindrical rods ($\lambda = 1.8$):

$$[\eta] = \frac{14}{15} + \frac{r^2}{15}\left[(\ln 2r - \lambda)^{-1} + 3(\ln 2r - \lambda + 1)^{-1}\right]$$

For very large axial ratio, this reduces to $[\eta] = 0.53r^2$.

Intrinsic viscosities for polymers and other macromolecules are usually expressed in terms of concentration variables other than ϕ, frequently as c in grams per deciliter. Early in the development of polymer science, it was assumed (Staudinger 1932) that intrinsic viscosity was proportional to molecular weight. The more general relation $[\eta] = KM^a$ was independently proposed by Mark (1938) and by Houwink (1940), also on empirical grounds. Representing the macromolecule as a spherical domain of variable porosity, Debye and Bueche (1948) showed that the range of the exponent a should be 0.5 to 1 for random coil molecules. If the molecules are more tightly coiled, due to incompatibility with solvent or to branching or internal crosslinks, the exponent a can approach the value 0 of Einstein's model. At the other extreme of a stiff, fully extended chain, the asymptotic formula for cylindrical rods shows that the upper limit for a is 2.

Theoretical study of concentrated systems is much more difficult than is the study at infinite dilution, which has told us so much about the intrinsic viscosity. Even to estimate the coefficient b of the quadratic term in the series expansion $\eta_r = 1 + [\eta]\phi + b\phi^2 + \ldots$. has proved to be a formidable task. The best current estimate (Batchelor and Green 1972) places b at 7.6 ± 0.8 for extensional flows, but the problem remains unsolved for shear flows. A very crude theoretical approach considers the suspension to behave as a simple liquid toward the addition of new particles. This leads to a functional equation $\eta_r(\phi_1 + \phi_2) = \eta_r(\phi_1) \times \eta_r(\phi_2)$, which is satisfied by a relationship originally proposed by Arrhenius:

$$\eta_r = e^{[\eta]\phi}$$

The Arrhenius relation is useful at moderate dilutions, but fails at high concentrations, predicting finite viscosity even when the volume fraction reaches 100%. In an attempt to take into account crowding of newly added particles by those already present, Mooney (1951) defined effective volume fractions

$$\phi_{2,1} = \frac{\phi_2}{1 - \phi_1/p} \quad \text{and} \quad \phi_{1,2} = \frac{\phi_1}{1 - \phi_2/p}$$

and satisfied the functional equation $\eta_r(\phi_1 + \phi_2) = \eta_r(\phi_{1,2}) \times \eta_r(\phi_{2,1})$ by

$$\eta_r = \exp \frac{[\eta]\phi}{1 - \phi/p}$$

Dougherty (1959) and Krieger (1967) modified Mooney's analysis to reduce the crowding correction. Their functional equation $\eta_r(\phi_1 + \phi_2) = \eta_r(\phi_1)[\eta_r(\phi_{2,1})] = \eta_r(\phi_{1,2})[\eta_r(\phi_2)]$ yielded as solution:

$$\eta_r = (1 - \phi/p)^{-[\eta]\phi}$$

Non-Newtonian Effects

An elastic body for which stress is proportional to strain is said to obey Hooke's law; the shear modulus of a Hookean solid is therefore a material constant, independent of the strain. For fluids, Newton's law of viscous flow postulates proportionality between stress and strain rate. The viscosity of a Newtonian liquid is therefore a material constant, independent of the strain rate. Many materials, like egg-white and mayonnaise, show both viscous and elastic behavior.

When an emulsion or a dispersion exhibits elasticity, the cause is almost always formation of long-range network structure, although it is possible in principle to deform individual particles elastically. Models for viscoelastic substances are frequently postulated to consist of various arrays of Hookean solid elements and Newtonian liquid elements. Such models and the systems they represent are said to be *linear viscoelastic bodies*, because their stress–strain relationship is expressible as a linear differential equation. The theory of nonlinear viscoelasticity is more difficult. There are several excellent references (especially Ferry 1980) on viscoelasticity of polymers; the methods they describe are directly applicable to food rheology.

Non-Newtonian fluids are those for which the viscosity depends on shear rate (or, equivalently, on shear stress). All materials that exhibit non-Newtonian behavior under ordinary conditions are colloidal or macromolecular. *Shear thinning*, in which the viscosity decreases with increasing shear rate, is a very common phenomenon. The converse, known as *shear thickening* or dilatancy, is less frequently encountered. A simple relationship between shear stress σ and shear rate $\dot{\gamma}$, which illustrates the gamut of fluid behavior, is the power law:

$$\dot{\gamma} = a\sigma^{\eta}$$

Here $1/a$ is the viscosity at unit shear stress and n is an index of flow behavior. When $n = 1$ we have Newtonian flow; when $n<1$ the fluid is shear-thinning; and when $n>1$ it is shear-thickening.

Certain shear-thinning materials (mayonnaise, gelatin) behave as elastic solids under low stresses, yet flow quite readily once the applied stress exceeds a critical value σ_y known as the yield stress. Two stress–strain rate relations useful to describe such behavior are due to Bingham (1922) and Casson (1959).

$$\dot{\gamma} = 0 \qquad \text{for} \qquad \sigma < \sigma_y$$

$$\dot{\gamma} = \eta_1^{-1}(\sigma^m - \sigma_y{}^m)^{1/m} \qquad \text{for} \qquad \sigma > \sigma_y$$

Here η_1 is the limiting viscosity at high shear stress and the exponent m takes on the value 1 in Bingham's equation and 1/2 in Casson's.

It may require measurable time after shearing has ceased for a shear-thinning fluid to recover its rest viscosity. This dependency of viscosity on recent shear history is called *thixotropy*, and is attributed to a reversible transformation in the state of aggregation or orientation of the particles. Steady-state non-Newtonian viscosity is determined by a competition between the recovery process and the disruptive effect of shear.

Eyring's theory of rate processes, when applied to viscous flow (Eyring 1936, Ree and Eyring 1955), predicts a size-dependent relaxation time τ for each kind of molecular or particulate flow unit in a fluid. At shear rates small compared to $1/\tau$, the flow unit will make a Newtonian contribution to the viscosity, while at very high shear rates ($\dot{\gamma}>>1/\tau$) the contribution will vanish. For ordinary fluids like water or alcohol, the relaxation time is so short that their contribution is Newtonian at all experimentally realizable shear rates, but for colloids the contribution is usually non-Newtonian. For a dispersion of uniform colloidal spheres in a Newtonian medium, the Ree–Eyring theory would predict a shear-dependent viscosity:

$$\eta = \eta_1 + (\eta_2 - \eta_1)\,\frac{\sinh^{-1}(\tau\dot{\gamma})}{\tau\dot{\gamma}}$$

Here η_2 is the viscosity at rest. Similar shear-thinning behavior is predicted by a theory based on doublet formation (Krieger and Dougherty 1959):

$$\eta = \eta_1 + \frac{\eta_2 - \eta_1}{1 + \sigma/\sigma_c}$$

Here σ_c is a size-dependent characteristic shear stress. In a more generalized version (Cross 1965), the adjustable parameter m usually takes on a value between 2/3 and 1.

$$\eta = \eta_1 + \frac{\eta_2 - \eta_1}{1 + (\dot{\gamma}/\dot{\gamma}_c)^m}$$

DIMENSIONAL ANALYSIS

Principles of Dimensional Analysis

Each physical variable has a dimensionality indicated by the units in which it is measured or expressed. In most cases, this dimensionality can be described by a product of a mass M, a length L, and a time T, each raised to an appropriate power. Thus the dimensionality of velocity is LT^{-1}, while that of density is ML^{-3}. Dimensionless groups are those products of physical variables (each raised to an appropriate power) that are dimensionless. The most elementary examples of dimensionless groups are simple ratios, like volume fraction or relative viscosity. An example familiar to all engineers is the Reynolds number $\mathrm{Re} = Du\rho/\eta$, which gives a criterion for the transition from laminar to turbulent flow for a fluid of viscosity η and density ρ flowing with average velocity u in a tube of diameter D.

A physical system knows nothing about the units in which we humans measure its properties. In a sense, each system carries its own internal standards of mass, length, and time. As a consequence, *all equations among physical variables can be written as equations among dimensionless groups*. Since the number of groups in the dimensionless equation is less than the original number of physical variables, the use of dimensional analysis reduces the order of complexity. If we start with V physical variables, all expressible in terms of U dimensional units, then the number of independent dimensionless groups which can be formed is $V-U$. (Independence means here that no dimensionless group in the set can be written as a product of the others.)

Dispersions of Uniform Rigid Spheres

The governing equation of colloid rheology relates the viscosity of an emulsion or dispersion to composition, temperature, shear rate, and shear history. The simplest case, that of a dispersion of uniform rigid noninteracting spheres in a Newtonian liquid (Krieger 1963), involves nine physical variables and one physical constant (Boltzmann's constant

k). By utilizing the combined variable kT, where T is the absolute temperature, the number of variables is held to nine, all expressible in terms of the three units of mass, length, and time.

$$\eta = f(\eta_0, \rho_0, R, \rho_p, N, \dot{\gamma}, kT, t)$$

(Where R is the radius, ρ_p the density, and N the concentration (number per unit volume) of the particles; ρ_0 is the density of the suspending liquid and t is the time.)

This means that the governing equation can be written in terms of six (nine minus three) dimensionless groups. Table 14.1 lists the nine physical variables, each with its dimensionality, while Table 14.2 gives a complete set of six dimensionless groups. For added convenience, the shear stress $\sigma = \eta\dot{\gamma}$ and a characteristic time $t^* = \eta_0 R^3/(kT)$ have been utilized. In terms of the groups defined in Table 14.2, the dimensionless equation takes the form

$$\eta_r = f(\phi, \sigma_r, t_r, \rho_r, \mathrm{Re})$$

Even a six-variable system may prove formidable to investigate. Fortunately, some of the variables take on limiting values in certain cases of interest. If we limit ourselves to low-shear ($\mathrm{Re} \to 0$) steady ($t \to \infty$) flow of neutrally buoyant ($\rho_r \to 1$) spheres, the dimensionless equation contains only three variables:

$$\eta_r = f(\phi, \sigma_r, \infty, 1, 0) = f(\phi, \sigma_r)$$

TABLE 14.1. PHYSICAL VARIABLES IN RHEOLOGY OF UNIFORM RIGID SPHERES

Symbol	Variable	Dimensionality
η	Viscosity of dispersion	$ML^{-1}T^{-1}$
η_0	Viscosity of medium	$ML^{-1}T^{-1}$
ρ_0	Density of medium	ML^{-3}
ρ_p	Density of particles	ML^{-3}
R	Radius of particles	L
N	No. of particles/volume	L^{-3}
$\dot{\gamma}$	Shear rate	T^{-1}
t	Time	T
kT	Thermal energy	ML^2T^{-2}

TABLE 14.2 DIMENSIONLESS GROUPS IN THE RHEOLOGY OF RIGID-SPHERE DISPERSIONS

Group	Symbol	Name
η/η_0	η_r	Relative viscosity
$4\pi R^3 N/3$	ϕ	Volume fraction
$\sigma R^3/(kT)$	σ_r	Reduced shear stress
ρ_p/ρ_0	ρ_r	Relative density
t/t^*	t	Reduced time
$\dot{\gamma} R^2 \rho_\sigma/\eta_0$	Re	Reynolds number

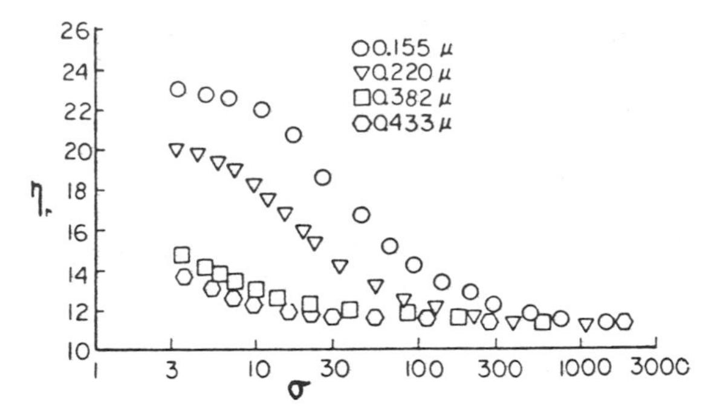

Fig. 14.3. Relative viscosities of 50% (by volume) dispersions of cross-linked polystyrene spheres of different sizes in benzyl alcohol at 30°C. *From Papir and Krieger (1970).*

Since dimensional analysis tells us nothing about the functional form of this three-variable equation, any further progress requires experimental or theoretical study. A simple experimental design might hold one of the two independent variables (ϕ or σ_r) constant, and study the effect on η_r of the other variable.

Such a study was carried out using polystyrene latex spheres of highly uniform size, dispersed in water (Woods and Krieger 1970) and in organic liquids of differing viscosities (Papir and Krieger 1970). Figures 14.3 and 14.4 show the relative viscosities of dispersions in benzyl alcohol of four different particle sizes, all at $\phi = 0.50$. The abscissa in Fig. 14.3 is the shear stress σ, while that in Fig. 14.4 is the dimensionless shear stress σ_r. Note that the four disparate curves of Fig. 14.3 superpose in Fig. 14.4, as is required by the dimensional analysis. Also plotted on Fig. 14.4 are relative viscosities of dispersions in *m*-cresol, together with a curve fitted to data on aqueous dispersions. Twenty-fold ranges in particle volume and in viscosity of the dispersing liquid are correlated on a single curve. Equations of the Krieger–Dougherty form were fitted to the data at all volume fractions up to 0.5, giving for $\eta_r = f(\phi,\sigma_r)$:

$$\eta_r = \eta_{1r} + \frac{\eta_{2r} - \eta_{1r}}{1 + 0.431\sigma_r}$$

where

$$\eta_{1r} = (1-\phi/0.68)^{-1.82}$$

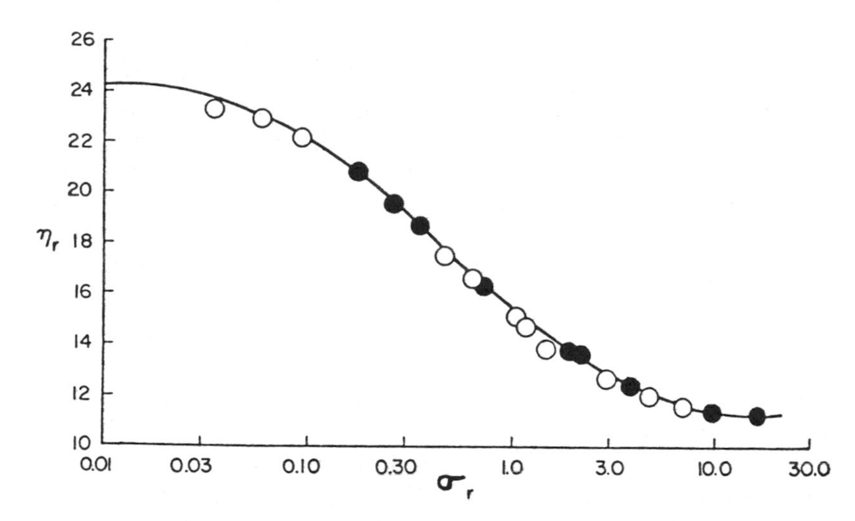

Fig. 14.4. Data of Fig. 14.3 replotted vs. reduced shear stress, together with data for 50% dispersions of the same particles in *m*-cresol (●). Solid line is fitted curve to data on 50% aqueous dispersions. (○) Benzyl alcohol.
From Papir and Krieger (1970).

and

$$\eta_{2r} = (1-\phi/0.57)^{-1.50}$$

Data taken on other colloidal dispersions may be tested against this master equation to determine whether their behavior corresponds to that expected of a uniform dispersion of noninteracting rigid spheres.

Extensions to Other Systems

For particles of ellipsoidal shape, two variables—the lengths of the semiaxes—are needed to define their size and shape. The extra physical variable requires an additional dimensionless group: the axial ratio. We would then seek a master equation for all colloidal dispersions of noninteracting rigid ellipsoids having the same axial ratio. Similar considerations would apply to rod-like or disk-like particles, where again two length variables are adequate to define size and shape.

The governing equation for deformable particles will contain their modulus of elasticity. Here again the addition of one variable requires the addition of a dimensionless group. Since the shear modulus G has the same dimensionality as does the shear stress, the ratio σ/G furnishes the additional dimensionless group. Dispersions of particles of differing elasticities should therefore behave alike when measured at the same values of σ/G. For emulsions, the viscosity η' of the inner phase determines flow within the particle, while the interfacial tension Γ determines its deformability. The two new dimensionless groups are the viscosity ratio η'/η_0 and a shear-deformation criterion $R\sigma/\Gamma$ (Taylor 1934).

The study of interacting particles introduces new physical variables. Any expression for the potential energy of interaction between two particles would be expected to contain a length parameter to indicate the range of the force and an energy parameter to indicate its intensity. Dividing the range parameter by the particle radius gives the first of the two new dimensionless groups required; the second can be obtained by dividing the energy parameter by kT. An example of particle interactions is the so-called electroviscous effect. Here electrostatic interactions are capable of increasing the viscosity of a dispersion by several orders of magnitude. Study of the effect of electrostatic interactions is greatly facilitated by the use of dimensional analysis (Krieger and Eguiluz 1976).

SUMMARY

Emulsions and dispersions, especially those encountered in the food industry, are complex systems that exhibit a wide range of rheological behavior. Considerable insight can be obtained by starting with the simplest cases and working forward to more complex systems. Just as the ideal gas serves as a starting point for the study of dense gases and the liquid state, the rigid-sphere dispersion is the starting point for studying colloid rheology. Ellipsoids, rods, and disks are models for anisotropic particles; elastic spheres and emulsions are tractable examples of systems of nonrigid particles. Semiempirical relations are available for the viscosities of concentrated dispersions and non-Newtonian fluids. Through the use of dimensional analysis, the number of independent variables in a rheological problem can be significantly reduced, and useful correlating principles can be developed.

BIBLIOGRAPHY

BATCHELOR, G.K., and GREEN, J.T. 1972. The determination of bulk stress in a suspension of spherical particles to order c^2. J. Fluid Mech. *56*, 401–427.

BINGHAM, E.C. 1922. Fluidity and Plasticity. McGraw-Hill, New York.

CASSON, N. 1959. A flow equation for pigment-oil suspensions of the printing ink type. *In* Rheology of Disperse Systems. C.C. Mill (Editor). Pergamon Press, London and New York.

CROSS, M.M. 1965. Rheology of non-Newtonian fluids: A new flow equation for pseudoplastic systems. J. Colloid Sci. *20*, 417–437.

DEBYE, P., and BUECHE, A.M. 1948. Intrinsic viscosity, diffusion and sedimentation rate of polymers in solution. J. Chem. Phys. *16*, 573–579.

DOUGHERTY, T.J. 1959. Some problems in the theory of colloids. Ph.D. Thesis, Case Institute of Technology, Cleveland, OH.

EINSTEIN, A. 1905. On the behavior of suspended particles in fluids at rest, as advanced by the kinetic-molecular theory of heat. Ann. Phys. *17*, 549–560.

EINSTEIN. A. 1906. A new determination of molecular dimensions. Ann. Phys. *19*, 271.

EINSTEIN, A. 1911. A correction to my work: A new determination of molecular dimensions. Ann. Phys. *34*, 591–592.

EYRING, H. 1936. Viscosity, plasticity and diffusion as examples of absolute reaction rates. J. Chem. Phys. *4*, 283–291.

FERRY, J.D. 1980. Viscoelastic Properties of Polymers, 3rd Ed. John Wiley & Sons, Inc., New York.

HOUWINK, R. 1940. Relation between viscometrically and osmotically determined degrees of polymerization of high polymers. J. Prakt. Chem. *157*, 15–18.

KRIEGER, I.M. 1963. A dimensional approach to colloid rheology. Trans. Soc. Rheol. *7*, 101–109.

KRIEGER, I.M. 1967. Flow properties of latex and concentrated solutions. *In* Surfaces and Coatings. R. Marchessault and C. Skaar (Editors). Syracuse University Press, Syracuse, NY.

KRIEGER, I.M., and DOUGHERTY, T.J. 1959. A mechanism for non-Newtonian flow in rigid-sphere dispersions. Trans. Soc. Rheol. *3*, 137–152.

KRIEGER, I.M., and EGUILUZ, M. 1976. The second electroviscous effect. Trans. Soc. Rheol. *20*, 29–45.

KUHN, W., and KUHN, H. 1945. The dependency of viscosity on flow gradients for highly dilute suspensions and solutions. Helv. Chim. Acta *28*, 97–128.

MARK, H. 1938. Der fester Körper. Hirzel, Leipzig.

MOONEY, M. 1951. The viscosity of a concentrated suspension of spherical particles. J. Colloid Sci. *6*, 162–170.

PAPIR, Y.S., and KRIEGER, I.M. 1970. Rheological studies on dispersions of uniform colloidal spheres. II: Dispersions in nonaqueous media. J. Colloid Interface Sci. *34*, 126.

REE, T., and EYRING, H. 1955A. Theory of non-Newtonian flow. I: Solid plastic systems. J. Appl. Phys. *26*, 793–799.

REE, T., and EYRING, H. 1955B. Theory of non-Newtonian flow. II: Solution sytems of high polymers. J. Appl. Phys. *26*, 800–809.

RUTGERS, I.R. 1962. Relative viscosity and concentration. Rheol. Acta *2*, 305–348.

SIMHA, R. 1940. The influence of Brownian movement on the viscosity of solutions. J. Phys. Chem. *44*, 25–34.

SIMHA, R. 1945. Influence of molecular flexibility on the intrinsic viscosity, sedimentation and diffusion of high polymers. J. Chem. Phys. *13*, 188–195.

STAUDINGER, H. 1932. Die hochmolekularer organischen Verbindungen. Springer Verlag, Berlin and New York.

TAYLOR, G.I. 1932. The viscosity of a fluid containing small drops of another fluid. Proc. R. Soc. London Ser. A *138*, 41–48.

TAYLOR, G.I. 1934. The formation of emulsions in definable fields of flow. Proc. R. Soc. London Ser. A *146*, 501–523.

WOODS, M.E., and KRIEGER, I.M. 1970. Rheological studies on dispersions of uniform colloidal spheres. I: Aqueous dispersions in steady shear flow. J. Colloid Interface Sci. *34*, 91–99.

15

Physical and Chemical Properties Governing Volatilization of Flavor and Aroma Components

C. Judson King[1]

INTRODUCTION

Many important flavor and aroma components are highly volatile, and are easily lost during processing, especially when water is removed by means such as evaporation and drying, which allow direct exposure to a purged vapor phase. Loss of these components is the reason for consumer reactions that most evaporated or dried foods tend to be flat and flavorless.

In recent years research into the mechanisms of loss of volatile components has yielded considerable insight into the phenomena controlling these losses and has provided scientific bases for devising and altering processes to increase retentions of volatile substances. As another application of the same knowledge, there are now effective ways of preparing "locked-in" flavor ingredients, which are themselves articles of commerce.

Understanding the mechanisms of loss of volatile flavor and aroma substances requires differentiating between equilibrium and rate phenomena. Volatiles retention can be controlled through either of these; however, equilibrium volatilities in most cases are high enough so that the only practical way to achieve good retention is through control of the rate of approach to equilibria, rather than through control of equilibrium itself.

[1]Department of Chemical Engineering, University of California, Berkeley, Ca 94720.

This chapter will not attempt to cover this large field of volatiles loss and retention comprehensively. Instead it will focus solely on the underlying physical and chemical properties, indicating how each may be measured, interpreted, predicted, and correlated. The discussion may be looked on as an updating of the portions of the earlier reviews by Bomben *et al.* (1973) and Thijssen (1971) relating to physical and chemical properties.

EQUILIBRIUM PROPERTIES

Definitions and General Behavior

The dominant equilibrium property is the *equilibrium partition coefficient* of a volatile component from the condensed phase into the vapor phase. Since volatiles loss tends to be most severe from aqueous solutions, this discussion is developed in terms of the condensed phase being liquid, with water a major constitutent. Equilibrium partition coefficients can be expressed in several ways, among which are

$$K_i = y_i/x_i \tag{1}$$

and

$$H_i = p_i/C_i \tag{2}$$

where K_1 is the mole-fraction-based partition coefficient for component i, H_i is the Henry's law constant for component i, y_i is the vapor-phase mole fraction, x_i is the liquid-phase mole fraction, p_i is the partial pressure of i in the vapor phase, and C_i is the concentration (moles/liter) of i in the liquid phase, all at equilibrium. Another form of partition coefficient is defined as the ratio of solute concentrations (moles or mass per unit volume) in the two phases.

Another important property is the *relative volatility* of the volatile component with respect to water, since this determines the relative proportions in which the volatile component and water come off during an equilibrium vaporization. This relative volatility, α_{iw}, is defined by the ratio of K_i to K_w. K_w, in turn, is defined as y_w/x_w, the ratio of vapor and liquid mole fractions for water.

$$\alpha_{iw} = \frac{K_i}{K_w} = \frac{y_i}{x_i}\frac{x_w}{y_w} \tag{3}$$

Activity coefficients in the liquid phase, γ_i, are defined by reference to the concept of an ideal solution:

$$p_i = \gamma_i x_i P_i^0 \tag{4}$$

where P_i^0 is the vapor pressure of pure component i at the temperature in question. The activity coefficient is a measure of the degree of compatability of i with the liquid phase—the tendency for intermolecular forces to develop between i and the major constituents of the liquid, in comparison with the strength of intermolecular forces among the major components of the liquid themselves. A similar definition may be made for the activity coefficient of water:

$$p_w = \gamma_w x_w P_w^0 \tag{5}$$

Activity coefficients may also be defined for the vapor phase, but these tend to be important only at pressures higher than those normally encountered in food-processing operations.

Putting Eqs. (4) and (5) together with Eqs. (1) and (3), and recognizing that by Dalton's law

$$y_i = p_i/P \tag{6}$$

where P is the total pressure, gives

$$K_i = \gamma_i P_i^0/P \tag{7}$$

and

$$\alpha_{ij} = \frac{\gamma_i P_i^0}{\gamma_w P_w^0} \tag{8}$$

For many important flavor and aroma compounds, the vapor pressure of pure substance is not greatly different from that of water. On the other hand, values of γ_i tend to be very large, with values of order 1000 being common. These very large activity coefficients stem from the fact that common volatile flavor and aroma components tend to be relatively nonpolar, and therefore relatively incompatible with a highly polar, aqueous solution in terms of intermolecular forces. Conversely, values of γ_w tend to be of order unity, since water is a major component of food liquids. By Eqs. (7) and (8), the high values of γ_i make K_i

and α_{iw} large. If P_i^0 is of the same order of magnitude as P_w^0, and γ_w is of order unity while γ_i is of order 1000, then α_{iw} will be of order 1000.

Effects of K_i and α_{iw} on Volatiles Loss at Equilibrium

In many evaporation processes, volatiles will be lost to a degree that corresponds to near-equilibrium being achieved between vapor and liquid. There are two simple situations that can be pictured for such an evaporation. In one ("simple equilibration"), the vapor and the liquid remain fully in contact with each other at all times. In that case it is easy to show that the fraction loss of i (denoted by f_i) is given by

$$f_i = \frac{1}{1 + (L/K_i V)} \tag{9}$$

where L and V are the molar amounts of liquid and vapor, respectively (King 1980). If water forms a very high mole fraction of both the liquid and the vapor phases, y_w, x_w, and K_w will all be approximately unity, and α_{iw} may be substituted for K_i in Eq. (9). If $\alpha_{iw} = 1000$, even $V/L = 0.01$ will give a high fraction loss (0.91) of i.

The other simple situation is a "batch" or Rayleigh evaporation, where bits of vapor are removed from contact with the liquid as soon as they are formed. This is a better representation for many industrial evaporators. For such an evaporation, it can be shown that when water is the major component of both the vapor and the liquid

$$f_i = 1 - \left(\frac{L}{L + V}\right)^{K_i} \tag{10}$$

Equation (10) is obtained by substitution into and rearrangement of Eqs. (3–17), of King (1980). Again, α_{iw} may be substituted for K_i in Eq. (10). If K_i and α_{iw} are equal to 1000, f_i rapidly approaches unity with only small fractions vaporized:

Fraction Vaporized	$L/(L + V)$	f_i, for $K_i = 1000$
0.001	0.999	0.632
0.002	0.998	0.865
0.005	0.995	0.993

0.01	0.990	1.000
0.02	0.980	1.000
.	.	.
.	.	.
.	.	.

Losses will be large even for $\alpha_{iw} = 100$, for which we have

Fraction Vaporized	$L/(L + V)$	f_i, for $K_i = 100$
0.002	0.998	0.181
0.01	0.990	0.634
0.02	0.980	0.867
0.05	0.950	0.994
0.10	0.900	1.000
.	.	.
.	.	.
.	.	.

Hence, an important conclusion is that most food volatiles will be lost almost completely during an equilibrium evaporation of a simple aqueous solution.

Relationship between Activity Coefficient and Solubility

For sparingly soluble compounds that are liquids in the pure state at the temperature in question, there is a simple relationship between the mole-fraction solubility in aqueous solution, x_i^*, and the activity coefficient in the same solution, γ_i:

$$\gamma_i \cong 1/x_i^* \tag{11}$$

Equation (11) follows from Eq. (4) with the assumption that no other component of the aqueous phase dissolves in pure, liquid i sufficiently to lower the equilibrium partial pressure for i below the vapor pressure of its pure state. Thus p_i in equilibrium with the aqueous solution (Eq. 4) must also be equal to P_i^0 if all three phases (the vapor and two liquids) are at equilibrium.

Measurements of solubilities of relatively insoluble flavor and aroma components in aqueous solutions thereby provide a method for determining γ_i at the solubility concentration. In many cases [see, e.g.,

Massaldi and King (1973) and Kieckbusch and King (1979A)] γ_i may be considered to be independent of concentration below x^*_i. Eq. (11) is not applicable if the component is a solid in the pure state and γ_i is referred to the liquid state.

Influence of an Extractive Liquid Phase

Another common situation is for a second liquid phase (e.g., an oil phase) to be present and to have significant extractive power for important volatiles. Such is the case for milk and citrus juices, and for coffee extract if coffee oil is present. In such a case writing Eq. (4) for both liquid phases and eliminating p_i yields

$$\gamma_i x_i = \gamma'_i x'_{i'} \tag{12}$$

where the prime marks on the terms on the right-hand side refer to the oil phase, and the unprimed terms, as before, refer to the aqueous phase. By analogy to Eq.(1), one can also define an oil-water *equilibrium distribution coefficient*, K_{ow}, as

$$K_{ow} = x'_i/x_i = \gamma_i/\gamma'_i \tag{13}$$

K_{ow} can also be called the partition coefficient between oil and water. For organic volatiles that are not highly polar, the volatiles will be relatively compatible with an oil phase, and γ'_i will be of order unity. Coupling this fact with the usually quite high values of γ_i means that K_{ow} can often be quite high, such as of order 1000.

A simple mass balance shows that the ratio of the amount of solute in the oil phase to the amount in the aqueous phase (R_{ow}) is given by

$$R_{ow} = K_{ow}(S/W) \tag{14}$$

where S and W are the molar amounts of oil and aqueous phases, respectively. The quantity on the right-hand side of Eq. (14) is known as the *extraction factor*. Since values of K_{ow} can be quite large, relatively small values of S/W can still be given substantial extraction. For example, with $K_{ow} = 1000$ and $S/W = 0.01$, 91% of the solute will have been taken into the oil phase at equilibrium.

When an aqueous solution is contacted with an extractive oil phase, a fraction of solute corresponding to Eq. (14) is taken out of the aqueous phase and into the oil phase at equilibrium. The resultant lower mole fraction of solute in the aqueous phase exerts a lower equilibrium partial pressure. Thus, extraction into an oil phase can reduce the loss of volatiles during a concentration or drying process, providing that the oil phase is sequestered from direct exposure to the vapor phase. Such is the case for spray drying, where addition of an emulsified oil phase has been shown to suppress losses of volatiles that are highly extracted into the oil (Zakarian and King 1982). On the other hand, in freeze drying an extracting oil phase tends to be directly exposed at interfaces left behind by subliming ice crystals, thereby causing nearly complete loss of extracted volatiles. Therefore, addition of an extractive oil phase can cause an increase in volatiles loss in freeze drying (Etzel and King 1980). Even in cases where the presence of an extractive oil phase increases volatiles retention, allowance must also be made for the fact that the oil suppresses the volatiles response in the equilbrium vapor over a reconstituted product (King and Massaldi 1974; Zakarian and King 1982).

Effect of a Solid Phase

A solid phase, present by itself or suspended in a liquid phase, can hold solutes through *adsorption* onto surface sites. There have been a number of studies of adsorption of volatile organics onto food-related substances. Representative among these are the work of Solms *et al.* (1973), Maier (1968, 1970A, 1975), and Issenberg *et al.* (1968). Whereas values of K_i and K_{ow} for fluid-phase equilibria tend to be relatively independent of solute concentration, as long as the solute concentration is low enough, partition coefficients for solutes between a solid surface and a fluid phase tend to be quite sensitive to solute concentration. The fraction adsorbed onto the solid surface, for a given ratio of solid surface area to fluid volume, usually becomes much higher as solute concentration decreases. This phenomenon can be attributed to the presence of a nonhomogeneous distribution of adsorption sites on the solid surface, with a few very strong sites dominating and holding the solute very strongly at very low solute concentrations.

When the solute concentration becomes high enough to saturate the surface, the amount of solute adsorbed usually tends to level off toward a certain constant value. Above this concentration, the partition coefficient for the solute becomes inversely proportional to the solute concentration in the fluid phase.

Effect of Complexation

Certain organic solutes can interact strongly through complexation in aqueous solution. Examples are cases where strong Lewis bases (e.g., amines) interact with strong acids (e.g., carboxylic acids). In such situations, the concentration of the free (uncomplexed) solutes in solution is diminished, and the equilibrium partial pressure of the solute(s), if volatile, decreases. The fraction of a solute that is complexed is determined by the equilibrium constant of the complexation reaction, and thereby depends on the relative proportions of the complexing reactants.

The effect of complexation in solution on volatiles equilibrium for food systems has not been studied much, although Solms *et al.* (1973) do discuss these and related phenomena. Complexation or adsorption onto solids are probably responsible for cases where the ratio of equilibrium partial pressure to solute concentration in solution drops off at low solute concentrations.

Methods of Measurements

By far the most common method for measuring vapor–liquid partition coefficients (K_i, H_i, and related quantities) is *vapor headspace analysis*, wherein a liquid solution is brought to thermal equilibrium at a controlled temperature, and the vapor phase in equilibrium with it is sampled and analyzed by gas chromatography. The liquid-phase concentration is determined simultaneously, either through direct analysis or through a mass balance based on the known composition of the feed liquid and the amounts of the two phases. This technique is described by Buttery *et al.* (1971) and by Kieckbusch and King (1979A).

Another method that has been used is monitoring the solute concentration in a gas phase over time as it equilibrates with a stagnant or gelled liquid. This technique is useful when the liquid is too viscous to be kept reliably mixed in a vapor headspace analysis.

Typical Experimental Results

Solubilities. Solubilities of sparingly soluble organic compounds in water are tabulated in a number of standard references. Coupled with pure-component vapor pressures, also from standard sources, these form one of the most extensive sources of experimental information on values of γ_i (Eq. 11) and K_i (Eq. 7).

Data are much less available for solubilities of sparingly soluble organic solutes in aqueous solutions that contain other major solutes. Among food-related systems, data have been reported for solubilities of *d*-limonene in 0–60% aqueous sucrose solutions (Massaldi and King 1973), for hexanol in 50–90% aqueous dextrin solutions (Grulke *et al.* 1980), and for various terpenes in 10% sucrose solution at 10°C (Smyrl and LeMaguer 1978).

Air–Water Partition Coefficient. Buttery *et al.* (1969, 1971) reported values of Henry's law constants (H_i) for families of alcohols, aldehydes, ketones, esters, and pyrazines in water, with no other solute present. Data for partitioning of other organic volatiles between air and otherwise pure water are given by Chandrasekaran and King (1971, 1972A) Massaldi and King (1973), Kieckbusch and King (1979A, B), and Grulke *et al.* (1980).

Influence of Major Solute(s) on Air–Water Partition Coefficients. Various major solutes ("dissolved solids") are present in food liquids and can have quite strong effects upon vapor–liquid partition coefficients for organic volatiles. Figure 15.1 shows results reported by Chandrasekaran and King (1972A) for activity coefficients (γ_i) of three organic volatiles, as influenced by increasing concentrations of fructose in aqueous solution. Two factors are apparent. (1) Activity coefficients (and hence partition coefficients, K_i) for different volatiles vary over more than three orders of magnitude, approaching 10^4 for *n*-hexanal in 60% sucrose solution. (2) The activity coefficient for a particular volatile compound increases substantially as fructose concentration increases. In some cases, the increase in activity coefficient approaches an order of magnitude as fructose concentration increases from 0 to 60%.

Not all major solutes cause activity coefficients and partition coefficients of volatile organics to increase as the concentration of the major solute increases. Massaldi and King (1973) noted that increasing sucrose concentration caused activity coefficients of *n*-hexyl acetate to increase, whereas activity coefficients for *d*-limonene and *n*-butylbenzene decreased. Figure 15.2 shows results obtained by Kieckbusch and King (1979B) for *n*-pentyl acetate in various carbohydrate solutions. The activity coefficient increases with increasing major-solute concentration in solutions of maltose, sucrose, and dextran, while it decreases with increasing major-solute concentration in solutions of maltodextrin, dextrin, and coffee extract. Data for the effect of major-solute concentration on activity coefficients and partition coefficients in other systems have been reported by McDowall (1959), Wientjes (1968), and Voilley *et al.* (1977).

Maier (1970B) made a systematic study of the influence on equilibrium partition coefficients of various volatile organics caused by addition of various nonvolatile solutes. The nonvolatile solutes examined included proteins (casein, gelatin, ovalbumin), simple sugars (fructose, sucrose, sorbitol), higher-molecular-weight carbohydrates (dextrin, pectin), methylcellulose, and starch. All these solutes served to suppress the volatilities of acetone and acetaldehyde. Suppression of volatilities of ethanol, ethyl acetate, and diethyl ether also occurred with a number of the added solutes, but addition of the simple sugars, oval-

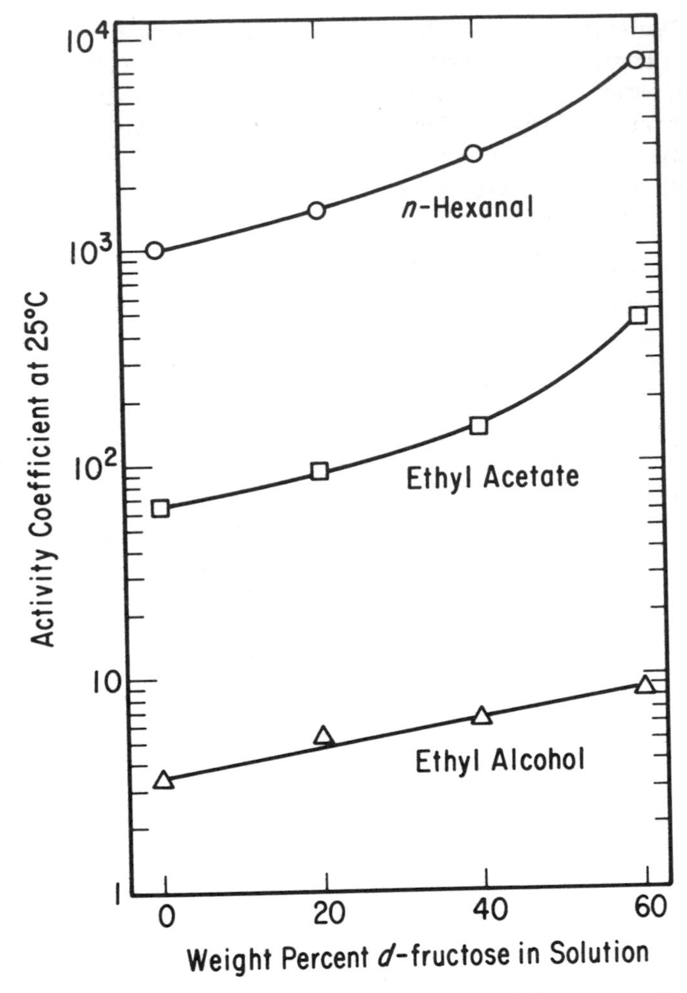

Fig. 15.1. Activity coefficients of trace organic volatile components in aqueous solutions of *d*-fructose at 25°C.
From Chandrasekaran and King (1972A).

bumin, starch, and casein served to increase volatilities of those solutes in several cases. The results observed for addition of simple sugars are in general agreement with the observations of Chandrasekaran and King (1972A), Massaldi and King (1973), and Kieckbusch and King (1979B) on similar systems.

These results probably reflect the conflicting effects of (1) intermolecular associations or even complexation between the volatile compound and the added solute, which suppress the volatility, and (2) salting-out effects from the added solute, which increase the volatility. The identi-

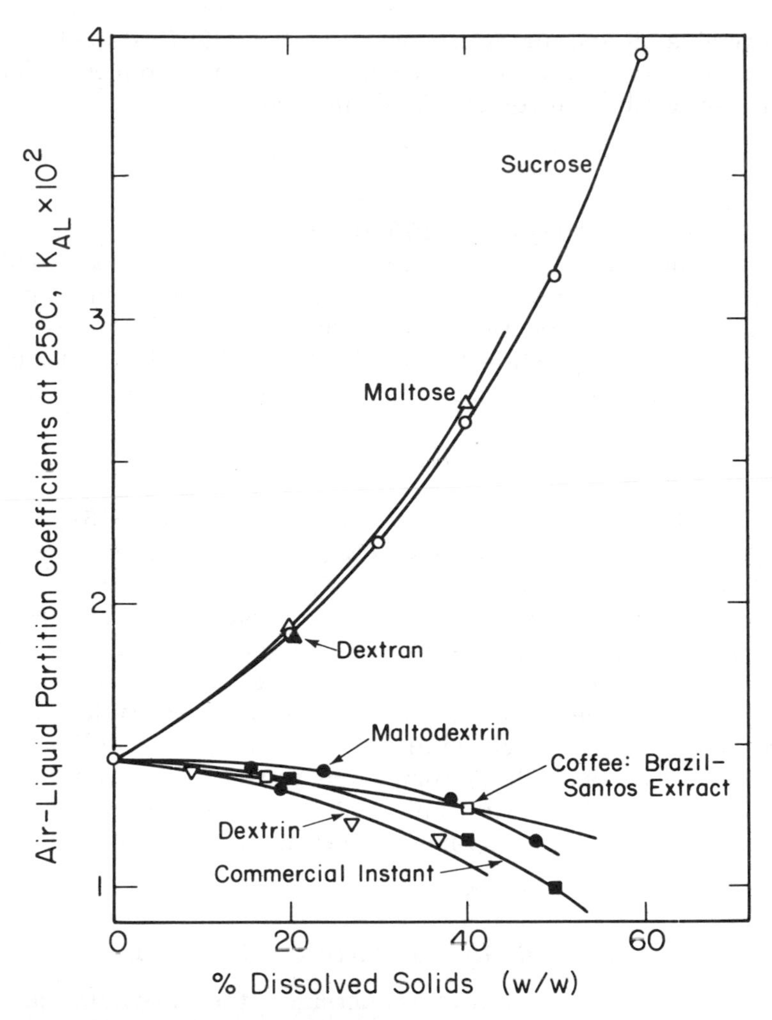

Fig. 15.2. Partition coefficients (concentration in gas/phase concentration in liquid phase) for *n*-pentyl acetate in aqueous solutions of various carbohydrates and coffee extract at 25°C.
From Kieckbusch and King (1979B).

ties of phenomena at play and which of them will dominate under different circumstances are still poorly understood, and are deserving of more research.

Oil–Water Partition Coefficients

Buttery *et al.* (1973) have made measurements of equilibrium partition coefficients for various aldehydes, ketones, and pyrazines between a vegetable oil and air. Kieckbusch and King (1979B) made similar measurements for a family of acetates. These measurements yield an oil-to-air partition coefficient, K'_i, defined as

$$K'_i = y_i/x'_i \tag{15}$$

at equilibrium. Buttery *et al.* (1973) report a concentration-based partition coefficient, which can be readily converted to K'_i, with knowledge of the molecular weight of the oil, or an assumption concerning it.

Values of the oil–water partition coefficient, K_{ow}, can be derived from K'_i and corresponding measured values of K_i for the same volatile compound in aqueous solution through the relationship

$$K_{ow} = K_i/K'_i \tag{16}$$

which is obtained by combination of Eqs. (1), (13), and (15).

Massaldi and King (1974) show how gas-chromatographic headspace analyses of oil–water systems with varying added quantities of a partitioning volatile substance can be used to determine the overall concentration of that volatile substance originally present, as well as the value of K_{ow}. They present results for the distribution of *d*-limonene between the lipid and aqueous phases in various orange juices. Radford *et al.* (1974) report experimental observations for the equilibrium distribution of various volatile compounds between the serum and pulp portions of centrifuged fruit juices. McDowell (1959) reports equilibrium data for the distribution of several substances between butterfat and water in cream.

Correlation and Prediction of Partition Coefficients

In recent years several effective molecular–thermodynamic methods have been developed for interpretation and correlation of experimentally measured partition coefficients. For present purposes the most useful of these is probably the UNIFAC method, which is based on summing contributions of individual functional groups in solute and solvent molecules in order to predict the activity coefficient for the solute in solution (Reid *et al.* 1977, Chapter 8). An extensive development of the UNIFAC method and tabulation of interaction parameters for various functional groups is given by Fredenslund *et al.* (1977), Skjold-Jørgenson *et al.* (1979), and Gmehling *et al.* (1982).

In its present state of development, the UNIFAC method is useful for prediction of activity coefficients of numerous organic solutes in water solution with good accuracy. With modified interaction parameters (Sorensen *et al.* 1979) it is also useful for approximate prediction of partition coefficients between water and an immiscible organic phase. However, there are not yet good methods for accounting for the effects of added major solutes such as carbohydrates or proteins on the activity coefficients, and experimental data must still be relied on in those cases.

The UNIFAC method and other related methods (Reid *et al.* 1977, Chapter 8) yield activity coefficients. For prediction of air–liquid partition coefficients, the activity coefficients must be used in conjunction with pure-substance vapor pressures (Eq. 7). An extensive compilation of vapor-pressure data is given by Boublik *et al.* (1973), and methods for prediction of vapor pressures are discussed by Reid *et al.* (1977, Chapter 6).

MASS-TRANSFER PROPERTIES

As has been pointed out previously, the relatively high volatilities of most important flavor and aroma compounds will cause quite high percentage losses during evaporative processing, unless mass transfer of the compound becomes sufficiently rate-limiting to retard the loss. Mass-transfer analyses of drying and concentration processes are usually rather complex. General mass-transfer considerations that need to be taken into account are developed in a number of references—for example, Sherwood *et al.* (1975) and King (1980, Chapter 11). More specific mass-transfer analyses have been presented by Thijssen and Rulkens (1968), Kerkhof and Schoeber (1974), Kerkhof and Thijssen (1977), Kieckbusch and King (1980), and Zakarian and King (1982) for spray drying; by Thijssen and Rulkens (1968), Bomben *et al.* (1973), King (1970, 1973), and Etzel and King (1980) for freeze drying; and by Chandrasekaran and King (1972B) for drying of slabs.

Under conditions leading to good volatiles retention, mass transfer within the material being dried becomes the rate-limiting factor. The transport property governing mass transfer within the material is the *molecular diffusivity*. If diffusion occurs within a stagnant, consolidated medium of simple geometry, the mass-transfer rate is predictable and correlatable through solutions to Fick's law of diffusion, written for the particular geometry. However, frequently other factors enter as well, such as development of internal voidage and/or irregular geometric shape and convection currents within the medium.

Salient Features

There are several common features of mass transfer properties in food liquids and related model systems that strongly affect volatiles retentions:

1. The diffusion coefficient for water in most carbohydrate and protein solutions, and in food systems in general, decreases substantially as the concentration of major solutes increases. Figure 15.3 shows data reported by Bomben *et al.* (1973) for diffusion of water in various food-related systems, while Fig. 15.4 shows data reported by Chandrasekaran and King (1972A) for diffusion of water in solutions of sucrose and glucose. The diffusion coefficient at high concentrations

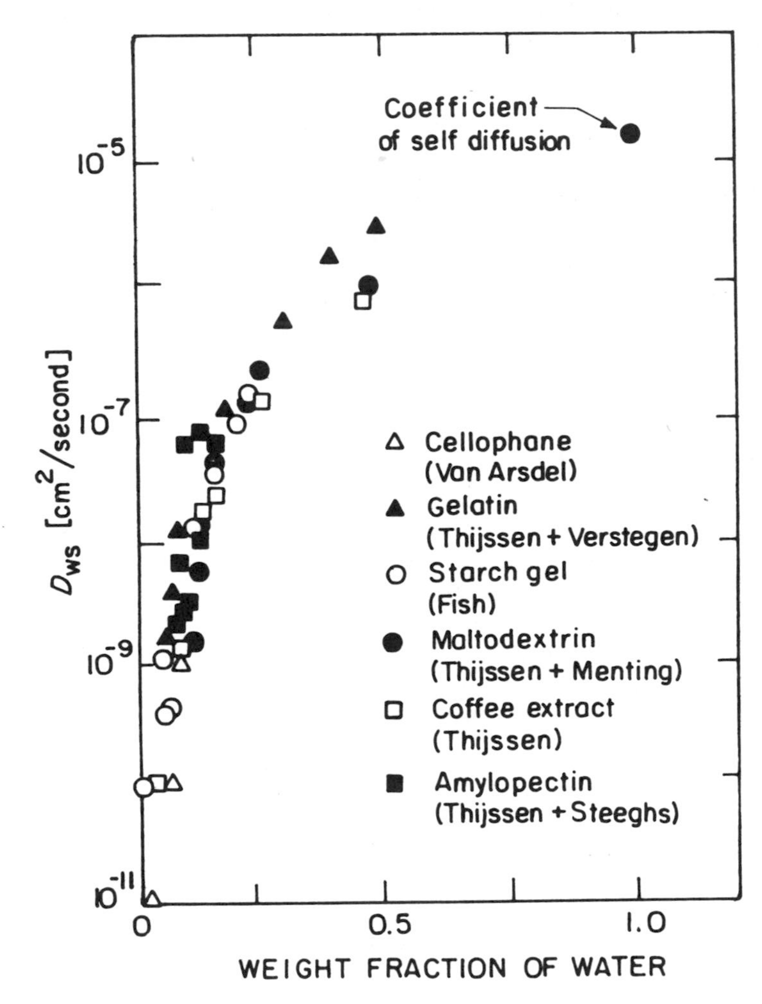

Fig. 15.3. Diffusivities of water in aqueous solutions of various food components and related substances, assumed to be at 25°C.
From Bomben et al. (1973).

of dissolved solids can be several orders of magnitude lower than that at low concentration.

2. As the concentration of major solutes increases, especially at high concentrations, the diffusion coefficients for dilute volatile organic substances drop off even more sharply than the diffusion coefficients of water. Figure 15.5 shows results presented by Bomben *et*

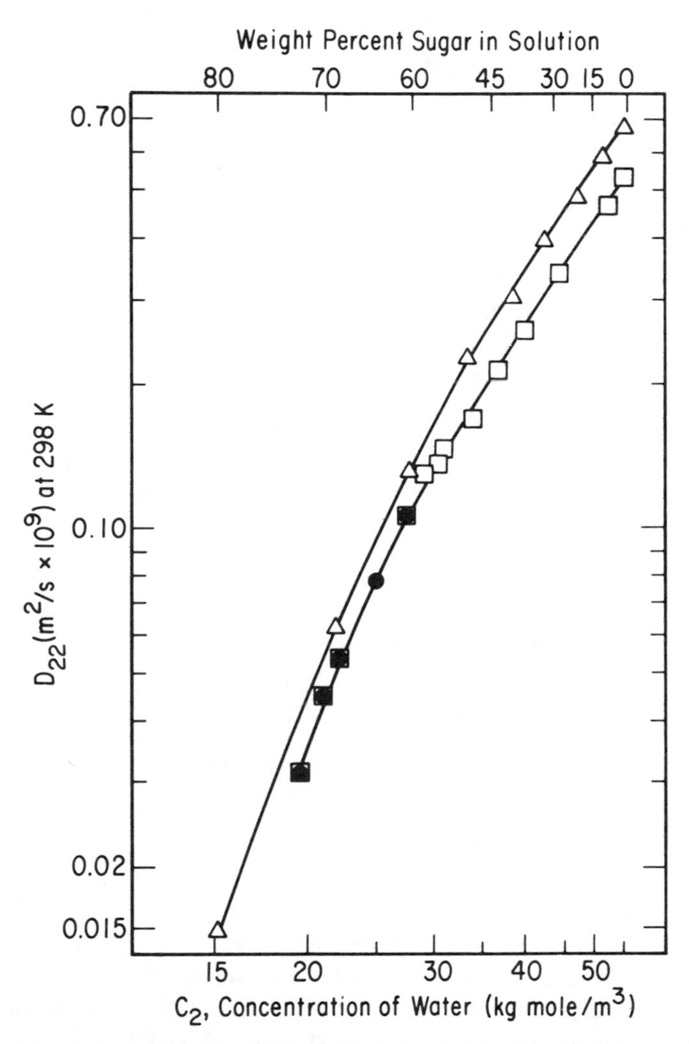

Fig. 15.4. Diffusion coefficients (D_{22}) of water in aqueous solutions of sucrose (●) (□), *d*-fructose and *d*-glucose (△) at 25°C.
From Chandrasekaran and King (1972A).

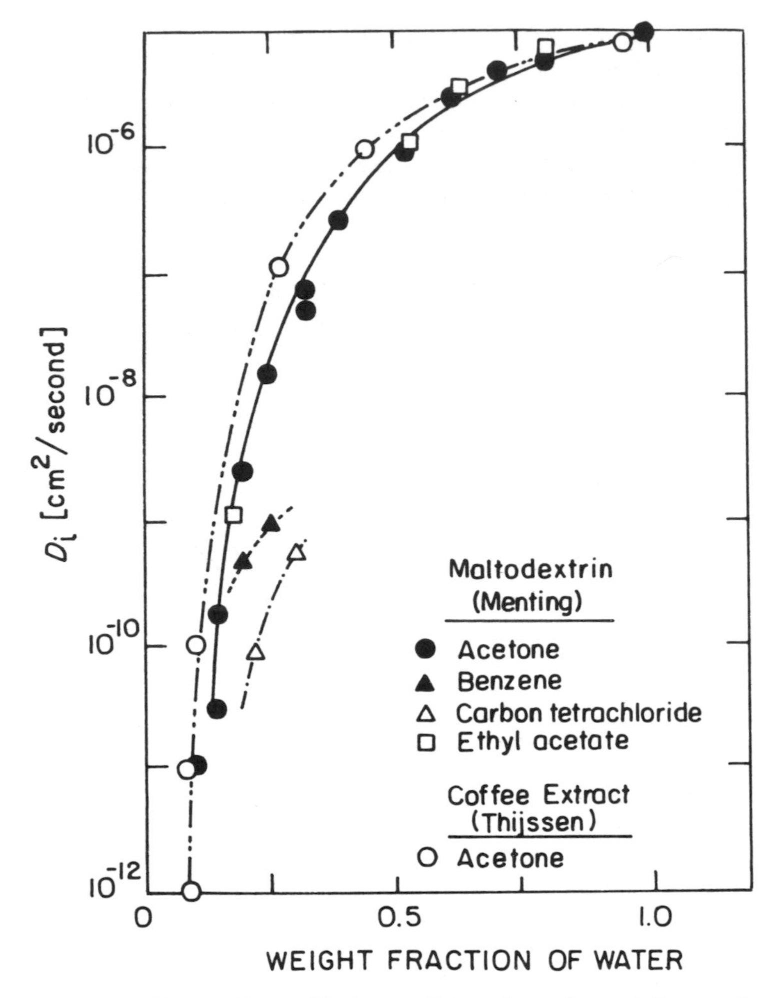

Fig. 15.5. Effective binary diffusion coefficients for various volatile organic solutes at high dilution in aqueous solutions of various food components and related substances, assumed to be at 25°C.
From Bomben et al. (1973).

al. (1973) for a variety of volatile solutes in solutions of maltodextrin and coffee extract. Figure 15.6 shows the ratio of diffusion coefficient for volatile organics to the diffusion coefficient of water, as reported by Chandrasekaran and King (1972A) for ethyl acetate in sucrose solutions, and as reported by Menting *et al.* (1970A) for acetone in maltodextrin solutions. These results, and ones similar to them, support the theory of selective diffusion (Thijssen and Rulkens 1968), which

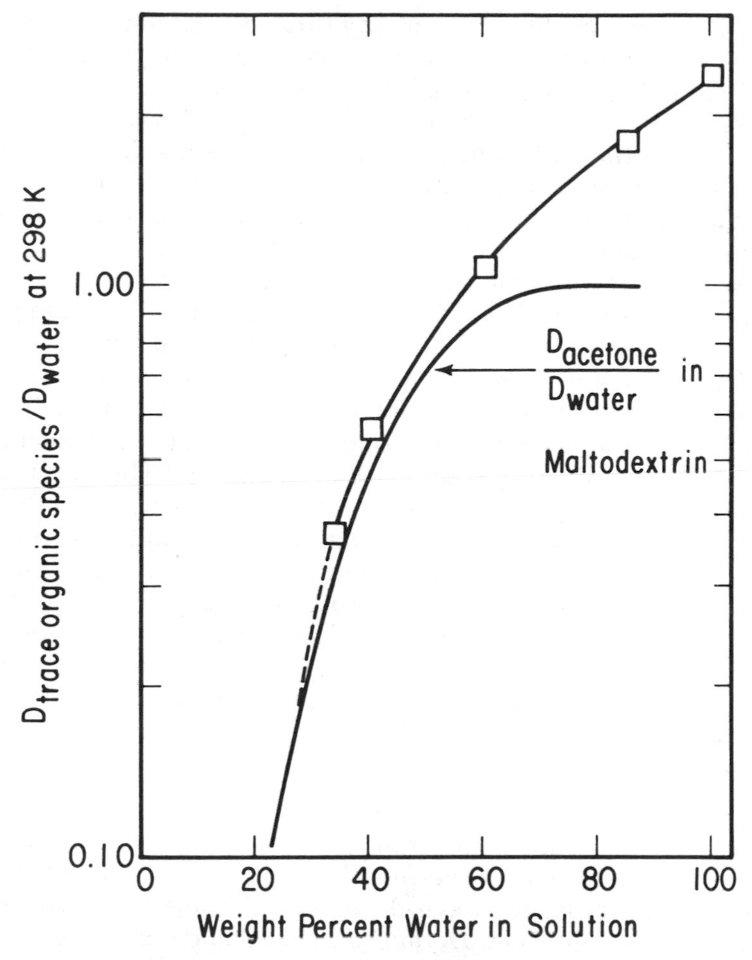

Fig. 15.6. Ratio of effective binary diffusion coefficient for volatile organic compound at high dilution (D_{11}) to binary diffusion coefficient for water (D_{22}) at 25°C. Data for the systems ethyl acetate–sucrose–water (□), and acetone–maltodextrin–water (Menting *et al.* 1970A).
From Chandrasekaran and King (1972A).

postulates that above some certain concentration of major dissolved solutes the volatile compounds are no longer sufficiently mobile to reach the surface and be lost. As a consequence of this very useful theory, a major goal in the design of concentration and drying processes is to establish conditions (e.g., fast drying) that lead rapidly to a high concentration of major solutes at the surface, thereby "sealing" the surface against further volatiles loss.

3. Reducing the diffusive mobility through a build-up of dissolved-solids concentration at the surface requires that the process have entered the falling-rate period of drying. In the falling-rate period the resistance to mass transfer inside the material becomes more rate-limiting than the mass-transfer resistance in the gas phase outside the material. Because of this, solute concentration gradients develop within the material, allowing the surface concentration of the major solutes to become substantially greater than the bulk average concentration.

To a first approximation, then, volatiles loss will slow down greatly or cease altogether once the drying process enters the falling-rate period. The time of entry into the falling-rate period depends on the diffusion coefficient of water, rather than the diffusion coefficients of volatiles. Hence we have the interesting situation that, to a first approximation, the loss of volatiles can be interpreted in terms of the diffusion coefficient for water (Menting *et al.* 1970B; Kerkhof and Thijssen 1977).

4. When an evaporative flux of water is present, the diffusional mass-transfer process controlling volatiles loss should properly be interpreted in terms of multicomponent diffusion concepts. Water, dissolved solids (major solutes), and volatiles should be accounted for separately, leading to a ternary diffusion analysis. The water flux influences volatiles transport in two ways—through convection of volatiles with the flux of water, and through the gradient in activity coefficient for a volatile component which results from the gradient of water concentration.

In the ternary diffusion analysis, there are three independent diffusion coefficients (Chandrasekaran and King 1972A). One is the effective binary diffusivity for the volatile compound (D_{11}); another is the binary diffusivity for the water-dissolved solids mixture (D_{22}); and the third is a cross coefficient (D_{12} or D_{21}). The other cross coefficient is not independent and is related to the other three through the Onsager relationships, which stem from the theory of irreversible thermodynamics.

Chandrasekaran and King (1972B) show how the ternary diffusion equations can be applied to the analysis of volatiles loss from a drying slab. The resulting concentration profiles for volatiles can exhibit an internal maximum.

Measurement Techniques

A common technique for the measurement of diffusion coefficients in liquid systems utilizes glass-diaphragm cells, which have two well-mixed compartments separated by a diaphragm of fritted glass. This technique is suitable for liquids of low-to-moderate viscosity but is impractical for liquids of high viscosity because of the difficulty of maintaining complete mixing in the cell compartments. Measurements may be made with good precision and reproducibility. However, the experiments do take several days or even weeks per cell per data point, and it is necessary to calibrate with a system of known diffusivity. Examples of the use of diaphragm cells are the work of Chandrasekaran and King (1972A) and Frey and King (1982).

Methods for measuring diffusivities in liquids of high viscosity generally involve immobilization of a precisely known amount of solution in a system with known and controlled dimensions. The rate of uptake or loss of a volatile solute into or out of the immobilized body of solution can then be followed over time, and the indicated diffusivity can be obtained by application of the simple penetration theory of mass transfer. This technique has been used by Menting *et al.* (1970A) and Grulke *et al.* (1980).

Experimental Results

Diffusivities of water in solutions containing various food-related solutes are given in Fig. 15.3 (Bomben *et al.* 1973; Menting *et al.* 1970A) and Fig. 15.4 (Chandrasekaran and King 1972A). These all show the common feature of a substantial drop-off of diffusivity with increasing solute content.

Diffusivities of acetone in aqueous solutions containing from 0 to 90% w/w maltodextrin are reported by Menting *et al.* (1970A). These and other related data are summarized in Fig. 15.3 and by Thijssen (1971). Chandrasekaran and King (1972A) report diffusivities (including cross coefficients for a ternary analysis) for ethanol and ethyl acetate in solutions containing 0–60% fructose, glucose, sucrose, or mixtures of these sugars. The effect of temperature was also determined, along with isolated measurements of diffusivities of *n*-butyl acetate and *n*-hexanal in sucrose solutions. Frey and King (1982) have measured diffusivities (D_{11} only) for ethyl, *n*-propyl, *n*-butyl, and *n*-pentyl acetates in sucrose solutions with concentrations up to 55% and have also determined the effect of temperature. Grulke *et al.* (1980) report effective binary diffusivities for *n*-hexanol in 50% and 70% dextrin solutions as a function of temperature.

Hikita *et al.* (1978) report diffusivities of oxygen in sucrose solutions for sucrose concentrations up to 54%. The diffusivity of oxygen drops off more sharply than that of water as the sucrose concentration increases.

Correlation and Prediction of Diffusion Coefficients

Reid *et al.* (1977, Chapter 11) outline the more commonly used empirical methods for correlation and prediction of diffusivities in binary liquid systems. The correlations are most effective for solutes present at high dilution, including cases of solutes in dilute solution in water. More uncertainty is involved when experimental or predicted diffusivities at high dilution are extended through available methods to predict diffusivities in more concentrated binary solutions. There are no reliable methods for prediction of effective diffusivities of dilute solutes in solutions that are themselves mixtures of other major components, nor are there any reliable methods for prediction of cross coefficients in ternary systems. Temperature changes are usually accounted for by postulating that the group $D\mu/T$, (where D is the diffusivity, μ is the viscosity, and T is the absolute temperature) is constant, unless experimental data are available for activation energies of diffusion in the system of interest.

Taking the system ethyl butyrate (volatile compound at high dilution)−lactose (major solute)−water as an example, one could predict the diffusivity of lactose in water at high dilution and the diffusivity of ethyl butyrate in water at high dilution from standard empirical correlations with some confidence, if experimental data are not available. However, only a first approximation could be made by existing methods for predicting the diffusivity of lactose in concentrated aqueous solution, and any method of prediction for the diffusivity of ethyl butyrate in this concentrated aqueous lactose solution would be highly uncertain. The latter two quantities should best be measured experimentally.

NOMENCLATURE

C	Concentration (moles/volume)
f	Fraction loss (Eq. 9)
H	Henry's law constant (Eq. 2)
K	Equilibrium partition coefficient (Eq. 1)
K_{AL}	Concentration-based partition coefficient (Fig. 15.2)
K'	Oil to air partition coefficient
L	Moles of liquid
p	Partial pressure
P	Total pressure
P^o	Vapor pressure
R	Molar ratio (Eq. 14)
S	Moles of oil phase
V	Moles of vapor
W	Moles of aqueous phase

x	Aqueous phase mole fraction
x'	Oil phase mole fraction
x^*	Solubility, expressed as mole fraction
y	Vapor phase mole fraction
α_{ij}	Relative volatility of i to j (Eq. 3)
γ	Activity coefficient in aqueous phase (Eq. 4)
γ'	Activity coefficient in oil phase

Subscripts

i	Component i, a volatile component
w	Water
ow	Between oil and water

BIBLIOGRAPHY

BOMBEN, J.L., BRUIN, S., THIJSSEN, H.A.C., and MERSON, R.L. 1973. Aroma recovery and retention in concentration and drying of foods. *In* Advances in Food Research. G.F. Stewart, E. Mrak and C.O. Chichester (Editors), Vol. 20. Academic Press, New York.

BOUBLIK, T., FRIED, V., and HALA, E. 1973. The Vapour Pressures of Pure Substances. Elsevier Publ. Co., Amsterdam.

BUTTERY, R.G., LING, L.C., and GUADAGNI, D.G. 1969. Food Volatiles. Volatilities of aldehydes, ketones and esters in dilute water solutions. J. Agric. Food Chem. *17*, 385–389.

BUTTERY, R.G., BOMBEN, J.L., GUADAGNI, D.G., and LING, L.C. 1971. Some considerations of the volatilities of organic flavor compounds in foods. J. Agric. Food Chem. *19*, 1045–1048.

BUTTERY, R.G., GUADAGNI, D.G., and LING, L.C. 1973. Flavor Compounds: Volatilities in vegetable oil and oil–water mixtures. Estimation of odor thresholds. J. Agric. Food Chem. *21*, 198–201.

CHANDRASEKARAN, S.K., and KING, C.J. 1971. Retention of volatile flavor components during drying of fruit juices. Chem. Eng. Prog. Symp. Ser. *67* (108), 122–130.

CHANDRASEKARAN, S.K., and KING, C.J. 1972A. Multicomponent diffusion and vapor-liquid equilibria of dilute organic components in aqueous sugar solutions. AIChE J. *18*, 513–520.

CHANDRASEKARAN, S.K., and KING, C.J. 1972B. Volatiles retention during drying of food liquids. AIChE J. *18*, 520–526.

ETZEL, M.R., and KING, C.J. 1980. Retention of volatile components during freeze drying of substances containing emulsified oils. J. Food Technol. *15*, 577–588.

FREDENSLUND, A., GMEHLING, J., and RASMUSSEN, P. 1977. Vapor–Liquid Equilibria Using UNIFAC. Elsevier Publ. Co., Amsterdam.

FREY, D.D., and KING, C.J. 1982. Diffusion coefficients of acetates in aqueous sucrose solutions. J. Chem. Eng. Data *27*, 419–422.

GMEHLING, J., RASMUSSEN, P., and FREDENSLUND, A. 1982. Vapor-liquid equilibria by UNIFAC group contribution. Revision and extension, 2. Ind. Eng. Chem. Process Des. Dev. *21*, 118–127.

GRULKE, E.A., GRAY, J.I., and OOSTING, E. 1980. Flavor Retention in model food systems. AIChE Symp. Ser., In press.

HIKITA, H., ASAI, S., and AZUMA, Y. 1978. Solubility and diffusivity of oxygen in aqueous sucrose solutions. Can. J. Chem. Eng. *56*, 371–374.

ISSENBERG, P., GREENSTEIN, G., and BOSKOVIC, M. 1968. Adsorption of volatile organic compounds in dehydrated food systems. I. Measurements of sorption isotherms at low water activities. J. Food Sci. *33*, 621–625.

KERKHOF, P.J.A.M., and SCHOEBER, W.J.A.H. 1974. Theoretical modelling of the drying behaviour of droplets in spray dryers. *In* Advances in Preconcentration and Dehydration of Foods. A. Spicer (Editor) pp. 349–397. Applied Science Publishers, London.

KERKHOF, P.J.A.M., and THIJSSEN, H.A.C. 1977. Quantitative study of the effects of process variables on aroma retention during the drying of liquid foods. AIChE Symp. Ser. *73* (163), 33–46.

KIECKBUSCH, T.G., and KING. C.J. 1979A. An improved method of determining vapor-liquid equilibria for dilute organics in aqueous solution. J. Chromatogr. Sci. *17*, 273–276.

KIECKBUSCH, T.G., and KING, C.J. 1979B. Partition coefficients for acetates in food systems. J. Agric. Food Chem. *27*, 504–507.

KIECKBUSCH, T.G., and KING, C.J. 1980. Volatiles loss during atomization in spray drying. AIChE J. *26*, 718–725; *27*, 528.

KING, C.J. 1970. Freeze Drying of Foodstuffs. CRC Crit. Rev. Food Technol. *1*, 379–451.

KING, C.J. 1973. Freeze drying. *In* Food Dehydration, 2nd Ed., Vol. 1, Chapter 6. W.B. Van Arsdel, M.J. Copley and A.I. Morgan, Jr. (Editors). 1, AVI Publishing Co., Westport, CT.

KING, C.J. 1980. Separation Processes. 2nd Ed. McGraw-Hill Book Co., New York.

KING, C.J., and MASSALDI, H.A. 1974. Aroma preservation during evaporation and drying of liquid foods: Consideration of mechanism and effects of more than one phase being present. Proc. 4th Int. Congr. Food Sci. Technol. *4*, 183–202.

McDOWELL, F.H. 1959. Steam distillation of taints from cream. Parts IV-VII. J. Dairy Res. *26*, 24–32, 33–38, 39–45, 46–52.

MAIER, H.G. 1968. Bindung von aromastoffen an lebensmitteln. Naturwissenschaften *55* (4), 180–181.

MAIER, H.G. 1970A. Zur bindung fluchtiger aromastoffe an lebensmitteln. Z. Lebensm. Unters. Forsch. *144* (1), 1–4.

MAIER, H.G. 1970B. Volatile flavoring substances in foodstuffs. Angew. Chem., Int. Ed. Engl. *9*, 917–926.

MAIER, H.G. 1975. Binding of volatile aroma substances to nutrients and foodstuffs. Proc. Int. Symp. Aroma Res., Zeist, Pudoc, pp. 143–157. Wageningen, Netherlands.

MASSALDI, H.A., and KING, C.J. 1973. Simple technique to determine solubilities of sparingly soluble organics: Solubility and activity coefficients of *d*-limonene, *n*-butylbenzene, and *n*-hexyl acetate in water and sucrose solutions. J. Chem. Eng. Data *18*, 393–397.

MASSALDI, H.A., and KING, C.J. 1974. Determination of volatiles by vapor headspace analysis in a multi-phase system: *d*-Limonene in orange juice. J. Food Sci. *39*, 434–437.

MENTING, L.C., HOOGSTAD, B., and THIJSSEN, H.A.C. 1970A. Diffusion coefficients of water and organic volatiles in carbohydrate-water systems. J. Food Technol. *5*, 111–126.

MENTING, L.C., HOOGSTAD, B., and THIJSSEN, H.A.C. 1970B. Aroma retention during the drying of liquid foods. J. Food Technol. *5*, 127–139.

RADFORD, T., KAWASHIMA, K., FRIEDEL, P.K., POPE, L.E., and GIANTURCO, M.A. 1974. Distribution of volatile compounds between the pulp and serum of some fruit juices. J. Agric. Food Chem. *22*, 1066–1070.

REID, R.C., PRAUSNITZ, J.M., and SHERWOOD, T.K. 1977. Properties of Gases and Liquids. 3rd Ed. McGraw-Hill Book Co., New York.

SHERWOOD, T.K., PIGFORD, R.L., and WILKE, C.R. 1975. Mass Transfer. McGraw-Hill Book Co., New York.

SKJOLD-JØRGENSEN, S., KOLBE, B., GMEHLING, J., and RASMUSSEN, P. 1979. Vapor-liquid equilibria by UNIFAC group contribution. Revision and extension. Ind. Eng. Chem. Process Des. Dev. *18*, 714–722.

SMYRL, T.G., and LEMAGUER, M. 1978. Retention of sparingly soluble volatile compounds during the freeze drying of model solutions. J. Food Process Eng. *2*, 151–170.

SOLMS, J., OSMAN-ISMAIL, F., and BEYELER, M. 1973. The interaction of volatiles with food components. Can. Inst. Food Sci. Technol. J. *6* (1), A10–A16.

SORENSEN, J.M., MAGNUSSEN, T., RASMUSSEN, P., and FREDENSLUND, A. 1979. Liquid-liquid equilibrium data. Their retrieval, correlation and prediction. Fluid Phase Equil. *2*, 297–309; *3*, 47–82.

THIJSSEN, H.A.C. 1971. Flavour retention in drying preconcentrated food liquids. J. Appl. Chem. Biotechnol. *21*, 372–377.

THIJSSEN, H.A.C., and RULKENS, W.H. 1968. Retention of aromas in drying food liquids. Ingenieur (The Hague) *80* (47), Ch45–Ch56.

VOILLEY, A., SIMATOS, D., and LONCIN, M. 1977. Gas phase concentration of volatiles in equilibrium with a liquid aqueous phase. Lebensm. Wiss. Technol. *10*, 45–49.

WIENTJES, A. 1968. The influence of sugar concentration on the vapor pressure of food odor volatiles in aqueous solutions. J. Food Sci. *33*, 1–2.

ZAKARIAN, J.A., and KING, C.J. 1982. Volatiles loss in the nozzle zone during spray drying of emulsions. Ind. Eng. Chem. Process Des. Dev. *21*, 107–113.

16

Expression-Related Properties

Henry G. Schwartzberg[1]

INTRODUCTION

Expression is the act of expelling liquid from a solid – liquid mixture by squeezing or compaction. The compaction is usually produced by movement of solid or perforated barriers pressing against surfaces of the mixture. However, centrifugal or gravitational forces, pneumatic pressure, osmotic pressure or internal intermolecular attraction can, in some cases, be used to induce compaction.

In the case of foods, such mixtures often consist of moist particles with air interspersed among them. During the initial stages of compaction the interparticle porosity decreases and air is expelled. Subsequently, liquid is forced out of the particles into the interparticle pores; and when these pores are full of liquid, or nearly so, outflow from the mixture begins. In subsequent discussion we will call the solid – liquid mixture, the "presscake" or simply the "cake," and the perforated or porous surface through which outflow occurs, the "media."

The fluid in the particles may be contained in relatively closed cells, which usually have to be ruptured to facilitate rapid fluid expulsion. Often, compaction itself generates fluid pressures within the cells, which induces rupture. The cell walls may be quite weak as in the case of oranges or grapes, or very strong as in the case of sugar cane.

In other cases, the particles are sponge-like and contain a communicating network of pores; and there is no need to induce cell rupture. The pore structure may be readily visible; or, in the cases of curds, gels, and flocs, the pores may be too small to be distinguishable by eye. Because of the fineness of such pore structures fluid expulsion may be quite slow.

[1]Department of Food Engineering, University of Massachusetts, Amherst, MA 01003.

Expression is also used to expel fluid from slurries and beds that contain fluid but no interspersed air. In such cases compaction induces immediate outflow.

When dilute slurries are subject to expression, a solids-rich and a fluid-rich layer may form because of settling and particle retention at the outflow surface. When such layers form, the process resembles pressure-induced filtration until the fluid-rich layer disappears and compaction of the solids-rich bed starts. This type of process has been extensively studied by Shirato and his coworkers (1969, 1970, 1971, 1977A,B).

CHARACTERISTICS

The bed and particle properties that influence expression, collectively lumped together, constitute a material's expression characteristics. These properties include the compaction stress vs. volume characteristics of the solid-like components of the presscake; the filtration resistance of the cake and the flow resistance of the cake-media interface; the relationship between the axial compaction stress and the concurrently generated transverse stress; the coefficient of friction between the moving presscake and adjacent surfaces; the pressures generated inside particle cells by their internal fluid content; the rupture strength of such cells; the relationship of specific bulk volume to pore volume and to expression yield, etc.

In many cases these properties are not wholly intrinsic but are partly response-related and depend somewhat on the way expression is carried out.

USES OF EXPRESSION

Expression is used to (a) recover fruit, vegetable, and meat juices; (b) recover vegetable oils from oil-rich seeds and animal and fish fats and oils from dry-rendering residues; (c) expel whey from cheese curds while consolidating the curds into solid blocks; (d) dewater fibrous webs, felts, and mats; (e) dewater moist processing wastes, by-products and processing intermediates; (f) control the cocoa butter content of cocoa; (g) recover colloidal protein contained in plant-cell juices; (h) recover sugar juice from sugar cane; (i) recover pectin liquors from cooked pomace; (j) dewater moist, combustible materials so as to enhance their fuel value; (k) facilitate subsequent imbibition of fluids by resilient porous solids; and (l) produce pulps and purees. Materials whose processing involves expression are listed in Table 16.1.

TABLE 16.1. MATERIALS FREQUENTLY SUBJECTED TO EXPRESSION

Sugar Cane	Tomatoes
Bagasse	Leaching Residues
Spent Sugar Beet Cosettes	Paper Webs
Peanuts	Cooked Seaweeds
Olives (for making olive oil)	Tofu
Cocoa	Protein-Fiber Residues
Rapeseed	from Wet Corn Milling
Copra and Coir	Potato Slurries
Corn Germ	Pectin Greens
Potato Processing Wastes	Spent Coffee Grounds
Alfalfa	Roasted Ground Coffee
Protein Concentrates	Grape Pits
Gluten	Winery Lees
Fish Meal	Spent Brewing Mashes
Grape Pomace	Cheese Curds
Apple Chunks	Stillage
Cooked Apple Pomace	Starch Sponges
Citrus Peel	Cassava
Citrus Fruit Halves	Sauerkraut
Whole Citrus Fruit	Silage
Citrus Pulp	Haylage

EXPRESSION EQUIPMENT

The most common types of expression or expression equipment are single screw expellers, double screw presses, cake presses, enclosed cylinder presses, filter presses with compactible frames, cheese presses, rack and cloth presses, inflatable-tube presses, belt presses, roll mills, dice mills, juice reamers, interlocking-finger extractors, overburden induced expression, centrifugally induced expression, and syneresis. Since information on expression equipment is readily available, the design and operation of this type of equipment need not be covered here.

PRESSING FORCES

Regardless of the type of expression equipment used, the major factors governing expression are the interaction between the extent and rate of compaction, the expressed liquid yield, and the forces required to achieve compaction. Pressures of up to 16,000 psi are used in oilseed expellers. Pressures up to 6000 psi are used in the nip of sugar cane roll mills; 500 – 1000 psi to dewater spent sugar beets, and 1000 – 2000 psi for spent coffee grounds. The maximum pressures for fruit juice expression are in the 200 – 450 psi range. At the low end of the scale, pressures as low at 2 – 5 psi are used to press semi-hard cheese and 6 – 12 psi for hard cheeses.

Three types of forces occur: the compaction stress required to reduce the volume of the presscake; the fluid pressure-drop required to propel expelled fluid through pores in the cake and through the outflow media; and friction between the presscake and surfaces of the cavity in which it is confined. The forces interact, so that fluid-pressure drops tend to increase compaction stress and friction tends to reduce it. The fluid pressure drop depends on local fluid velocities and on the filtration resistance of the presscake. This resistance, in turn, depends on the presscake's particle size distribution and degree of compaction. The local fluid velocities depend on local rates of change of specific volume. The local compaction stress depends on local specific volume, the structural characteristics of the solid itself, and pressure exerted by fluid occluded in the solid. Obviously it is impossible to analyze these interactions and measure the factors that affect them without the use of mathematical models for the expression process. The expression properties occur as parameters in these models.

DEFINITIONS

To facilitate discussion and to identify various aspects of expression behavior, special definitions will be used. The insoluble solids that cannot pass through the outflow media will be called the "marc." Fine solids, which the fluid carries through the media and whose passage is undesirable, will be called "dregs." The terms are not mutually exclusive since fines, which might pass through the media, will at times be trapped in the cake and form part of the marc. Solids that are extruded through the media by compaction pressure, when such extrusion is not desired, will be called "foots." Migratable fines are particles that migrate to some extent because of fluid flow in the cake. These include the dregs. Sometimes the inclusion of certain fines in the expelled juice is desired, e.g., chloroplasts and other protein-rich organelles in alfalfa juice. These will be called "juice fines."

"Bound water" is water so strongly bound to the marc that it cannot be expelled at all. Under certain circumstances the marc will break up completely due to the dissolution of the protopectin holding it together or because of comminution and will readily be extruded through openings larger than some critical size. The converted marc in such cases will be called "pulp."

While the volume and height of the presscake change during pressing, the dry mass of marc provides a convenient invariant frame of reference for following pressing. The local bulk volume per unit mass of dry marc will be designated by the symbol v. Its units in the S.I. system are m^3/kg dry marc, or in cgs terms (cm^3/g dry marc). The average value of v or total cake volume divided by the total mass of dry marc is $\bar{v}$. Flow within the pores of the presscake will be called the internal flow, or when no ambiguity results, simply the "flow." Flow through the media will be called outflow. Flows from pores within the particles into the interparticle pores will be called exudation. Transfer of fluid into the particle will be called imbibition. Expression can usually be divided into an induction period, during which air is expelled from the presscake pores and the pores gradually fill with exuded liquid, and an outflow period. If the presscake does not contain air, outflow commences as soon as pressing starts and there is no induction period. Presscakes may initially contain several types of fluid: occluded fluid contained within the particles; cling on the surface of the particles; seepage, non-uniformly distributed cling, which concentrates at the bottom of the cake; and saturation fluid, which completely fills the interparticle pores.

The volume of the cake can be divided into several parts: the matrix, or space occupied by the marc alone; the solid volume, which is occupied by the marc and occluded fluid together; the cells, the space occupied by occluded fluid; and the pores, the volume of the interparticle space. The external porosity, ε, is the fraction of the local space occupied by pores. The internal porosity, ψ, is the fraction of the local volume occupied by cells. ε and ψ have counterparts, $\bar{\varepsilon}$ and $\bar{\psi}$, which are based on the pore volume and cell volume, respectively, for the entire cake.

During outflow there is a plane or surface in the cake, the no-flow surface, at which internal flow starts, and the fluid velocity is zero. The no-flow surface may also be the ram surface, the moving surface that compacts the presscake. The ram surface may actually be part of a ram, or it may be the expanding root surface of a screw, the surface of an inflatable tube, or the surface of a belt being pressed by rollers. If the cake is pressed from both sides by surfaces through which outflow occurs, the no-flow surface will lie somewhere between these surfaces. In cylindrical cage presses in which the internal flows are radially directed, the no-flow surface degenerates into the axis of the cylinder.

COORDINATE SYSTEMS

The mass of marc between a unit area of outflow surface and the corresponding area on the no-flow surface will be called the "total specific load" and designated by W. Corresponding areas on the no-flow surface and outflow surface are circumscribed by stream tubes, which originate at the no-flow surface and pass through the outflow surface. If the no-flow and outflow surfaces are opposing planes, the corresponding no-flow and outflow areas are parallel and equal in size. For outwardly directed radial flow the corresponding no-flow area is smaller than the outflow area. The reverse is true for inwardly directed flow, such as occurs during expression in solid bowl centrifuges. During juice expression in roll mills, the flow stream lines are curved, the outflow surface is not well defined, and the situation is quite complicated.

In most cases the no-flow and outflow surfaces will be symmetric or parallel, and the space between them can be divided into strata by a series of similar surfaces. The mass of marc per unit area of outflow surface between the outflow surface and one of these surfaces will be designated by w. The fraction of the marc between the particular surface and the outflow is w/W. This fraction remains invariant as pressing proceeds even though the surface moves and the space between the surface and the outflow surface contracts.

In the case of roll mills it *may* be preferable to use Eulerian coordinates to describe the marcs movement and juice flow. If this is done, there is a steady-flow of marc through the constant w surfaces, but the actual amount of marc between two such surfaces remains constant. In most other cases it is preferable to use Lagrangian coordinates, which follow the marc as it moves.

When the constant w surfaces are parallel, the cake depth, the distance from the outflow surface, and the no-flow surface is given by

$$Z = \int_0^W v \, dw \tag{1}$$

$Z/W = \bar{v}$. We use the convention that w increases as one moves away from the outflow surface. Thus $w = 0$ at that surface and $w = W$ at the no-flow surface. The local depth, z, is given by Eq. (1) with the upper limit W replaced by w.

The overall compaction rate dZ/dt for the parallel w surface case is given by

$$\frac{dZ}{dt} = \int_0^W \left(\frac{\partial v}{\partial t}\right) dw \tag{2}$$

YIELDS

Let x_i denote the initial mass fraction of a component i in a presscake and y_i indicate the initial mass of i per unit mass of marc.

$$y_i = x_i / x_m \tag{3}$$

where x_m is the weight fraction of marc.

The partial volume of a component in a mixture is the increase in mixture volume divided by the increase in mixture mass that occurs when a differentially small amount of the component is added to the mixture. The partial volume of presscake components are V_s, the partial volume of the juices solutes; V_M the partial volume of the marc; and V_W the partial volume of the solvent in which the solutes are dissolved. In estimating pressing yields it will be assumed that these partial volumes remain constant and do not depend on composition or pressure. The critical volume, v_c, is the bulk specific volume of the cake when all the interspersed air is expelled but all other components remain. Based on the foregoing assumptions

$$v_c = V_M + y_S V_S + y_W V_W \tag{4}$$

η, the juice yield based on initial cake weight, is given by

$$\eta = \frac{(v_c - \bar{v}_f)\,[\rho_j]}{1 + y_S + y_W} \tag{5}$$

where $\bar{v}_f$ is the final value of $\bar{v}$ and ρ_j is the juice density. For estimation purposes

$$\eta = \frac{(v_c - \bar{v}_f)\,(y_S + y_W)}{(1 + y_S + y_W)\,[y_S V_S + y_W V_W]} \tag{6}$$

Obvious corrections can be made to account for the presence of dregs or foots.

η_j the recovery factor or yield based on the initial juice content is given by

$$\eta_j = \frac{(v_c - \bar{v}_f)\,\rho_j}{y_S + y_W} \tag{7}$$

or for estimation purposes

$$\eta_j = \frac{(v_c - \bar{v}_f)}{y_S V_S + y_W V_W} \tag{8}$$

$(\eta_j)_{corr}$ a corrected recovery factor that accounts for bound water is given by

$$(\eta_j)_{corr} = \frac{(v_c - \bar{v}_f)}{y_S V_S + (y_W - b) V_W} \tag{9}$$

where b is the binding factor, or amount of water bound per unit mass of marc.

INTERNAL FLOW VARIATION

After outflow commences, or in any region of the cake where the pores are saturated, a volume balance when the constant w surfaces are parallel, yields

$$\frac{\partial U}{\partial w} = \frac{\partial v}{\partial t} \tag{10}$$

where U is the superficial fluid velocity relative to the solids and t is time. Such flows occur in belt presses, rack and frame presses, compactible frame filter presses, and in enclosed cylinder presses where outflow only occurs through the bottom of the cylinder or the face of the piston.

If the constant w surfaces are concentric cylindrical surfaces and $\partial v/\partial t$ and v are independent of w

$$\frac{\partial U}{\partial w} = \frac{\partial v/\partial t}{[1 - 4I/D_o]^{1/2}} + \frac{2Uv}{[D_o - 4I]} \tag{11}$$

where I stands for v integrated with respect to dw between the limits 0 and w, and D_o is the diameter of the outflow surface. Such flows occur in inflatable tube presses, in screw presses where the screw's root diameter progressively increases, and in cylindrical cage presses where outflow occurs only through the sides of the cylinder.

In certain cases (e.g., cylindrical cage presses and screws with diminishing pitches) compaction occurs at right angles to the direction of flow; and the outflow area progressively decreases. This progressively increases both w and W. Because of friction between the outflow surface and the cake sliding across it and cross-flow, both w and W may vary somewhat in the direction of compaction but not in exactly parallel fashion. The constant w surfaces will no longer be invariant, but the w/W surfaces may remain roughly parallel and provide a convenient frame of reference.

LOCAL VALUES OF U

In the simple case where w/W surfaces are parallel, $\bar{v} < v_c$, and compaction occurs in the direction of outflow:

$$U = \int_W^w \left(\frac{\partial v}{\partial t}\right) dw \tag{12}$$

If $\partial v/\partial t$ is constant

$$U = U_o\left[1 - \frac{w}{W}\right] = -\left(\frac{dZ}{dt}\right)\left[1 - \frac{w}{W}\right] \tag{13}$$

where U_o the outflow value of U equals $-dZ/dt$, the compaction or ram-surface movement rate relative to the outflow surface.

If constant-rate, uniform compaction occurs at right angles to parallel w/W surfaces and Z' is the cake thickness in the compaction direction.

$$U = U_o\left[1 - \frac{w}{W}\right] = \frac{-Z_o}{Z'}\left(\frac{dZ'}{dt}\right)\left[1 - \frac{w}{W}\right] \tag{14}$$

where Z_o is the constant cake thickness in the outflow direction. Further

$$U = \frac{-Z_o(dZ'/dt)}{Z'_c + (dZ'/dt)t}\left[1 - \frac{w}{W}\right] \tag{15}$$

where Z'_c is the value of Z' at which outflow starts. Since dZ'/dt is negative, U increases with time. This contrasts with uniform compaction in the direction of flow, where U remains constant at fixed w/W planes.

For uniform compaction in the direction of flow when w/W surfaces are cylinders, $\bar{v} < v_c$ and the inside diameter D_i expands at a constant rate

$$\mathrm{U} = \frac{U_o\,[1 - (w/W)]}{[(1 - \mathrm{w/W}) + (\mathrm{D_i}/D_o)^2(w/W)]^{1/2}} \tag{16}$$

or

$$\mathrm{U} = \frac{(D_i/D_o)(dD_i/dt)\,[1 - (w/W)]}{2\{1 - [1 - (\mathrm{D_i}/D_o)^2](w/W)\}^{1/2}} \tag{17}$$

since D_i increases with time, U increases at all w/W surfaces except $w/W = 1.0$, where U is always zero.

For constant rate uniform compaction at right angles to cylindrical w/W surface, when D_i/D_o is constant and $\bar{v} < v_c$ Eq. (16) still applies; but upon substituting for U_o, one obtains

$$U = \frac{[(D_o^2 - D_i^2)/4D_o Z'](dZ'/dt)}{\{1 - [1 - (D_i/D_o)^2](w/W)\}^{1/2}}[1 - (w/W)] \qquad (18)$$

Equation (18) is useful for screw presses with decreasing pitches. Then (dZ'/dt) is the rate of change in pitch. For cage presses, where $D_i = 0$,

$$U = -\left(\frac{D_0}{4Z'}\right)\left(\frac{dZ'}{dt}\right)\left[1 - \left(\frac{w}{W}\right)\right]^{1/2}$$

as in the case of Eq. (15), Z' in Eq. (18) and (19) can be replaced by $Z'_c + (dZ'/dt)t$.

NONUNIFORM COMPACTION

When the compaction is nonuniform because v varies because of wall friction and flow-induced drag, the expressions for U become very much more complicated. Therefore, when carrying out tests to determine expression parameters it is preferable to use compaction and flow arrangements that minimize such complication. Thus, axial compaction and axial flow, where the w/W planes are parallel and compaction occurs in the w direction, is probably best for this purpose. Even in this case, complications cannot be wholly eliminated.

COMPACTION STRESS EVALUATION

The expression test unit shown in Fig. 16.1 provides both axial compaction and axial flow. As the ram advances, the presscake in the test cavity is compacted. When $\bar{v} < v_c$, outflow occurs through the perforated plate and connecting channel at the bottom of the unit. If the unit is mounted in an Instron compression tester as shown in Fig. 16.2 constant compaction rates and a record of ram force vs time can be obtained. The Instron can provide compaction rates ranging between 0.5 and 200 mm/min. Cake heights can be read to within 0.1 mm from dials on the control panel.

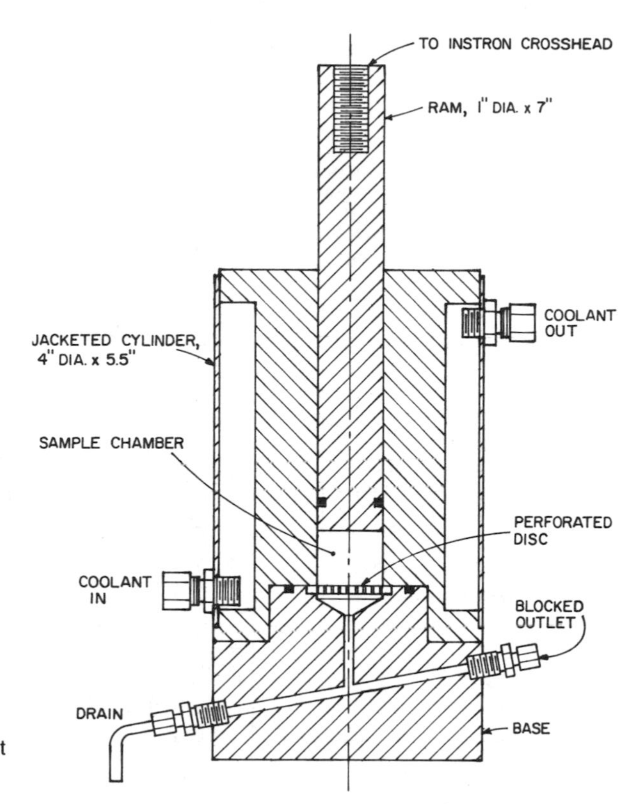

Fig. 16.1. Expression test unit.

By adjusting an interlock the ram advance can be stopped whenever a tolerable, chosen maximum force is reached. Alternatively, another interlock can be used to stop the ram when a chosen cake height is reached. The compaction force decays after the ram motion stops. A minimum force interlock can be used to restart the ram advance when the force decays below a chosen value. In general it is wise to use the maximum force interlock when running expression tests. This will prevent damage to the equipment due to the development of excessive forces, which may occur when stroke length control alone is used.

A typical ram force vs time record for an expression test is shown in Fig. 16.3. Moist, spent coffee grounds were used in the test shown; but the record is similar to that usually exhibited by other materials. However, if cells rupture occurs during pressing, the ram force will decrease momentarily before it increases again. Sometimes several rupture inci-

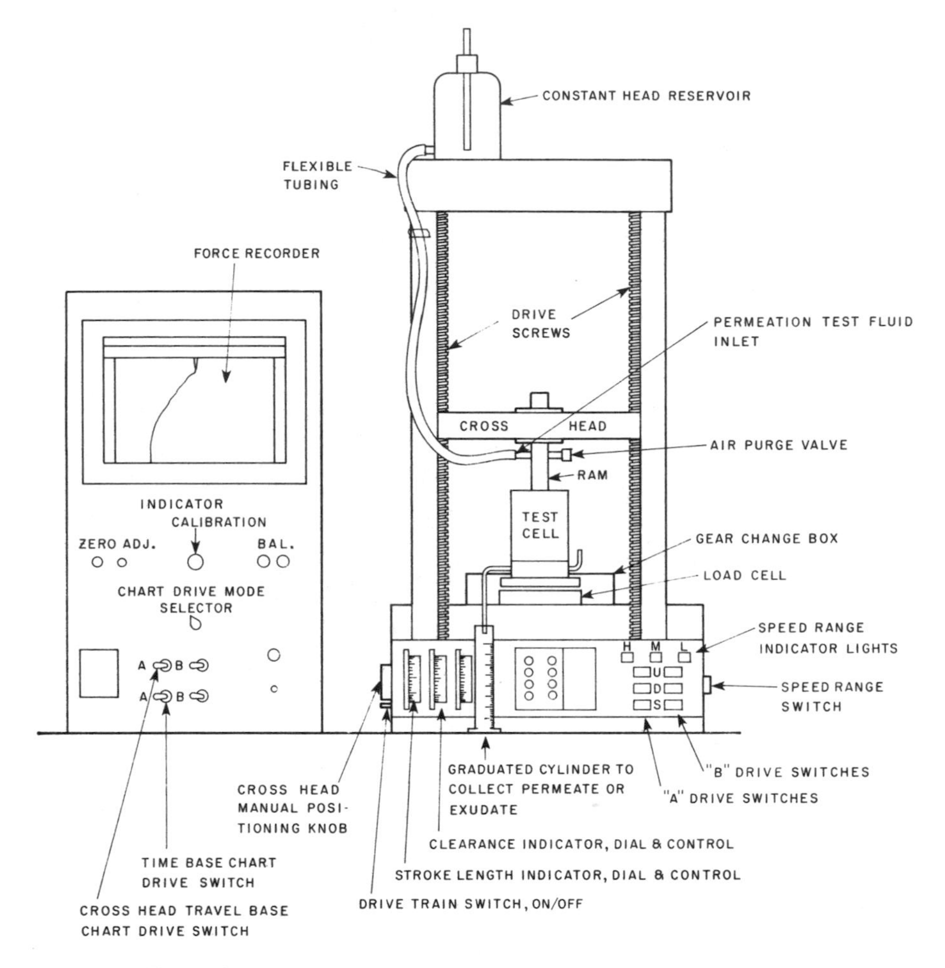

Fig. 16.2. Expression test set-up.

dents occur during pressing. The force vs time trace is jagged immediately following rupture. When the force rises at a very rapid rate it may be difficult to detect rupture incidents unless an electronic oscilloscope is used instead of a mechanical recorder.

Data Transformation

Excluding instrument overshoot or undershoot, the peak in the force vs time record corresponds to the instant the ram is stopped. The decay in force after that time represents relaxation. Since a uniform rate of

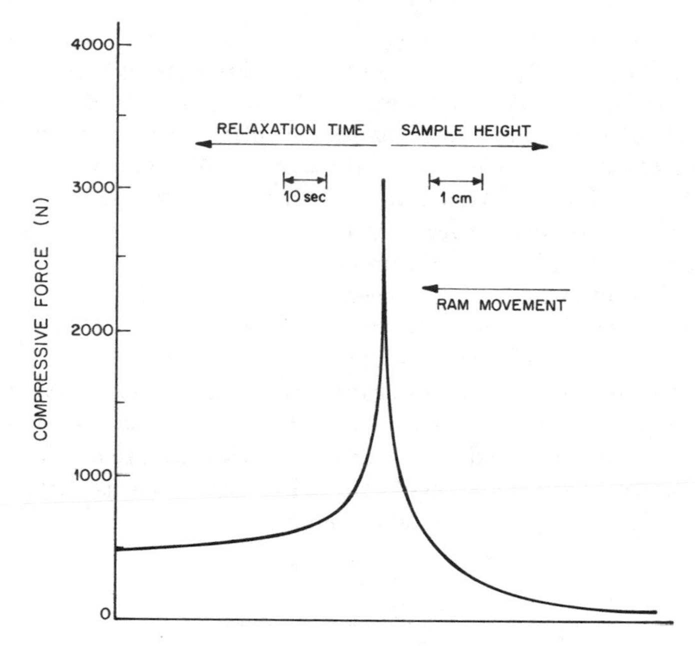

Fig. 16.3. Ram force vs. cake height and relaxation time.

ram advance is used, the record prior to stopping is also a record of force vs ram position. Since the cake height, Z, varies linearly with ram position the record can be converted into a plot of force vs cake height. By subtracting the ram extension at the end of pressing from the ram extension at the outflow surface, Z_f, the final cake height is obtained. Cake heights corresponding to chosen forces on the record will be given by

$$Z = Z_f - \Delta L\, r \tag{20}$$

where ΔL is the distance in the direction of chart movement between the peak force and the chosen force, and r is the ratio of ram speed to chart speed. Values of the relaxation time, θ, can be obtained as follows

$$\theta = \Delta L_R/(dL/dt) \tag{21}$$

where ΔL_R is the chart distance in the direction of chart movement between the peak force and the chosen partly relaxed force, and (dL/dt) is the chart speed.

The ram force, F, can be converted into the ram pressure P_1 by dividing F by $\pi D^2/4$, where D is the ram diameter. When P_1 is plotted vs Z for different cake load weights, P_1 increases much more rapidly for small loads than for large loads. This difference is greatly reduced, but not eliminated, when Z is divided by W yielding $\bar{v}$; and P_1 is plotted vs $\bar{v}$. Plots of this type are shown in Fig. 16.4 for constant loads of spent coffee grounds pressed at compaction rates ranging between 0.5 mm/min to 200 mm/min.

Two distinct regions can be seen in these plots. When $\bar{v} > v_c$, log P_1 rises linearly as $\bar{v}$ decreases. Shortly after $\bar{v}$ becomes smaller than v_c, log P_1 begins to rise more rapidly. As the compaction rate increases, this log P_1 rise becomes much more rapid. Since outflow occurs when $\bar{v} < v_c$ and its rate increases as the compaction rate increases, it appears very likely that the increased rate of rise of log P_1 is due to outflow induced pressure. As will be shown later this inference is supported by independent fluid pressure measurements.

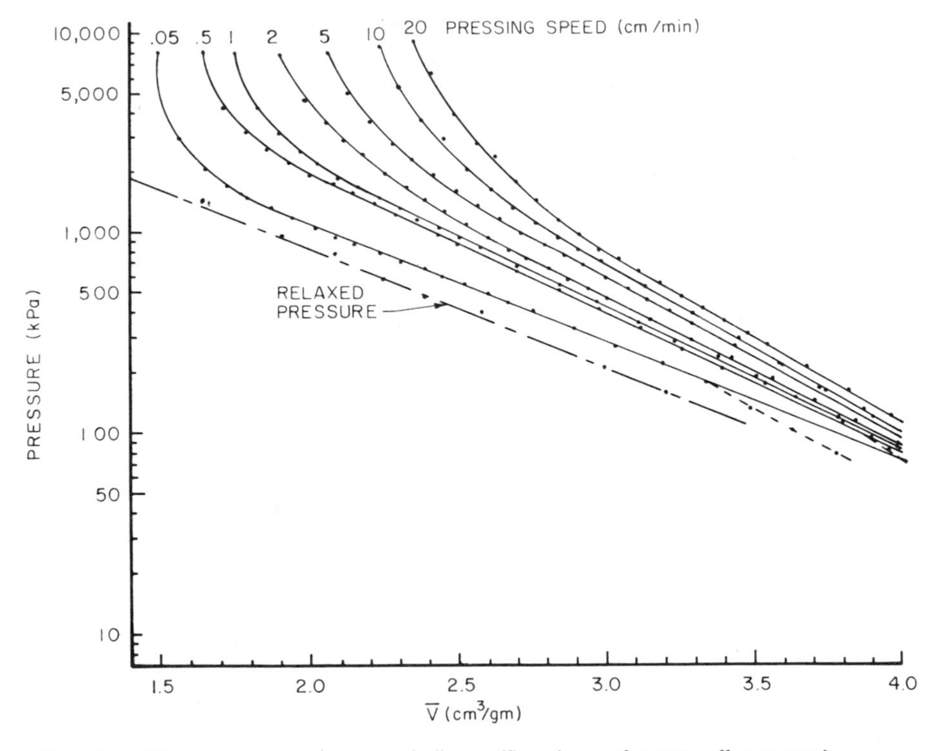

Fig. 16.4. Ram pressure vs. average bulk specific volume of spent coffee grounds.

Little or no internal flow occurs when $\bar{v} > v_c$. Therefore it is likely that in this $\bar{v}$ range P_1 equals P_{s1}, the solid compaction stress at the ram surface. It appears reasonable to suppose that extrapolation of the straight line region of the log P_1 vs $\bar{v}$ plot into the outflow region (i.e., when $\bar{v} < v_c$) will indicate the value of log P_{s1} during outflow. If this is so, $P_1 - P_{s1}$, provided by extrapolating P_1 to given values of $\bar{v}$ should be equal to ΔP_f, the corresponding fluid pressure drop across the presscake. When such $P_1 - P_{s1}$ values are compared with corresponding measurements of ΔP_f, qualitative agreement is obtained, but the accuracy of agreement is relatively poor. Possible reasons for this will be discussed later.

The preoutflow ln P_{s1} vs $\bar{v}$ data can be correlated as follows

$$\ln(P_{s1}/P_{1c}) = k_1\,[v_c - \bar{v}] \tag{22}$$

where P_{1c} is the value of P_{s1} when $\bar{v} = v_c$.

INTERNAL PRESSURE, CELL RUPTURE, AND CLING

Much of the solid compaction stress is caused by internal pressure exerted by occluded fluid rather than by the marc itself. This can be demonstrated by comparing values of P_{1c} for alfalfa samples that have been passed through a hammer mill various numbers of times with those for unmilled alfalfa. P_{1c} for the unmilled alfalfa is 300 psi (2070 kPa). For one-pass alfalfa P_{1c} is 51 psi (350 kPa), and for two- to five-pass alfalfa it is 27 psi (186 kPa). The cell rupture induced loss in compaction strength and reduction in P_{1c} is readily apparent. Impact or shear-induced cell rupture prior to expression can lead to marked increases in juice yield. For example, the juice yield for uncomminuted alfalfa pressed to 1200 psi (8300 kPa) was 18% compared to an average of 57% for milled alfalfa. Though very little yield difference is obtained at 1200 psi as the number of mill passes is varied from one to five, differences are obtained at low pressures. The yield data in Table 16.2 for Schwartzberg *et al.* (1978) pressing tests for macerated alfalfa illustrate this. The tendency of pressed alfalfa to imbibe more water as the number of mill passes increases also indicates increased cell rupture.

The large effect of occluded fluid on P_{s1} suggests that the v_c value used to predict P_{s1} in Eq. (22) should be based on v_c', the v_c value obtained when no cling or pore liquid is present prior to pressing. Liquid, which lies outside the particles does not contribute to internal pressure and hence should not affect P_{s1}. Therefore, v_c in Eq. (22) should be replaced by v_c' whenever possible.

TABLE 16.2. JUICE YIELDS AND IMBIBITION RATIOS VS. COMMINUTOR PASSES FOR ALFALFA PRESSED BY HAND IN A POTATO RICER

Passes through Fitzpatrick comminutor	Recovery factor $\left(\frac{\text{Juice expelled}}{\text{Initial juice content}}\right)$	Imbibition ratio $\left(\frac{\text{Grams of water imbibed}}{\text{Grams of press alfalfa}}\right)$
1	0.366	1.17
2	0.394	1.39
3	0.434	1.39
4	0.451	1.42
5	0.461	1.46

The consequences of such replacement can be seen from the P_{1c} values for spent coffee grounds listed in Table 16.3. This data was obtained by Schwartzberg *et al.* (1977A) for expression tests carried out with different sizes of spent grounds. It is known that cling and undrainable seepage increase as particle size decreases. Correlations such as that of Dombrowski and Brownell (1954), which are based on capillary rise effects in porous beds, can be used to estimate the amount of such cling and seepage. If the presscake feed is well mixed before pressing the seepage shows up as cling. The cling for a well-mixed presscake feed should be inversely proportional to the particle diameter squared.

The data in Table 16.3 suggest a number of points. (1) It is preferable to base comparison pressing tests on equal dry weight loads rather than on equal as-is loads. (2) If cling is neglected and P_{1c} is based on v_c rather than on v_c' a much lower value of P_{1c} will be obtained. These P_{1c} values will not truly indicate the compaction stress contribution produced by internal pressure exerted by the occluded liquid. Such pressures should tend to prevent cell walls from buckling and therefore should also increase the compaction stress contribution of the marc itself. (3) Smaller particles probably exhibit smaller values of P_{1c} because more cells in small ground particles are initially ruptured than in large particles. Based on the size of cells in roasted coffee (i.e., roughly 30 – 40 μm),

TABLE 16.3. P_{1c} VALUES BASED ON v_c AND v'_c FOR DIFFERENT SIZES OF SPENT COFFEE GROUNDS[1]

Average particle diameter (mm)	Average moisture content (% as-is)	W (g/cm^3)	v_c (cm^3/g)	v_c' (cm^3/g)	P_{1c} (psi) Based on v_c	v_c'
3.57	50.6	2.39	3.002	3.002[2]	1530	1530
1.79	65.6	1.78	3.927	3.002[2]	563	1065
0.72	74.4	1.32	5.459	3.002[2]	110	507
0.30	77.3	1.18	6.224	3.002[2]	34	308

[1] Pressing load 26.2 g of moist grounds.
[2] Based on assumption that the amount of cling was negligibly small for the 3.57 mm grounds.

it is estimated that 50 – 60% of cells in the 0.30 mm grounds are ruptured. The corresponding figure for the 3.57 mm grounds is 5.0 – 6.5%.

CLING DETERMINATION

It is difficult to eliminate cling completely in the case of extracted, ground particles. Unless the particles are wiped with a towel or blotter the minimum amount of cling will usually be about 4 – 5% of the solids weight, even for 2 – 3 mm particles. One can determine the amount of cling by rapidly mixing a known weight S of the moist particles with a known weight E of an aqueous solution containing a known concentration C_0 of a readily measurable solute. Preferably, the solute should exert a low osmotic pressure. Small samples of the liquid should be withdrawn from the stirred mixture every 5 or 10 sec for a period of 1 min. The solute concentration of these samples should be measured and plotted vs $\sqrt{t}$, the time elasped after the start of mixing. The initial part of the concentration vs $\sqrt{t}$ plot should be a straight line which can be extrapolated to $t = 0$. If the intercept concentration is C_1, G the amount of cling per unit weight of as-is solid will be

$$G = \left[\frac{C_0}{C_1} - 1 \right] \frac{E}{S} \tag{23}$$

Since the amount of cling is SG, the weight of cling-free solid is $S - SG$. Therefore the weight of cling per unit weight of cling-free solid is $G/(1 - G)$. To maximize accuracy, the E/S ratio should be small, but even so the technique is not very precise.

RUPTURED CELLS

The fraction of the cells in the particles that is ruptured can be estimated by an analogous method. A known weight of particles is mixed with a known weight of a solution containing a high molecular weight solute, such as gelatin or dextran blue, which cannot diffuse into intact cells. The concentration of this solute is sampled frequently during the first minute of mixing and the extrapolated zero time concentration is used to determine G as previously described. The mixture is stirred until equilibrium is reached. This may take several hours. The mixture weight is brought back to its initial value by the addition of solute-free solvent so as to compensate for any evaporation and then the equilibrium concentration C_2 of the solute is measured. F_R the fraction of the cells which are ruptured is given by

$$F_R = \left(\frac{E}{x_{WF}\, S(1 - G)}\right)\left[\frac{C_0}{C_2} - 1\right] - \frac{G}{(1 - G)x_{WF}} \tag{24}$$

where x_{WF} is the weight fraction of free water in the cling-free solid. The value of x_{WF} can be determined by carrying out the preceding test using a readily measurable solute that can diffuse into all the cells. For such solutes

$$x_{WF} = \left(\frac{E}{S(1 - G)}\right)\left[\frac{C_0}{C_2} - 1\right] - \frac{G}{(1 - G)} \tag{25}$$

Since the bound water weight fraction $x_{WB} = x_W - x_{WF} = bx_m$, where x_W is total weight fraction of water (as measured by drying tests) and x_m is the weight fraction of marc. x_{WF} as determined from Eq. (25) can be used to determine x_{WB} and the water-binding factor b. The factor can then be used in Eq. (9) to obtain the corrected juice recovery factor.

EFFECT OF PRESSING SPEED PRIOR TO OUTFLOW

When pressing is stopped, the ram pressure relaxes and ultimately reaches an equilibrium value, P_r. A plot of log (P_r) vs $\bar{v}$ yields a straight line, and this line extends into the outflow region. In this region relaxation is no doubt largely caused by progressive decreases in the fluid pressure that sustained flow during active pressing. Indeed, independent measurements indicate that fluid pressure drops at a relatively slow rate following the end of ram movement.

The difference between P_1 and P_r increases as pressing speed increases, and $\bar{v}$ decreases even in the absence of outflow. What causes this difference to increase has not yet been wholly resolved. Observations of cake movement in transparent compression cells show that no detectable cake movement occurs after the ram stops. Therefore, the hypothesis that relaxation is caused by equalization of local compaction stresses due to shifts in the cake appears to be untenable.

RELAXATION

It is more likely that relaxation is caused by the relief of internal pressure by exudation. There is no reason why such exudation should not continue after the ram stops. Measurements of the pore fluid pressure in the cake following outflow sometimes show that local fluid pressures continue to rise after the ram motion stops (Huang 1979).

These pressure rises could be due to exudation which simultaneously causes a reduction in internal pressure and a rise in pore pressure. If particles have to deform more rapidly as pressing speed increases, and such deformation is resisted because of the presence of neighboring fluid-filled cells, internal pressures should rise to a greater extent as pressing speed increases. This internal pressure is presumably relieved by exudation after the ram motion stops. After such relief occurs the compaction stress will revert to common P_r regardless of the prior pressing rate and prior value of P_{s1}. Some of the available evidence is consistent with this hypothesis, but some is not. Both types of evidence are reviewed below.

Molecular Weight Cutoff

Holtz and Ohlson (1981) found that the cutoff molecular weight for diffusive transfer of corn syrup oligosaccharides into intact cells in spent coffee grounds was 1800 – 1960. This corresponds to an equivalent cell wall pore diameter of roughly 20 – 30 ⟨ngstroms. The thickness of such cell walls as observed by electron microscopy is roughly 10 μm (100,000 ⟨ngstroms). Thus, the length to diameter ratio of channels in the cell walls should be roughly 3300 – 5000. Based on the use of structural models to analyze the ratio of the diffusivity of small solutes in moist coffee grounds to the corresponding diffusivity in free water, it appears likely that such channels or their equivalents occupy 4 – 5% of the cell wall area. Flow through such a sparse array of fine-long channels would be very slow even if the internal pressure provides a large driving force. Hence pressure relaxation may be quite slow.

The difference between the partially relaxed pressure P and P_r exhibits a peculiar time dependency. After the first few tenths of a second log $(P - P_r)$ decreases linearly when plotted vs the cube root of time (Schwartzberg *et al.* 1977A). Such relaxation, which is depicted in Fig. 16.5, also occurs for radial flow (Ngoddy 1976). It is exhibited by a wide variety of materials, including some with open cells (e.g., sponges) and some with cells that rupture easily, (e.g., apples). The existence of such relaxation for open-celled materials is certainly not consistent with the exudation theory. However, in the case of these materials, relaxation was only measured after very high levels of compaction. Hence, the cell openings may have been largely blocked.

As shown in Table 16.4 the rate of relaxation decreased markedly as $\bar{v}$ decreases. R is the slope of the straight-line portion of the $\ln(P - P_r)$ vs $\theta^{1/3}$ curve. It is possible that cell wall openings become progressively blocked as $\bar{v}$ decreases. Though P_1 increases as $\bar{v}$ decreases, R

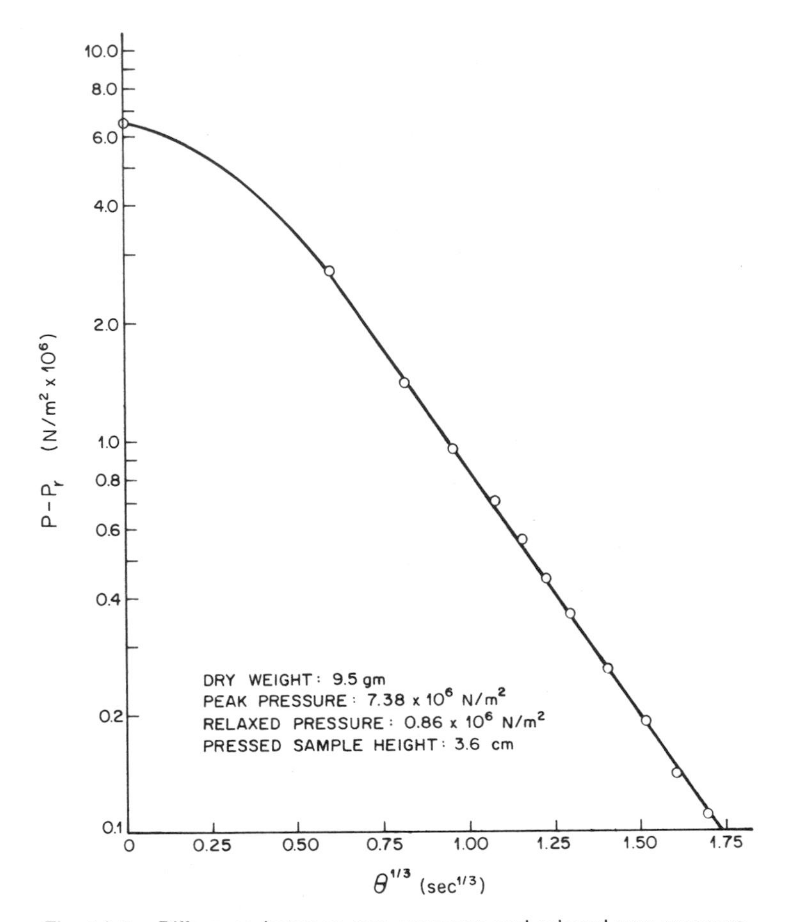

Fig. 16.5. Difference between ram pressure and relaxed ram pressure vs. relaxation time of coffee grounds.

does not truly correlate with P_1, as can be seen from the constant P_1 data in Table 16.4. In other tests carried out to determine the maximum possible degree of compaction and maximum possible juice yield, P_1 was permitted to cycle automatically between a fixed maximum value of roughly 1150 psi and a slightly lower value, roughly 1000 psi. The process was continued for several days until the successive reductions in $\bar{v}$ became so small that it did not appear worthwhile to continue. During this time the relaxation rate progressively decreased until it took several hours for P_1 to drop to 1000 psi.

Grounds containing 50% sugar solution were prepared by repeatedly shaking spent grounds in fresh portions of the 50% solution. When these grounds were pressed, R was roughly 2.7 times smaller that R for

TABLE 16.4. RELAXATION CONSTANT VS. $\bar{v}$ FOR COFFEE GROUNDS

Condition	$\bar{v}$ (cm^3/g)	R ($sec^{-1/3}$)	$-dZ/dt$ (mm/min)
As-is load weight 2.2 grams, P_1 increasing	8.42	−6.93	25
	7.37	−5.76	25
	6.31	−3.32	25
	4.74	−2.72	25
	3.17	−1.70	25
As-is load weight 25 grams, varying P_1 approximately constant at 1145 psi	2.05	−1.44	200
	1.97	−1.09	100
	1.86	−0.97	50
	1.71	−0.72	20
	1.55	−0.62	10
	1.47	−0.60	5
Grounds containing 50% Sugar solution, as-is load weight 25 g P_1 approximately constant at 1145 psi	2.58	−0.58	50

regular spent grounds at comparable values of $\bar{v}$. The sugar solution also increased $P_1 - P_r$ at $\bar{v} = v_c$ roughly 3.4-fold compared to regular grounds at $\bar{v} = v_c$. Fifty percent sucrose is about 15 times as viscous as water. If $P_1 - P_r$ and the rate of relaxation are truly related to exudation, the rate of relaxation should decrease and $P_1 - P_r$ should increase as the viscosity of the exuded fluid increases. Since this occurred, it appears to support exudation as a contributing factor to both the $P_1 - P_r$ difference and relaxation.

EXUDATION

Abularach (1979) attempted to measure the extent of exudation both prior to and during outflow. He filled the interparticle pores of various presscakes with a Dextran blue solution and retained it in place by using a fine pore outflow line positioned slightly higher than the top of the cake in the expression test unit. After pressing started, this solution and any exudate mixed with it flowed out of the unit at a volumetric rate equal to the volumetric displacement rate of the ram. Since the Dextran blue could not diffuse in the closed cells (as demonstrated by prior testing), any reduction in its concentration would have to be due to exudation. He collected the expressed liquid at evenly spaced time intervals by using a fraction collector and measured the Dextran blue concentration by spectrophotometry.

He then used the concentration and ram displacement measurements to determine the rate and extent of reduction in the cell volume and $\bar{\psi}$ (i.e., the rate and extent of exudation), and the rate and extent of reduction in the pore volume and $\bar{\varepsilon}$. Initially very little, if any, exudation occurred, and almost all of the reduction in cake volume occurred because of reductions in $\bar{\varepsilon}$. After the onset of outflow, exudation increased rapidly. Exudation started earlier with materials with weak cell walls such as hydrolyzed coffee grounds and apples (see Fig. 16.6). This early exudation is almost certainly due to more rapid cell wall rupture. Additional evidence indicates that cell wall rupture finally occurred in all cases.

Toward the end of pressing the interparticle pore space stopped decreasing and the reduction in presscake volume was almost wholly due to reductions in cell volume. As a consequence $\bar{\varepsilon}$ actually increases toward the end of the pressure. $\bar{\varepsilon}$ apparently never went lower than 0.25. The $\bar{\varepsilon}$ measurements and corresponding measured values of fluid pressure drop indicate that equations such as the Kozeny – Carman equation or the Ergun equation cannot be successfully used to predict flow pressure drop in the cake.

The lack of detectable exudation during the early stages of pressing is not consonant with the hypothesis that internal pressure is responsible

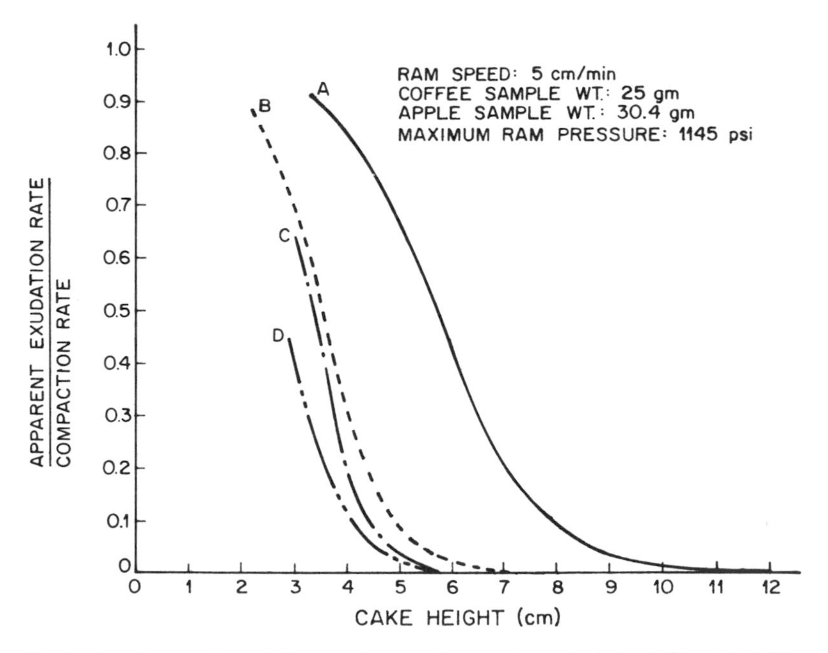

Fig. 16.6. Apparent exudation rate/compaction rate vs. cake height. Curve A, 0.25 in. apple cube; curve B, hydrolyzed coffee grounds; curve C, drip grind spent coffee; curve D, drip grind spent coffee with 50% sugar solution.

for the difference between P_{s1} and P_r. On the other hand, because of the small size and sparcity of cell wall channels, the exudation prior to cell wall rupture may have been too small to detect by the techniques that Abularach used. In subsequent work, using a transparent test unit, exudation was observed at the top of the presscake early in the pressing cycle. Because of the time lag before such exudation appeared in the outflow, Abularach did not detect it until much later in the pressing cycle.

MODIFIED TEST UNIT

The test unit shown in Fig. 16.7 was used in conjunction with the Instron to measure fluid pressure in the presscake, friction between the presscake and the wall of the pressing chamber, and the force acting on the outflow surface. In the absence of flow pressure drop across the cake – media interface this force should be solely due to the compaction stress at the bottom of the cake. As in the case of P_1, the outflow surface pressure P_o was calculated by dividing this force by $\pi D^2/4$.

The frictional force was measured by mounting a load cell in one of the three evenly spaced legs that supported the yoke, which in turn supported the wall section of the test unit. The other two legs consisted of rugged screws, which were adjusted to make sure the yoke was level. Because of this arrangement the load cell measured one-third of the frictional force acting on the test unit wall. This force was converted into ΔP_W, the reduction in compaction stress due to wall friction, by multiplying the force on the pillar load cell by $12/\pi D^2$.

Pressure transducers mounted in the wall of the test unit communicated with the presscake cavity via disc like annular slits in the test unit wall assembly. These units sensed the fluid pressure P_f as it developed; and with the aid of a four pen strip chart recorder provided a record of P_f vs time. The record could readily be converted into a P_f vs $\bar{v}$ record prior to the end of ram motion and a P_f vs θ record after the end of ram motion. The system was adjusted so that the bottom transducer measured the fluid pressure just above the cake media interface.

WALL FRICTION AND RESIDUAL STRESS

When log (ΔP_W) and log (P_o) were plotted vs $\bar{v}$ graphs very similar to the log (P_1) vs $\bar{v}$ plot were obtained. By a line of reasoning similar to that previously used it can be shown that P_o prior to outflow represents P_{so}, the compaction stress at the bottom of the cake. ΔP_W was roughly twice as great as P_o (i.e., P_{so}) at $\bar{v} = v_c$, but this ratio decreased markedly

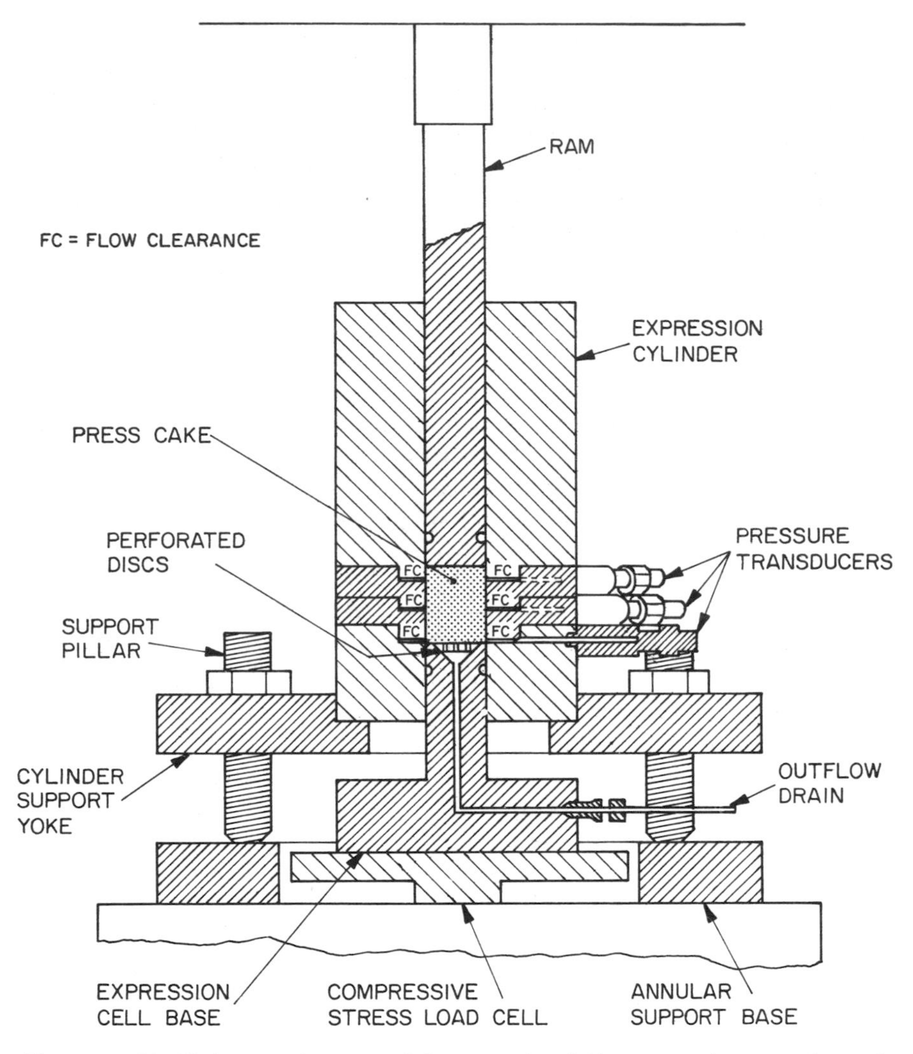

Fig. 16.7. Modified expression test unit for measuring fluid pressures in presscake, wall friction, and compaction stress at outflow surface.

during outflow. When pressing stopped, ΔP_W was only roughly 30% as large as P_o. The slope of log (P_{so}) vs $\bar{v}$ was markedly greater than that of log (P_{s1}) vs $\bar{v}$; and that of log (ΔP_W) was much less than that of log (P_{s1}). Within the accuracy of the measurements ΔP_W plus P_{so} equalled P_{s1} prior to outflow. These data indicate that, due to wall friction, compaction was very nonuniform. This conclusion was later supported by visual observation of the presscake movement.

Friction Calculation

If one assumes, as did Shirato *et al.* (1968), that the compaction stress produces a proportional transverse stress in the cake, the ratio of proportionality being B, this transverse stress will produce a differential frictional force equal to $BfP_s\pi D\,dz$, where f is the coefficient of friction. But, by a force balance $BfP_s\pi D\,dz = \pi D^2\,dP_s/4$.This differential equation can be integrated between $z = 0$, the outflow surface, and $z = Z$ the ram surface yielding

$$\ln(P_{sl}/P_{so}) = 4\,BfZ/D = 4\,Bf\bar{v}W/D \tag{26}$$

Since Eq. (26) shows that the difference between $\ln(P_{s1})$ and $\ln(P_{so})$ will diminish as $\bar{v}$ decreases, it is consistent with the relative slopes of $\ln(P_{so})$ vs $\bar{v}$ and $\ln(P_{s1})$ vs $\bar{v}$.

There should be three parallel relationships (Eq. 22),

$$\ln(P_{so}/P_{oc}) = k_o\,(v_c - \bar{v}) \tag{27}$$

and

$$\ln(P_s/P_{sc}) = k_s\,(v_c - v) \tag{28}$$

where P_s is the local stress in the cake, P_{sc} is its value when the local bulk specific volume v is v_c, P_{oc} is the value of P_{so} when $\bar{v}$ equals v_c, and k_o is the slope constant for $\ln(P_{sc})$ vs $\bar{v}$ and k_s is the corresponding constant for $\ln(P_s)$ vs v. It can be shown that

$$k_s = \frac{4BfW/D}{\ln[1 - (BfW/Dk_o)]} = \frac{4BfW/D}{\ln[1 + (BfW/Dk_1)]} \approx \frac{k_o + k_1}{2} \tag{29}$$

and

$$\ln(P_{sc}) = \ln(P_{oc}) + v_c\,[k_o - k_s] = \ln(P_{1c}) + v_c\,[k_1 - k_s] \tag{30}$$

Equation (28), which correlated the local compaction stress with the local bulk specific volume, provides a true measure of the presscake's compaction characteristics, whereas Eq. (22) and (27) are basically, equipment-dependent correlations. However, through the use of Eq. (26), which permits Bf to be determined, and Eqs. (22) and (27) in conjunction with Eqs. (29) and (30), one can determine values for the constants k_s and P_{sc}, which are used to correlate P_s in terms of v. Conversely, by applying Eqs. (29) and (30) in rearranged form one can determine k_1 and P_{1c} in new circumstances once P_{sc}, k_s, and Bf are known. This provides a method for determining P_{s1}, the compaction stress at the ram surface vs $\bar{v}$ in new circumstances and hence provides a scaleup method and basis for design which was heretofore lacking.

If k_s is truly constant, Eq. (29) indicates that k_1 should decrease slightly as W increases, which is in agreement with our observations. Value of Bf, k, and P_{sc} vs pressing speed for spent drip grind coffee are shown in Table 16.5.

TABLE 16.5. FRICTION PARAMETERS AND COMPACTION STRESS CONSTANT VS. PRESSING SPEED FOR SPENT DRIP COFFEE

Pressing speed (mm/min)	Bf	k (g/cm^3)	P_{sc} (kPa)	P_{sc} (psi)
200	0.144	2.15	643	93.3
100	0.140	2.10	540	78.3
50	0.149	2.05	432	62.7
20	0.119	1.87	383	55.5
10	0.113	1.78	342	49.6
5	0.111	1.72	313	45.4

FLUID PRESSURES

A sample plot of fluid pressure at the three transducer locations vs cake height is shown for drip grind coffee in Fig.16.8. In this case it can be seen that the maximum fluid pressure drop (1000 psi) almost equaled the entire pressure exerted by the ram, roughly 1200 psi. The maximum value of the fluid pressure drop, ΔP_f, decreased slightly as the pressing speed decreased, e.g., from 1000 psi at a pressing speed of 200 mm/min to 500 psi at 5 mm/min. But the final value of $\bar{v}$ concurrently dropped from 2.05 cm^3/g at 200 mm/min to 1.45 cm^3/g at 5 mm/min. Thus the juice yield increased from 0.40 at 200 mm/min to 0.66 at 5 mm/min. This clearly indicates the importance of fluid pressure drops and internal and outflow rates on yield. (Recall that based on Eq. (13) U and U_o are proportional to the ram speed.)

It is particularly noteworthy that in the case examined the fluid pressure in the cake was high and relatively uniform. Hence in this case most of the pressure drop occurred across the cake – media interface, and not the cake itself. When fines were screened out of the coffee

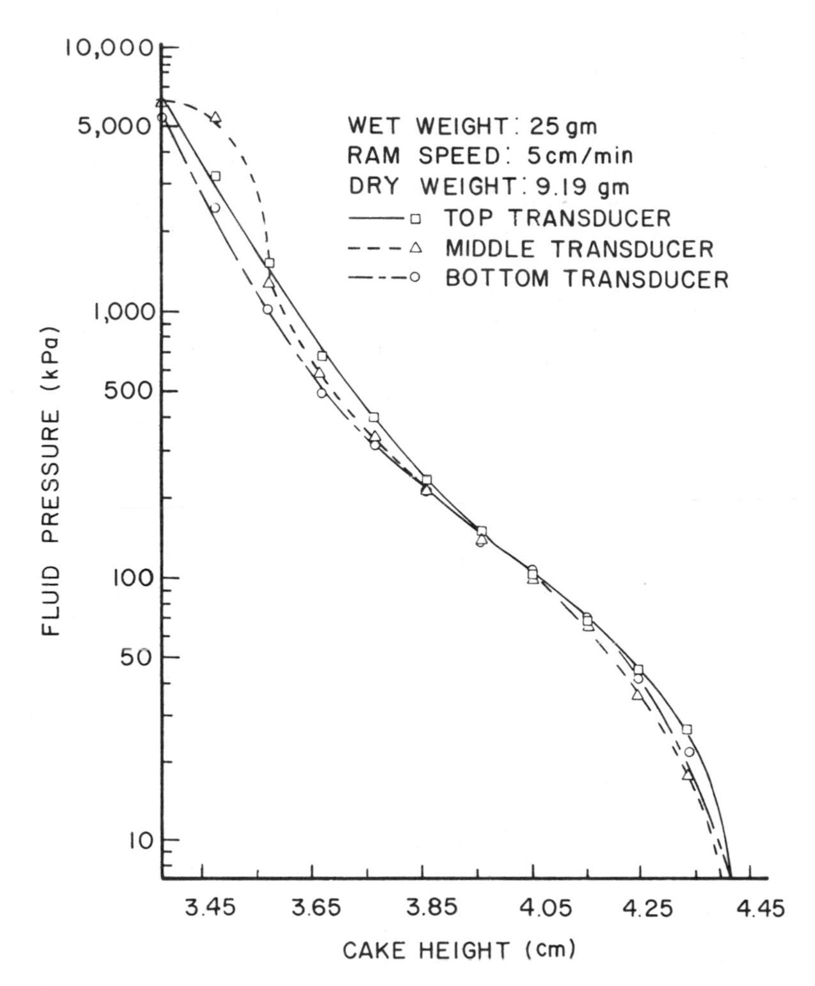

Fig. 16.8. Fluid pressure in presscake vs. cake height on spent drip coffee grinds.

used, the fluid pressure drop decreased markedly and the juice yield increased markedly at all pressing speeds. Further, except at the highest pressing speed, most of the fluid pressure drop occurred across the cake rather than across the outflow surface. Consequently, compaction stress and wall friction were the major factors limiting further juice recovery. With −16+20 mesh grounds (i.e., drip grind coffee with the fines removed), compaction stress and wall friction accounted for two-thirds of the ram pressure at a 200 mm/min pressing speed and 95% of the ram pressure at 5 mm/min. As the particle size increased, ΔP_f decreased further. However, since P_s increases as particle size increases there should be, at least in the case of spent coffee grounds, an optimum particle size in terms of maximizing either yield or processing rate. Further, it appears that mixtures of large particles and fines may provide the worst possible situation; the large particle will increase P_s, and the fines will increase ΔP_f by blinding the outflow media, thereby increasing ΔP_f. The tendency of fines to blind the outflow media could be reduced by increasing the perforation size of the media; but this, in turn, would produce more dregs. This situation suggests that there is an optimum perforation size which depends on the particle size distribution of the cake and the cake's ability to inhibit fines migration. The optimization involves balancing the disadvantages of added dregs in the juice vs the advantage of greater yield as the perforation size is increased.

Radial Flow

The importance of fluid velocity was further demonstrated by the work of Ngoddy (1976), who used a radial flow test unit in the Instron setup previously described. He obtained dryer cakes and higher juice yields than were obtained at comparable final ram pressures in the axial flow cell. Radial flow cells should provide more outflow area and lower values of U_o than axial flow cells as long as $D/Z < 4$ at the end of pressing. In Ngoddy's case his final D/Z values were about 0.6. One can maximize the outflow surface area/cake volume ratio by using either long, narrow diameter cells or very short wide diameter cells. Since short, wide diameter cells will minimize ΔP_W they should be the preferred alternative in terms of minimizing pressing force and increasing yield. However, excessively thin cakes may permit more migratable fines to escape, thus increasing the dregs level.

Sugar Beet Pressing

The fluid pressure curves for spent sugar beet strips (cossettes) undergoing pressing exhibited some peculiarities not seen with coffee grounds. The pressures in the middle of the beet presscake rose smoothly during the beginning of outflow, then followed a highly jagged plateau and did not exhibit any further sustained rise. After the ram stopped, these pressures decreased slowly but smoothly. On the other hand, just above the outflow surface the fluid pressure rose smoothly during the entire period of outflow and continued to rise appreciably for roughly 7 sec after the ram motion stopped. It then decreased smoothly. The jagged portions of the P_f curve corresponded roughly to a jagged portion of the ram pressure curve. However, the ram pressure continues to rise jaggedly during most of the period. Just before the ram stopped, the sustained rise in ram pressure ceased; but the high pressure interlock was tripped by one of the jagged peaks in ram pressure. The combined fluid pressure and ram pressure behavior suggest repeated incidents of cell rupture and breakthroughs of internal flow blockages during pressing.

When sugar beet cossettes, which had been cooked and extracted at 100°C, were subjected to pressing, the ram pressure became jagged at lower pressures, around 70 psi vs around 300–500 psi for beets extracted at 60° – 70°C; and the fluid pressure drop was roughly 2.5 times as great as that for the 60° – 70°C beets. Moreover, the fluid pressure curves for the 100°C extracted beets were jagged at both the middle of the cake and the outflow surface. There were one or more jagged plateaus in fluid pressure, each followed by further jagged rises in pressure. With the 100°C beets, 82 – 95% of the fluid pressure drop occurred across the outflow surface, whereas with the 60° – 70°C beets only 70% of the pressure rise occurred across this surface. It appears that cooking or extraction at temperatures where protopectin is converted to pectin (i.e., temperatures higher that 75°C) leads to easier cell rupture, higher fluid pressure drops and greater blinding of the outflow surface.

FILTRATON RESISTANCE

The fluid pressure drop across the presscake is related to the internal flow rate U and the filtration resistivity α as follows

$$dP_f/dw = \alpha\mu U \tag{31}$$

The integral counterpart of this equation is

$$(P_{fl} - P_{fo})/W = \bar{\alpha}\mu\bar{U} \tag{32}$$

where $\bar{\alpha}$ is the average resistivity of the cake, μ is the viscosity, and $\bar{U}$ is the effective internal flow rate in the cake. While reasonably simple rigorous procedures exist for determining $\bar{\alpha}$ from α in the case of filtration, no such procedure is available in the case of expression. In most cases of filtration, excepting those involving concentrated slurries of highly compactible solids, U is approximately equal to $\bar{U}$ in all sections of the cake. By contrast, U varies strongly during expression, increasing from $U = 0$ at the no-flow surface to $U = U_o = -dZ/dt$ at the outflow surface. Because this increase is not necessarily linear, it is difficult to evaluate $\bar{U}$. Further $\bar{\alpha}$ and $\bar{U}$ are interdependent during pressing; and the effective value of $\bar{U}$, as opposed to its arithmetic mean value depends on how α varies with w/W. Alternatively, one could use the arithmetic mean value of U in Eq. (32) and determine the mean effective value of $\bar{\alpha}$ from dynamic pressing tests.

In the past, α has usually been calculated from permeation flow rates and pressure drops measured at static relaxed conditions, which follow active compaction. Usually $\bar{\alpha}$ has been determined as a function of P_r, but in the work reported herein $\bar{\alpha}$ was determined as a function of $\bar{v}$. In contrast to most former work, Murry and Holt (1967) determined $\bar{\alpha}$ for shredded sugar cane by using dynamic compaction force vs $v_c/\bar{v}$ measurements for linear flow transverse to the direction of compaction. The procedure was apparently necessary because of the highly anisotropic nature of sugar cane. However, Murry and Holt apparently did not subtract the P_s and ΔP_w contributions in calculating ΔP_f from the pressing force measurements.

Usually, to measure $\bar{\alpha}$, a known, axial-flow fluid pressure drop ΔP_f is imposed at a chosen static value of $\bar{v}$ and the corresponding value of U is measured. Since the cake is not undergoing active compaction, U is constant throughout the cake. $\bar{\alpha}$ is calculated as follows

$$\bar{\alpha} = \frac{\Delta P_f}{W \mu U} - \frac{R_m}{W} \tag{33}$$

where μ is the fluid viscosity and R_m is the flow resistance of the outflow media. R_m is usually measured by permeation rate tests in the absence of the presscake. Then

$$R_m = \Delta P_f / \mu U_m \tag{34}$$

where U_m is the value of U in the absence of the cake. Since cake – media interactions often greatly increase R_m when the cake is compacted, this procedure is highly suspect.

Values of log ($\bar{\alpha}$) vs $\bar{v}/v_c'$ obtained in this manner are plotted in Fig. 16.9 along with log ($\bar{\alpha}$) vs $\bar{v}/v_c'$ values determined from Murry and Holt's work. Because of the large number of curves, the curves are numbered; and the materials represented by curves, the source of the data involved, and the corresponding calculated values of v_c' are listed in Table 16.6.

A highly compressed log $\bar{\alpha}$ scale is used, and though the change in ordinate sometimes appears small, $\bar{\alpha}$ invariably increased very rapidly toward the end of pressing.

The data shown in Fig. 16.9 and similar data illustrate a number of interesting points: (1) The $\bar{\alpha}$ vs $\bar{v}/v_c'$ values of nominally similar mate-

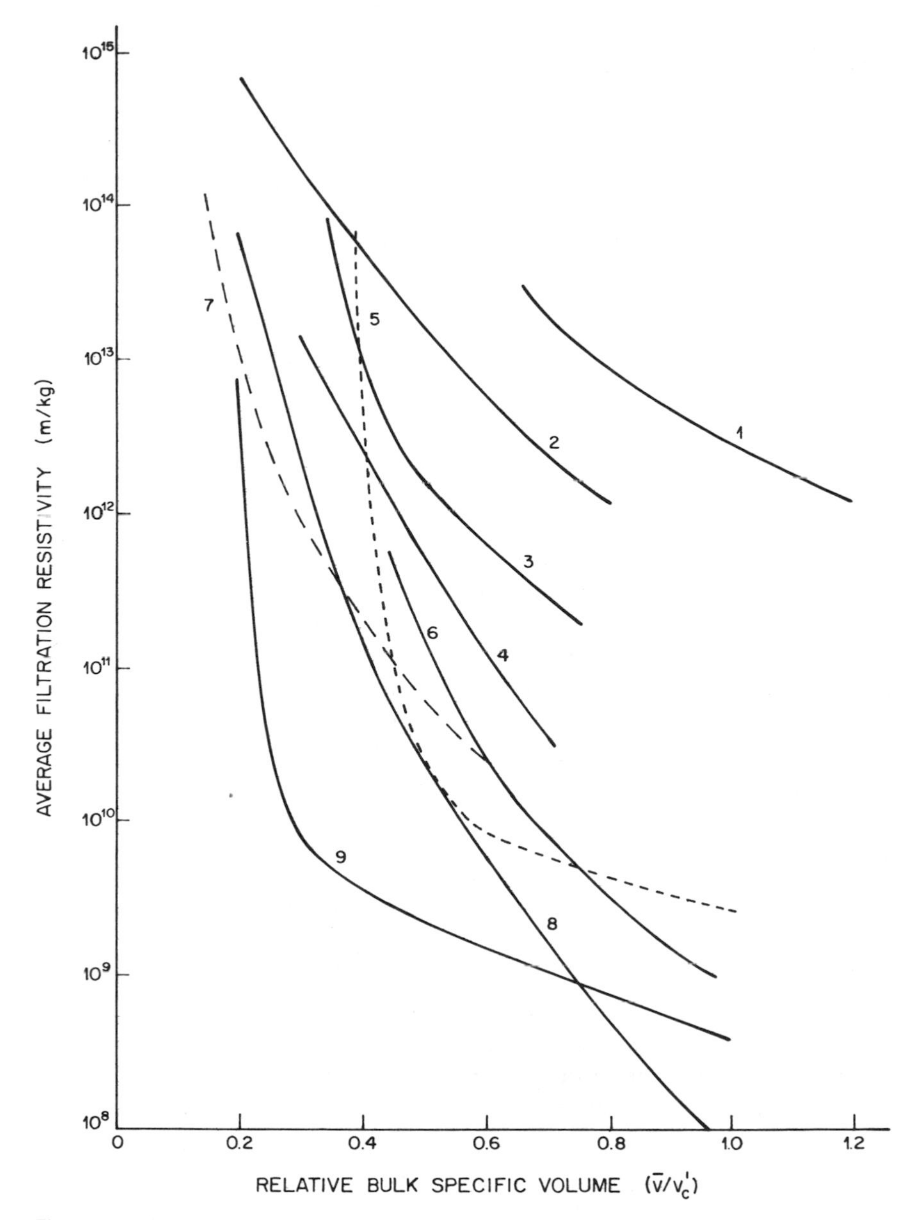

Fig. 16.9. Average filtration resistivity vs. $\bar{v}/v'_c$. See Table 16.6.

rials can be radically different. This is shown by the data for Kehura apples and Jonathan apples, although the experimental technique used to measure $\bar{\alpha}$ probably differed in these two cases. The various $\bar{\alpha}$ vs $\bar{v}/v'_c$

TABLE 16.6. MATERIALS, v'_c VALUES, AND DATA SOURCES FOR log $\bar{\alpha}$ vs. $\bar{v}/v'_c$ CURVES SHOWN IN FIG. 16.9

Curve number	Material	v'_c (cm^3/g marc)	Source
1	Spent drip grind coffee	3.00	Pollak (1978)
2	Jonathan apples	47.4	Körmendy (1979)
3	Milled alfalfa	9.27	Schwartzberg *et al.* (1978)
4	Spent sugar beet cossettes	19.65	This work
5	Spent −16+30 mesh coffee	3.00	Huang (1979)
6	Spent −4+8 mesh coffee	3.00	Schwartzberg *et al.* (1977A)
7	Spent milled alfalfa	9.65	Schwartzberg *et al.* (1978)
8	Shredded sugar cane	6.91	Murry and Holt (1967)
9	Kehura Apples	47.4	Meparishvili and Zhvaniya (1972)

curves for different samples of coffee also illustrate this point. (2) $\bar{\alpha}$ increases as particle size decreases. Murry and Holt (1967) indicate that $\bar{\alpha}$ for sugar cane also changes as the amount of shredding prior to pressing increases. They indicate that the bulk density of the shredded cane may provide a convenient basis for correlating this $\bar{\alpha}$ variation. (3) The presence of small amounts of fines can radically increase the apparent value of $\bar{\alpha}$. For example, compare curves 2 and 5. $\bar{\alpha}$ for spent drip grind coffee is roughly 1000 times greater than $\bar{\alpha}$ for −16+30 mesh coffee even though their mean particle sizes are not radically different. However, the drip grind coffee contains a small amount of very fine material (7% −30+50 mesh solids, and 7.3% − 50 mesh solids). Probably, the fine material blinds the outflow media. This is consistent with the P_f vs $\bar{v}$ measurements for drip grind coffee, where almost all of the fluid pressure drop occurred across the outflow media. (4) Trapped fine particulate solids from the juice can greatly increase $\bar{\alpha}$. Thus $\bar{\alpha}$ for freshly milled alfalfa is roughly ten times higher than $\bar{\alpha}$ for alfalfa for which these solids have been removed by repeated sequences of low pressure expression and imbibition.

Resistance Measurement Reproducibility

It should be noted the $\bar{\alpha}$ measurements are highly nonreproducible. Kormendy (1979) repeated his permeation measurements eight times. The $\bar{\alpha}$ values calculated from these measurements varied as much as 4.6-fold at some $\bar{v}/v'_c$ and at least a 2.2-fold $\bar{\alpha}$ variation occurred at all $\bar{v}/v'_c$ values. Tests by Pollak (1978) indicated roughly similar $\bar{\alpha}$ variability for spent drip ground coffee. In both Kormendy's and Pollak's work the coefficient of variation decreased as $\bar{v}/v'_c$ decreased. Huang (1979) attempted to measure the effect of W on the apparent value of $\bar{\alpha}$ and noted considerable variation; but because of the inherent variability of $\bar{\alpha}$ measurement one cannot be sure that the changes he observed are induced by the change in W.

The measurements of Schwartzberg *et al.* (1977A, 1978A), Pollak (1978), Huang (1979) and those reported for the first time here were carried out in a unit similar to that shown in Fig. 16.1. The permeating liquid passed through the channel in the ram, the perforations in its tip, the presscake and the outflow media. Because relatively thick cakes and a small diameter were used, the compaction was probably relatively nonuniform in the axial direction due to friction-induced variation in P_s and v. Flow-induced drag forces probably also caused axial variations in P_s and v. Such variations can be minimized by using large diameter cells, as Kormendy did, small cake thicknesses, and very slow pressing rates. However, the use of large diameter cells would have limited the maximum ram pressure that the Instron could have provided; and the use of small cakes thickness would have limited the accuracy with which $\bar{v}$ could be measured, particularly at the low $\bar{v}$ values which occur near the end of pressing.

Outflow resistance probably represents a large fraction of the apparent values of $\bar{\alpha}$ show in Fig. 16.9. Thus the true values of $\bar{\alpha}$ are probably less than indicated in Fig. 16.9.

FLOW PRESSURE DROP PREDICTIONS

If one assumes that the filtration resistivity values shown in Fig. 16.9 really represent $\bar{\alpha}_{APP}$, apparent values of $\bar{\alpha}$, then

$$\bar{\alpha}_{APP} = \bar{\alpha} + R_m/W \tag{35}$$

Huang (1979) attempted to use such values of $\bar{\alpha}_{APP}$ to predict values for ΔP_f for expression tests involving the same coffee grounds for which $\bar{\alpha}_{APP}$ was measured. He used the equation

$$\Delta P_f = \bar{\alpha}_{APP}\, \bar{U} \mu W \tag{36}$$

to calculate the predicted values of ΔP_f using $\bar{\alpha}_{APP}$ values corresponding to chosen values of $\bar{v}$. Since the effective value for $\bar{U}$ in the case of uniform compaction is $-(dZ/dt)/2$, and that for compaction where the flow resistance occurred primarily at the outflow surface would be

$-(dZ/dt)$, Huang used these two choices for $\bar{U}$. With spent drip grind coffee, moderately good agreement between the predicted and observed values of ΔP_f was obtained fairly often. In most cases better agreement was obtained by setting $\bar{U}$ equal to $-(dZ/dt)/2$. In the case of $-16+20$ mesh grounds, much poorer agreement between the predicted and measured values of ΔP_f was obtained. This no doubt stems from the greater steepness of the $\bar{\alpha}$ vs $\bar{v}$ curve for $-16+30$ mesh grounds (curve 5) in Fig. 16.9 as compared to that for drip grind coffee (curve 1). Because of the steepness of curve 5 even miniscule errors in determining $\bar{v}$ will cause very large errors in determining $\bar{\alpha}_{APP}$ and ΔP_f. Thus, the poor ΔP_f agreement for the $-16+20$ mesh grounds is readily understandable.

CAKE VOLUME AND JUICE YIELD PREDICTION

Since the log $\bar{\alpha}_{APP}$ vs $\bar{v}$ curve is very steep, it should be possible to predict $\bar{v}$ very accurately if the corresponding value of $\bar{\alpha}_{APP}$ is known. Such values of $\bar{\alpha}_{APP}$ can be obtained from measured values of ΔP_f by rearranging Eq. (36) and using whichever of $-(dZ/dt)/2$ or $-(dZ/dt)$ appears more suitable for $\bar{U}$. The $\bar{\alpha}$ values calculated in this fashion are then used to look up corresponding values of $\bar{v}$ on log $\bar{\alpha}_{APP}$ vs $\bar{v}/v_c$ curves. $\bar{v}$ values calculated in this fashion were compared by Huang (1979) with corresponding measured values of $\bar{v}$. The measured and calculated $\bar{v}$ values agreed within 20% in almost all cases for the drip grind spent coffee grounds. The average difference between the final predicted and measured values of $\bar{v}$ was 8.3% when the $\bar{U}=-(dZ/dt)/2$ choice was used. When $\bar{U}=-(dZ/dt)$ was used, the corresponding average difference was 16%. The average difference between the predicted and observed value of $\bar{v}$ for $-16+20$ mesh grounds was 9.2% at $\bar{U}=-(dZ/dt)$ and 12.5% at $\bar{U}=-(dZ/dt)/2$. Considering that, $\bar{\alpha}_{APP}$ vs $\bar{v}$ measurements are subject to reasonably large experimental errors, the agreement is not bad.

If $\bar{v}_f$ the final value of $\bar{v}$ can be predicted, juice yields and juice recovery factors can be predicted through the use of Eq. (5)–(9). Schwartzberg *et al.* (1977B, 1978A) noted that for milled alfalfa P_{s1} is very small, hence they assumed that the final value of P_1 was approximately equal to ΔP_f. Using this assumption and taking $\bar{U}$ as $-(dZ/dt)$ they calculated $\bar{\alpha}_{APP}$ and $\bar{v}_f$ for milled alfalfa at different pressing speeds. These $\bar{v}_f$ values were used to calculate juice yields. When the calculated and measured juice yields were compared, the estimated juice yield ranged from 2% (too low at the lowest pressing speed 5 mm/min) to 38% (too high at the highest pressing speed

200 mm/min). The error progressively increased with pressing speed and no doubt resulted from neglecting P_{s1}, which also progressively increases with pressing speed. Therefore, Schwartzberg *et al.* (1978B) utilized a correction based on linear proportionality between the error and the pressing speed. They scaled this proportionality based on the juice yield error at the highest pressing speed. When this was done the average magnitude of the juice yield error was reduced to 2.6%.

Schwartzberg *et al.* (1979A) subsequently developed an iterative procedure for predicting $\bar{v}_f$ and the juice yield when $\bar{P}_{s1}$ is not negligible, but P_{s1} and $\bar{\alpha}_{APP}$ vs $\bar{v}$ correlations or charts are available. By using Eq. (35) the method can be readily extended to the case when $\bar{\alpha}$ and R_m are known as functions of $\bar{v}$. If P_{s1} is likely to be small compared to ΔP_f, an initial value for $\bar{v}_f$ is obtained by setting the chosen final value of P_1, the ram pressure, equal to ΔP_f and using the previously described procedure to calculate $\bar{v}$ from P_f. This $\bar{v}$ value is used in Eq. (22) to estimate P_{s1}. Then ΔP_f is set equal to $P_1 - P_{s1}$ and a new value of $\bar{v}_f$ is calculated. This in turn is used to calculate a new value of P_{s1}. The procedure is continued until successive calculated values of $\bar{v}_f$ do not differ by more than some maximum tolerable percentage, e.g., 0.1%. The final value of $\bar{v}_f$ is then used to calculate η the juice yield by means of Eq. (5) or (6). The k_1 and P_{1c} values needed to calculate P_{s1} for the W values used during production can be obtained from k_s and P_{sc} values obtained in small scale tests by using Eq. (29) and (30).

In order to scale-up $\bar{\alpha}_{APP}$ it is necessary to know both R_m and $\bar{\alpha}$ as a function of $\bar{v}$. Separate values of $\bar{\alpha}$ and R_m can be obtained by carrying out permeation tests in a test unit with a pressure transducer whose inlet slit is positioned just above the outflow surface. By setting ΔP_m, the pressure measured by this transducer, equal to ΔP_f in Eq. (34), one can determine R_m. By carrying out permeation tests at successive values of $\bar{v}$ one can obtain both $\bar{\alpha}$ and R_m as functions of $\bar{v}$. In estimating $\bar{v}_f$ for scaled up operation when $\bar{\alpha}$ and R_m are known it would probably be best to use $(\bar{\alpha}_{APP})_{MOD}$, a modified $\bar{\alpha}_{APP}$, and $\bar{U} = -(dZ/dt)/2$

$$(\bar{\alpha}_{APP})_{MOD} = [\bar{\alpha} + (2R_m/W)] \tag{36}$$

If $\bar{\alpha}$ and R_m are known as functions of $\bar{v}$, $(\bar{\alpha}_{APP})_{MOD}$ can be readily determined as a function of $\bar{v}$. The reason for these choices of $(\bar{\alpha}_{APP})_{MOD}$ and $\bar{U}$ is that $\bar{U}$ in the cake will most often be approximated best by $-(dZ/d\theta)/2$, whereas U_o the flow rate determining the pressure drop across the media is equal to $-(dZ/d\theta)$. Equations (16) – (19) can be used to calculate appropriate approximate values of $\bar{U}$ in the cake and U_o for the various radial flow and transverse flow cases. In cases where the radial flow occurs across layers of cake that are thin relative to the outflow diameter, the flow can be treated as linear axial flow as a reasonable approximation.

If it appears that P_{s1} will be larger than ΔP_f at the end of pressing, the order of iteration used to determine $\bar{v}_f$ should be reversed. That is, P_{s1} should be set equal to the maximum allowed value of P_1, and the first estimate of $\bar{v}$ should be obtained from Eq. (22). This $\bar{v}$ estimate should then be used to calculate $(\bar{\alpha}_{APP})_{MOD}$ and ΔP_f. The calculated value of ΔP_f should be then subtracted from P_1 to get a revised estimate of P_{s1}. The above procedure is then repeated until successive calculated values of $\bar{v}_f$ agree within some tolerable percentage.

FLUID VISCOSITIES

In carrying out ΔP_f calculation it is often assumed that μ can be treated as constant. However, when pressing juices out of fruit this may not be the case. For example, Körmendy (1979) measured μ for Jonathan apple juice as a function of the incremental juice yield $\Delta\eta$. His μ vs $\Delta\eta$ values are shown in Table 16.7.

TABLE 16.7. APPLE JUICE VISCOSITY VS. INCREMENTAL JUICE YIELD

Juice yield range	Viscosity (cP)
0 – 20%	2.34
20 – 40%	2.79
40 – 60%	3.02
60 – 80%	3.09

This sequence of viscosities probably results from the progressive mixing and displacement of dilute intercellular juice by stronger intracellular juice during the course of pressing.

In pressing pectin liquor out of cooked apple pomace, the grade of the pectin liquor progressively decreases during the course of pressing (Pilnik 1981). This presumably occurs because the pectin is created by solubilizing protopectin, which lies in the middle lamella between the cells. This pectin is progressively diluted by pectin-free juices or juices of low pectin content pressed out of the cells. Since the pectin liquor viscosity tends to increase as its grade increases, the pressing of pectin liquor out of pomace is likely to be accompanied by a progressive decrease in juice viscosity.

Whatever case occurs, $\bar{v}_f$ estimates should be calculated using μ values appropriate to the end of pressing.

MORE DETAILED MODELS

More detailed models of the flow local compaction and pressure transfer processes in the presscake should permit more accurate estimations of $\bar{v}_f$ and juice yield. These models can be generated by setting up differential equations pertaining to the force balances and local flow rates in the presscake. The local force balance during outflow is

$$\frac{dP_s}{dw} = \frac{-dP_f}{dw} + \frac{4BfP_s v}{D} \tag{37}$$

Substituting from Eq. (31) for dP_f/dw

$$dP_s/dw = -\alpha\mu U + 4BfP_s v/D \tag{38}$$

But, $dP_s/dw = (dP_s/dv)(\partial v/\partial w)$; P_s can be found from Eq. (28); α is a function of v similar to those shown for $\bar{\alpha}$ vs $\bar{v}$ in Fig. 16.8; and U can be obtained from Eq. (12). Over reasonably wide ranges of v, the α vs v relationship can be correlated by an equation of the form

$$\alpha = \alpha_o v^n \tag{39}$$

If the equations just cited are substituted in Eq. (38) there is obtained

$$-P_{sc}k \exp[-kv]\left(\frac{\partial v}{\partial w}\right) = -\alpha_o v^n \mu \int_W^w \left(\frac{\partial v}{\partial t}\right) dw + (4Bfv/D) P_{sc} \exp[k(v_c - v)] \tag{40}$$

If v_o, the value of v at $w = 0$ is known, values of $(\partial v/\partial w)$ obtained from Eq. (40) can be used in the following equation to obtain v as a function of w.

$$v - v_o = \int_0^w \left(\frac{\partial v}{\partial w}\right) dw \tag{41}$$

Equations (40) and (41) have to be solved simultaneously because $\partial v/\partial w$ is a function of v.

The solution to Eqs. (40) and (41) must satisfy the constraints

$$\int_0^W v\, dw = Z = Z_0 + \left(\frac{dZ}{dt}\right) t \tag{42}$$

and

$$\int_{W}^{0} \left(\frac{\partial v}{\partial t}\right) dw = -\frac{dZ}{dt} \tag{43}$$

Thus, even when the compaction rate (dZ/dt) is constant, the solution of Eq. (40) will be a function of time. Because it is nonlinear, Eq. (40) must be solved by numerical techniques. Alternative differential equations derived by Shirato *et al.* (1971), Kormendy (1974), Schwartzberg (1977), and Schwartzberg *et al.* (1977A; 1979A) also require numerical solution when the compaction rate rather than the compaction pressure is prescribed.

Numerical Solutions

Solutions to Eq. (40) can be obtained as follows. The integral term in Eq. (40) is initially replaced by $-(dZ/dt)\,[1 - w/W]$. This will automatically satisfy the integral constrain (Eq. 43). A value for v_o, v at the bottom of the presscake, is selected. Equations (40) and (41) are then numerically integrated from $W = 0$ to $w = W$ simultaneously, thereby obtaining v as a function of w. The local values of v are then integrated with respect to w to obtain a value of Z. A second slightly smaller value of v_o is selected and the above process is repeated. The time interval Δt between the successive value of v_o is then determined by dividing the difference between the second and first Z values by dZ/dt. Note: Both the Z value difference and dZ/dt are negative. The difference between the first and second local v values at corresponding values of w is then divided by Δt to obtain values of $(\partial v/\partial t)$. These are then progressively substituted in Eq. (40); and Eqs. (40) and (41) are again numerically integrated simultaneously. This provides new sets of v values, which in turn can be used to get new Z values and Δt values. The process is repeated until convergence is obtained for each of the sets of Z values corresponding to the two chosen values of v_o.

The value of v_1 the bulk specific volume at the top of the presscake is obtained by integrating Eq. (41) from 0 to W. Actually, this has already been done as part of the preceding calculation. The values for v_1 are then substituted into Eq. (28) in order to find P_{s1}. The calculated local values of $(\partial v/\partial t)$ are substituted into Eq. (12) in order to find local values of U. Again, this process has already been done as part of the preceding calculations. P_{fo} the fluid pressure just above the outflow surface is then set equal to $-R_m\mu(dZ/dt)$. In carrying out this last calculation it is assumed that a R_m vs v correlation is available. Local values of P_f in the cake are then determined by substituting for α in Eq. (37) and integrating it from $w = W$ to w, using the previously determined local values of U and v.

$$P_f = P_{fo} + \int_W^w \alpha_o v^n U \, dw \quad (44)$$

When w is set equal to 0, P_f equals P_{f1} the fluid pressure at the top of the cake. $P_1 = P_{s1} + P_{f1}$. The values of P_1 corresponding to the chosen pair of values of v_o are compared to $(P_1)_{max}$, the maximum allowable value of P_1. If the P_1 are greater than $(P_1)_{max}$, a larger set of v_o values is chosen; and if they are smaller than $(P_1)_{max}$ a smaller set of v_o values is chosen.

Volume Adjustment

A reasonable procedure for choosing the new values of v_o is to multiply the old values by the calculated values of $P_1/(P_1)_{max}$. The preceding procedure is then repeated in its entirety, and the new calculated values of P_1 are compared with (P_1). If convergence within an acceptable error limit is not obtained, the entire process is repeated again and again until convergence does occur. If the successive values of P_1 exhibit oscillating divergence or converge slowly, the method of Wegstein (Perry and Chilton 1973) can be used to achieve and accelerate convergence. If Z_f is the value of Z at which convergence occurs $\bar{v}_f = Z_f/W$. Once $\bar{v}_f$ is known, the juice yield can then be obtained by substituting $\bar{v}_f$ in Eqs. (5) or (6).

Obviously, the procedure just described is far more complex and time consuming than the previously discussed yield estimation procedure. However, the complex procedure provides a better basis for testing the validity of the proposed expression model and a basis for determining when the simpler yield estimating procedure can be validly used. Tests for validity of the proposed model can be based on comparisons of the experimentally measured variation of v, U, and P_f vs w/W. The predicted behavior of v, U, and P_f vs w/W depends on the fraction of the fluid pressure drop that occurs across the outflow media. It also depends on relative magnitude of the wall friction, which tends to increase P_s and decrease v in the direction of increasing w, and the fluid pressure drop across the cake, which tends to increase v and decrease P_s in the direction of increasing w.

If the outflow resistance is very high compared to the internal flow resistance, U will usually tend to decrease linearly as w/W increases. However, at high pressing speeds, U will decrease at a greater than average rate as w/W increases near the outflow surface. Because of the opposing effects of fluid drag and wall friction, v may first decrease as w/W increases and then increase again at higher w/W values; but if the internal fluid pressure drop across the cake is small, v will increase monotonically as w/W increases. Evaluation of the types of compaction, flow pressure drop, and frictional behavior that can occur, indicates that the short cut yield estimation methods should work best when outflow media resistance is the main factor limiting the expression yield. When the filtration resistance of the cake itself is large relative to the outflow resistance, the simple yield prediction method should be reasonably accurate as long as the wall friction effects are small; but as the wall friction effects and cake filtration resistance increase, the short-cut procedure will show a tendency to overpredict yields. This occurs because during outflow ΔP_w is no longer adequately accounted for by the equation

$$\Delta P_w = P_{s1} [1 - \exp(-4Bf\bar{v}W/D)] \tag{45}$$

Equation (45) is truly valid only before outflow, but it has been implicitly used during outflow in the short-cut yield prediction procedure. Equation (45) is not valid during outflow because the fluid pressure drop tends to increase P_s in the flow direction, thereby partly counterbalancing the friction induced reduction in P_s. Since the local values of P_s are higher than would have been anticipated on the basis of friction alone, ΔP_w as a whole will be greater than predicted by Eq. (45). Since friction is counterproductive in general, large enough flow cross sections should be used to reduce friction to negligible proportions. This will not only insure more efficient operation, but will also provide improved yield predictability.

TRANSPARENT CELL STUDIES

The transparent cell shown in Fig. 16.10 was constructed in order to provide local v measurement and a means of calculating local values of U. The cell was operated in the Instron in the same manner as the other cells, but motion pictures were taken of the cell, its contents, and an adjacent clock during pressing. The presscake was loaded into the cell in equal segments separated by perforated metal wafers that contrasted optically with the presscake. The extent of compaction of the cake segments and local average value of v could be determined as a function of time by reading the clock and measuring the distance between the wafers on the motion pictures. The fluid velocity increase across each segment is $\Delta Z_s/\Delta t$, where ΔZ_s is the decrease in segment thickness

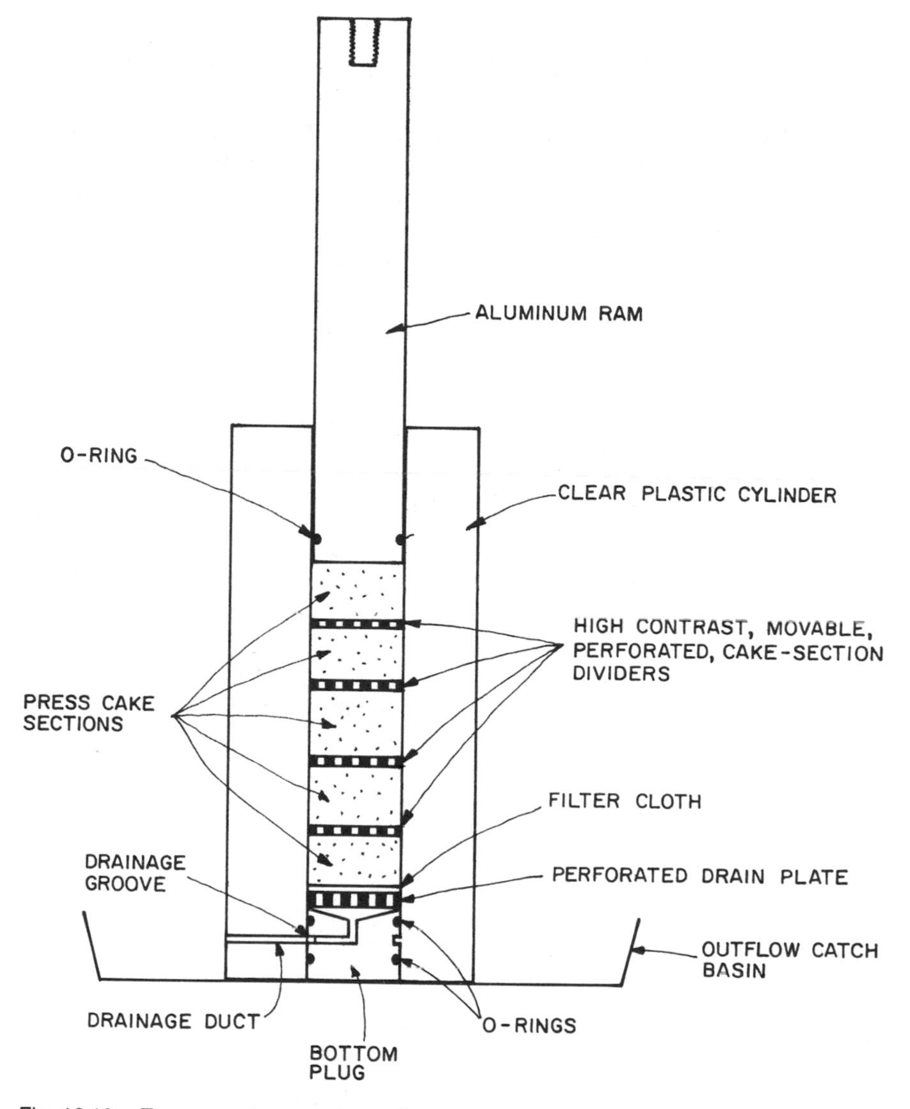

Fig. 16.10. Transparent expression unit.

during time interval Δt. Since $U = 0$ at the ram surface, U_n the value of U leaving the nth segment counting outward from the ram surface is given by

$$U_n^i = \frac{1}{\Delta t} \sum_{i=1}^{N} (\Delta Z_s)_i \tag{46}$$

where the maximum N is the total number of segments when $\bar{v} < v_c$, or in case $\bar{v} > v_c$, the number of the last segment in which the local average of v is less than v_c.

A plot of the relative extent of compaction Z_s/Z_{so}, where Z_{so} is the initial segment height is shown for spent drip grind coffee in Fig. 16.11. It can be seen that the top cake layer initially compacts faster than the lower layers and that v decreases monotonically from the top layer to the outflow layer. This agrees with our expectations since R_m is very large for spent drip grind coffee.

Values of U were calculated from Eq. (46) and are plotted vs w/W in Fig. 16.12. At low pressing speeds U increased linearly as w/W decreased; but, as anticipated, at high pressing speeds U increased at a greater than average rate when w/W was small.

FLUID PRESSURES IN THE CAKE

The v and U behavior just described is in fairly good agreement with the mathematical model of the compaction process; but because it was obtained using a presscake that induced a high value of R_m it does not provide an adequate test of the model. The fluid pressure profile in the cake, particularly when R_m is small, should provide such a test. Predicted values of P_f vs w can be obtained from Eq. (44). Based on Eq. (44) P_f should monotonically increase as w increases.

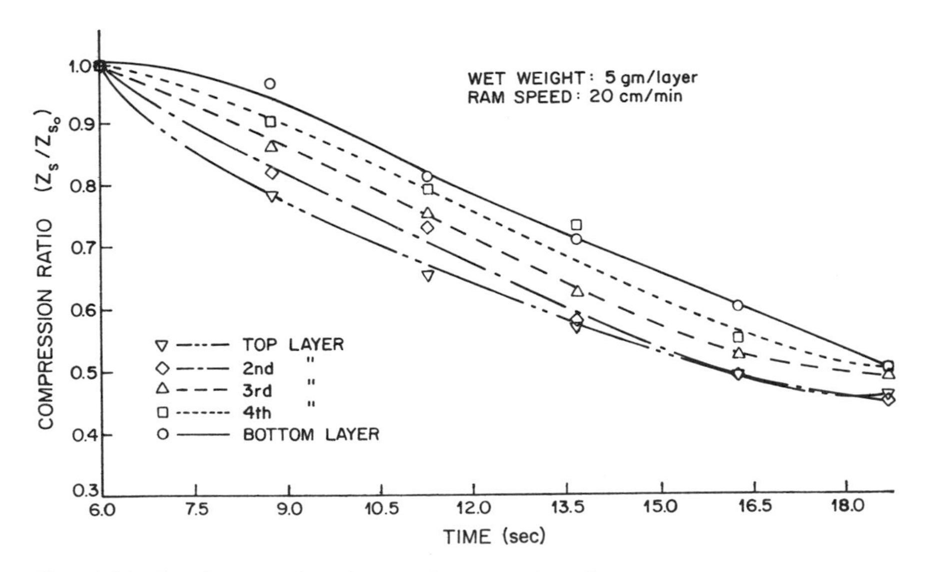

Fig. 16.11. Local compaction of presscake segment vs. time.

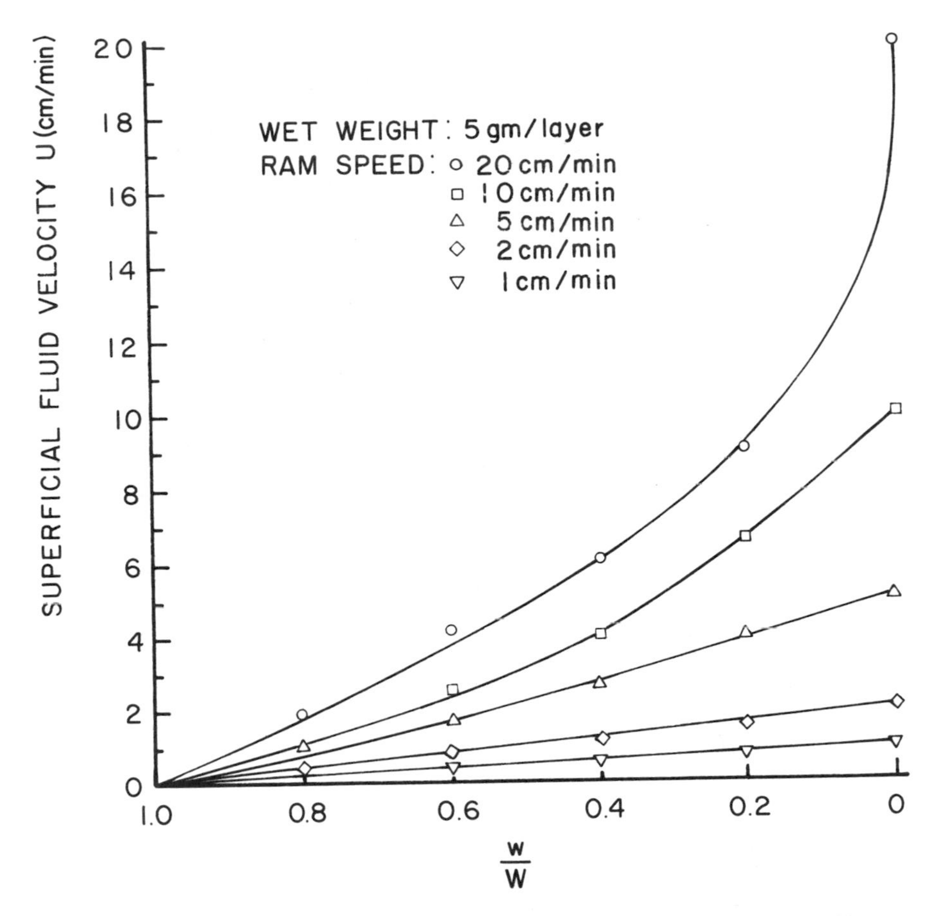

Fig. 16.12. Local fluid velocity vs. w/W as a function of pressing speed on spent drip coffee grounds.

Fig. 16.13 is plot of log P_f vs (transducer elevation/cake height). While P_f initially increased monotonically with distance above the outflow surface, a hump in the P_f profile ultimately developed. Similar humps were observed in most of our fluid pressure measurements. While the reproducibility of the fluid pressure measurements is poor the characteristic features of the profile including the hump usually remain the same. Since the observed P_f humps are in conflict with the expression model used herein and with all other expression models of which the author is aware, efforts were made to make sure the humps were not artifacts generated by the use of improper experimental procedures.

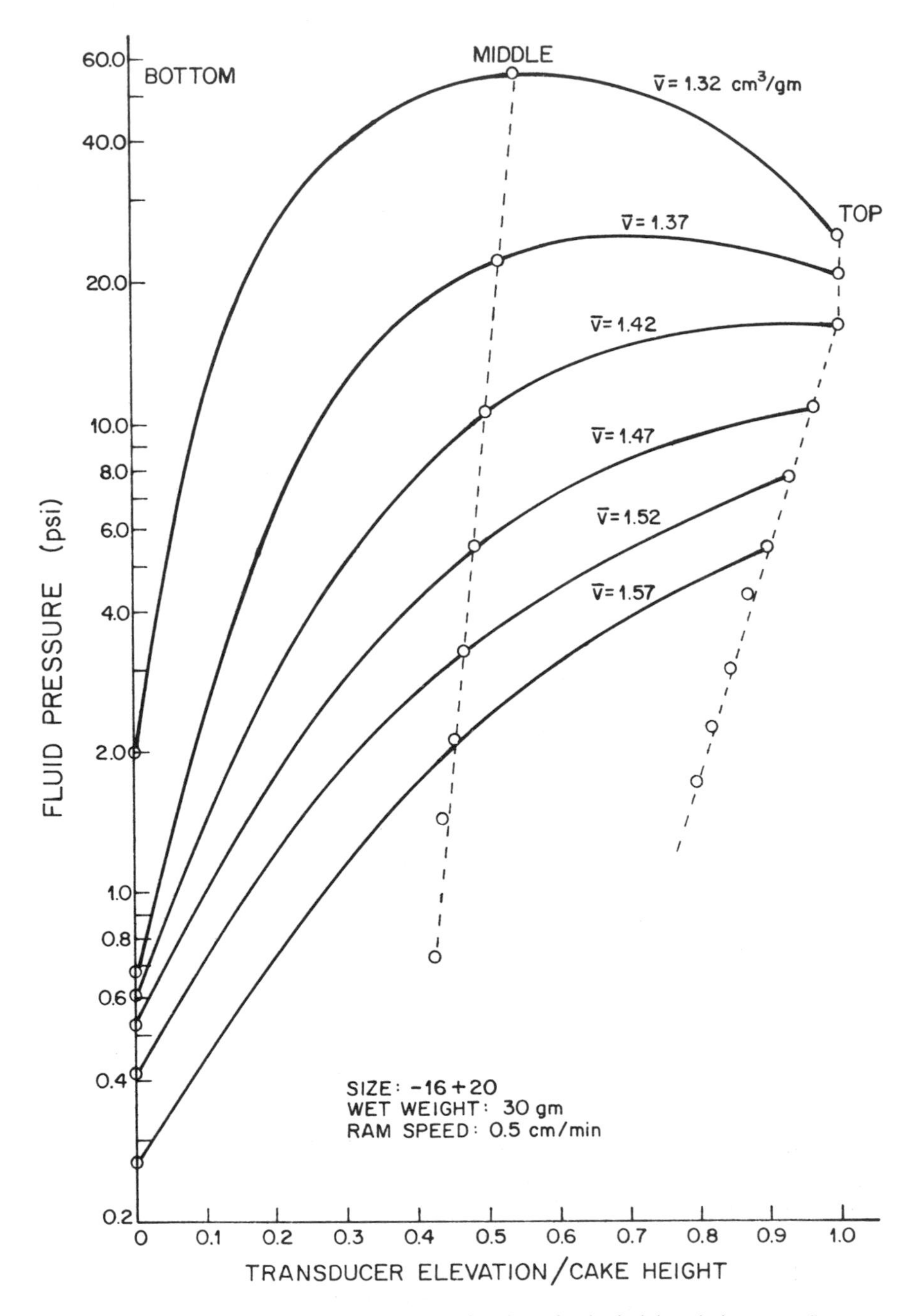

Fig. 16.13. Fluid pressure in presscake vs. fraction of cake height relative to outflow surface as a function of bulk specific volume on spent drip coffee grounds.

The possibility that fluid leakage or entrapped air compaction could have generated the P_f humps was eliminated by making sure that no such leakage occurred and by displacing all air by filling all the pores in the presscake and all channels in the test unit with water prior to pressing. Since the pressure transducers could not check P_f in the cake itself, the possibility that the fluid pressures might be abnormal near the cylinder's wall could not be eliminated or tested. Abnormal P_f profiles might be generated by a shear field in a thin fluid layer between the presscake and the wall. This fluid would be stationary at the wall but would move with the presscake at the cake surface. The rate of movement of the presscake relative to the wall goes from $-dZ/dt$ at the ram tip to zero at the outflow surface. Thus the film shear rate must decrease in going from the rim tip to the outflow surface. This decrease in shear might give rise to an increase in pressure in the direction of cake movement.

Alternatively, the humps might truly represent the nature of the P_f change in the cake. It is usually assumed that P_1, the solid stress in the ram, is immediately transformed into $P_{s1} + P_{f1}$ at the ram − cake interface. Since the cake particles also have a solid surface there is no intrinsic reason why the particles cannot similarly convert part of their compressive stress into fluid pressure. The process might be analogous to peristaltic pumping. One possible cause for this behavior might be changes in the exudation contribution to P_s at various levels in the cake.

These hypotheses and other possible explanations for the existence of the P_f humps should be critically tested. If the postulated P_s to P_f transfer is validated, the mechanistic basis for transfer should be identified. This could provide means for calculating the extent of transfer as a function of w.

CONCLUSION

Many of the factors affecting expression can be identified, measured, and correlated by using procedures described herein. These factors can be utilized in mathematical models that permit the prediction of juice yields, ram pressures, cake heights, local specific volumes, and local fluid velocities. Methods for improving expression yields can be readily inferred through the use of the models. In some cases relatively simple methods can be used to predict juice yields; however, in other cases the use of these simple methods is suspect. The fluid pressure in the presscake apparently behaves in a manner not predicted by known expression models. The causes for this fluid pressure behavior have not yet been clearly identified.

NOMENCLATURE

B	Transverse compaction stress/axial compaction stress
b	Mass of water bound per unit mass of marc
C	Tracer solute concentration
C_0	Initial value of C
C_1	Intercept equilibrium value of C
C_2	Equilibrium value of C
D	Diameter of presscake cake
D_i	Inside diameter of pressing unit
D_o	Outside diameter of pressing unit
E	Mass of solution mixed with solid in cell rupture and free water evaluation tests
F	Ram force
F_R	Fraction of cells ruptured
f	Coefficient of friction between presscake and pressing unit
G	Mass of cling per unit mass of as-is presscake
I	Integral of v with respect to dw between the limits 0 and W
i	Number of cake segments from ram surface
k_1	Stress coefficient at ram surface
k_o	Stress coefficient at outflow surface
k	Local stress coefficient
dL/dt	Chart speed
ΔL	Chart travel length on force recorder chart between peak force and chosen force during active pressing
ΔL_R	Chart travel length on recorder chart between peak force and chosen force during relaxation
N	Number of presscake segments in which flow is occurring
n	Exponent in filtration resistivity correlation
P	Pressure or compaction stress
P_1	Ram pressure
P_{1c}	Ram pressure when $\bar{v} = v_c$
P_o	Pressure at outflow surface
P_{oc}	Outflow surface pressure when $\bar{v} = v_c$
P_f	Fluid pressure
P_{f1}	Fluid pressure at ram surface
P_{fo}	Fluid pressure at outflow surface
ΔP_f	Fluid pressure drop
ΔP_m	Fluid pressure drop across outflow media
P_r	Relaxed ram pressure
P_s	Local compaction stress
P_{sc}	Value of P_s when $v = v_c$
P_{s1}	Value of P_s at ram surface
P_{so}	Value of P_s at outflow surface
ΔP_w	Change in P_s due to wall friction
R	Ram pressure relaxation constant
R_m	Flow resistance of outflow media
r	Ram speed/chart speed
S	Weight of solids in cell-rupture and free-water evaluation tests
t	Time

U	Internal superficial velocity of fluid relative to presscake solids
U_i	Value of U leaving ith cake segment
U_m	Fluid velocity through outflow media in the absence of the presscake
U_o	Outflow velocity
$\bar{U}$	Effective mean value of U
V_m	Partial volume of marc
V_s	Partical volume of solutes
V_w	Partial volume of water
v	Local bulk specific volume, local volume of cake per unit mass of marc
$\bar{v}$	Average value of v for whole presscake
v_c	Initial value of v excluding the volume occupied by air
v'_c	Initial value of v excluding the volume occupied by air and cling
$\bar{v}_f$	Final volume of v
v_o	Volume of v at the outflow surface
v_1	Volume of v at the ram surface
W	Mass of marc between outflow and no flow surfaces/outflow surface area
w	Mass of marc between a chosen level above the outflow surface and the outflow surface/outflow surface area
x_m	Weight fraction of marc
x_s	Weight fraction of solutes
x_w	Weight fraction of water or other solvent
x_{wb}	Weight fraction of bound water
x_{wf}	Weight fraction of free water
y_s	Mass of solute/mass of marc
y_W	Mass of water/mass of marc
Z	Cake height
Z_f	Final value of Z
$-dZ/dt$	Compaction rate
Z_o	Original value of Z
Z_s	Height of cake segment
Z_{so}	Original height of cake segment
ΔZ_s	Change in Z_s during time interval Δt
z	Distance above outflow surface
Z'	Cake thickness transverse to direction of flow
$Z_{0'}$	Original value of Z'
Z'_c	Value of Z' at which outflow starts
α	Local filtration resistivity of cake
$\bar{\alpha}$	Average value of α for entire cake
$\bar{\alpha}_{APP}$	Value of $\bar{\alpha}$ including the outflow resistance
$(\bar{\alpha}_{APP})_{MOD}$	Value of $\bar{\alpha}_{APP}$ modified so as to take into difference between flow rate through the media and the mean effective flow rate in the cake
α_o	Coefficient in α correlation
ε	Fraction of cake volume occupied by interparticle pores
$\bar{\varepsilon}$	Average value of ε for whole cake
η	Juice yield, expelled fluid mass/original cake mass
$\Delta\eta$	Change in η during a pressing interval
η_j	Juice recovery factor expelled fluid mass/original fluid mass in cake
$(\eta_j)_{CORR}$	A corrected value of η_j which takes into consideration the nonexpressibility of bound water
θ	Relaxation time

μ	Viscosity
ρ_j	Juice density
ψ	Fraction of cake volume occupied by fluid contained in cells within the presscake solids
$\bar{\psi}$	Average value of ψ for entire cake

BIBLIOGRAPHY

ABULARACH, V.H. 1979. A study of cake porosity in constant rate expression. M.S. Thesis. University of Massachusetts, Amherst.

DOMBROWSKI, H.S., and BROWNELL, L.E. 1954. Residual equilibrium saturation of porous media. Ind. Eng. Chem. *46*, 1207 – 1216.

HOLTZ, E., and OHLSON, I. 1981. Investigating the use of spent coffee grounds as low-cost dialysis membranes for separating low and high molecular solutes in counter-current extractors. Report of research done at the University of Massachusetts, Amherst. (Submitted to the University of Lund, Sweden.).

HUANG, B.W. 1979. Flow induced pressure drop during expression. M.S. Thesis. University of Massachusetts, Amherst.

KESSLER, H.G. 1981. Food Engineering and Dairy Technology. Kessler Verlag, Freising, West Germany.

KÖRMENDY, I. 1974. Contributions to the three dimensional pressing theory and its one dimensional application. Acta Aliment. Acad. Sci. Hung. *3*, 93 – 110.

KÖRMENDY, I. 1979. Experiments for the determination of the specific resistance of comminuted and pressed apple against its own juice. Acta Aliment. Acad. Sci. Hung. *4*, 321 – 342.

KEARNEY, R.D. 1979. Applications of helical conveyor centrifugal in the food industry. Paper presented at the 87th National Meeting, Am. Inst. Chem. Eng., Boston.

MEPARISHVILI, S.G., and ZHVANIYA, G.G. 1972. Cited by I. Kormendy. Experiments for the determination of the specific resistance of comminuted and pressed apple against its own juice. Acta Aliment. Acad. Sci. Hung. *3*, 321 – 342.

MURRY,C.R., and HOLT, J.E. 1967. The Mechanics of Crushing Sugar Cane. Elsevier Publishing Co., Amsterdam.

NGODDY, P. 1976. The expression of fluid from biological material, the radial flow case. Report of research done at the University of Massachusetts, Amherst.

PERRY, J.H., and CHILTON, C.H. 1973. Chemical Engineers Handbook, 5th Edition, pp. 2 – 54. McGraw-Hill Book Company, New York.

PILNIK, W. 1981. Personal communication. Wageningen, The Netherlands.

POLLAK, N. 1978. Study of the feasibility of gas aided expression. M.S. Thesis. University of Massachusetts, Amherst.

ROWSE, J.A. 1962. Production of apple juice and vinegar stock. U.S. Pat. 3, 042,528, July 3.

SCHWARTZBERG, H.G. 1977. Expression of fluid from cellular plant material. Proc. 2nd Pacific Chemical Engineering Congress (PACHEC 77), 675 – 680, Am. Inst. Chem. Eng., New York.

SCHWARTZBERG, H.G., ROSENAU, J.R., and RICHARDSON, G. 1977 A. The removal of water by expression. *In* Water Removal Processes. AIChE Symp. Ser. 73, pp. 177 – 190. C.J. King and J.P. Clark (Editors). Am. Inst. Chem. Eng., New York.

SCHWARTZBERG, H.G., COLLYER, T., and WHITNEY, L.F. 1977B. Improved protein juice recovery by combined imbibition and expression. ASAE Paper No. 77-6504, 1977 Winter Meeting. Am. Soc. Agric. Eng., Chicago, IL.

SCHWARTZBERG, H.G., COLLYER, T., and WHITNEY, L.F. 1978A. Alfalfa protein recovery by combined imbibition and expression. *In* Food, Pharmaceutical and Bioengineering 1976/77. AIChE Symp. Ser. 74, pp. 166 – 174. G.T. Tsao (Editor). Am. Inst. Chem. Eng., New York.

SCHWARTZBERG, H.G., HUANG, B.W., ABULARACH, V., and ZAMAN, S. 1978B. Force requirements for water and juice expression from cellular plant material. Paper presented at the National Meeting, Am. Inst. Chem. Eng., Boston.

SCHWARTZBERG, H.G., HUANG, B.W., ABULARACH, V., and ZAMAN, S. 1979A. Expression of water and juice in plants. *In* Food Process Engineering, Vol. 1, Food Processing Systems. P. Linko, Y. Mailku, J. Olkku and J. Larinkari (Editors). Applied Science Publishers Ltd., London.

SCHWARTZBERG, H.G., HUANG, B.W., ABULARACH, V., and ZAMAN, S. 1979A. Mechanical dewatering of food systems. Paper 284, presented at the 39th Annual Meeting of the Institute of Food Technologists, St. Louis, MO.

SHIRATO, M., ARAGAKI, T., MORI, R., and SAWAMOTO, K. 1968. Prediction of constant pressure and constant rate filtrations based upon an approximate correction for side wall friction in compression permeability cell data. J. Chem. Eng. Jpn. *1*, 86 – 90.

SHIRATO, M., SAMBUICHI, M., KATO, H., and ARAGAKI, T. 1969. Internal flow mechanisms in filter cakes. AIChE J. *15*, 405 – 409.

SHIRATO, M., MURASE, T., NEGAWA, M., and SENDA, T. 1970. Fundamental studies of expression under constant pressure. J. Chem. Eng. Jpn. *3*, 105 – 112.

SHIRATO, M., ARAGAKI, T., ICHIMURA, K., and OOTSUJI, N. 1971. Porosity variation in filter cake under constant pressure filtration. J. Chem. Eng. Jpn. *4*, 172 – 177.

SHIRATO, M., MURASE, T., and HAYASHI, N. 1977A. Filter cake dewatering by mechanical expression. Proc. 2nd Pacific Chemical Engineering Congress (PACHEC 77), pp. 667 – 574. Am. Inst. Chem. Eng., New York.

SHIRATO, M., MURASE, T., HAYASHI, N., MIKI, K., FUKUSHIMA, T., SUZUKI T., SAKAKIBA, N., and TAZIMA, T. 1977B. Fundamental study on screw expression. Kagaku Kogoku Robunshu *3*, 303 – 310.

TAYLOR, D. W. 1960. Fundamentals of Soil Mechanics. John Wiley & Sons, New York.

TERZAGHI, K. 1954. Theoretical Soil Mechanics. Springer-Verlag, Berlin and New York.

17

Structure and Structure Transitions in Dried Carbohydrate Materials

James M. Flink[1]

INTRODUCTION

The generation of solid low molecular weight carbohydrate from its aqueous solution involves a number of events, and for each event the factors of temperature, moisture, and time are of major significance. The particular values for these factors, together with the particular sequence of events utilized will be important in determining the physical state of the solid phase obtained, and thus its properties.

In particular, the influence of the factors of temperature, moisture, and time will be noted in the concepts of equilibrium vs nonequilibrium conditions, system concentrations, viscosity levels, diffusivity values—the last two strongly influencing mass transport. (An additional factor, heat transport, can also be considered, but will for the most part not be included in this chapter.)

At this point, one can name four "states" in which the solid carbohydrate can be obtained, depending on process conditions. These are generally labeled as crystalline, amorphous, vitreous, and glassy. The crystalline state involves a three-dimensional ordered array of molecules in which there is periodicity and symmetry. All molecules are equivalent with respect to their binding energy, which results in the breakdown of the crystal occurring at a fixed, unique energy level (i.e., temperature). The difference between amorphous, vitreous, and glassy carbohydrate appears to be less definite. It seems in most cases that vitreous and glassy are used to describe the same condition, while the difference

[1] Department for the Technology of Plant Food Products, The Royal Veterinary and Agricultural University, Copenhagen, Denmark.

between them and amorphous may reflect the method of preparation. In general, it can be mentioned that the noncrystalline "states" can be characterized in a manner similar to the crystalline, in that the molecules are interacting with each other and form an extended three-dimensional network. However, for noncrystalline materials this network is not periodic and symmetric, and all molecules are thus not equally involved in the binding of the network. This means that with input of energy, there is an increasing number of bonds that are broken, giving a temperature range over which the network breaks down. It should also be noted that while the total binding energy is not equal for crystalline and noncrystalline states, the binding energy of the individual bond is the same. For the case of carbohydrates, interaction is through hydrogen bonds. Lastly, it can be mentioned that of the solids obtainable, the crystalline state is the stable form, e.g., that which develops when equilibrium conditions exist during formation of the solid phase. The noncrystalline materials are inherently unstable, being formed under nonequilibrium conditions. Given the opportunity, they will change into the more stable crystalline configuration.

Historically, the carbohydrate materials that have been most often examined are sucrose and lactose (low MW) and starch and cellulose (high MW). This chapter will consider low MW carbohydrates, which, when referred to collectively, will be called sugar. Of the variety of processes available for conversion of sugar in solution to the solid state, three (crystallization, drying, and freezing) will be considered here.

THE WATER – CARBOHYDRATE RELATIONSHIP IN SOLUTION

In investigating the preparation of solid carbohydrate from solution, the obvious starting point is to examine the relationship of water and carbohydrate in the starting material, the solution. This could be approached in a number of ways; the one chosen here explores information available on the properties of sucrose–water solutions, especially regarding the binding of water by sucrose. Hydrogen bonds (H bonds) between hydrogen atoms and the oxygen atoms of hydroxyl groups and ether linkages are the major mode of interaction for the various combinations of interest in the sucrose–water system (that is, water – water, sucrose – water, and sucrose–sucrose). A review of sucrose–water associations by Allen *et al.* (1974) serves as a major reference source for this section.

The process of dissolving a solute, such as sucrose in water, involves the breaking of H bonds, both in the water (so as to form a "hole" into which the sucrose can be placed) and between sucrose molecules. Based on investigations by numerous researchers, water in the liquid state is concluded to exist as a highly H-bonded material, this internal organization being responsible for many of the interesting properties of liquid water. The high degree of H bonding in water can be noted from data of Dack (1975) on cohesive energy density (ced), which is the measure of total intermolecular bonding energy and internal pressure (P_i, the measure of polar and nonpolar interactions) for a number of solvents. The difference between ced and P_i (ced$-$P$_i$) gives a measure of the intermolecular hydrogen bonding. Water has the highest ced and the lowest P_i, which results in a difference that is twice that of the next solvent on the list (formamide).

Sugar Bonding Behavior

Allen *et al.* (1974) note that the sucrose molecule contains 11 oxygen-containing groups, eight being in OH groups (Fig. 17.1). This means that there is a maximum of 11 H-bond sites available. In the solid (crystalline) state, four sucrose oxygen groups are taken up with two intramolecular H bonds (three OH and one ring oxygen) with the others being available for intramolecular H bonds to hold the crystal together. Both Jeffrey (1972) and Allen *et al.* (1974) have noted that the intramolecular H bonds are important for determining the conformation of sucrose in the solid state. Allen *et al.* (1974) note that if the intramolecu-

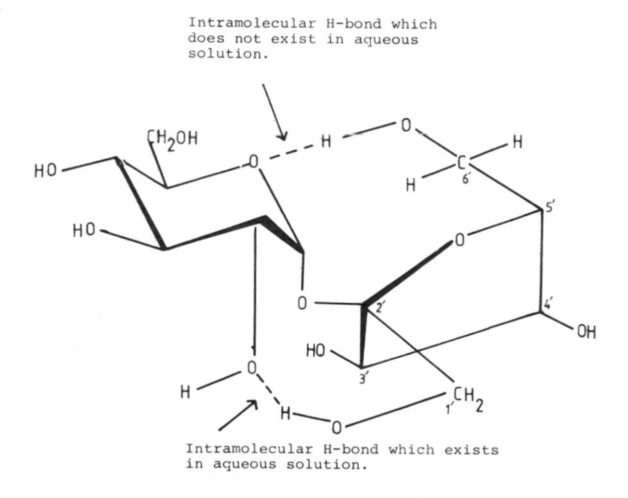

Fig. 17.1. Structure of sucrose.

lar H bonds are present in solution, they would tend to hold the molecule in a spherical form, reducing the sucrose–water interaction level. They deduce that since this interaction level is quite high (demonstrated, e.g., by hindrance to crystallization) it seems likely that the intramolecular H bonds do not exist in solution.

Very recently, Bock and Lemieux (1982) have indicated that one of the intramolecular H bonds of the crystal structure (OH_{1f}–OH_{2g} in Fig. 17.1) exists in aqueous solution. The other intramolecular H bond of the crystal structure (OH_{6f}–OH_{5g}) does not exist in aqueous solution, and it was further noted that the OH_{6f} unit is presumably solvated by water. [They noted that in dimethyl sulfoxide (DMSO) the second intramolecular H bond does exist, but with a different orientation of the C-6 fructose hydroxymethyl group than that which exists in the crystal.]

Neal and Goring (1970) describe a hydrophobic bonding that has been observed for some carbohydrates. For example, glucose has certain surfaces that are hydrophobic in character and thus the interaction of glucose in water is a balance of the effect of the —OH groups and the hydrophobic surfaces. For maltose, which has two rings with these hydrophobic surfaces, the conformation in aqueous solution is dependent on an intramolecular "hydrophobic bond," which more properly is an adoption of conformation such that two relatively hydrophobic surfaces are oriented to face each other. This gives the molecule the highest hydrophilic character possible. Cellobiose, on the other hand, which cannot rotate its rings relative to each other, cannot shield these surfaces. While Neal and Goring (1970) did not investigate sucrose, Bock and Lemieux (1982) recently have postulated that the internal bonding of sucrose in aqueous solution results in a sizable surface of the molecule being able to partake in hydrophobic bonding.

As noted above, the process of dissolving sucrose in water results from the breaking of sucrose–sucrose and water–water H bonds and developing sucrose–water H bonds. Studies by Arshid *et al.* (1956) showed that in solution the —OH groups of sugar are solvated. They also indicated that there was evidence that sugar–water bonds can be stronger than water–water bonds, but more recently James and Frost (1974) using infrared (IR) analysis observed that the forces acting on water molecules do not change greatly by addition of sucrose, indicating that the forces in the sucrose–water bond are similar to the water–water bond.

Another view regarding the H bonding of sugar in solution has been presented by Jeffrey and Lewis (1978). On the basis of quantum-mechanical calculations, they noted that "cooperativity" in H bonding means that for carbohydrates, where H bonding is the important cohesive interaction, the molecules will tend to orient themselves such that the hydroxyl groups can be bonded in chains. This is due to the fact that it is more favorable for OH groups to be both a H-bond donor and acceptor. The anomeric OH (C-1) is only a weak H-bond acceptor and thus will act as a chain stopper. This behavior can have an influence on the eventual organization of sucrose in the aqueous solution.

Hydration of Sucrose

From the above, it can be seen that in solution there is a high level of H-bond interaction between sucrose and water, with hydroxyl groups of sucrose being solvated by the water molecules. To quantitatively evaluate the relationship of sugar in solution, the degree of this interaction can be investigated. This has been done in a number of studies in which bound water or unfreezable water is measured. "Bound" and "unfreezable" water are terms used to describe water present in a solution which does not act in a manner similar to pure water, this generally being referred to as "free" water in this context. (The difference in the terminology bound and unfreezable generally reflects the measuring method used.) Allen *et al.* (1974) noted that in dilute solution there is a dynamic equilibrium between bound and free water molecules. A term used to quantify the degree of binding is the "hydration number," which is the number of molecules of bound water per molecule of sucrose. The concentration dependence of hydration number indicates that as sucrose concentration increases, the amount of free water available in the solution and the amount of bound water decrease. For a wide range of concentrations, there appears to be an equilibrium between free and bound water, such that both species exist. [Similar behavior was observed by Ross (1978) for sorption isotherms of intermediate moisture foods.] Hydration numbers in the literature range from 21 to 1, but for the most part values of 4 to 7 are reported. The hydrate level most often quoted for 20°C is 5 moles of water per mole of sucrose, which corresponds to about 0.26 g water per gram of sucrose (g/g). The hydration number tends to decrease with increasing temperature or concentration. (One can note that hydration numbers of 4, 6, or 7 correspond to moisture levels of 0.21, 0.32 and 0.37 g/g, respectively.)

Sucrose–Water Binding Measurements

Some recent measurements of sucrose–water binding follow. Maltini (1977) investigated the freezing of sucrose–water solutions. For samples with initial moisture contents up to 31% water, no recrystallization peak was observed under warming in the differential scanning calorimeter (DSC). Thus, the unfrozen water level is 0.31 g/g, which calculates to a hydration level of about 5.9. In comparing the bound (unfreezable) water content with the BET monolayer for water adsorption, Almasi (1979) noted that as temperature is decreased, these two values approach each other. For starch (the only carbohydrate tested) they were equal at − 30°C and a moisture level of 0.26 g/g, corresponding to a hydration level of 5.2 per disaccharide (s) unit of the starch. From a study on the activity coefficient vs concentration of sucrose in water, Bertran and Ballester (1971) found that the bound water level (moles water/mole sucrose) falls from 6 at 0°C to 2 at 55°C, with the value at 10–20°C being 4. Suggett (1976) calculated hydration numbers for various sugars in solution at 5°C and a concentration of 2.8 *M* (monosaccharide basis). He obtained 6.6 for sucrose, 5.0 for maltose, and 3.7 for glucose. Values for the glassy phase of the frozen material was 6.3 for sucrose. He noted that in both cases the same H-bond forces are active and deduces that water in the glassy phase must correspond to the water most closely associated with solute.

Relationship between Hydration and Sugar Concentration

In a survey of the state of water in foods, Karel (1975) discussed a number of topics of relevance here, including the relationship of bound (unfrozen) water and the BET monolayer. Unfrozen water (ranging from 0.2–0.4 g/g) is generally two to four times the BET monolayer. He noted that in the literature, information on the amount of bound water per functional group showed that the aliphatic hydroxyl group has one water per hydroxyl, which gives a binding level similar to that found for water binding by sugars.

Steinbach (1977), in studying freezing of sucrose solutions, indicated that sucrose hydrates are formed (Table 17.1). For solutions starting between 3 and 10% sucrose the hexahydrate will form, while between 10 and 50% the pentahydrate forms. The cryosolvate point is about 80% solids, corresponding roughly to 0.25 g/g unfrozen water.

The great bulk of the data noted above indicates that in solution, between four and six water molecules are strongly interacting with the sucrose molecule. As this would correspond to a solids content of about 80% (somewhat over saturation), it would seem likely that the condition in solution would be one of individual sucrose molecules, each with a sheath of tightly bound water, surrounded by a sea of water. It has been indicated (Allen *et al.* 1974) that this is not always the case, and that results of a number of experiments indicate that in solutions at concen-

TABLE 17.1. HYDRATE IN FROZEN SUCROSE SOLUTION

Initial Solution Concentration (%)	Hydration Number (moles water/mole sucrose)
2 <	10–12
3–10	6
10–50	5
55	3.5
57	2.5
65–73	2
77–83	1

Source: Steinbach (1977)

trations greater than 30–40%, the system changes from a solution of solvated sucrose in water to a sucrose–water phase, in which all the water is involved in H bonds with sucrose, either directly or indirectly. This tight water shielding reduces the tendency for sucrose molecules to self-associate (i.e., form nuclei for crystallization). They note that some structural changes occurring in sucrose solutions require a period of time to occur. This influences, for example, the onset of crystallization in supersaturated solutions, where, presumably, time is required to form protonuclei (clusters of sucrose molecules with some water molecules present). These observations can give grounds for considering the equilibration times necessary for changes to occur in concentrated solution. Because reaction rates will vary with temperature and concentration in solution, there is the question of how much time will be required to form or interconvert the different hydrates or sucrose–water structures under changing conditions. It is possible that the crystallization behavior of sucrose, in which a time is required for the supersaturated solution to form nuclei naturally, indicates something of the natural time scale for changes in the sucrose–water relationship in concentrated sucrose solution. This point will be considered further later.

Sucrose Clusters

Bowski *et al.* (1971), in analyzing their results of a study on the hydrolysis of sucrose with invertase, reached a similar conclusion that a sucrose clustering was present. From an analysis of the enzyme kinetics at high sucrose concentrations, they could determine the amount of free vs bound water. They found that the bound water level falls with increasing concentration of sucrose, having values ranging from 6.5 at 20% sucrose (w/v) to 3.7 at 60–70% sucrose. As their values deviated from the "theoretical" bound water value reported by Einstein (seven water molecules per sucrose), they suggested that at the higher concentrations, sucrose molecules may be incompletely hydrated and thus able to interact with each other, forming clusters.

Relationship Between Hydration and Concentration

Biswas *et al.* (1975) investigated the relationship between hydration and concentration in solution. They found the hydration level to be greater than 9 g/g for a 1% concentration and decreasing to approach a 1 g/g level for concentrations above 30%. These hydration levels are much higher than those reported by others. They considered that there could exist an equilibrium between hydrated and nonhydrated species. The increase of hydration number with decreasing solute concentration is considered to be due to multiple layers of water around the binding sites. These multiple layers disappear as the solute concentration increases, presumably due to the water molecules being taken up as the primary binding by the new solute molecules.

Mathlouthi *et al.* (1980) and Mathlouthi and Luu (1980) conducted studies of the solvated properties of glucose, fructose, and sucrose with laser Raman spectroscopy, the results being presented in terms of the different Raman spectra band regions. Shifts in the spectrum at 1460 cm^{-1} at a sugar concentration of 20–40% for glucose and about 35% for sucrose indicated modification of the association of water and sugar. This is referred to as a kind of pre-nucleation step and is noted to agree with a number of results in the literature that described a change in the relationship of sucrose and water at a concentration of 30–40%. The same change from solvated sucrose to sucrose–sucrose associations referred to above is noted in Allen *et al.* (1974). On the basis of band shifts, it is assumed that the association produces aggregrates that have a compact network comparable to a crystalline network. (They also make reference to an article by Voilley *et al.* regarding the acceleration of crystallization of glucose if the solution is centrifuged, presumably due to the closer association of the aggregates after centrifugation.) It is noted that this behavior is not found with fructose. Further information at the 1640 cm^{-1} band supports these observations. This band for the water shift indicates that at sucrose concentrations above 30%, the change to a sucrose–sucrose relationship acts to structure the surrounding water and to reinforce the water–water interactions. This behavior is not common for all sugars, and for fructose it appears that the fructose–water bonding has the same energy levels as the fructose–

fructose or water–water H bonds, a fact that explains some of fructose's properties.

A similar change in solution behavior at sucrose concentrations above 30% was also observed by Sugisawa and Shiraishi (1974). In a study on volatile retention in freeze drying, they found significantly higher retention when the initial concentration of sucrose was above 30%. In a concurrent investigation of electrical conductivity of the solution, a maximum for conductivity at 30% sucrose was noted. These effects were attributed to the formation of an associated sucrose phase (micellar) at a concentration of 30%. Addition of electrolytes to the solution acted to move the conductance maximum to a lower sucrose concentration.

Summation

In this first part, it has been noted that water is a strongly hydrogen-bonded liquid, but that the introduction of sucrose into the water results in a rearrangement of bonding conditions, such that sucrose–water H bonding is established. It appears that one of the intramolecular H bonds of the sucrose crystal is not broken, however. It has been noted that the relationship of sucrose to water in solution is dependent on the sucrose concentration. This has been expressed by the change in hydration numbers as the sucrose concentration increases. It is tempting to consider that this reduction in water binding indicates that the unoccupied hydroxyl groups have become involved in interractions other than with water, and that these could be sucrose–sucrose bonds, signifying the beginning stages of the formation of the solid sucrose structure.

It was also noted that at sucrose concentrations below 30%, the solution apparently consists of individual sucrose molecules that are strongly solvated by water. There are indications that, at concentrations above 30%, the status of sucrose and water changes such that hydrated sucrose is able to undergo associations that have also been referred to as sucrose clusters. These clusters have been associated with the beginning stages of the development of the solid material. It seems that most studies on the binding of sucrose and water in solution investigate the static case, that is, where the sucrose concentration is fixed from the beginning of the experiment. The question of system dynamics, which is of major importance, has not been commented on. It is immediately apparent that the rate of development of sucrose–water associations will be highly significant in determining if such associations develop in systems where sucrose concentration changes with time (see below).

MODES OF SEPARATION OF CARBOHYDRATE AND WATER

As seen in the previous section, the sucrose–water solutions can exist with the sucrose either fully hydrated with solvent or with some form of sucrose clustering, depending on the carbohydrate concentration in the solution. To form the dry solid carbohydrate, it is necessary to have a process by which the carbohydrate molecules can be separated from the water solvent. In this discussion, the following three methods for preparing solid carbohydrate will be noted: (1) crystallization of carbohydrate, (2) a rapid air drying (such as spray drying), and (3) freezing (crystallization of water) followed by freeze drying.

Each of these processes differs in concept, though the desired result (separation of carbohydrate from water to give solid carbohydrate) is the same. However, the properties and appearance of the solid carbohydrate formed will differ for the various processes. Following a further analysis of the glassy state, the differences in these processes and how they can affect the solid carbohydrate will be discussed.

Conditions for Glass Formation

As noted in the Introduction, solids can be obtained in two basic states, crystalline or glassy (also referred to as amorphous or vitreous). Generally, the solid phase is formed from the melt by cooling, while either cooling or increase in concentration is used to form the solid phase from solution. For crystal-forming materials, the basic factor that will determine whether the eventual solid is crystalline or glassy will be the extent to which the molecules are able to attain equilibrium configurations. Low molecular mobility due to high sample viscosity (which can result from many factors) will result in the inability to attain equilibrium configurations, and consequently in glass formation. The glass is the metastable state and will tend to convert to the crystal, if the molecules attain adequate mobility.

A number of investigators have considered the causes for formation of the glassy state. Kauzmann (1948), in what is essentially the bible of the glassy state, noted that thermodynamically, the presence of the metastable state indicates that there are free energy barriers, which hinder the attainment of the stable state (see below).

Meares (1957, 1965) described the glass transition for pure polymer systems (i.e., no solvent) in terms of a balance between energy and entropy, with the molecule's thermal energy being one side of this balance and the system free volume ("holes") available for molecular movement associated with the thermal energy being the other side. Above the glass transition temperature T_g, both the energy and entropy terms adjust to changing temperature so that volume vs Gibbs energy is minimum and the system is at equilibrium (Fig. 17.2). The molecule's movements (rotation, translation, vibration) are such that they overlap and thus free volume exists throughout the system. Free volume is the volume unoccupied by the "solid matter" of the molecules and represents the volume available for free movement of the molecules. On cooling at temperatures above T_g, liquids contract, and as the molecule's kinetic energy decreases, the free volume available for movement decreases as well. As long as free volume extends throughout the system, the molecules are able to change structure, and molecular associations leading to crystal formation are possible. However, at the point that the free volume no longer extends throughout the system, the structure that exists at that moment is "frozen in" and the glassy state is established. System viscosity is said to become too high for the volume relaxation to follow the temperature lowering and nonequilibrium conditions enter into consideration, as there is space in the system, which should not exist at equilibrium. It appears that for polymers at T_g, the free volume is about 0.025. Below T_g, no further adjustment of geometrical structure can occur, and the system retains the structure present at T_g. (On rewarming, the molecules take up some thermal energy and movements increase. When the range of the movements begins to overlap

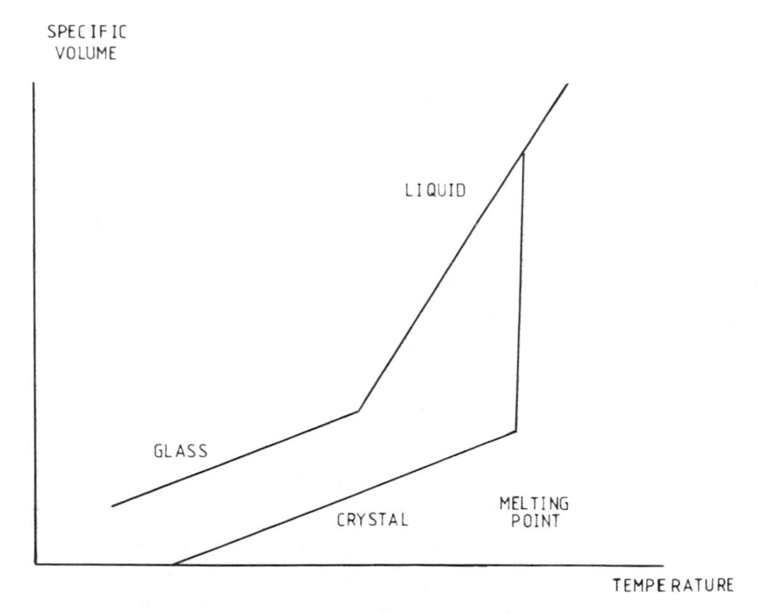

Fig. 17.2. Specific volume dependence on temperature for glass transition vs crystallization.

and extend throughout the system, then the system takes on liquid-like properties again, and T_g has been passed.)

Factors that influence free volume requirements, such as molecular weight (MW), steric hindrance, diluents, etc., will alter T_g. In particular, as diluents have a high free volume, which is added to the molecule's free volume, the effect of diluents on T_g is large.

Williams *et al.* (1955) showed that the mathematical analyses used for glass transition phenomena in polymers can also be used for low molecular weight organic glasses, such as sucrose. They demonstrated the existence of a unique equation that relates various properties of a material near T_g to the difference in temperature from T_g. This indicates that the behavior for all glasses is based on the same physical principle; that is, that as the glass transition is approached, the relative free volume decreases sharply and that this is the primary cause for increase in viscosity. The universal WLF function indicates that the nature of the volume change and its effect on rates of molecular rearrangement is independent of molecular structure. Williams *et al.* (1955) also showed that the change of free volume with temperature is the difference between the thermal expansion coefficients above and below T_g, which for essentially all materials gave a fractional free volume of 0.025 at T_g.

Lesikar (1975) described some of the hydrogen bonding aspects of the glass transition. Factors tending to reduce fluidity in a system will tend to increase T_g. From an analysis of the ratio of H-bond energy to thermal energy in a supercooled liquid, he concluded that H-bonding coordination of the molecules must be almost complete at the time of the glass transition. The picture developed has a molecule of liquid vibrating with respect to its neighbors, with flow occurring by jumps into neighboring holes. With the high H-bonding levels, as presumably exist in a supercooled liquid, such jumps are energetically unfavorable and flow is hindered (i.e., the glassy state is established).

Kauzmann (1948) noted that in thermodynamics, the presence of a metastable state implies that there are free energy barriers between the metastable and stable states that hinder attainment of the stable state. He thus discusses the cause of the glass transition in terms of failure to achieve the stable, crystalline state directly. In a supercooled liquid, there exist at least two free energy barriers, one to crystal nucleus formation and one to crystal growth. The free energy barrier to nucleation results from the melting point of small crystals being lower than large crystals. Thus to form a nuclei, it is necessary to form a crystallite that has a free energy that is higher than the same amount of liquid. This implies that at any time small crystallites are formed which then dissolve. The free energy barrier to crystal growth acts to prevent movement of a molecule at the crystal–liquid interface from going from the liquid to the crystal. At temperatures below freezing, hindrance to rotation of the molecule is active in limiting crystal growth. Hindrance of molecular rotation will be similar to dielectric relaxation behavior of a material. The barrier to nuclei formation falls with temperature, and it is thought that at a temperature 0.65 times the melting temperature ($T = 0.65\ T_m$), crystallites will form. The remaining barrier will then be the growth barrier.

Crystallization of Sucrose

The separation of sucrose from water in a crystallization process first involves the removal of the solvent (water) through evaporation so as to increase the sucrose concentration in the system. Under these conditions, the sucrose molecules will eventually associate and form crystals. Many studies have been made on sucrose crystallization, owing to its commercial importance. In this chapter, reviews by Lees (1965) and Smythe (1971) have been quite valuable. It can be noted at this time that sucrose generally does not crystallize directly from the anhydrous melt, rather it forms the glass. Crystalline sucrose can, however, form from the humidified glass; this will be discussed further later. In any case, recrystallization of the anhydrous melt is not a normal method for separation of crystalline sucrose from the sucrose–water solution.

Chalmers (1964) presented a thermodynamic analysis of nucleation and crystal growth. Nucleation theory builds on the concept of the presence of clusters. Thermodynamic functions are used to describe the condition of these clusters, particularly if they attain a stable configuration called a nucleus. Pidoux (1972) studied the viscosity of supersaturated solutions. The results indicated that there exist clusters in solution (called proto-nuclei), which over an induction period capture other molecules or groups of molecules to form nuclei. It is the growth of these units which results in the increase of system viscosity. Van Hook (1971) interpreted the comparable levels for the energy of activation of nucleation, viscosity, and diffusion, and the major influence that viscosity has on nucleation to indicate that nucleation in sucrose solution can be considered a diffusion-controlled process. He also indicated that viscosity could be so high that nucleation can be inhibited.

Lees (1965) described the importance of supersaturation on the crystallization process of sucrose. The solubility (temperature–composition) diagram for sucrose is generally presented with two or three regions that describe different degrees of stability for sucrose in solution for different levels of supersaturation. The definition of supersaturation (S) is based on weight of sucrose per weight of water compared to the same value at saturation. It is shown that at values of $S = 1.0-1.3$ existing crystals will grow, but new crystals do not normally form. At $S >$ 1.3, spontaneous formation of new crystals may occur. There are many factors that enter into these considerations, such as impurities and methods of preparation, so that the actual S values for various phenomena can be somewhat variable. It can be noted, though, that higher solution concentrations are required to initiate nucleation than to achieve the growth of crystals. It was also noted that most crystallization theories are based on low values of supersaturation and that action at very high concentration may not follow the theories.

Smythe (1971) has reviewed crystal growth information, both with respect to the influence of various process conditions on growth of crystals, and on the structure of the sucrose crystal. It is noted that in the crystal, the packing is determined mainly by the hydrogen bonds of the hydroxyl groups. There are seven hydrogen bonds, five intermolecular and two intramolecular. At each crystal face, there are a number of hydroxyl groups that are not involved in hydrogen bonds and these will be the sites for further crystal growth. Crystal growth involves the mass transfer of molecules from the bulk phase to the surface sites, and the incorporation of the molecule onto the site. The rate of crystal growth will depend on both factors, though under different conditions, one or the other can be considered limiting. On the basis of activation energy similarities, it appears that at low temperatures, it is the surface incorporation step that is limiting, while at higher temperatures it is diffusion of sucrose to the crystal surface that limits growth. Van Hook (1973) investigated the mechanism of growth at low temperatures and reports that the surface integration step is limiting. The rate constant for this step appears to be dependent only on temperature. Allen *et al.* (1974) considered the intramolecular H bonds present in the sucrose crystal to be absent in solution, while Bock and Lemieux (1982) recently indicated that in aqueous solution one of the intramolecular H bonds found in the crystal is present. It would seem likely that the incorporation step of crystal growth involves molecular orientations or rearrangements which give the formation of the required intramolecular H bonds, this then giving the sucrose molecule the proper H-bond orientation for crystal propagation.

From the above, it can be noted that there are many factors that must be fulfilled if one is to obtain crystalline sucrose. The relationship of concentration of sucrose relative to water (expressed as supersaturation) and temperature must be carefully controlled to allow the various associations (nuclei formation, crystal growth) to develop without hindrance. If the system is not controlled adequately, the evaporation of water at high temperature will result in amorphous sucrose when the concentrate is cooled. Also, rapidly cooling the concentrated solution so that viscosity rises significantly can result in the formation of amorphous sucrose.

Regarding amorphous sucrose, Lees (1965) notes that generally it should contain, at most, 2% water, which is equivalent to a hydration number of 0.4. This means that there are many hydroxyl groups available for bonding. According to the standard definition of supersaturation, a hydration number of 0.4 corresponds to $S = 2450$, which would be a highly unstable material with respect to nucleation and crystal growth.

Rapid Air Drying

In a somewhat analogous situation to the crystallization methods noted above, drying also involves the evaporation of water to obtain the separation of the water from the sucrose. The major difference is that drying is usually conducted at high rates of water removal, such that the solution quickly attains high sucrose concentrations. Under these conditions, the intermolecular associations noted above do not apparently have time to occur and the orderly crystalline structure is not obtained. The product will be in the glassy state. Bushill *et al.* (1965, 1969) have investigated the properties of powder obtained from spray drying milk or a lactose solution. They noted that the powder was generally amorphous, though on occasion some crystalline α-monohydrate was found. Varshney and Ojha (1977) showed that both spray drying and roller drying of milk-based baby foods would give amorphous lactose. Thus, it appears that the rate of drying or temperature–time–moisture content behavior in a roller dryer is adequate to give amorphous lactose.

Herrington (1934) showed that dry amorphous lactose could be obtained by cooling a concentrated lactose solution and then vacuum drying the viscous mass.

Roetman and Van Schaik (1975) and Roetman (1979) studied the effect of precrystallizing lactose in milk and its effect on the product obtained after spray drying. Lactose in milk, whey, and water (i.e., pure solution) were spray dried before and after precrystallization. The general structure of the spray-dried products was amorphous, though if a precrystallization step were used, lactose crystals in an amorphous matrix were obtained. For the milk or whey, however, the major part of the amorphous fraction was not due to the lactose, but rather to the other components.

Freezing (Crystallization of Water) Followed by Freeze Drying

In freezing, water is first separated from the solution by crystallization to ice. The ice formed is later removed by sublimation. The remaining water that is associated with the sucrose (unfrozen water) is then removed through an evaporation/desorption process. In contrast to the earlier mentioned evaporation processes, evaporation/desorption in freeze drying can occur at a combination of low temperature and high concentration. The development of the final structure can conceivably come at any stage of the overall process.

During freezing of sucrose–water solutions, the unfrozen material undergoes an increase in concentration as water is converted to ice. In accordance with the terminology of Bellows and King (1973), this unfrozen phase is called the concentrated amorphous solute (CAS) phase, though a part of the discussion will consider whether it is completely amorphous. It can be noted that the previous discussion refers to measurements of unfrozen water as a means to determine the sucrose–water association level. This information is also relevant here.

Consideration of freezing as a mode for separation of sucrose and water and its effect on structure can begin with a general description of what occurs during freezing. Ideally, as the temperature of a solution is lowered below the freezing point (i.e., supercooled), nucleation of water molecules occurs and ice crystals form and grow. This removal of water from the solution as ice results in a more concentrated solution, which exists in equilibrium with ice crystals at a lower temperature. As further energy is removed from the system (i.e., temperature is further lowered), the nucleation/crystal growth continues until one of a number of conditions occurs such as, no further reduction in temperature, insufficient water being available for further crystallization, or system viscosity becoming so high that the low mass transport rate of water from solution to the crystals prevents further ice formation. If the temperature reaches a low enough level, slowly enough, the CAS will generally attain the unfreezable water level noted earlier. The formation of the ice crystals itself has a definite effect on *overall* freeze-dried structure; the porosity of the final product will depend on the orientation and size of the ice crystals since following sublimation, the ice crystal locations form the pores of the dry material. Overall structure is thus primarily decided in the freezing step, as long as structural collapse does not occur in the freeze-drying steps. In this discussion of the freezing step, the conditions present in the CAS were not considered. The status in the CAS is very important regarding the "microstructure" of the dry product. The behavior of the CAS under the influence of varying temperature has been investigated. This can be combined with information regarding associations occurring in aqueous solutions and leads to the question, "can these associations form and develop under the freezing or freeze drying process and contribute to the microstructure of the dry CAS?"

A part of the behavior of the system under freezing depends on the degree of formation of ice and subsequent concentration of the CAS. Lusena and Cook (1954) presented information relating the relationship of heat removal (cooling) and heat production (crystal formation) to whether ice forms or not. Lusena (1955) indicated that solutes at higher concentrations affect ice crystal growth rate by apparently blocking the diffusion pathways of water to the growing ice crystal. The presence of solutes also tends to reduce the nucleation temperature in proportion to the depression of the freezing point. Thus, if the heat removal rate is high and/or the heat production rate low, then at high solute concentrations, where crystallization can be hindered, the system can cool so rapidly that further crystallization will not occur due to nonequilibration of the system (as described earlier under glass formation). Lusena (1955) showed that this situation can exist with sucrose solutions as the concentration increases, and thus, while the solution becomes more viscous due to the lower temperature, its composition is unchanged because no additional ice has been formed.

MacKenzie (1966) presented an interesting overall view of CAS behavior. His later review (MacKenzie 1975) on structure changes in freeze drying is referred to here. From thermal analysis studies for the sucrose–water system, he demonstrated that there are a number of changes in system associations with rise in temperature. Diagrams were prepared containing a number of characteristic system temperatures—Tg (glass transition temperature), Tc (collapse temperature), Tam ("ante-melting" temperature), Td, Tr (devitrification or recrystallization temperature),Tim (incipient melting temperature), Tm (melting temperature of ice), Ts (solubility temperature), and Te (eutectic temperature)—as a function of system composition. Tam and Tim were considered to relate to the solute phase after devitrification, though others later have associated it with ice crystallization in the CAS (see below). The effect of freezing of sucrose solutions followed by rewarming was described. During freezing, the melting (Tm) curve eventually crosses the Tg curve, and the CAS hardens to a glass. Upon rewarming, at T_g the CAS softens (i.e., undergoes the glass transition) before any other phenomena are observed. The temperature region between T_g and T_r is thus one of constant composition in the CAS, but with decreasing viscosity. T_g is characterized by a viscosity of about 10^{13} cP, while collapse first comes after the CAS achieves a lower viscosity (10^7-10^{10} cP). For recrystallization to occur, it is apparently necessary to have a still lower viscosity.

Simatos *et al.* (1975A, B) described the freezing of solutions and concentration of the CAS that results. They note in particular that the real concentration of the CAS will depend on the freezing rate and the ability of the remaining water to diffuse from the CAS to the ice crystals. This will have an effect on the measured glass transition temperature of the CAS, as well as the size of an eventual devitrification (recrystallization) peak on the DSC on rewarming (the devitrification resulting from crystallization of freezable water that had remained unfrozen in the CAS owing to hindered crystallization during the cooling). If all the freezeable water does freeze during the initial cooling, such as with slow cooling, then the CAS glass transition will be the maximum T_g possible for that CAS obtainable by a freezing process. (It can be remarked here that the T_g's of the CAS obtained by freezing should be related to the T_g's measured for dried sample that has been humidified.) The ultimate CAS concentration must represent the absolute minimum of water–sugar interactions which exists in the frozen state.

Couach *et al.* (1977) and Le Meste and Simatos (1980) examined the "ante-melting" (AM) phenomenon in more detail, and its relationship to incipient melting (IM). The basic concepts regarding freezing and the concentration of the CAS have been noted above. AM is considered to be the melting of the small ice crystals that form in the CAS during devitrification of the CAS, while IM is the melting of the larger ice crystals that formed during the initial freezing. It was noted for the sucrose system that CAS T_g is significantly increased in the region between devitrification, where CAS water is converted to ice, and ante-melting, where these ice crystals melt again to give an increasing CAS water content. Devitrification thus results in an increase in system viscosity over a narrow range of increasing temperature, CAS viscosity decreasing again at T_{am} due to the melting of the newly formed ice crystals in the CAS. It was noted (Couach *et al.* 1977) that choosing freeze drying temperatures above or below T_{am} will have a definite effect on both the process and on product properties, this probably being due to the very different water concentrations in the CAS. It was further observed that high molecular weight additives had no effect on this thermal behavior or on solute mobility at low temperatures, but rather influenced the macroviscosity of the CAS network and thus gross structural collapse.

Bellows and King (1973) studied the concentration behavior that occurred in the CAS under freezing. Since the CAS exists in the presence of ice crystals, in the equilibrium state there would be a unique relationship between system temperature and CAS concentration. Some of their measurements indicated that the equilibrium state is not achieved, and that ice formation ceases when the CAS concentration reaches 72%. That this value is lower than equilibrium could also indicate that the CAS does not have a uniform concentration. Luyet and Rapatz (1969) had reported that the CAS does not have a uniform concentration, but instead that the surface layers of the CAS that are in contact with ice will be at the equilibrium concentration, such that further ice crystallization does not occur at the surface layers. Due to the high viscosity and resultant low water diffusivity in the CAS, there can result a steep water concentration gradient in the CAS. Due to the high viscosity of the system, this metastable state can last for a very long time.

Maltini (1977) described the freezing of sucrose solutions at a wide range of concentrations. DSC analysis indicated that at concentrations above 69% solids, no ice formed on freezing, and that freezing resulted only in a glass transition, with Tg decreasing as the sample moisture content increased. At medium concentrations, a recrystallization phenomenon was noted, while at lower concentrations (less than 59%) other thermal transitions are observed as well. On the basis of DSC and microscopic observations, it was concluded that the CAS formed had a large concentration gradient. When the temperature rises, the CAS first undergoes an ice recrystallization, which is then followed by a melting. As long as the major ice phase (that formed in the initial freezing) does not melt, however, these actions will not be associated with a physical movement of the water from its original location in the CAS; rather, the recrystallization and melting will occur solely in the amorphous phase. It was especially noted that if the transitions occur in the CAS without transport of water, they will appear to be reversible upon repeated coolings and warmings. Ito (1970, 1971) also studied the freezing of sucrose solutions. He noted that a $3M$ (about 81%) sucrose solution will not produce ice at any temperature and that the final CAS composition is essentially independent of the initial solution concentration.

In the above studies, the changes occurring in the CAS under varying temperature conditions have been investigated. It has been shown that in most cases, the freezing process results in a metastable concentrated

CAS phase, where the mass transport of water from the CAS becomes limited and thus the ultimate water concentration will be below the value corresponding to the highest T_g for freezing. This unfrozen water is able to recrystallize on warming, and these crystals will melt at a temperature that is lower than the melting point for the crystals formed during freezing. Figure 17.3 summarizes these comments.

EQUILIBRIUM FREEZING

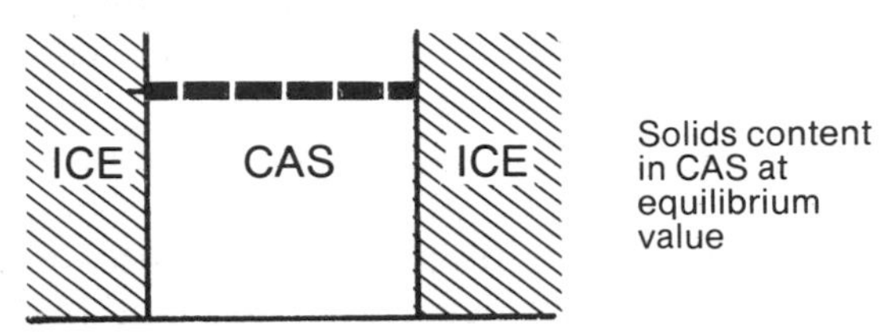

DSC will show glass transition (Tg) and melting of ice (Tm). Tg is highest value which can be found for the frozen system.

NON-EQUILIBRIUM FREEZING

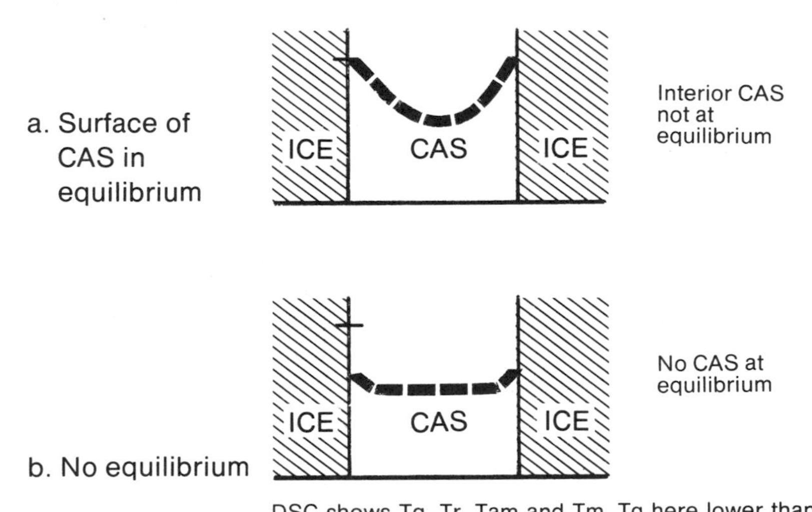

DSC shows Tg, Tr, Tam and Tm. Tg here lower than Tg with equilibrium freezing. (If water in CAS cannot redistribute from CAS to ice, can get cyclic behavior between Tr and Tam depending on thermal treatment).

Fig. 17.3. Distribution of solute concentration in the CAS for equilibrium and non-equilibrium freezing processes, and resultant thermal response in DSC analysis.

The main interest of the previously mentioned studies has been the water component, with the influence of the sucrose primarily being related to the temperature of the glass transition. It should be noted that the sucrose concentrations that ultimately exist in the CAS are high, generally being in the range associated with saturation or supersaturation in sucrose crystallization studies (Fig. 17.4). Samples that have been prepared in which ice cannot form in freezing (for example, Ito's 81% or Maltini's 83%) are definitely above the supersaturation concentration considered unstable relative to crystallization. Thus, as part of the overall view of the CAS during freezing and freeze drying, the associations that can exist between sucrose molecules should be considered.

Measuring Phase Behavior of Sucrose. Steinbach (1977), Zimmer (1979), and Weisser (1979) have reported on a series of studies involving various methods for measuring the phase behavior of sucrose under freezing [refractometer and nuclear magnetic resonance (NMR)]. As previously noted, Steinbach (1977), using a refractometer method, indicated that the sucrose hydrate that forms in solution remains stable

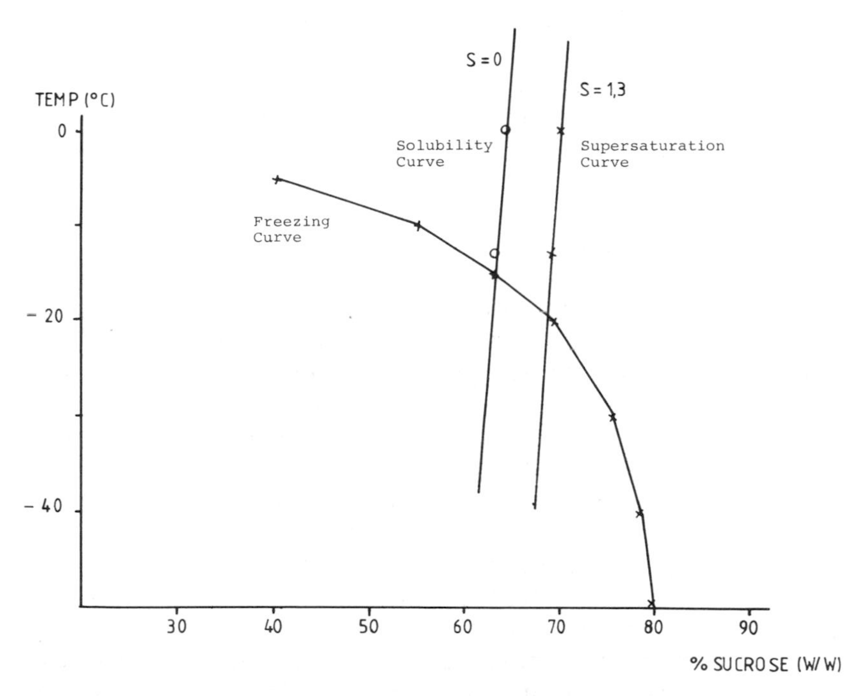

Fig. 17.4. Freezing curve for sucrose in water, with calculated supersaturation curve.

during freezing, the hydration level depending on the initial concentration of sucrose in the solution. For example, at concentrations from 10 to 50% in the initial solution, the pentahydrate would be present as the CAS. He also showed that the ultimate concentration of solids in the CAS is 80% (reached at −38°C), which corresponds well with the pentahydrate being the remaining species in the CAS. Based on the enthalpy change that occurs at the cryosolvate temperature, he implies that it is at this temperature that the hydrate freezes, and that in warming from a lower temperature, the phenomenon generally called "ante-melting" results from the melting of the five water molecules and the sucrose of the hydrate. This contrasts with the interpretation of Couach *et al.* (1977). Zimmer (1979) refined Steinbach's measurements by use of NMR techniques. Studies showing crystallization and melting of the sucrose in the hydrate would seem to indicate that during freezing, sucrose exists in two bonding relationships, one in which the sucrose is in a liquid environment and the other where sucrose is considered solid (termed by Zimmer as crystalline). The formation of the crystallized sucrose would occur concurrently with the formation of ice, and it was noted that the presence of both phases could be observed until almost all sucrose was "crystallized," at which time sucrose crystallization is nearly complete. It is questionable whether the solid state described is crystalline in the traditional view, since freeze drying of such a system should then yield crystalline sucrose, which has not been reported. However, Zimmer (1979) and Steinbach (1977) could be determining a different type of bonding of the sucrose from that which exists in a liquid environment, and which is similar to solid sucrose, perhaps that associated with some type of sugar glass.

At this point it is perhaps well to recall the likelihood for associations of sucrose molecules or a sucrose–water phase in aqueous solution, especially when the solution concentration is higher than 30–40%. One point to consider concerns the freezing of a sucrose solution whose initial concentration is at these levels; What happens with respect to the sucrose-water phase? For solutions that start at concentrations lower than the 30–40% level, the potential for development of the sucrose–water phase as solution concentration increases during the freezing process must depend on the relative rates of the concentration step (and resultant viscosity increase) and the rate of association to form the sucrose–water phase. There has been little work done in the area of the state of the solid material in the CAS or in freeze-dried material.

Simatos and Blond (1975) noted that while the structure is essentially vitreous, it also appears to contain small crystals. The presence of these crystallites is not dependent on thermal treatments to the frozen material, these apparently just giving further freezing of water, perhaps with a leveling of concentration gradients in the CAS. While this recrystallization of water should result in further saturation of the CAS, it was felt that the CAS phase is too viscous to allow crystal growth. Freeze-drying conditions were noted to have little effect on crystallite formation, while the type material being processed does have a major influence on the final structure obtained.

A number of the references on the crystallization behavior of sucrose could also be considered regarding the question of eventual structure formation in the CAS during the concentration phase. Pidoux (1972) indicated that clusters (called proto-nuclei) exist in supersaturated solution and that these can capture other clusters or molecules to form nuclei. While an induction period for formation of the nuclei was noted, it can be asked if the proto-nuclei themselves exist in the frozen material and/or later in freeze drying. Kauzmann (1948) noted that the free energy barrier to nucleation results from the melting point of small crystals being lower than large crystals, which means that to form a nucleus, it is necessary to form a "crystallite," which has a free energy higher than the same amount of liquid. While Kauzmann does not mention it, this implies that at any time there are small crystallites formed in solution, which then dissolve. It remains to be seen whether those crystallites will be able to dissolve at the moment of solidification of the total system, or (even though they are subcritical in size) will they remain at the moment of solidification and thus be found in the dried product.

It has been noted that the range of concentrations of solids of the CAS would be high enough that a traditional discussion of crystallization could be considered.

To summarize a few of the earlier references, Lees (1965) discussed supersaturation in solution. The CAS concentration can reach up to $S = 1.3$ if nonequilibrium mass transport arises (Fig. 17.4). Van Hook (1971) indicated that viscosity has a major effect on nucleation and that at high viscosity nucleation can be inhibited. Van Hook (1973) also indicated that at low temperatures, it is the orientation step at the surface that hinders crystal growth. While it may be questionable whether nucleation can occur, it is definite that crystal growth is limited. This would tend to give higher supersaturations in the CAS because of retention of sucrose in the CAS rather than it being deposited on any eventual nuclei.

Mentioned previously are a number of material-associated changes occurring during freezing and rewarming processes that can also occur during freeze drying and could thus affect solid structure. Ice crystallization occurs simultaneously with concentration of the CAS, and the relative rates of a number of steps will be of importance in determining which of a number of possible pathways are followed.

It seems that the glass transition temperature (T_g) marks the formation of a H-bonded solid structure from the less-organized liquid. Thus, if one freeze-dries at a temperature below T_g, the dried product should have the solid structure that was fixed by the glassy state. If the sample is freeze-dried at a temperature above T_g, then the extension of free volume throughout the system will result in the loss of the glassy structure that was "frozen in" when free volume became inadequate to allow further rearrangements. There appears to be a temperature range between T_g and T_c where, while the "liquid-like" characteristics for the very viscous material are reestablished, the viscosity is still so high that there is essentially no mobility for the sucrose molecules. Thus, while the microstructure bonding of the glass (sucrose–sucrose and sucrose–water molecules) is altered above T_g, the gross overall structure is still relatively fixed. If the actual water content in the CAS is higher than the minimum unfreezable water content associated with the CAS's solute, this "excess" water will first become mobile at T_r, and recrystallization can occur. (If the CAS had the minimum unfreezable water content possible, there would be no excess water and thus no recrystallization would be possible.) The recrystallization of the excess water of the CAS actually results in an increase in sample viscosity owing to the resultant reduction in the unfrozen water content of the CAS. This recrystallized water then first begins to melt at T_{am}, and it is around here that T_c is found, at a system viscosity much lower than that of T_g. At T_c, the overall structure is no longer stable and the sample will be able to undergo viscous flow ("collapse"). In this case the final structure obtained can be quite variable, since the concentration will still be high though viscosity is much lower than that present in a glass. It is possible therefore in this case to obtain some crystalline material in the dry sample, because the reduced viscosity can allow the proto-nuclei to capture other clusters or molecules to form nuclei.

In these cases, it is assumed that the glassy state forms on freezing and that with the exception of collapse during freeze drying, the formation of nuclei or the growth of crystals is not possible. When considering the situation of forming the crystalline state under the freezing or freeze-drying process, it must also be remembered that to form crystalline sucrose, it is necessary both to form intramolecular H-bonds, one of which does not exist in solution, and to obtain the proper orientation for the intermolecular H-bonds. Since in the hydrated state the sucrose molecule does not contain one of the necessary intramolecular H bonds, the surface incorporation step in crystallization from solution could be thought to be related to the formation of this intramolecular bond. The surface incorporation step is limiting at low temperature, and failure to form the second H bond would be responsible for the failure to give crystalline sucrose in freeze drying. In freeze-drying, once the residual water has been removed in the desorption steps, a structure would be established that involves H bonds between sucrose molecules, but only one intramolecular bond. It is conceivable that the addition of small amounts of water (e.g, by humidification) could give the sucrose molecule mobility without its attaining full hydration, which could allow reorientation of the molecules (recrystallization, see next section). In these circumstances, it can be considered that the intramolecular bonds are the directors of an eventual recrystallization structure.

Sucrose–Water Phase. Still to be considered are the sucrose–water phases, which form in solution at sucrose concentrations above 30–40%, or the hydrates, which have been found in freezing studies. Together with the "unfrozen" water data presented earlier, they would indicate that there are about five H-bonding groups (out of 11) that are ultimately associated with water molecules. If any of these five groups are required for formation of the second intramolecular H bond of the sucrose crystal, then it appears obvious that water blockage of this group would mean that freeze-dried sucrose cannot contain a crystalline phase. Bock and Lemieux (1982) have indicated that one of the groups that is involved in one of the intramolecular H bonds is hydrated in aqueous solution, but it is not known if it is one of the five groups comprising the "unfrozen" water.

At lower solution concentrations, there will be more water molecules associated with each sucrose. When solutions with a concentration above 40% are frozen, it can be assumed that if sufficient time is allowed for the cluster phase to form, then a structure for the CAS is already present in the solution. When freezing solutions of lower initial concentration, however, it can be questioned if clusters will be able to form before the CAS becomes so viscous that the clusters are unable to develop. Under these freezing conditions, it could also be possible that, in addition to the failure to form clusters, the water content in the CAS associated with the sucrose might not be at the minimum level possible when the CAS solidifies. This would result in a material in which there is water available for recrystallization. Recrystallization of CAS water would result in a change in the bonding relationships of the sucrose molecules because a number of free H-bonding groups would appear on the sucrose molecule, which had been previously associated with the now recrystallized water. The ability to develop new bonding relationships would depend on the mobility of the sucrose molecules, but the potential for such bonding is present. With the high concentrations that exist in the CAS, and the numbers of H-bonding groups available for bonding, it could well be that the clusters in the CAS are associated in a structure which, when formed under controlled conditions, frees the highest amount of water possible for formation of ice. This would be a CAS in which there would be no recrystallization.

SOLID CARBOHYDRATE STRUCTURE TRANSITION AND ITS RELATIONSHIP TO FOOD PRODUCTS AND PROCESSES

Solid Carbohydrate Structure Transition

The properties of solid carbohydrates will depend on the method of preparation, particularly regarding its structure (crystalline or amorphous). The largest effect on solid carbohydrate properties results from sorption of water, especially for the recrystallization of amorphous carbohydrate.

Nissan (1976A, B; 1977A, B) investigated how the change in mechanical properties for H-bond-dominated solids (mechanical properties that are determined by H bonds) is related to the destruction of the H bonds. Two modes of action for the addition of water are defined, one at water contents below the BET monolayer, where each added water dissociates one H bond, and the second at water contents above the BET monolayer where H-bond cooperativity is active and one water acts to dissociate one H bond, but where that breakage disrupts more H bonds in the structure. Nissan reports that the cooperativity index for cellulose is about seven. Bond breaking is defined as a cellulose–cellulose H bond being replaced by a cellulose–water H bond, or by an intramolecular cellulose H bond. To and Flink (1978B) showed that this same analysis could be applied to low molecular weight carbohydrates such as sucrose.

Effect of Water Sorption. There have been many investigations of the effect of water sorption on the properties of dried carbohydrates, in particular, the recrystallization of amorphous carbohydrates.

Kargin (1957) noted that glucose glass sorbs water only on the surface due to the low diffusivity of water in the glass. As the surface hydrates, a saturated solution forms and the glass softens, giving higher water diffusion. Thus, in tightly packed sugar glasses, water sorption is a surface phenomenon, and on the usual weight basis, sorption will be higher the higher the specific surface area. Lees (1965) similarly described the graining of boiled sweets, which is recrystallization of the amorphous sugar.

Makower and Dye (1956) studied the recrystallization of spray-dried sucrose occurring during storage at various relative humidities. An induction period prior to the start of recrystallization, which depended on the relative humidity (RH), was noted. According to Lees (1965) seeding studies at low humidity indicated that the induction period could be related to the formation of nuclei. The course of recrystallization was measured by the loss of water from the sample and by X-ray diffraction. Upon recrystallization, sucrose lost all the water that had been sorbed.

Guilbot and Drapron (1969) give moisture sorption isotherms of amorphous and crystalline sugars. The crystalline sugars first begin to sorb after attaining a minimum water activity (a_w) level, while the amorphous sugars sorb at any water activity. For example, in Fig. 17.5 crystalline maltose first starts sorbing water at a_w=0.25 and immediately forms the monohydrate. Amorphous maltose sorbs from a_w greater than zero and appears to recrystallize at a_w = 0.52, at which time it loses water down to the monohydrate level. Tests on a series of amorphous sugars showed all to have approximately the same isotherm, at least up to their recrystallization points, probably indicating that the same sites (—OH groups) are available for uptake of water for all sugars. As these sugars recrystallized, they lost water to become their respective hydrates, though an anhydrous sugar will lose all the previously sorbed water. At high enough a_w the sugars will eventually go into solution.

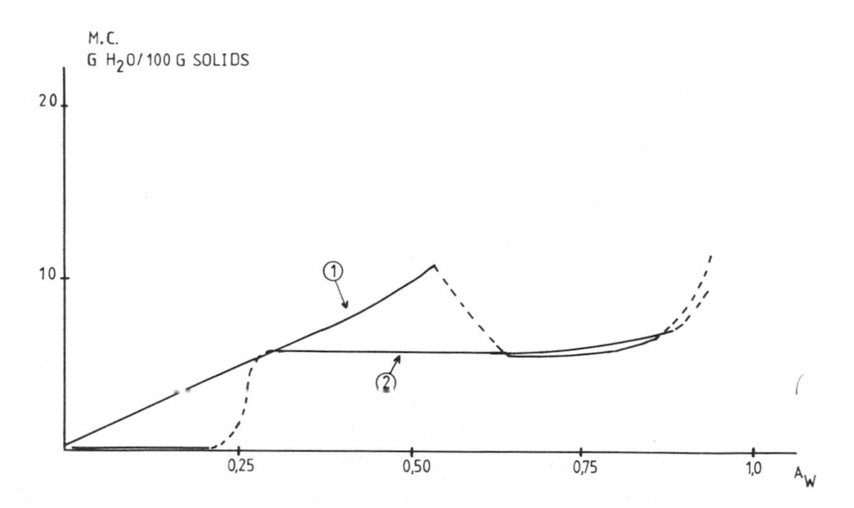

Fig. 17.5. Water vapor sorption by amorphous (curve 1) and crystalline (curve 2) maltose.
From Guilbot and Drapron (1969).

They also observed that the temperature for a solid–liquid transition of the amorphous sugars was much lower than the melting temperatures of the respective crystalline sugars. The similarity of this transition temperature (collapse temperature) for the various amorphous sugars tested indicated that their bonding energy and degree of bonding is similar in the amorphous state.

Sloan and Labuza (1974) also presented many sorption isotherms, including those for amorphous and crystalline sucrose. Amorphous sucrose begins sorbing water already at $a_w = 0.10$, with the major rise in moisture content at $a_w = 0.30$. Crystalline sucrose remains essentially dry (i.e., about zero moisture content) until about $a_w = 0.84$, and then begins to absorb considerably, meeting the amorphous sucrose sorption curve at about $a_w = 0.90$. The isotherms do not reflect any recrystallization effect, which probably means that the time was too short for recrystallization to occur.

Roth (1976, 1977) demonstrated that mechanical crushing of sugar crystals produces an amorphous surface capable of recrystallization after sorbing water. Measurements showed that the amorphous layer on the surface comprises about 2% of the sugar mass. Recrystallization of sucrose was noted at $a_w = 0.42$ at 20°C, at which point there was caking and liquifaction due to recrystallization. (It should be noted that the sample moisture content corresponding to $a_w = 0.42$ will be very low, as only 2% of the mass is sorbing water and being responsible for the sample structure changes.) Subsequent isotherms on the recrystallized sucrose are normal for the crystalline material and were repeatable, indicating no further change after the recrystallization was completed.

Simatos and Blond (1975) noted that freeze-dried sucrose adsorbs water at high rates while the glass produced from the melt is slow to sorb water, this being due to the difference in specific surface areas for the two samples. Differential thermal analysis (DTA) of freeze-dried sucrose showed a glass transition at 45°C, a devitrification at 91°C, and melting of a crystalline phase at 180°C. Glassy sucrose produced from the melt has a glass transition at about 60°C and a shallow, broad melting peak at about 160°C. They concluded that freeze-dried sucrose contains sucrose nuclei that crystallize upon rewarming, these crystals then melting at the normal melting point. X-Ray diffraction did not indicate a crystalline structure, but electron diffraction revealed some organization in freeze-dried sucrose. The size of the structured regions must be small and it could be questioned whether these are crystal nuclei, small crystals, or from the organized regions in the glass.

Iglesias *et al.* (1975) prepared freeze-dried sucrose in two different ways: (1) as a free powder and (2) supported in the pore space of porous ceramic cylinders. Following humidification, the equilibrium moisture level at each a_w was higher for the supported samples, presumably due to the availability of greater surface area for sorption. Recrystallization was delayed or prevented in the presence of the cylinder. The porous nature of the cylinder was felt to hinder the rearrangement of the sucrose molecules. Further, it was noted that as temperature increases, the a_w for recrystallization decreases. In a followup study, Iglesias and Chirife (1978) showed that addition of macromolecular compounds to sucrose solution prior to freeze drying acted to hinder eventual recrystallization when the dry powders were humidified. They also indicated the existence of an induction period before recrystallization begins, which was presumed to be related to formation of nuclei. Because the induction period depends on the presence of additives, the delay could be due to increased viscosity or to interactions of the additives with other components of the system. The additives were also carbohydrates (i.e., with similar —OH groups), and it is possible that the sucrose is bound to the polymer. Thus, as the sucrose obtains mobility due to water uptake, its nearest neighbor is not another sucrose, but rather a carbohydrate unit of the polymer with which it is capable of interacting. It thus takes a longer time for a sucrose molecule to find other sucrose molecules and form the stable crystal. Another explanation for the presence of an induction period may be shielding of the sucrose by the polymer, a form of encapsulation.

Collapse Temperature of Freeze-Dried Sugars. Tsourouflis *et al.* (1976) investigated collapse temperatures of freeze-dried sugars and sugar mixtures as a function of moisture content and blend composition. It was shown that by addition of high molecular weight materials, the collapse temperature of sensitive low molecular weight materials can be effectively raised. They demonstrated the existence of common temperature–moisture content relationships for collapse behavior of dried products, stickiness in spray drying, and collapse during freeze drying (Fig. 17.6). These curves are quite useful on a practical level, but To and Flink (1978B) showed that they are theoretically incorrect.

To and Flink (1978A, B, C) expanded the earlier studies on collapse of freeze-dried carbohydrates. The influence of moisture, molecular weight, and various additives on second-order transitions in sugars was determined. Freeze-dried sucrose was shown to have a glass transition at 52°C, a devitrification at 112°C, and a melting of the crystals formed at 192°C. Freeze-dried maltose, on the other hand, demonstrated only a glass transition at 100°C (Fig.17.7). Since maltose forms a monohydrate crystal, it is possible that the lack of a recrystallization peak in the DTA

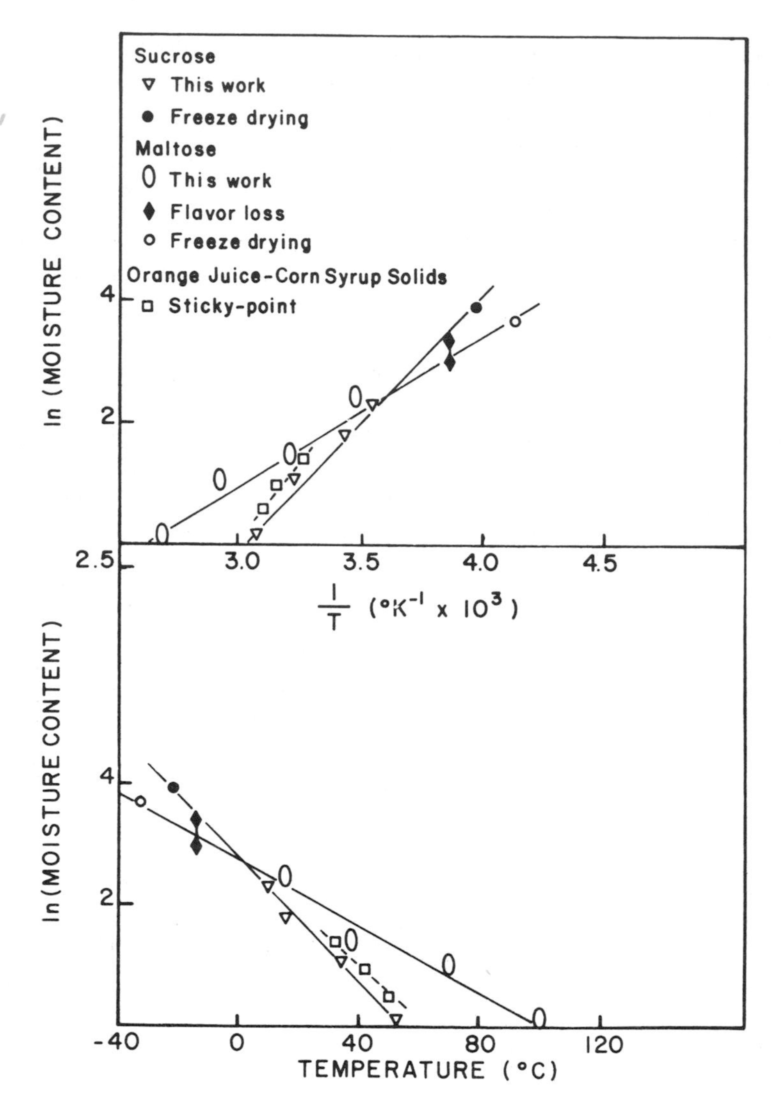

Fig. 17.6. Influence of moisture content on the collapse temperature of carbohydrates.
From Tsourouflis et al. (1976).

is due to the fact that only heating was used to cause collapse. In this case, the anhydrous state of the sucrose crystal allows the formation of the crystal under the action of heat alone, while formation of a maltose crystal would also require the presence of adequate water to develop the monohydrate.

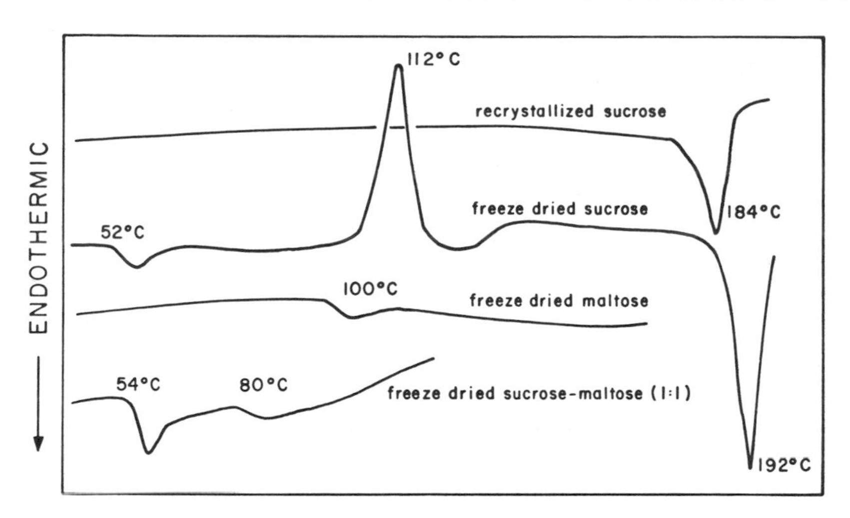

Fig. 17.7. Structure transitions in freeze-dried carbohydrates as determined by DTA. *From To and Flink (1978A).*

Collapse was shown to occur stepwise over a range of temperature (20°–40°C); that is, it is possible to hold a sample at a fixed temperature in the collapse range and apparently have a stable partial collapse of the system. As noted earlier, the glassy state has nonperiodic and nonsymmetric bonding and thus amorphous CAS generally will have different levels of internal bonding prior to freeze drying, which would carry over to the freeze-dried product. Thus, considering the inhomogeneity of structure (which is expected in an amorphous solid), the stability of partial collapse is not surprising as this merely reflects the situation where only a part of the matrix has adequate conditions for flow to occur.

The effect of moisture is related to the BET monolayer. Based on Nissan's theory of the H-bond breaking effect of water (see above), log (T_c) vs moisture content of sucrose was shown to give two straight line portions with a break point at the BET monolayer (Fig.17.8).

The collapse behavior of blended systems of two components showed two types of behavior. When the freeze-dried system contained components that could form a crystalline phase, then total collapse was associated with the melting of the crystalline phase, though partial collapse of the amorphous component could occur prior to crystallite melting. For blends that freeze dry as amorphous solids, collapse temperature showed a nonlinear relationship with composition. Freiser and Tummala (1975) indicated that the glass transition temperatures for mixtures generally showed nonlinear behavior when the mixture composi-

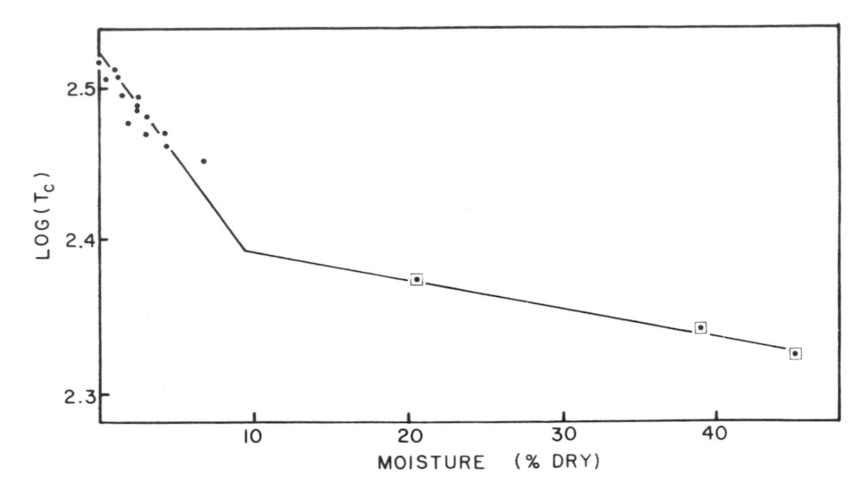

Fig. 17.8. Influence of moisture content on the collapse temperature of sucrose. *From To and Flink (1978B).*

tion is expressed on a weight % basis. When volume % or mole % is used to quantify the mixture compositon, a linear relationship is obtained, which supports the concept that volume change is the critical factor in the glass transition.

The collapse temperature was shown to be inversely proportional to the molecular weight for a homologous series of malto-oligosaccharides.

In all the above conditions, collapse behavior could be calculated using equations similar to those developed in the literature for glass transitions. It seems likely, therefore, that collapse and glass transitions are closely related. While it is tempting to think of collapse as the macroscopic manifestation of a glass transition in the freeze-dried material where the subsequent viscous flow occurs over a time frame so that it is visible, it is valuable to remember that the glass transition is reversible (if no recrystallization has occurred), while collapse has not been shown to be reversible.

It can be seen that the amorphous solid carbohydrate is a metastable material, and that through the action of temperature and moisture, the metastable state will convert to the stable crystalline state with important changes in the properties of the solid. It has also been noted that additives (especially high molecular weight components) will hinder the conversion from the metastable to stable states. The remainder of this chapter notes some consequences of the above-mentioned structure transformations.

Relationship of Carbohydrate Structure Transition to Food Products and Processes

White and Cakebread (1966; 1969) have described conditions for formation of glassy carbohydrates in foods, the relationship of the glassy state in food products to their properties, and the changes that can occur if the glassy state is allowed to become unstable. They describe changes that can occur in boiled sweets (distortion, stickiness, or graining), milk powders (caking and lumpiness), ice cream (sandiness), and freeze drying (liquification and caking). The primary change in all these cases is the viscous flow of the carbohydrate, which in many cases will be followed by its recrystallization. It can also be noted at this point that Tsourouflis *et al.* (1976) describe a number of areas for which structure transformation (called "collapse") would have major consequences. In this section, some of the consequences of these structure transitions will be noted. A number of categories can be mentioned (Table 17.2), but the following discussion will be limited to collapse in freeze drying, encapsulation of oils and flavors, crystal forms in milk products, and agglomeration, caking and flow.

TABLE 17.2. SOME PROCESS AND PRODUCT AREAS INFLUENCED BY CARBOHYDRATE STRUCTURE

Collapse during freeze drying
Encapsulation of flavors and oils
Crystal forms in dry products
Agglomeration, caking, and flow of powders
Graining of boiled sweets
Stickiness in spray drying
Shrinkage in freeze drying
Sandiness in ice cream

Collapse in Freeze Drying. A review of the status of information on collapse phenomena during freeze drying was presented by MacKenzie (1975). Since that time further information regarding the thermal transitions that can result in collapse during freeze drying has appeared; some of these (most notably the work by Simatos and co-workers) have been noted above. In addition, it can be mentioned that MacKenzie's earlier work (MacKenzie 1966) still remains a valuable source of information regarding collapse during freeze drying. Collapse during freeze drying results from a CAS viscosity, which is inadequate to support the weight of the CAS and withstand surface forces present. This loss of structure can occur at either the interface or in the "dry" layer.

Bellows and King (1973) indicated that both drying rate and aroma retention during freeze drying were dependent on collapse behavior. While drying rate would increase with increase in ice layer temperature, due to higher heat transfer rates, the onset of collapse resulted in no further improvements in drying rate. This was presumably due to blockage of the pores which would normally be the pathways for removal of water vapor. Thus, it may be concluded that the onset of collapse will serve as a natural limit to heat input, as further heat input will not give any further improvement in drying rate. Aroma retention was observed to decrease essentially linearly with the estimated degree of collapse of the surface. Ettrup Pedersen *et al.* (1973) examined the influence of chamber pressure on retention of flavor for freeze-dried coffee and were able to show that when the chamber pressure corresponded to an ice front temperature above coffee's collapse temperature, there was a significant loss of volatile compounds. It was also noted that the color of the coffee darkened significantly at this point, due to the changed light reflectance properties of the collapsed surface. In that study, overall particle geometry was maintained, collapse being limited to the surfaces of the CAS.

Swinnen *et al.* (1974) and Tobback *et al.* (1979) studied the effect of mixed systems of glucose and salt on collapse behavior during freeze drying. Tobback *et al.* (1979) also demonstrated that aroma retention was dependent on the stability of the CAS, though in this case, conditions which lead to recrystallization of the salt are detrimental. At low salt concentrations, where recrystallization does not occur under freeze drying, aroma retention is high. At higher concentrations, where salt recrystallization occurs, the aroma level is reduced. At still higher salt concentration, where salt eutetics form on freezing and no recrystallization occurs during freeze drying, aroma retention was at the same level as when recrystallization occurs during freeze drying. It thus seems that the disruption of the integrity of the amorphous structure, due to the presence of salt crystals, is responsible for the reduction in the volatile retention level from that obtained with a fully amorphous CAS. This observation indicates also that the presence of a eutectic phase, even though it is stable with respect to structure changes due to the higher viscosity associated with the crystallites, will give reduced retention of volatile.

Couach *et al.* (1977) observed that when freeze drying sucrose, choosing the process temperature to be above or below the ante-melting temperature of the CAS resulted in a discontinuity for the residual moisture, diffusivity, and specific enthalpy. As noted earlier, ante-melting is the beginning of melting of the ice crystals that formed through recrystallization of excess unfrozen water present in the CAS. It is likely that the liquid formed at ante-melting is located internal to the CAS and cannot easily diffuse out to the surface. Under this condition, the local moisture content in the CAS will be quite high, and collapse will result. This collapse can be local in nature, depending on the amount of water involved in the original recrystallization. To and Flink (1978C) have also indicated that sample composition can result in structure changes during freeze drying. It was shown that when acetone–sucrose blends were freeze-dried, the relationship of acetone to sucrose concentrations was important in determining whether the sucrose would be obtained in the amorphous or crystalline state. While collapse of the sample appears to give a higher probability of sucrose crystallization during freeze drying, it was not essential to have collapse for sucrose crystallization to occur. The sucrose crystallization resulted in complete loss of the acetone.

Seager *et al.* (1978) also investigated the effect that the presence of organic volatiles at relatively high concentrations (e.g., 5 – 10%) had on the structure of freeze-dried pharmaceutical preparations. (The solute phase was not identified specifically.) Their results showed that the various volatiles would influence the overall structure obtained, but to different degrees. The type of volatile would also influence the ease of drying and the drying time.

Lerici *et al.* (1975) investigated the influence of adding various pectins on the collapse of a glucose–fructose blend. Pectin concentration in the mixture and degree of esterification of the pectin were considered. They showed that by adding small quantities of pectin, collapse of the system during freeze drying could be hindered. Since DSC analysis did not indicate any changes in thermal properties of the frozen system due to the presence of the pectin, it was felt that the pectin acts as a physical support, increasing the viscosity of the CAS. Degree of esterification and pectin concentration both showed an effect, with higher degree of esterification and higher concentration giving better protection against collapse. In both these cases, the conditions giving higher solution viscosity gives more protection. This can indicate that in fruit juices where pectin is present as a natural component, the concentration of pectin and degree of esterification could be important factors in determining the ease of freeze drying and the subsequent stability of the dry product. Tsourouflis *et al.* (1976) also investigated the effect of adding higher molecular weight gums and starch hydrolyzates to orange juice on the structural stability of the dry product. It was shown that small amounts of added gums (arabic, tragacanth, or karaya) could cause significant increases in the collapse temperature.

Stern and Storrs (1969) obtained a patent for a process by which high-carbohydrate liquids (apple juice, grape juice, and honey were named) could be more easily freeze dried (i.e., without collapse). In this process, either corn syrup solids or lactose were added to the juice in order to improve the structure of the dried material. The product with added solids was shown to dissolve more readily and to have a lower hygroscopicity. This improvement in product quality was due to prevention of the collapse of the product during drying.

Encapsulation of Oils and Flavors. There have been a number of studies in which the development and retention of structure in freezing and drying processes have been shown to be important in the entrapment and retention of volatile flavor compounds in carbohydrate-based systems. A systematic review of the results from the research group of Karel and Flink has been presented by Flink (1975B). Other studies building on the entrapment of organic volatiles in sucrose systems have been reported by Sugisawa *et al.* (1973), Sugisawa and Shiraishi (1974), Kojimi *et al.* (1975), and Seager *et al.* (1978). More recently, similar studies have been conducted on the influence of matrix structure on protection of encapsulated lipids from environmental interactions, such as would cause lipid oxidation (Gejl-Hansen and Flink 1976, 1977, 1978; Gejl-Hansen 1977; To and Flink 1978C). As noted, these drying methods generally result in the attainment of an amorphous carbohydrate phase, and it has been shown (Flink and Gejl-Hansen 1972; Karel and Flink 1973; Flink 1975B) that the organic compounds are entrapped in the amorphous carbohydrate matrix, generally in the form of liquid droplets. These droplets are not able to interact with the environment (such as loss of volatiles through volatilization or oxidation of the lipid through oxygen diffusion) as long as the physical structure of the matrix remains unaltered. A number of factors can be involved in the disruption of the structure of the dry material, these generally being increase of water content, increase in temperature, or changes in structure occurring in the drying process, such as collapse during freeze drying. In all cases, an alteration of structure results in a loss of the protective properties of the matrix material.

Flink (1975B) reviewed the effect of moisture and temperature on the loss of structure of freeze-dried material and retention of flavor compounds. Additional studies by To and Flink (1978A, B, C) have been mentioned earlier in this section. The influence of moisture content on volatile loss during storage was shown to change at the BET monolayer (Table 17.3). It is seen that if moisture uptake is below the BET monolayer—this being well below the hydration number of the carbohydrate—

TABLE 17.3. CONTENT OF 2-PROPANOL RETAINED IN FREEZE DRIED MALTOSE[1] AFTER 20 HR HUMIDIFICATION AT 25°C

Water Activity a_w	Moisture content (g/100 g)	2-Propanol content (g/100 g)
0	0.42	2.60
0.11	2.41	2.62
0.20	3.93	2.53
0.32	5.21	2.44
0.52	8.04	1.80
0.75	23.15	0.70
1.00	91.70	0.06

[1] BET monolayer value = 5 g/100 g; Unfrozen water value ~ 25 g/100 g.

there is no volatile loss and no structural change. As was noted by To and Flink (1978B), the analysis of collapse based on Nissan's theory results in a mechanism change at the BET monolayer value, at which point further uptake of water results in cooperativity of breakdown of H bonds holding the carbohydrate structure together. This results in large changes in structure and structure stability at moisture levels above the BET monolayer, even though this is well below the hydration number (bound or unfreezable water) that had been discussed earlier in relation to the glassy state.

Chirife and Karel (1974) showed that temperature and other H-bond disrupting solvents (besides water) could also be responsible for loss of structure and resultant loss of volatile from freeze-dried material. This was followed up with microscopic studies by To and Flink (1978C) in which it was shown that the degree of collapse of the freeze-dried matrix was linearly related to the degree of loss of volatile, whether this collapse was caused by moisture uptake or heating. It thus appears certain that the loss of volatile is directly related to the loss of structure. It can be noted that as the material remains amorphous, recrystallization is not taken into consideration in this test. It has been noted earlier that there can be moisture content requirements that must be met if collapsed amorphous carbohydrates are to undergo recrystallization. Thus, heating of a carbohydrate that forms the monohydrate crystal, such as maltose, in the absence of water can cause collapse without subsequent recrystallization. Sucrose, on the other hand, which forms an anhydrous crystal, can recrystallize by the action of heat alone, as was shown by To and Flink (1978A, C). Collapse through moisture uptake will potentially give a different final structure, as the water of crystallization is available. Recrystallization of the collapsed amorphous matrix results in significantly faster loss of entrapped volatile than collapse alone, since the formation of the crystal structure does not allow for the presence of the volatile. This has been shown by Flink (1975B) for the volatile loss behavior of humidified lactose, in which the initial humidification phase, the collapse phase, and recrystallization phase are recognizable. Volatile loss rate changes markedly at the point of onset of recrystallization. To and Flink (1978C) also showed that when acetone containing freeze-dried sucrose was heated, collapse of the sucrose was rapidly followed by sucrose recrystallization and volatilization of the acetone, resulting in a rapid, complete loss. Gerschenson *et al.* (1981) have also recently investigated the influence of structural collapse (induced by both heating and humidification) on volatile retention of freeze-dried tomato juice, and particularly noted the influence of pectin on both collapse and volatile retention. Their results showed that the loss of the volatile substances is related to collapse, in that addition of pectin results in a rise of both the collapse temperature and the temperature at which volatile loss occurs.

As noted above, the influence of structure of freeze-dried oil-in-water emulsions on the protection of the lipid phase have been investigated by Gejl-Hansen (1977) and To (1978). In these studies, it was also shown that the retention of an amorphous matrix is critical to the protection of the lipid phase. During freezing, there is a separation of the oil phase into that which is capable of being encapsulated and that which is located on the surface, and thus not protected. In addition, during the separation step, there can exist oil droplets that are not completely encapsulated, and in fact penetrate the surface of the CAS matrix. While these various possibilities for the oil–matrix relationship make the system complex, the completely encapsulated lipid is stable as long as the amorphous matrix structure is retained. The effect of collapse of maltose based emulsions results in contact between the oil phase and the environment (Gejl-Hansen and Flink 1978). Collapse appears to result in spillage of some of the droplets from the matrix to the surface, where they form pools of oil. They are then readily available for oxidation. To and Flink (1978C) also showed that the temperature dependence for oxidation of dry emulsions depends on the collapse temperature of the matrix. If the test temperature is lower than the collapse temperature of the matrix, then oxidation is hindered. When the test temperature is above the collapse temperature of the matrix oxidation of the material occurs.

Gejl-Hansen (1977) studied the effect of water sorption on the retention of lipid encapsulation properties of freeze-dried maltose and maltodextrin-based emulsions. In the case of maltose-based emulsions, it was shown that at a_w values above the BET monolayer, the maltose matrix will undergo collapse due to uptake of water and that some of the encapsulated oil that was previously unavailable for oxidation will be released as surface fat. It was shown further by optical microscopy that at the higher a_w values, the collapsed maltose system had further undergone recrystallization, and that the encapsulated lipid was expelled completely from the solids. This lipid would then be located on the surface and be available for oxidation. Maltodextrin, which can undergo collapse but not recrystallization, tended to show retention of the lipid droplets, presumably due to the high viscosity present in the collapsed media. Addition of gelatin to emulsions prior to freeze drying would result in a more stable system. It appeared from SEM analysis that the gelatin can coat oil droplets, and, thus, even though the system had undergone collapse, the oil droplets remained encapsulated in the collapsed matrix. This would give a significant difference in lipid protection from the behavior following collapse without gelatin in the system.

Crystal Forms in Milk Products. De Vilder (1975) investigated the influence of lactose crystallinity in whey products on hygroscopicity and cakiness of the dry powder. Lactose crystallization was initiated by the addition of seed crystals to the whey prior to the drying process. It was shown that the presence of crystals in the final product resulted in lower hygroscopicity and less caking tendency. Berlin and Anderson (1975) noted that recrystallization of lactose in cottage cheese whey occurs over the a_w range $0.26 - 0.49$. At the same time that recrystallization occurs, lactic acid is lost and an increased permeability of the powder to nitrogen and organic solvents occurs.

Saltmarch and Labuza (1980) showed that the degree of crystallization in whey powder influenced the water sorption behavior, giving products that range from hygroscopic to nonhygroscopic. Recrystallization appears to cause the rejection of other whey components from the lactose phase, in many cases making them available to the environment.

Warburton and Pixton (1978A) conducted water sorption studies for spray-dried skim milk and lactose. Lactose was amorphous below $a_w =$ 0.42, with recrystallization occurring over the range $a_w = 0.42–0.52$ (7–9.5% water). SEM analysis showed some small prism-shaped crystals in the amorphous matrix of freshly dried skim milk samples. After long-term storage, these can develop into tomahawk shaped α-hydrate crystals. Samples originally without crystals developed a different type of crystal structure (many, small prism crystals) after humidification at intermediate a_w levels, while humidifying at high a_w results in tomahawk crystals.

Roetman (1979) investigated the effect of pre- or postcrystallization of lactose on the structure of the spray-dried whey product. Precrystallization involves the formation of lactose crystals by concentration and seeding of the solution prior to drying, while postcrystallization involves humidification of the product to induce recrystallization of the amorphous lactose obtained following spray drying. Major differences of structure were noted for the two processes. With precrystallization, the product contained tomahawk-shaped crystals imbedded in an amorphous matrix of the other components of the whey product. Postcrystallization resulted in the recrystallization of the lactose of the amorphous matrix as thin needles and a breakdown of the overall matrix structure. These two different structures could be expected to have vastly different porosities and permeabilities, and thus the stability of sensitive components such as fats could be very different for the two products.

Heldman *et al.* (1965) discussed the influence that the amorphous or crystalline state of lactose had on the ease of spray drying and hygroscopicity of the final product. It was postulated that spray-dried lactose exists as a glass since the moisture content is below the monohydrate moisture content. The amorphous lactose will readily adsorb water, and at a moisture content above 5% (presumably at room temperature), recrystallization of the lactose can occur, the ultimate moisture content depending on the crystal structure obtained (monohydrate or anhydrous). Also the ease of spray drying (extent of moisture removal) could depend on the extent of lactose crystallization in the product and the type that forms, since it is more difficult to remove water from the monohydrate. Berlin *et al.* (1968) also indicated that the presence of lactose crystals in a material could have an influence on the removal of the water of crystallization. Varshney and Ojha (1977) showed that the a_w for recrystallization of lactose in spray-dried and roller-dried milk-containing baby food depended on the temperature, with higher temperature giving recrystallization at lower a_w.

Bushill *et al.* (1965; 1969) observed that spray-dried and freeze-dried milk and lactose are amorphous. With humidification at a_w above 0.50, they take up water and then recrystallize with loss of the sorbed water. In milk, lactose generally forms the α-monohydrate crystal, while in lactose solution, the lactose forms the α-monohydrate and an anhydrous β-lactose. In some cases a more complex anhydrous combination of α-lactose and β-lactose forms.

Roetman and Van Schaik (1975) presented information on the β-lactose/α-lactose ratio for amorphous lactose in spray-dried and freeze-dried milk and whey products. It was shown that mutarotation takes place in solution prior to drying, during the drying process, and further in the final dried product during storage. The rate of mutarotation depends on the sample temperature and moisture content, and in this respect has a behavior similar in concept to the collapse phenomena. It was also noted that mutarotation will also occur when the amorphous lactose undergoes recrystallization.

Otsuka *et al.* (1976, 1978) have investigated the effect of sorption of water on the volume expansion of amorphous lactose tablets. This is an area of interest in the pharmaceutical industry. Water sorption studies with α- and β-anhydrous lactose showed different behavior for the two materials. α-Anhydous lactose sorbs water at low a_w and eventually forms the α-monohydrate, resulting in a marked volumetric expansion of the tablets. β-Lactose is less hygroscopic at $a_w < 0.9$, and tablet expansion was not observed. As a_w approaches saturation, β-lactose sorbs a large amount of water and gradually formed the α-monohydrate. Otsuka *et al.* (1978) further investigated this phenomena. At low a_w, they noted some hygroscopic swelling, but this did not lead to crystallization. At high a_w, the lactose formed supersaturated solutions on the surfaces and recrystallization of the lactose occurred, leading to a rapid expansion of the tablets. Because recrystallization releases the sorbed water, the process is autocatalytic, the eventual result being complete recrystallization.

Agglomeration, Caking and Flow. Carbohydrate structure and structure transitions are very important in the area of agglomeration, caking, and flow of powder materials. A factor of major significance, even at low water activities, is the "wetting" of the surfaces such that the particle surfaces can undergo viscous flow into one another. This can be stabilized by eventual recrystallization, giving crystal bridges between particles, and will tend to result in a fusion of the particles due to carbohydrate structure transitions caused by the action of both heat or moisture, both of which have been discussed by Lees (1965) with respect to confections.

Quast and Teixeira Neto (1976) presented sorption isotherms for various amorphous and crystalline sugar products. The a_w levels at which caking was observed are given, and it can be noted that the water activities for caking of the amorphous products are much lower than for the crystalline. For the amorphous products, slight caking was already noted at $a_w = 0.25-0.30$. They also gave information on the influence of water uptake on the rheological properties of cashew nuts, and on the flowability of coffee products. Soluble coffee solids, with its high sugar content showed a loss of free-flowing ability and agglomeration tendency at relatively low a_w values. These differences in caking and flow behavior are very important in determining the packaging requirements for maintaining product quality during distribution. Amorphous products place stricter requirements on the properties of packaging materials.

Peleg and Mannheim (1973) investigated the effect of conditioners on the cohesiveness and compressibility of powdered sucrose in the dry or humidified (a_w = 0.52) condition. This a_w is near the level at which amorphous sucrose undergoes recrystallization. As noted above, Roth (1977) showed that grinding of sucrose gave the crystal an amorphous surface which constitutes about 2% of the crystal's mass. It is this surface layer that is responsible for the surface-related behavior of the sucrose particles. Their results indicated a relationship between caking and the cohesion of powdered sucrose. Smaller particles had higher cohesion and caking, which can be explained by the higher surface to volume ratio.

Moreyra and Peleg (1981) have conducted an investigation of the effect of a_w on the bulk properties of a number of powders, including powdered sucrose. While the powdered sucrose is referred to as a crystalline material, it is well to recall again that the surface is apparently amorphous and is thus labile to collapse and recrystallization behavior. Their results showed that the behavior of sucrose changes markedly at about $a_w = 0.6$, which could be related to transitions in the structure of the surface, first with collapse and eventually recrystallization. The sorption isotherm does not show a recrystallization occurrence, though the fact that only 2% of the mass is involved at a local moisture content of about 5 g water/100g solids would probably make any measurement difficult (see also Roth 1977).

Peleg and Mannheim (1977) also investigated the caking properties of onion powder and showed that at ambient temperature, caking does not occur at $a_w < 0.4$, while at $a_w > 0.45$, the time at which caking was observed was inversely proportional to the a_w. At these levels, anti-caking agents are ineffective in preventing caking. The extent to which this is related to the recrystallization phenomena, in which different caking behaviors would occur at water activities above and below the recrystallization level is not discussed.

Hollenbach *et al.* (1982) investigated the caking characteristics of ground sugar and the effect of adding four anti-caking agents. Again, as noted above, ground sugar has an amorphous surface that controls the particle's interaction behavior. The effectiveness of anti-caking agents depends on either their ability to hinder contact of the humidified amorphous surfaces, or through an action in which they hinder the recrystallization of the amorphous material, thus preventing the formation of the crystal bridges between particles. They observed that the anti-caking agents adhered to the surfaces of the particles, though it was unclear what mechanism was active. Because a certain moisture uptake of moisture is required for recrystallization, the effectiveness of anti-caking agents could well be dependent on the a_w levels in the system (e.g., above or below the a_w giving recrystallization over a measurable time period).

Warburton and Pixton (1978B) also discussed the influence of water uptake on the quality of dried milk products, especially with respect to caking and lumpiness. They described these changes as due to lactose recrystallization, and that this resulted in protein insolubilization and poor rehydration properties for the milk powder. This would result in a significant loss of value, as the product would only be suitable for animal feeding.

Masters and Stoltze (1973) discussed the methods used for agglomeration of dry powders. Included was a curve giving the relationship of moisture content and temperature for the successful agglomeration of coffee powder. They note specifically that the conditions at the surface of the particle must be carefully controlled, so that the amorphous material undergoes only a limited collapse at the surface. The overall structure of the particle must be maintained for successful agglomeration. Their description concerns particles as a unit, and does not discuss the behavior on a molecular level. While it is not reported whether the interparticle bridges that form are themselves amorphous or crystalline, it seems likely that due to the redrying step in an agglomeration process, the bridges are maintained in the amorphous state. This contrasts with the behavior that occurs in caking, where there is adequate time for the material to recrystallize after viscous flow. This difference is quite important as recrystallization at the surface of an agglomerate could eventually carry throughout the whole piece, resulting in major changes in product quality. This would not be acceptable for an agglomerated product.

Hamano and Sugimoto (1978) discussed caking and free flow for spray-dried powdered soy sauce. It was shown that there were three caking behaviors. At $a_w < 0.2$, the material remained powdery, while in the range of a_w between 0.2 and 0.33, there was slight caking (i.e, agglomerates) that could be easily broken up to give a free-flowing material. At a_w above 0.4, the powder showed high hygroscopicity and was not free flowing. Powdered soy sauce undergoes an amorphous to crystalline transition at a moisture content above 2%, which corresponds to an a_w of 0.2. Referring to data from Flink (1975A), they felt that this indicated that the amorphous to crystalline transition is important in caking.

In the previous section, the importance of retention of amorphous structure and the consequences of undergoing an amorphous to crystalline transition have been presented for a number of areas related to food product quality. The importance of retention of amorphous structure and the hindrance of viscous flow of an amorphous structure has been shown to be important in the areas of agglomeration and caking, and in the retention of encapsulated materials, such as flavors and oils. Also, subsequent recrystallization of the collapsed amorphous material can have very significant effects, leading to complete loss of flavor components, and it is presumably responsible for the hard caking of high carbohydrate powders.

SUMMARY

Factors affecting the formation of structure in carbohydrate systems has been investigated from the point of view of (1) that which occurs in solution and (2) the factors that affect the separation of the solid carbohydrate from the aqueous solvent under various water removal processes. The solid carbohydrate is able to form in either a stable crystalline or a metastable amorphous state, where the metastable amorphous form will transform to the crystalline under suitable conditions. The structure obtained under the water removal process has a significant influence on the properties of the food material. In addition, the transition from the metastable amorphous state to the stable crystalline state will have a large effect on the properties and quality of food products. Thus, knowledge regarding the conditions leading to the production of desired structures and to the retention of these structures is important. These have been discussed in the latter part of this chapter along with examples of the influence of structural transitions on product quality.

BIBLIOGRAPHY

ALLEN, A.T., WOOD, R.M., and McDONALD, M.P. 1974. Molecular association in the sucrose-water system. Sugar Technol. Rev. *2*, 165 – 180.

ALMASI, E. 1979. Dependence of the amount of bound water of foods on temperature. Acta Aliment. *8*, 41 – 56.

ARSHID, F.M., GILES, C.H., and JAIN, S.K. 1956. Studies on hydrogen-bond formation. Part IV. The hydrogen-bonding properties of water in non-aqueous solution and of alcohols, aldehydes, carbohydrates, ketones, phenols, and quinones in aqueous and nonaqueous solutions. J. Chem. Soc. A *1956*, 559 – 569.

BELLOWS, R.J., and KING, C.J. 1973. Product collapse during freeze-drying of liquid foods. AIChE Symp. Ser. *132*. 69, 33 – 41.

BERLIN, E., and ANDERSON, B.A. 1975. Reversibility of water vapor sorption by cottage cheese whey solids. J. Dairy Sci. *58*, 25 – 29.

BERLIN, E., ANDERSON, B.A., and PALLANSCH, M.J. 1968. Water vapor sorption properties of various dried milks and wheys. J. Dairy Sci. *51*, 1339 – 1344.

BERTRAN, J.F., and BALLESTER, L. 1971. The heat of crystallization and activity coefficients of sucrose in saturated water solutions. Int. Sugar J. *73*, 40 – 43.

BISWAS, A.B., KURSAH, C.A., PASS, G., and PHILLIPS, G.O. 1975. The effect of carbohydrates on the heat of fusion of water. J. Solution Chem. *4*, 581 – 590.

BOCK, K., and LEMIEUX, R.U. 1982. The conformational properties of sucrose in aqueous solution: Intramolecular hydrogen-bonding. Carbohydr. Res. *100*, 63 – 74.

BOWSKI, L., SAINA, R., RYU, D.Y., and VIETH, W.R. 1971. Kinetic modeling of the hydrolysis of sucrose by invertase. Biotech. Bioeng. *13*, 641 – 656.

BUSHILL, J.H., WRIGHT, W.B., FULLER, C.H.F., and BELL, A.V. 1965. The crystallization of lactose with particular reference to its occurrence in milk powder. J. Sci. Food Agric. *16*, 622 – 628.

BUSHILL, J.H., WRIGHT, W.B., FULLER, C.H.F., and BELL, A.V. 1969. The crystallization of lactose with particular reference to its occurrence in milk powder. *In* Proc. 1st Int. Congr. Food Sci. Technol., 1962. J.M. Leitch (Editor), Vol I, pp. 237 – 245. Gordon and Breach Science Publ., New York.

CHALMERS, B. 1964. Principles of Solidification. John Wiley and Sons, New York.

CHIRIFE, J., and KAREL, M. 1974. Effect of structure disrupting treatments on volatile release from freeze-dried maltose. J. Food Technol. 9, 13 – 20.

COUACH, M., MOREIRA, Th., LE PEMP, M., BONJOUR, E., and SIMATOS, D. 1977. Physical state of sugar solutions at low temperature in connection with structure collapse in freeze drying. Int. Inst. of Refrig. Proc. of Commissions C1-C2 Meeting, Karlsruhe-Ettlingen 1977, pp. 475 – 485.

DACK, M.R.J. 1975. The importance of solvent internal pressure and cohesion to solution phenomena. Chem. Soc. Rev. *4*, 211 – 229.

DE VILDER, J. 1975. Influence of lactose crystallization in concentrated whey on the hygroscopicity and caking of whey powder. Rev. Agric. (Brussels) *28*, 963 – 975. (French)

ETTRUP PETERSON, E., LORENTZEN, J., and FLINK, J. 1973. Influence of freeze-drying parameters on the retention of flavor compounds of coffee. J. Food Sci. *38*, 119 – 122.

FLINK, J.M. 1975A. Application of freeze drying for preparation of dehydrated powders from liquid food extracts. *In* Freeze Drying and Advanced Food Technology. S.A. Goldblith, L. Rey, and W.W. Rothmayr (Editors), pp. 309 – 329. Academic Press, New York.

FLINK, J.M. 1975B. The retention of volatile components during freeze drying: a structurally based mechanism. *In* Freeze Drying and Advanced Food Technology. S.A. Goldblith, L. Rey, and W.W. Rothmayr (Editors), pp. 351 – 372. Academic Press, New York.

FLINK, J., and GEJL-HANSEN, F. 1972. Retention of organic volatiles in freeze-dried carbohydrate solutions: microscopic observations. J. Agric. Food Chem. *20*, 691 – 694.

FREISER, R.G., and TUMMALA, R.R. 1975. Glass transition temperatures of glass mixtures determined by DTA. Glass Technol. *16*, 149.

GEJL-HANSEN, F. 1977. Microstructure and stability of freeze dried solute containing oil-in-water emulsions. Sc.D. Thesis. Mass. Inst. Technol., Cambridge, Massachusetts.

GEJL-HANSEN, F., and FLINK, J.M. 1976. Application of microscopic techniques to the description of structure of dehydrated food systems. J. Food Sci. *41*, 483 – 489.

GEJL-HANSEN, F., and FLINK, J.M. 1977. Freeze-dried carbohydrate containing oil-in-water emulsions: Microstructure and fat distribution. J. Food Sci. *42*, 1049 – 1055.

GEJL-HANSEN, F., and FLINK, J.M. 1978. Microstructure of freeze dried emulsions: Effect of emulsion composition. J. Food Proc. Preserv. *2*, 205 – 228.

GERSCHENSON, L.N., BARTHOLOMAI, G.B., and CHIRIFE, J. 1981. Structural collapse and volatile retention during heating and rehumidification of freeze-dried tomato juice. J. Food Sci. *46*, 1552 – 1556.

GUILBOT, A., and DRAPRON, R. 1969. Evolution of the state of organization and affinity for water of various carbohydrates as a function of relative humidity. Bull. Int. Inst. Froid Annexe. *9*, 191 – 197. (French)

HAMANO, M., and SUGIMOTO, H. 1978. Water sorption, reduction of caking and improvement of free flowingness of powdered soy sauce and miso. J. Food Proc. Preserv. *2*, 185 – 196.

HELDMAN, D.R., HALL, C.W., and HEDRICK, T.I. 1965. Equilibrium moisture of dry milk at high temperatures. Trans. ASAE *8*, 535 – 541.

HERRINGTON, B.L. 1934. Some physico-chemical properties of lactose. J. Dairy Sci. *17*, 501 – 518.

HOLLENBACH, A.M., PELEG, M., and RUFNER, R. 1982. Effect of four anticaking agents on the bulk characteristics of ground sugar. J. Food Sci. *47*, 538 – 544.

IGLESIAS, H.A., and CHIRIFE, J. 1978. Delayed crystallization of amorphous sucrose in humidified freeze dried model systems. J. Food Technol. *13*, 137 – 144.

IGLESIAS, H.A., CHIRIFE, J., and LOMBARDI, J.L. 1975. Comparison of water vapor sorption by sugar beet root components. J. Food Technol. *10*, 385 – 391.

ITO, K. 1970. Freeze drying of pharmaceuticals. On the change in the macroscopic appearance during freezing and the critical temperature for freeze drying. Chem. Pharm. Bull. *18*, 1509 – 1518.

ITO, K. 1971. Freeze drying of pharmaceuticals. Eutectic temperature and collapse temperature of solute matrix upon freeze drying of three component systems. Chem. Pharm. Bull. *19*, 1095 – 1102.

JAMES, D.W., and FROST, R.L. 1974. Structure of aqueous solutions. Structure making and structure breaking in solutions of sucrose and urea. J. Phys. Chem. *78*, 1754 – 1755.

JEFFREY, G.A. 1972. Conformational studies in the solid state: Extrapolation to molecules in solution. *In* Carbohydrates in Solution. Adv. Chem. Ser. pp. 177 – 196. R.F. Gould (Editor), Am. Chem. Soc., Washington, DC.

JEFFREY, G.A., and LEWIS, L. 1978. Cooperative aspects of hydrogen bonding in carbohydrates. Carbohydr. Res. *60*, 179 – 182.

KAREL, M. 1975. Physico-chemical modification of the state of water in foods—a speculative survey. *In* Water Relations of Foods. pp. 639 – 656. R. Duckworth (Editor), Academic Press, New York.

KAREL, M., and FLINK, J.M. 1973. Influence of frozen state reactions on freeze-dried foods. J. Agric. Food Chem. *37*, 16 – 21.

KARGIN, V.A. 1957. Sorptive properties of glasslike polymers. J. Polym. Sci. *23*, 47 – 55.

KAUZMANN, W. 1948. The nature of the glassy state and the behavior of liquids at low temperatures. Chem. Rev. *43*, 219 – 256.

KOJIMA, T., TABATA, K., KIMURA, K., and SUGISAWA, H. 1975. The retention of volatile flavors in foods. Part VIII. On the moisture content of carbohydrate solution and the dried product. J. Food Sci. Technol. *22*, 594 – 597. (Japanese, English summary)

LEES, R. 1965. Factors affecting crystallization in boiled sweets, fondants and other confectionery. British Food Manufacturing Res. Assoc. (Leatherhead, Surrey, England) Sci. Tech. Surveys No. 42.

LE MESTE, M., and SIMATOS, D. 1980. Use of electron spin resonance for the survey of "ante-melting" phenomenon, observed in sugar solutions by differential scanning calorimetry. Cryo-letters. *1*, 402 – 407.

LERICI, C.R., MALTINI, E., and POLESELLO, A. 1975. The role of pectin in the freeze-drying of sugars model solutions. Proc. 16th Int. Congr. Refrig. (IRR), Moscow. Paper C1.39.

LESIKAR, A.V. 1975. Glass transitions of organic solvent mixtures. Phys. Chem. Glasses *16*, 83 – 90.

LUSENA, C.V. 1955. Ice propagation in systems of biological interest. III. Effect of solutes on nucleation and growth of ice crystals. Arch. Biochem. Biophys. *57*, 277 – 284.

LUSENA, C.V., and COOK, W.H. 1954. Ice propagation in systems of biological interest. II. Effect of solutes at rapid cooling rates. Arch. Biochem. Biophys. *50*, 243 – 251.

LUYET, B., and RAPATZ, G. 1969. Mode of distribution of the ice and the nonfrozen phase in frozen aqueous solution. Biodynamica. *10*, 293 – 317.

MAC KENZIE, A.P. 1966. Basic principles of freeze-drying for pharmaceuticals. Bull. Parenter. Drug Assoc. *20*, 101 – 129.

MAC KENZIE, A.P. 1975. Collapse during freeze drying-qualitative and quantitative aspects. *In* Freeze Drying and Advanced Food Technology, pp. 277 – 307. S.A. Goldblith, L. Rey and W.W. Rothmayr (Editors). Academic Press, New York.

MAKOWER, B., and DYE, W.B. 1956. Equilibrium moisture content and crystallization of amorphous sucrose and glucose. J. Agric. Food Chem. *4*, 72 – 77.

MALTINI, E. 1977. Studies on the physical changes in frozen aqueous solutions by DSC and microscopic observations. IIR Proc. Commissions C1-C2 Meeting, Karlsruhe, Germany.

MASTERS, K., and STOLTZE, A. 1973. Agglomeration advances. Food Eng. *45*(2), 64 – 67.

MATHLOUTHI, M., and LUU, D.V. 1980. Laser-Raman spectra of *d*-glucose and sucrose in aqueous solution. Carbohydr. Res. *81*, 203 – 212.

MATHLOUTHI, M., LUU, C., MEFFROY-BIGET, A.M., and LUU, D.V. 1980. Laser-Raman study of solute-solvent interactions in aqueous solutions of *d*-fructose, *d*-glucose and sucrose. Carbohydr. Res. *81*, 213 – 223.

MEARES, P. 1957. The second-order transition of polyvinyl acetate. Trans. Faraday Soc. *53*, 31 – 40.

MEARES, P. 1965. Polymers: Structure and Bulk Properties, pp. 251 – 275. Van Nostrand Reinhold, New York.

MOREYRA, R., and PELEG, M. 1981. Effect of equilibrium water activity on the bulk properties of selected food powders. J. Food Sci. *46*, 1918 – 1922.

NEAL, J.L., and GORING, A.I. 1970. Hydrophobic folding of maltose in aqueous solution. Can. J. Chem. *48*, 3745 – 3747.

NISSAN, A.H. 1976A. Three modes of dissociation of H bonds in hydrogen-bond dominated solids. Nature *263*, 5580.

NISSAN, A.H. 1976B. H-bond dissociation in hydrogen bond dominated solids. Macromolecules *9*, 840 – 850.

NISSAN, A.H. 1977A. The elastic modulus of lignin as related to moisture content. Wood Sci. Technol. *11*, 147 – 151.

NISSAN, A.H. 1977B. Density of hydrogen bonds in H-bond dominated solids. Macromolecules *10*, 660 – 662.

OTSUKA, A., WAKIMOTO, T., and TAKEDA, A. 1976. Moisture sorption and volume expansion of lactose anhydrate tablets. Yakugaku Zasshi *96*, 351 – 355. (Japanese, English sumary).

OTSUKA, A., WAKIMOTO, T., and TAKEDA, A. 1978. Moisture sorption and volume expansion of amorphous lactose tablets. Chem. Pharm. Bull. *26*, 967 – 971.

PELEG, M., and MANNHEIM, C.H. 1973. Effect of conditioners on the flow properties of powdered sucrose. Powder Technol. *7*, 45 – 50.

PELEG, M., and MANNHEIM, C.H. 1977. The mechanism of caking of powdered onion. J. Food Proc. Preserv. *1*, 3 – 11.

PIDOUX, G. 1972. Relationship between viscosity and nuclei formation in sucrose solutions. Zucker *25*, 170 – 173. (German, English summary).

QUAST, D.G., and TEIXEIRA NETTO, R.O. 1976. Moisture problems of foods in tropical climates. Food Technol. *30*(5), 102 – 105.

ROETMAN, K. 1979. Crystalline lactose and the structure of spray-dried milk products as observed by scanning electron microscopy. Neth. Milk Dairy J. *33*, 1 – 11.

ROETMAN, K., and VAN SCHAIK, M. 1975. The β/α ratio of lactose in the amorphous state. Neth. Milk Dairy J. *29*, 225 – 237.

ROSS, K.D. 1978. Differential scanning calorimetry of nonfreezable water in solute-macromolcule-water systems. J. Food Sci. *43*, 1812 – 1815.

ROTH, D. 1976. Production of the amorphous state during grinding and recrystallization as a cause of agglomeration of powdered sucrose and processes to prevent this. Dissertation, University of Karlsruhe, Germany.

ROTH, D. 1977. The water vapor sorption behavior of icing sugar. Zucker *30*, 274 – 284. (German, English summary.)

SALTMARCH, M., and LABUZA, T.P. 1980. Influence of relative humidity on the physicochemical state of lactose in spray-dried sweet whey powders. J. Food Sci. *45*, 1231 – 1236, 1242.

SEAGER, H., TASKIS, C.B., SYROP, M., and LEE, T.J. 1978 – 1979. Freeze drying parenterals containing organic solvent. Manuf. Chem. Aerosol News. Nov. 1978, pp. 43 – 46; Dec. 1978, pp. 59 – 64; Jan. 1979, pp. 40 – 45; Feb. 1979, pp. 41 – 45, 48.

SIMATOS, D., and BLOND, G. 1975. The porous texture of freeze dried products. *In* Freeze Drying and Advanced Food Technology. pp. 401 – 412. S.A. Goldblith, L. Rey and W.W. Rothmayr (Editors). Academic Press, New York.

SIMATOS, D., FAURE, M., BONJOUR, E., and COUACH, M. 1975A. The physical state of water at low temperatures in plasma with different water contents as studied with differential thermal analysis and differential scanning calorimetry. Cryobiology *12*, 202 – 208.

SIMATOS, D., FAURE, M., BONJOUR, E., and COUACH, M. 1975B. Differential thermal analysis and differential scanning calorimetry in the study of water in foods. *In* Water Relations of Foods. pp. 193 – 209. R. Duckworth (Editor). Academic Press, New York.

SLOAN, A.E., and LABUZA, T.P. 1974. Investigating alternative humectants for use in food. Food Prod. Dev. *8*(9), 75, 78, 80, 82, 84, 88.

SMYTHE, B.M. 1971. Sucrose crystal growth. Sugar Technol. Rev. *1*, 191 – 231.

STEINBACH, G. 1977. Phase equilibria in frozen solutions from refractometric measurements of freezing curves. *In* Freezing, Frozen Storage and Freeze-Drying of Biological Materials and Foodstuffs. Int. Inst. Refrig. Annex, Paris, Vol. 1, pp. 53 – 66. Paris.

STERN, R.M., and STORRS, A.B. 1969. Use of a high carbohydrate additive in processing apple and grape juice. U.S. Patent 3,483,032. Dec. 9.

SUGGETT, A. 1976. Molecular motion and interactions in aqueous carbohydrate solutions. III. A combined nuclear magnetic and dielectric-relaxation strategy. J. Solution Chem. *5*, 33 – 46.

SUGISAWA, H., and SHIRAISHI, K. 1974. The retention of volatile flavors in foods. Part II. The formation of micellar colloid and its contribution to flavor retention in the dried solution of sugars. J. Food Sci. Technol. *21*, 524 – 528. (Japanese, English summary).

SUGISAWA, H., KOBAYASHI, N., and SAKAGAMI, A. 1973. The retention of volatile flavors in food. Part I. Flavor retention in the dried solution of carbohydrates. J. Food Sci. Technol *20*, 364 – 368. (Japanese, English summary).

SWINNEN, J., TOBBACK, P.P., and MAES, E. 1974. Study by differential-scanning-calorimetry of the stability of model systems (sugar-salt-solutions) in the frozen state. Proc. 4th Int. Congr. Food Sci. Technol. Vol II. pp. 43.50.

TO, E.C. 1978. Collapse, a structural transition in freeze-dried matrices. Sc.D. Thesis, Mass. Inst. Technol. Cambridge, Massachusetts.

TO, E.C., and FLINK, J.M. 1978A. "Collapse," a structural transition in freeze dried carbohydrates. I. Evaluation of analytical methods. J. Food Technol. *13*, 551 – 565.

TO, E.C., and FLINK, J.M. 1978B. "Collapse," a structural transition in freeze dried carbohydrates. II. Effect of solute composition. J. Food Technol. *13*, 567 – 581.

TO, E.C., and FLINK, J.M. 1978C. "Collapse," a structural transition in freeze dried carbohydrates. III. Prerequisite of recrystallization. J. Food Technol. *13*, 583 – 594.

TOBBACK, P.P., SWINNEN, J., and MAES, E. 1979. Relation between interstitial fluid stability and volatiles retention in freeze-dried model systems (sugar-salt solutions). Int. J. Refrig. *2*, 206 – 210.

TSOUROUFLIS, S., FLINK, J.M., and KAREL, M. 1976. Loss of structure in freeze-dried carbohydrate solutions: Effect of temperature, moisture content and composition. J. Sci. Food Agric. *27*, 509 – 519.

VAN HOOK, A. 1971. Nucleation. Sugar Technol Rev. *1*, 232 – 238.

VAN HOOK, A. 1973. Sucrose crystallization-mechanism of growth. Z. Zuckerind. Boehm. *23*, 499 – 503.

VARSHNEY, N.N., and OJHA, T.P. 1977. Water vapor sorption properties of dried milk baby foods. J. Dairy Res. *44*, 93 – 101.

WARBURTON, S., and PIXTON, S.W. 1978A. The moisture relations of spray dried skimmed milk. J. Stored Prod. Res. *14*, 143 – 158.

WARBURTON, S., and PIXTON, S.W. 1978B. The significance of moisture in dried milk. Dairies Ind. Int., April.

WEISSER, H. 1979. NMR-techniques in studying bound water in foods. Personnal Communication, Institute of Food Processing Technology, Univ. of Karlsruhe, D-7500 Karlsruhe, Germany.

WHITE, G.W., and CAKEBREAD, S.H. 1966. The glassy state in certain sugar-containing food products. J. Food Technol. *1*, 73 – 82.

WHITE, G.W., and CAKEBREAD, S.H. 1969. The importance of the glassy state in certain sugar-containing food products. Proc. 1st Int. Congr. Food Sci. Technol., 1962. Vol I, pp. 227 – 235. J.M. Leitch (Editor). Gordon and Breach Science Publ., New York.

WILLIAMS, M.L., LANDEL, R.F., and FERRY, J.D. 1955. The temperature dependence of relaxation mechanisms in amorphous polymers and other glass-forming liquids. J. Am. Chem. Soc. *77*, 3701 – 3706.

ZIMMER, M. 1979. Investigation on sugar solutions of the freezing of foods through pulsed NMR. Diplomarbeit Thesis, Univ. of Karlsruhe, Karlsruhe, Germany.

Index

F